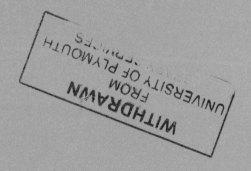

4

SERIES IN THEORETICAL AND APPLIED MECHANICS

Edited by R.K.T. Hsieh

SERIES IN THEORETICAL AND APPLIED MECHANICS

Editor: R. K. T. Hsieh

Published

Volume 1: Nonlinear Electromechanical Effects and Applications
by G. A. Maugin

Volume 2: Lattice Dynamical Foundations of Continuum Theories
by A. Askar

Volume 3: Heat and Mass Transfer in MHD Flows
by E. Blūms, Yu. Mikhailov, and R. Ozols

Volume 4: Mechanics of Continuous Media
by L. Sedov

Volume 5: Inelastic Mesomechanics
by V. Kafka

Volume 6: Design Technology of Fusion Reactors
edited by M. Akiyama

Volume 8: Mechanics of Porous and Fractured Media
by V. N. Nikolaevskij

Volume 9: Aspects of Non-Equilibrium Thermodynamics
by W. Muschik

Volume 10: Fragments of the Theory of Anisotropic Shells
by S. A. Ambartsumian

Volume 11: Diffusion Processes during Drying of Solids
by K. N. Shukla

Mechanics of Continuous Media

Volume 1

L. I. Sedov
Institute of Mechanics
Moscow University

Translated from the Russian, edited by J. P. Nowacki
Polish Academy of Sciences

World Scientific
Singapore • New Jersey • London • Hong Kong

Author

L I Sedov
Institute of Mechanics
Moscow University
119899 Moscow
Russia

Series Editor-in-Chief

R. K. T. Hsieh
Department of Mechanics, Royal Institute of Technology
S-10044 Stockholm, Sweden

Published by

World Scientific Publishing Co. Pte. Ltd.
P O Box 128, Farrer Road, Singapore 912805
USA office: Suite 1B, 1060 Main Street, River Edge, NJ 07661
UK office: 57 Shelton Street, Covent Garden, London WC2H 9HE

MECHANICS OF CONTINUOUS MEDIA — Volume 1

ISSN 0218-0235
ISBN 9971-50-728-5 (Set)
ISBN 981-02-3981-5 (Vol. 1)
ISBN 981-02-3982-3 (Vol. 2)

This book is printed on acid-free paper.

Printed in Singapore by Uto-Print

Foreword to the first edition

An understanding of Nature and the solution of many practical technical problems require the construction of new models for a thorough and more detailed description of microscopic and macroscopic, mechanical and, in general, physical objects, of interactions and phenomena, etc.

Experience and the intrinsic essence of science indicate that answers to questions, relating to the structure of matter, to problems in astrophysics, to essential properties of complicated interactions in bodies of the living organic and inorganic world, are in many ways linked to certain general universal concepts, ideas, laws, representations and methods.

By now, an enormous amount of scientific information has been accumulated; theories have been developed and experimental data on the behaviour of physical fields, on the motion and equilibria of gases, plasmas, fluids and rigid deformable bodies collected.

A lucid presentation of the foundations and internal relationships between different theories and observed effects will assist a thorough understanding of the real state of science, a true assessment of known and developing scientific achievements, and a measure of orientation in the wealth of information. All these aspects require attention before one can arrive at a sound basis for further scientific development.

In an exposition of the foundations of mechanics and physics, it is vital to pursue the following attitude which must be clearly emphasized throughout. All leading and useable concepts and relationships have a definite meaning only within the framework of a certain set of models which are constructed for the scientific description and study of interesting classes of real phenomena; this statement also applies to such fundamental concepts as space, time, force, temperature, entropy, etc. For example, it is well known that the idea of the Newtonian force is meaningless for certain interactions, described by quantum mechanics; it is quite insufficient for the description of mechanical interactions in contemporary, complicated models of continuous media. "Universal ideas" on entropy and temperature are senseless and not required in models of analytical mechanics.

Thus, one cannot speak of basic concepts and regularities outside sets of very wide or very narrow classes of models, which either have already been introduced explicitly, or may yet have to be proposed, or are implied implicitly in theories and by experimental observations. Failure to take into account this situation, excessive generalizations and isolation from the essence of a matter, may lead to pointless arguments on the fundamental nature of force, entropy, etc. In definite, correctly introduced models, the necessary clarity is attained and any misunderstandings which may arise are readily dispersed. However, obviously, all models reflect reality only in an approximate manner and only within certain confined limits; more refined definitions, more complicated mechanisms, new models or simplifications in a known sense of existing models, all these processes form part of the continuing process which is associated with scientific progress.

Obviously, the study of mechanics is especially useful not only from the point of view of already known applications, but also from that of future problems which will require investigation and lead to new applications.

In recent years, the need for the introduction of a course in continuum mechanics at more advanced levels at universities and technical high schools has become clear; it is needed as a general base for the theoretical development of thermodynamics, electromagnetism, hydrodynamics, gas dynamics, elasticity, plasticity, creep and of many other groups of phenomena of physics and mechanics. The generality and, at the same time, inseparability of these branches, which at a first glance may appear to be autonomous subdivisions of mechanics and physics, bring about the need to consider them as a single unit.

The author has designed this course in continuum mechanics in the light of these aims, following many years of lecturing at Moscow State University. Notes of these lectures were initially published in rotaprint in the years 1966–1968.

The course consists of two volumes. Volume I deals with universal mathematical methods and concepts, the foundations of thermodynamics and electrodynamics; the author has found that a short exposition of the foundations of thermodynamics and electrodynamics proves useful not only to the specialist working in the field of mechanics. In addition, the first volume establishes the basic physical equations and supplementary relations on strong discontinuities as well as initial, boundary and other conditions, explains important properties of approximate methods arising, for example, in the linearization of problems. Thus, the first volume prepares the foundations for the construction of concrete models of continuous media and exposes typical elements of schematic methods for the formulation of concrete problems.

Volume II is devoted to concrete models and theories in hydrodynamics, gas dynamics, elasticity and plasticity. It presents typical problems within the framework of classical models and establishes the most important regularities which occur in wide classes of motions and processes.

Fundamentally, the deductive style of this presentation and the choice of the material and its development are linked closely to any endeavour to prepare a logical skeleton of the theory of motion of continuous media and of its main goals: namely, to establish for the reader a minimum of factual information, limited to simpler examples with greatest practical significance which are the favourite ground for a detailed understanding of the essence of continuum mechanics and of the principal known effects arising in the motion of continuous media. In other words, it is the author's aim to obtain with the least amount of necessary information the best possible understanding.

Certain very important branches of hydromechanics and of the mechanics of rigid bodies are not included in this course, because there exist special courses and books where the corresponding theories are developed in detail. For example, this is true with regard to two-dimensional motions of liquids and gases, unsteady gas motion, surface gravity waves on fluids, dimensionality and similarity, boundary layers and turbulence, plasticity, creep and many other topics.

The foundations for this course were laid in notes of my lectures, taken by V. V. Rozantseva and M. E. Eglit, who gave me much help and suggested important improvements of the text following very fruitful discussions which took place during the compilation and redaction of this course. I wish to express my deep gratitude to V. V. Rozantseva and M. E. Eglit for their enormous effort which ensured the publication of this book.

I also wish to thank V. P. Karlikov for his great assistance in the compilation of the text relating to hydrodynamics and to D. D. Ivlev for help rendered with the work on the plane theory of elasticity.

In the preparation and editorial work on the sections dealing with general gas dynamics and hydraulic machines, I was greatly assisted by F. M. Bam-Zelikovich, G. Yu. Stepanov and A. Ya. Cherkez, and thanks are due to them.

I likewise wish to thank my Assistant E. I. Sveshnikova and many of my collaborators and students who helped with the preparation of the first version of this course.

Supplementary literature on Continuum Mechanics is given at the end of Volume II.

L. I. Sedov

Moscow, December 1968.

Foreword to the third edition

In recent years the indispensability of introducing a systematical course in continuum mechanics as an extension of the general mechanics, and as a basis of particular chapters of the dynamics of deformed material media and fields as well as phenomenological physics has been well understood in the scientific circles. Such courses are initiated everywhere in the scientific programmes of universities and technical high schools. As a result a considerable number of monographs and textbooks devoted to mechanics of continuous media has appeared in the last decade.

The proposed course is constructed on rational mathematical and physical grounds inherent to mechanics being natural exact science, directed to cognition of phenomena of nature and creating the basis for various actual domains of techniques. In the course purely formal constructions are reduced to the minimum. Theoretical model conceptions of the investigated objects of the studied phenomena and schematized expression of typical problems have been explicitly shown in the text.

A number of theories become comprehensible in the course of time of almost a half-century of the author's experience both in the theoretical studies and lectures at universities as well as in the run of research work in several scientific institutes realizing investigations for industry in various domains. It also explains certain complexity of contents and separation – in the author's opinion – the most useful theoretical methods and problems being essential for contemporary education in mechanics.

In a series of cases the problems of such a type are introduced for studying. We rank to them simple formulation of tensors as the objects inseparable from base vectors, with the corresponding mastering in tensor calculus; the general theory of symmetry; the theory of tensor functions of several tensor arguments; systematical expounding of the conception of penderomotive forces; forces which are exerting an influence on a material medium from the side of electromagnetic field; an extension of the theory of flow of fluid and gas around bodies; general theory of properties and principal characteristics of gas and hydraulic engines; foundations and applications of the theory of dimensionality and physical similarity; phenomenological thermodynamic theory of construction of models with internal degrees of freedom; thermodynamics of models of plastic bodies, and many other questions and problems in this field.

In the third edition the noticed misprints have been corrected. A series of supplements and specifications as well as certain necessary explanations have been introduced, too.

L. I. Sedov

Moscow, December 1975.

Foreword to the fourth edition

During the last ten years since the first edition of this course considerable progress in a number of fundamental theories of mechanics of continuous media has been achieved. Accordingly, in the actually proposed new fourth edition of the course certain substantial supplements have been introduced, the setting forth of several parts of the text have been improved; a series of methodological remarks which should help in understanding and acquiring the studied subjects with greater precision, and interpretation of the matter on the level of the last achievements in science were made.

In particular, the experience in teaching of mechanics of continuous media, taking advantage of this course; shows that the proposition of giving the definition of the notion of the vector and the tensor with the compulsory explicit introduction of base vectors, which appears to be quite adequate to the essence of the matter, does not cause any difficulties but contrariwise; it significantly helps the readers and participants, in a relatively short time, to master and acquire fully the essence of many geometrical and mechanical theories.

Necessity for the introduction of concomitant and in general arbitrary reference frames of an observer, and in this connection, a concept of covariance of laws of nature are stressed, with the required elements in detail, in the book. It is worth to note once more that the fundamental equations of mechanics and physics possess the property of covariance with respect to arbitrary moving and deforming reference systems as well as corresponding coordinate systems not only in the general theory of relativity, as some people think, but in Newtonian mechanics, too. However, it is necessary to take only into account the presence of corresponding postulates of a space and a time, and, moreover, the presence in some cases (as in Newtonian mechanics) of the chosen reference systems, as, for instance, global or local inertial systems being used not only for setting equations and formulation of the problem but also for introducing physical characteristics, similar to "absolute acceleration". Analogous characteristics are introduced in the theory of relativity by means of the chosen characteristic local inertial reference systems or in particular examples, just as in Newtonian mechanics, with the help of the chosen global reference systems.

For instance, all the equations in Newtonian mechanics will have a covariant form if an expression for the absolute acceleration and all the derivatives with respect to time, having by definition an important physical meaning, are expressed through the corresponding quantities in an arbitrary moving reference system. For an acceleration vector, taken with respect to an inertial reference system one should use the following formula:

$$a_{\text{abs}} = a_{\text{con}} + a_{\text{trans}} + a_{\text{add}} \,. \tag{1}$$

Here a_{abs} denotes the acceleration vector of an individualized point M in relation to an arbitrary convected reference system, a_{con} stands for the acceleration vector in relation to the inertial system of point M' of the convected system, coinciding at a given instant with point M, whereas a_{add} is a vector of additional generalized Coriolis acceleration occurring due to the presence of the velocity of point M in relative motion, vortex tensors and strain rate tensors of convected motion. Here in a general case, the main role fulfills a convected coordinate system.[1)] In special cases vectors entering into a general formula (1), may have a partial form for example, for a plane or rectilinear motion, etc.

Using the formula (1) and the analogous formulae for the characteristic scalars, vectors and tensors all the equations both in the theory of relativistic as well as in Newtonian mechanics take the covariant form. It means that all physical relations for each given phenomena can be written over in a universal form by means of identical tensorial formulae in arbitrarily chosen coordinate systems. These formulae will contain the components of a matrix

[1)] See Sedov L. I., *On the addition of motion relative to deformable reference systems*, PMM, Vol. 42, No. 1, 1978.

tensor, which are the functions of coordinates, different in various used coordinate systems.

However, covariance both in the theory of relativity, as well as in Newtonian mechanics, and in general in each theory, does not mean that the physical phenomena are perceived identically by various observers. Covariant equations written out in the components, in an arbitrary reference system in explicit form depend greatly not only on the phenomena considered but also on a fixed observer and, moreover, on the used concrete coordinate system! It appears not only in the theory, but also in the experimental results obtained by the observers.

Together with the concept of covariance, one may introduce the concept of form-invariance in the case when all the equations in components, in explicit form are identical for different observers moving each with respect to the other and associating with the different coordinate systems. In other words the question consists of the transformation of coordinates preserving all the numerical functions for formulated physical relations. Form-invariance in general cases is satisfied for particular transformation only.

As it is a well-known fact, in Newtonian mechanics the form-invariance takes place for the global Galilean–Newtonian transformations; in STR takes place or occurs for the global Lorenz transformations,[1)] whereas in GTR in each point of the four-dimensional pseudo-Riemannian space takes place or occurs only locally for arbitrary finite Lorenz transform.

Practice shows that the remarks made above are useful for construction of various models of media and fields.

The recent state in the scientific literature devoted to macroscopic thermodynamics, in the problem of construction of new models in which irreversible and non-equilibrium processes are considered is connected with the necessity of reconsideration of established traditions and conceptions. The corresponding more accurate definitions have been introduced in the first and the second editions of the course. In the edition proposed the new more precise definitions have been added to the formulae here and there.

Such type of questions was discussed with A. G. Kulikovskii and M. E. Eglit giving lectures on mechanics of continuous media to the students at Moscow State University (with noteworthy, creative contribution to thermo-

[1)] The formulae for Galilean–Newtonian and Lorenz transformations differ; they are very simple in Cartesian coordinates; they may be written in any arbitrary coordinate systems. GTR – the General Theory of Relativity, STR – the Special Theory of Relativity.

dynamics and to some other chapters). Their advice reflected in subsequent critical analysis and improvement of the text of the proposed edition.

The substantial supplements were made in the domain of electrodynamics; in working out of equations for penderomotive moments, in models of magneto-hydrodynamics and electro-hydrodynamics – the branches of mechanics and physics which have been widely developed and which found numerous applications.

In the second volume, some may find an extended elaboration of general thermodynamic theory of modelling of nonlinear elastic media with taking into account finite deformations together with various forms of stress tensors.

Certain supplements have been made in the 19 paragraphs of Chapter VIII, Volume II, concerning the theory of non-steady motion of gases and steam bubbles in fluid with taking part of condensation and vaporisation of fluid.

In arrangement of this part of course invaluable help showed N. S. Khabeev, to whom I would like to express my deep gratitude.

The paragraph 5 of Chapter II comprises additional exposition of construction of the structure of Fermi coordinates.

In the first volume, on the pages 505–510 new interpretation of the basic variational equation as the generalization of the equation of energy has been presented, whereas on the page 491 the generalization of the concept of tensor functions has been given.

Similarly as in the previous editions of the above-discussed material the great and complex works, with the supplementary editor's duties of the whole text have been made by V. V. Rozantseva, M. E. Eglit and A. G. Kulikovskii, to whom I desire to express my acknowledgements and gratitude.

L. I. Sedov

Moscow, September 1982.

Contents

Volume 1

Mechanics of Continuous Media

Volume 1

Chapter I

Introduction

§1. Continuum mechanics and its methods

The topic of continuum mechanics. Continuum mechanics, a large branch of mechanics, is devoted to the study of the motion of gaseous, liquid and rigid deformable bodies. In theoretical mechanics, one studies the motions of material points, of discrete systems of material points and of absolutely rigid bodies. In continuum mechanics, with the aid of and on the basis of methods and observations developed in theoretical mechanics, one considers motions of material bodies which fill space continuously in a continuous manner and the distances between the points of which change during motion.

Beside ordinary material bodies, such as water, air or iron, one investigates in continuum mechanics also special media, i.e., fields of electro-magnetism, radiation, gravitation, etc.

Many different motions of liquids, gases or rigid deformable bodies may be described such as are encountered in the study of material phenomena and in the solution of many technical problems.

Many motions of deformable bodies are encountered directly and may be explained to a necessary degree by relying on elementary, personal, day to day experience. Such a complex of observations gives one a feeling of reality and "common sense" which often makes it possible to correctly forecast and establish mechanical events.

However, in complicated situations, special development and concentration of systematic experimentation are required and special methods for theoretical and experimental studies must be found. Such investigations have led to the foundation and development of the science of continuum mechanics.

It is easy to state examples when everybody can immediately offer methods of solution of some of the most important practical problems of motions of deformable bodies. For instance, one knows the answers to questions such as how to proceed when pouring water from one barrel into another, how to preserve most simply warm air in a room, how to protect oneself against wind and rain, etc. Other problems may be formulated when solutions can only be found on the basis of special knowledge. Examples are: The speed of escape of compressed gas from a balloon, the motion of a cyclone in the atmosphere, methods for lowering the air resistance of an aircraft or the water resistance of a ship, the design of a metal, 500 metres high, television tower, the construction of a bridge with a distance between neighbouring spans bigger than 1 kilometre, the effect of changes in the size of a propeller on an aircraft, the pressure distribution and the motion of the air during the explosion of a bomb, etc.

It is immediately obvious that there are very many tests and problems which cannot be solved satisfactorily with the aid of known experimental and theoretical results. The solution of new complex problems, which are of vital practical importance or which are posed in the development of the sciences, constitutes the subject of contemporary research.

Examples of such problems are the reduction of the resistance of water to solids moving through it at speeds of the order of 100 m/sec, the creation and preservation of plasmas with temperatures of several millions of centigrades, the explanation of and peculiarities in the behaviour of heavily loaded materials at high temperature (such as plasticity and creep); the determination of the forces arising in shock waves, the design of supersonic passenger aircraft, the understanding of the general circulation of the atmosphere, weather forecasting, the mechanical processes in plants and living organisms, the evolution of stars, the phenomena occurring in the sun and another cosmic and cosmological problems.

At present, the advance of science and engineering in these directions is closely linked to and determined by research in progress; nevertheless, side by side with exact scientific and engineering observations, an important role is played by the sense, talent, intuition and mechanical feeling of designers and engineers. Such facilities are developed as a result of long experience. It must not be thought that all machines, aircraft, ships, etc. can be designed, stressed and analyzed beforehand in all details. Today, many achievements of engineering come about in the same manner in which the Vikings constructed their ships some thousand years ago. At that time, mechanics did not exist as a science, not even in an initial state; nevertheless, the Vikings built good ships.

However, in modern times, engineering has developed to such an extent that one cannot avoid utilization of the results of the sciences and of accumulated systematic experience. It is impossible to imagine contemporary industry without mechanization, and engineering without the support of the existing scientific basis.

Problems of continuum mechanics. Some of the more important problems, which are being studied today, will now be discussed.

Problems of the interaction between gases, liquids and solids in motion. The forces exerted by fluids on solids are determined by the fluid motion. Therefore the study of the motion of solids inside liquids is directly linked to the study of fluid motion.

The development of the problem was stimulated especially by engineering tasks arising through the motion of aircraft, helicopters, dirigibles, missiles, rockets, ships, submarines, etc., and through the need for the construction of various propelling devices such as propellers, airscrews, etc.

The motion of fluids and gases along tubes and, more generally, inside different machines. Here the laws of interaction between fluids and walls have fundamental importance; in particular, one studies problems of resistance of movable and immovable walls, of irregularities in velocity distributions, etc., which have real significance for pipe-lines, pumps, turbines and other hydraulic machines.

Filtration, i.e., the motion of fluid through soil and other porous media. For example, in soil, one observes constantly motion of water which must be taken into consideration in the building of foundations of structures (dams, bridge supports, hydroelectric stations), in the construction of earth tunnels, etc. Filtration is also very important for the oil industry.

Hydrostatics, i.e., the equilibrium of fluids and bodies floating inside or on the surface of a fluid, equilibrium figures of rotating fluid masses under the influence of Newtonian gravitational forces.

Wave motion. Propagation of waves in solids; waves on the surface of the ocean; ship waves; propagation of waves along canals and rivers; tides; seismic processes; sound waves; the general problem of noise in different media, etc. All surrounding media (fluids, gases, rigid bodies and various fields) are constantly in states of vibration and of different perturbed motions which propagate in time and space. It is immediately clear that these phenomena play a very important role in life and are essential for the solution of many engineering problems.

Unsteady motions of gases with chemical transformations in the presence of shocks, detonations and combustion, for example, in the flow of air, in the cylinders of piston engines or in the chambers of jet engines etc.

Protection of solids against burning and melting as they enter with large velocities into the denser layers of the atmosphere.

The theory of turbulent motion of gases and fluids, which in reality represents very complex irregular motion with a stochastic character, pulsating about some regular mean process which in many problems under consideration have great practical significance. An overwhelming number of motions of gases and fluids in stars and cosmic clouds, in the atmosphere of Earth, in rivers, canals, in conduits and other technical structures and machines have a turbulent character. Hence it is clear that the theory and experiments, arising in the study of turbulence, are enormously important. Studies of turbulence at the present time are still not sufficient for an understanding of many peculiarities and relationships of such complex motions which occur in nature.

Problems of the description of the motion of highly compressed fluids and gases with special consideration of the complicated physical properties of different media in such states, especially in the presence of high temperatures. There exist interesting and important branches of industry in which one must deal with bodies subject to high pressures (of the order of many thousands and millions of atmospheres), for example, in the artificial manufacture of diamonds, in the application of explosives for the punching of machine parts and in many other problems.

In *strongly diluted gases*, there occur also very important phenomena. For the study of different processes, arising in the motion of media in strong vacua in laboratory experiments, in cosmic space, in the atmospheres of planets and stars, one must also apply the methods of continuum mechanics.

Problems of magneto-hydrodynamics and the study of motions of ionized media, i.e., plasmas for which the interaction of a gas with an electromagnetic field is taken into account, have today acquired first priority, attention and technical significance. In particular, such phenomena must be studied for the construction of magneto-hydrodynamic generators of electric current in which the energy of the motion of plasmas is directly converted into the energy of electric current. Likewise one must note that the solution of problems of the utilization of thermo-nuclear energy is closely linked to the solution of problems of the behaviour of high temperature plasmas in strong magnetic fields.

The science of weather forecasting, i.e., of meteorology, represents itself to a significant degree a study of the motion of the air masses of the atmosphere of Earth. It is an important branch of continuum mechanics which is closely connected with a multitude of other branches of physics.

Basic problems of astrophysics and cosmology are studied within the framework of continuum mechanics. They touch questions of the internal structure of stars and of the structure of their photosphere, of the motion of nebulae and cosmic clouds, of eruptions and explosions of variable stars, of oscillations of Cepheids and, finally, the fundamental problem of the development of galaxies and of the structure and evolution of the universe.

A significant part of continuum mechanics deals with the study of the motion and equilibrium of "solid" deformable bodies. The *theory of elasticity* is basic for the construction of any kind of building and all possible machines. At the present time, special attention is given to those branches of mechanics which concern the study of complicated elastic properties of bodies and take into account inelastic effects in rigid bodies such as *plasticity* linked to phenomena of residual deformations, *creep* connected with gradual build up of deformation under constant external loading and with high temperature strength of machine parts (phenomena of creep manifest themselves in long period after-effects in many structures and at very high temperatures even during short time intervals).

Great significance have the study of different forms of *fatigue of materials*, taking into consideration phenomena of hereditary effects in processes of motion and equilibrium of bodies.

With the appearance and utilization of the new *polymer materials* one must necessarily give consideration to their internal physical structure which may be changed into phenomena which are of practical interest.

Finally, great importance attaches to work dealing with the general problem of *strength and failure* of structures made of different materials. This important practical problem has hitherto not found satisfactory solution.

Mention may still be made of *mechanical* problems which arise in the motion of viscous types of mixtures, i.e., in the motion of sand, snow and different soils, alloys, liquid solutions, suspensions and emulsions, fluids with polymer fillers, etc. There are interesting *problems of cavitation* which characterize the formation and disappearance in moving fluids of bubbles and large cavities, filled with gases and vapours of fluids.

It must be emphasized that recently problems of the technology of manufacture in *chemical plants* are based on mechanical investigations of motions of corresponding continuous media.

Important *new contemporary theories* in which problems of the interaction of powerful *laser beams* with different bodies are studied are problems of non-linear optics, of the interaction of moving bodies with electromagnetic fields. Such interactions in macroscopic scales are essentially linked up with effects which are described within the frame work of *quantum mechanics.* An analogous position is encountered in the description of macroscopic properties of bodies which are linked to motions at very high temperatures or when one takes into consideration magnetization and electric polarization.

Recently, very many studies in *biological mechanics* have been initiated; in particular, mechanical models have been constructed which are used to describe the propagation of impulse in nerves, the mechanism of dislocation of viruses, microbes and other small organisms in different media, the swimming of fishes, the motion of blood in living organisms and phenomena of the shortening of muscles.

Methods of continuum mechanics. This book presents a theoretical course of continuum mechanics. It deals with the mathematical methods for the study of the motion of deformable bodies. These methods will now be summarized.

A number of mechanical concepts are introduced which characterize and uniquely determine the motion of continuous media. These concepts must be defined in terms of numbers or other mathematical concepts, examples of which are velocity fields, pressure fields, temperatures, circulations, etc. In the sequel, these concepts and many other mechanical characteristics of the motion of continuous media will be introduced.

In continuum mechanics, methods for the reduction of mechanical problems to problems of mathematics are developed, i.e., to problems of finding certain values or numerical functions with the aid of different mathematical operations.

Besides, one of the most important aims of continuum mechanics is the discovery of several properties and the determination of the laws of motion of deformable bodies. In the sequel, there will be introduced several laws on the forces, which fluids exert on bodies moving through them, some relationships between pressures and velocities governing a rather wide class of important motions, rules holding between external loads and corresponding deformations, etc.

It should be noted that the very solution of concrete problems of continuum mechanics with the aid of mathematical operations also, as a rule, belongs to continuum mechanics. That circumstance is explained by the fact that even in the simplest cases mathematical formulations of problems of the mechanics of continuous media are arrived at which are very difficult and effectively insoluble by contemporary tools of mathematics. Therefore one must modify the formulation of a problem and find approximate solutions on the basis of various mechanical conjectures and arguments.

Under the influence of continuum mechanics, a number of fields of mathematics underwent great developments. For instance, continuum mechanics had great influence on the development of certain sections of the theory of functions of a complex variable, of boundary value problems for partial differential equations, of integral equations, etc.

Analogues of certain problems of continuum mechanics which were discovered in detailed studies of other branches of mechanics and physics were very useful.

It turned out that in many cases different problems of continuum mechanics and mathematical methods are closely interlinked. For example, the study of the motion of liquids in pipes serves to clarify certain basic facts of the motion of fluids around aircraft wings. The methods of solution of problems of the flow around aircraft wings have a lot in common with the mathematical methods of solution of problems of filtration of liquids through soil. Again, many results of the theory of the motion of gas in tubes may be used in the study of various problems of wave motion in channels, etc.

At first, one will be far away from a study of the problems which have just been mentioned. It will be necessary to prepare a lot of material of a general mechanical and mathematical nature. At first, one will not have the feeling that one is dealing with or directly approaching an investigation of problems which are connected with phenomena encountered in nature and engineering. A reference to the history of continuum mechanics might offer here some consolation. More than 100 years had to pass before the mathematical methods of continuum mechanics brought success with practical problems in the theory of the motion of liquids and gases.

The aim of this book is to make the reader familiar with the foundations of continuum mathematics which are sufficient and necessary for specialized studies of different concrete problems.

§ 2. Fundamental hypotheses

(a) The structure of real bodies and the hypothesis of continuity.

In studies of the motion of bodies one must rely heavily on the properties of real solids. As is well known, all bodies represent accumulations of various kinds of molecules and atoms. Sometimes a body may be ionized, i.e., it may consist of electrons, ions (atoms and molecules with an excess or deficiency in the number of electrons) and neutral particles. Consider next some data on elementary particles which are known from physics.

Data on elementary particles. The radius of the nucleus of an atom is of the order 10^{-13} cm. The radius of the molecule of hydrogen is 1.36×10^{-8} cm, i.e., the radius of a nucleus is much smaller than the radius of a molecule and, at the same time, the basic mass of a substance is concentrated in the nucleus: The mass of an electron is 9.1066×10^{-28} g, the mass of a proton is 1.6724×10^{-24} g.

Under normal conditions (temperature 0 °C, atmospheric pressure at sea level), one cubic centimetre of air contains $N = 2.687 \times 10^{19}$ molecules. If one considers a cube with edges 10^{-3} cm long, which lies quite often beyond the limits of accurate length measurement in engineering, it still contains 27×10^9 particles. At a height of 60 km, which is today's "ceiling" of aircraft, one has $N = 8 \times 10^{15}$ particles per cm^3. In interstellar space, where gases are strongly rarefied, $N = 1$ per cm$^3 = 10^{15}$ particles per km^3. The distance of a kilometer is very small compared with characteristic cosmic distances, so that even interstellar gas may be considered to be a medium with very large numbers of particles in small volumes.

There is no atmosphere on the moon, but one has there $N=10^{10}$ particles per cm^3, i.e., one has on Earth 2.7×10^9 times more particles than on the moon. Such intense vacua have not been achieved in practice under laboratory conditions on Earth. In such a vacuum, in many cases, materials in contact will fuse.

For iron (Fe), one has $N=8.622\times10^{22}$ particles per cm^3 and the density $\rho_{Fe}=7.8$ g/cm^3, while the density of its nuclear substance is $\rho_{\text{nucl. Fe}}=1.16\times10^{14}$ g/cm^3, so that $\rho_{Fe}/\rho_{\text{nucl. Fe}}=7\times10^{-14}$.

It is seen that the volume occupied by a body is, as a rule, much bigger than that in which its matter is concentrated. Hence, all bodies, in essence, consist of emptiness and, at the same time, practically small volumes of space, occupied by solids, always contain large numbers of particles.

Atoms and molecules are constantly in chaotic motion.

Under normal atmospheric conditions, the mean velocity of a hydrogen molecule is 1.692 m/sec (which is greater than the velocity of modern passenger aircraft). All the time, molecules collide with each other; the free path length in the normal atmosphere is $l = 11.2\times10^{-6}$ cm. For oxygen, one has $v_{\text{mean}} = 425$ m/sec and $l = 6.5\times10^{-6}$ cm, i.e., a molecule collides every second 6.54×10^9 times.

The interaction of particles. Particles interact in a definite manner. In gases, one has only collisions. In fluids and solids, particles come close to each other and interacting forces exist between them.

The forces which ensure the strength and elasticity of a body are electric in origin and, generally speaking, follow Coulomb's law. As regards nuclear forces and the forces of weak interaction, they only appear in nuclear reactions, when particles interact with each other over short distances. In order to bring particles together, one requires enormous amounts of energy which may arise through the chaotic motion of particles at high temperatures of many millions of centigrades.

On physico-chemical processes of importance for continuum mechanics. On the basis of a knowledge of the electric forces of interaction between particles, one may develop a theory of deformable solids. Hence electrodynamics assumes here great importance. In the construction of material continua for the modelling of real bodies, one must take into consideration their various structural peculiarities. Bodies may be gaseous, liquid, rigid, crystalline with different phases. At high temperatures, states arise in which matter may be considered simultaneously to be a gas, liquid or solid.

Besides its structure, the nature of matter and the properties of the components of mixtures, solutions and alloys play important roles.

In many cases, there arise mechanical problems relating to the motion of bodies with qualitative changes of components and of relative compositions. For instance, problems of the motion of gases, subject to nuclear and chemical reactions, and, in particular, to combustion, disassociation, reassociation, ionization, are of this type.

Processes of phase change may play important roles in motions of material bodies; for example, there may occur condensation, evaporation, melting, hardening, polymerization, recrystallization, etc.

In order to study motions of material continua, one must introduce internal stresses. In bodies with discrete molecular structure, internal stresses are statistical averages which depend on direct forces of interaction between molecules, situated on either sides of a cross-section under consideration, and on transfer of macroscopic momentum through that section as a result of the thermal motion of molecules.

The viscosity of gases can be explained by the thermal motion of molecules aligning the macroscope motion of neighbouring gas particles. Thus, properties of internal stresses in material bodies are determined by their molecular structure, by the forces of interaction between molecules and atoms, which take place at very short distances, and by thermal motion, characterized by the temperature.

In a similar way, heat conduction can be explained. For any two adjacent particles of a medium which are in contact, there occurs an interchange of energy either by collisions or directly by exchange of fast and slow molecules. Statistically, the mean energy of thermal motion characterized by the temperature tends to a state of equilibrium.

The mechanism of diffusion in mixtures can also be explained by molecular kinetic processes of mixing of molecules as a result of thermal motion.

The effect of radiation is somewhat complex owing to quantum effects of energy exchange levels in the system of a molecule or atom, or of the nucleus of an atom, and also owing to accelerated motions of charged particles. In many cases, radiation, which may be considered as emission of photons, is closely linked to chaotic thermal motion of molecules and atoms and depends essentially on the temperature which defines possible excitations of energy at collisions of particles. A study of the motions of material bodies at high temperatures must take into account effects of energy exchange and variations in temperature owing to absorption and dissipation of light energy.

Electric polarization and magnetization, connected with the regularly ordered structure of elementary particles in bodies, may also have essential significance for various motions of material bodies.

Mechanisms of internal interaction in solids, in materials with complicated molecular structures, in bodies with very high density, at comparably low temperatures and in other cases may be very complicated and, in general, cannot be described within the framework of Newtonian mechanics. In order to understand the corresponding interactions in many cases, one must utilize the concepts and laws of quantum mechanics.

One of the main tasks of physics is to find for the phenomena, listed above, macroscropical laws based on a thorough analysis of the physical microscopic mechanisms and properties of elementary particles.

Statistical and phenomenological approach. Note that the complex structure of molecules and their retention of the electric forces of interaction are not always known. It seems that mechanics should be developed on the basis of the representation of a material body as an accumulation of elementary particles. However, it is impossible to follow up the motion of each elementary particle, owing to their large numbers and a lack of knowledge regarding their forces of interaction. It is very important to note that, as a rule, it is even not essential to know the motion of each elementary particle.

In practice, one requires to know only some mean, overall or global characteristic.

One of the general methods of approach to the study of the behaviour of material media is the statistical method developed in physics.

It employs the theory of probability and introduces means of characteristics of huge assemblies of particles. Statistical methods always introduce additional hypotheses concerning the properties of particles and their interactions. Note that in many cases there does not exist a basis for the construction of such methods. In those cases when they are constructed, usually they do not provide effective means of solution of problems, owing to the excessive complexity of the corresponding equations.

Another general method of approach to the study of the motions of material bodies is the construction of phenomenological macroscopic theories based on general laws and hypotheses, obtained by experiment. Macroscopic theories provide effective means of the solution of practically important problems and results achieved with their aid agree with experiments.

In what follows, a phenomenological macroscopical theory of material media will be developed.

The hypothesis of a continuous medium. The concept of a continuous medium must first of all be defined. All bodies consist of individual particles; however, there are very many such particles in any volume which is essential for the ensuing considerations, and it is possible to assume that a medium fills space in a continuous manner. Water, air, iron, etc., will be considered to be bodies, which fill completely a part of space.

Not only common material bodies can be considered to be continuous continua, but also fields, for instance, electromagnetic fields.

In particular, such an idealization is necessary, because one wants to employ the apparatus of continuous functions, differential and integral calculus in the study of the motion of deformable bodies.

(b) On space and time

Space is understood to be a collection of points, determined with the aid of numbers which are called coordinates.

Metric space. In essence, continuous metric spaces will be considered, i.e., such spaces for which the distances between its points are defined. As an example of metric space, one has ordinary three-dimensional space the points of which are defined by a single Cartesian coordinate system (common to the entire space) and the distance between two points (x_1, y_1, z_1) and (x_2, y_2, z_2) of which are defined by the well-known formula

$$r = \sqrt{\{(x_1 - x_2)^2 + (y_1 - y_2)^2 + (z_1 - z_2)^2\}} . \tag{2.1}$$

Euclidean space. When can one introduce a single Cartesian coordinate system for an entire space? For the sake of simplicity, consider two-dimensional space. Obviously, one can always introduce for the entire plane a single Cartesian coordinate system. On the surface of a sphere, the curvature of which is not zero, this cannot be done, i.e., it is impossible to introduce on the sphere a coordinate system such that the distance between two arbitrary points in it is defined by a formula of the type (2.1). However, in the neighbourhood of each point on the sphere, one may approximate with the aid of a Cartesian coordinate system located in the local tangent plane. For three-dimensional spaces, it is also not always possible to introduce a simple, Cartesian coordinate system suitable for an entire space.

In the sequel, only such spaces will be considered for which it is possible to introduce a single Cartesian coordinate system. Such spaces are referred to as Euclidean spaces and the corresponding mechanics is called Newtonian mechanics. Experience shows that within not very large scales one may assume with a high degree of accuracy that real physical space is Euclidean.

Absolute time. The concept of time is linked to experience and is indispensable in mechanics. Any mechanical phenomenon is always described from a point of view of an observer. Time, generally speaking, may depend on the system of measurement of the observer.

Assume that time flows independently for all observers—in a train, in an aircraft, in an auditorium. Hence, use will be made of absolute time—an idealization, which is convenient for the description of reality only if the effects of the theory of relativity can be negelected.

Hence, it will be assumed that one is dealing with the motion of continuous media—of continua in Euclidean space—and absolute time will be used. Thus, three fundamental hypotheses have been introduced above with the aid of which the theory of motion of deformable bodies will be developed. Derivations from this theory, based on these hypotheses, very often, but not always, agree with experiment. If necessary, this model of space and time can be refined and generalized. However, all further generalizations will rest heavily on Newtonian mechanics, based on the fundamental hypotheses above. The essence of these hypotheses will become clearer as the theory is developed further.

Chapter II

Kinematics of a deformable medium

§1. Lagrangian approach to the study of the motion of a continuous medium

Coordinate systems. Motion is always determined with respect to some reference system — coordinate system. A correspondence between numbers and points in space is established with the aid of a coordinate system. For three-dimensional space, three numbers x^1, x^2, x^3 correspond to points; they are called the coordinates of the point. Lines along which any two coordinates remain constant, are called coordinate lines (fig. 1). For example, the line for which x^2 = const., x^3 = const. defines the coordinate line x^1, along which different points are fixed by the values of x^1; the direction of increase of the coordinate x^1 defines the direction along this line. Three coordinate lines may be drawn through each point of space. At each point, the tangents to the coordinate lines do not lie in one plane and, in general, they form a non-orthogonal trihedron.

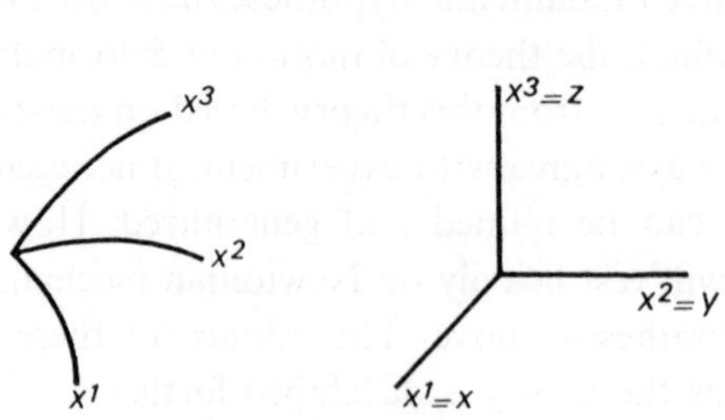

Fig. 1. Curvilinear and Cartesian system of coordinates.

If the coordinate lines x^i are straight, the system of coordinates is rectilinear; if not, the system is curvilinear. It will be seen below that curvilinear coordinate systems are necessary and essential in continuum mechanics.

Notations of coordinates and time. Let the symbols x^1, x^2, x^3 denote coordinates in any system which may also be Cartesian, the symbols x, y, z only orthogonal Cartesian coordinates and t the time.

Motion of a point. A point moves relatively to the coordinate system x^1, x^2, x^3, if its coordinates change in time:

$$x^i = f^i(t) \qquad (i = 1, 2, 3)\,. \tag{1.1}$$

The moving point coincides with different points of space at different instants of time. The motion of a point is known, if one knows the functions (1.1), referred to as the law of motion of the point.

Motion of a continuum. A continuous medium represents a continuous accumulation of points. By definition, knowledge of the motion of a continuous medium implies knowledge of the motion of all points (a study of the motion of a volume of a continuous body as a whole is, in general, insufficient).

On the individualization of the points of a continuum. For the above purpose, one must treat individually distinct points, which from a geometrical point of view are completely identical points of the continuum. It will be seen below that the laws of individualization used in the theory are determined, generally speaking, by the fact that the motion of each point of a continuous medium is subject to certain physical laws. Individual points of a medium may be allotted, for example, the values of their initial coordinates. Let the coordinates of points at the initial moment t_0 be denoted by a, b, c or ξ^1, ξ^2, ξ^3 and the coordinates of points at an arbitrary instant of time by x^1, x^2, x^3.

Laws of motion of a continuum. For any point of a continuum, specified by the coordinates a, b, c, one may write down the law of motion which contains not only functions of a single variable, as in the case of the motion of a point, but of four variables: the initial coordinates a, b, c and the time t.

$$\left.\begin{aligned} x^1 &= x^1(a, b, c, t)\,, \\ x^2 &= x^2(a, b, c, t)\,, \\ x^3 &= x^3(a, b, c, t) \end{aligned}\right\} \quad \text{or} \quad x^i = x^i(a, b, c, t)\,, \tag{1.2}$$

If in (1.2) a, b, c are fixed and t varies, then (1.2) describes the law of motion of one selected point of the continuum. If a, b, c vary, and t is fixed, then (1.2) gives the distribution of the points of the medium in space at a given instant of time. If a, b, c and t vary, then one may interpret (1.2) as a formula which determines the motion of the continuous medium and, by definition, the functions (1.2) yield the law of motion of the continuum.

Lagrangian coordinates. The coordinates a, b, c or ξ^1, ξ^2, ξ^3 (or sometimes definite functions of these variables), which individualize the points of a medium, and the time t are referred to as Lagrangian coordinates.

The fundamental problem of continuum mechanics consists of the determination of the functions (1.2).

In the sequel, the concept of the law of motion will always be used explicitly or implicitly.

For the sake of generality, note that a medium represents an accumulation of points, but it need not be a material body. For instance, it may sometimes be agreed to represent by points in a plane the prices of some products and to study by the methods of the kinematics of continuous media the motion of prices in economics.

Also, and this is done frequently, one may study the laws of displacement in space of different states of motion of material particles, and not of the particles themselves. For example, waves are observed on a rye field in windy weather, and one may speak of the displacements in space of the highest points or of the depressions of the rye surface, and not of the ears themselves.

Thus, in kinematics, a continuous medium may be conceived as an abstract geometrical object, and not merely as a material body. The motion of a medium can be governed by various laws. When considering the motion of a material body, these may be known physical laws or, when speaking, for example, of prices, they may be mathematical laws of economics which are at that time applicable.

Continuity of the functions specifying laws of motion. In a study of the mechanics of deformable media, one will wish to rely on the apparatus of differential and integral calculus. Therefore assume that the functions entering into the laws of motion of a continuum are continuous and possess continuous partial derivatives with respect to all variables. This assumption is general enough, but it limits greatly the class of phenomena which may be studied.

In fact, water, for example, may splash. Then particles, which lie infinitely close to each other at one instant, will not be close to each other at the following instant of time. It is impossible to describe such phenomena and retain the assumption of continuity of the laws of motion. At a later stage, it will be seen that in many cases the assumption of continuity of a motion must be dropped and that one has to consider also motions which are such that they themselves or their derivatives suffer discontinuities on some distinct surfaces. Motions of this sort, for example, shock waves, will be considered in the sequel. However, note that the investigation of discontinuous motions will still be based on the theory of continuous motions.

Single-valuedness of the laws of motion. On the grounds of physical considerations, assume that at every instant of time $t=\text{const.}$ the functions $x^i = x^i(\xi^1, \xi^2, \xi^3, t)$ are single-valued. As is known, in that case the Jacobian

$$\Delta = \begin{vmatrix} \dfrac{\partial x^1}{\partial \xi^1} & \dfrac{\partial x^1}{\partial \xi^2} & \dfrac{\partial x^1}{\partial \xi^3} \\[2ex] \dfrac{\partial x^2}{\partial \xi^1} & \dfrac{\partial x^2}{\partial \xi^2} & \dfrac{\partial x^2}{\partial \xi^3} \\[2ex] \dfrac{\partial x^3}{\partial \xi^1} & \dfrac{\partial x^3}{\partial \xi^2} & \dfrac{\partial x^3}{\partial \xi^3} \end{vmatrix}$$

cannot be identically equal to zero.

If $\Delta \neq 0$ then the Equation (1.2) can be solved for ξ^i and the solution represented in the form of single-valued continuous functions:

$$\xi^i = \xi^i(x^1, x^2, x^3, t). \tag{1.3}$$

General properties of continuous mappings. The set of values x^i constitutes in space a region D which is occupied by the solid at a given instant of time t. If the coordinates ξ^1, ξ^2, ξ^3 are looked upon as the values of x^1, x^2, x^3 at some other instant of time t_0, then the region D_0 of the variables ξ^1, ξ^2, ξ^3 corresponds to the volume occupied by the solid at time t_0.

In this case, the laws of motion (1.2) and (1.3) can be considered as a single-valued and continuous mapping of D on to D_0.

As is well known, the general topological properties of such mappings consist of the fact that any volume V_0 is transformed into a volume V, a surface S_0 into a surface S, a curve L_0 into a curve L, while a closed surface and a closed curve are transformed into a closed surface and a closed curve, respectively (fig. 2). For example, a volume cannot be transformed into a single point, since then the condition of single-valuedness would be violated; similarly, a closed curve cannot be transformed

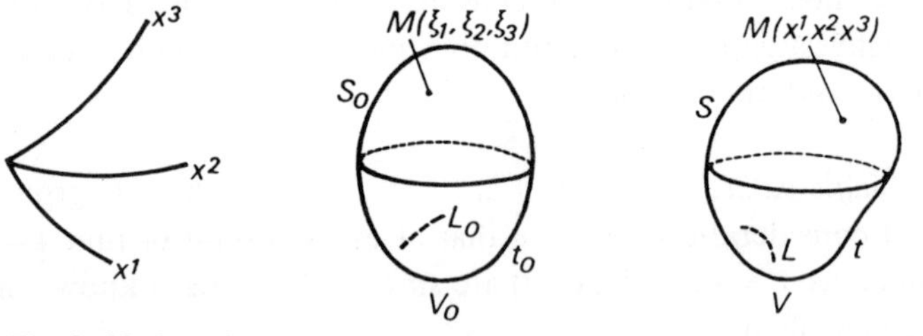

Fig. 2. Motion of a continuum. At $t = t_0$, $x^1 = \xi^1$, $x^2 = \xi^2$, $x^3 = \xi^3$.

into an open curve, since then the condition of continuity would be violated.

Reference frame. The motion of a continuum, like any other motion, is always determined with respect to some system of coordinates x^1, x^2, x^3 — the reference frame of the observer. This coordinate system can be chosen arbitrarily. By its very definition, its choice depends on the observer. In practice, it is often linked to Earth; however, it can be attached to the Sun, a star, an aircraft, a coach, etc. By definition, the system may be movable or immovable.

In Newtonian mechanics, special physical significance attaches to motions relative to inertial coordinate systems, translating with respect to each other with constant velocity. The presence of such coordinate systems (closely linked to the postulate of the Euclidean nature of physical space and the postulate of absolute and equal real time for different points) is a basic postulate of Newtonian mechanics.[1)]

All physical laws in Newtonian physics are normally formulated in inertial coordinate systems and do not depend on the choice of the inertial coordinate system. This is the celebrated principle of Galilei–Newton. In practice, in life, one may choose as inertial coordinate system a Cartesian coordinate system in which distant stars may be assumed to be fixed.

The concomitant system. At the same time, in the case of motion of continuous media, one must introduce still a concomitant coordinate system. Side by side with the coordinates x^1, x^2, x^3, the Lagrangian coordinates ξ_1, ξ_2, ξ_3 of the individual points may be conceived as different coordinates of the same points of space in the region D. The corresponding system of coordinates ξ_1, ξ_2, ξ_3 in the same space constitutes a moving, deforming, curvilinear coordinate system which is referred to as concomitant coordinate system. Hence, if at the initial instant t_0 one chooses in the

1) In the special theory of relativity, the existence of inertial coordinate systems is also postulated; they are interlinked by Lorenz transformations; however, physical space is presented as the four-dimensional pseudo-Euclidean space of Minkowski (the fourth coordinate being connected with proper time). In this theory, one may also use for the observer, describing relative motion, any moving coordinate system.

In the general theory of relativity, any coordinate system, which moves with respect to another coordinate system, is assumed to have equal rights, and physical space is not given, but defined, however, on the assumption that physical space is four-dimensional and Riemannian, and for small volumes the laws of the special theory of relativity are fulfilled.

It is interesting to note that, as a result, solutions of corresponding problems are obtained and that multiply connected, in a topological sense empty (i.e., without mass and charge) Riemannian space resembles in a known sense Euclidean space with gravitational and electric fields, conditioned by the presence of masses and charges. (see, Wheeler J. A., "Neutrinos, Gravitation and Geometry", Bologna 1960.)

medium some coordinate lines ξ_1, ξ_2, ξ_3 consisting of points of the continuous medium, then at subsequent instants of time they become together with points of the continuum coordinate lines of the concomitant system. However, if they were chosen to be straight lines at the initial instant, then, generally speaking, they will be curved at subsequent instants (fig. 3).

Hence, a coordinate system, linked to the particles of a continuous medium, varies with time. The choice of such a coordinate system at any given instant of time is arbitrary; however, at subsequent instants, it is already determined, since it is "frozen into" the medium and deformed together with it. Such a coordinate system, which is fixed in a medium, has been defined earlier as a concomitant system. All points of a continuum are always at rest with respect to the moving concomitant system, since, by definition, their coordinates ξ^1, ξ^2, ξ^3 do not vary in the concomitant system.

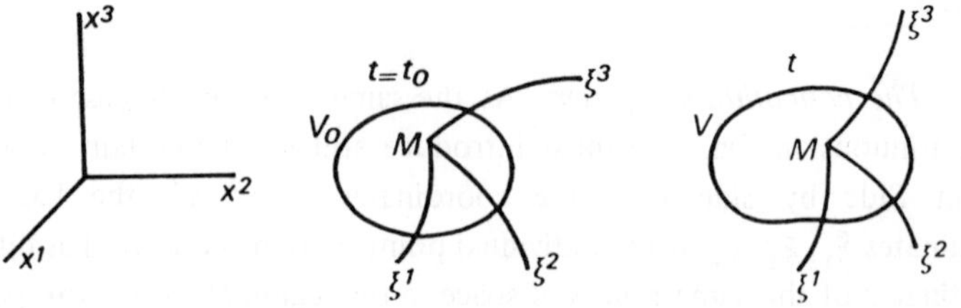

Fig. 3. x^1, x^2, x^3 = reference frame, ξ^1, ξ^2, ξ^3 = concomitant Lagrangian system.

However, the system itself moves, expands, contracts, bends, etc. The concept of the concomitant coordinate system is a generalisation of the coordinate system of a rigid body, used in theoretical mechanics, to the case of continuous media.[1)]

Thus, always in considering the motion of a continuous medium, one must individualize points and, consequently, use Lagrangian coordinates. Therefore, in investigations of motions of continuous media, one must

[1)] Obviously, for each coordinate system and, in particular, for the reference system of an observer, one may always introduce an imagined idealized medium for which the coordinate system under consideration is concomitant.

For the same medium one can introduce different concomitant coordinate systems. Then, sometimes, one can separate the concept of the reference frame of an observer as of the same invariant object, from the concept of the generally used coordinate system, which may differ for the observer in the given reference system.

always postulate the existence of a reference system x^1, x^2, x^3 with respect to which a motion is studied, and that of a concomitant coordinate system.

In the Lagrangian approach to the study of motions of continuous media, one uses ξ^1, ξ^2, ξ^3 and t as independent variables. In other words, it operates on the complete history of every individual point of the medium separately. In practice, such a description is often too detailed and too complicated; however, it is always implied in the formulation of physical laws. Besides the concept of laws of motion, one must still introduce for the description of the motion of continuous media certain other concepts, in particular, those of the velocity and acceleration of particles of a continuous medium.

Velocity. Let some point of a medium at the instant t occupy the point M, and at the instant $t+\Delta t$ the point M', and let $\overline{MM'}=\Delta \boldsymbol{r}$.

By the quantity $\Delta \boldsymbol{r}$ is understood the infinitesimal oriented displacement of the individual point of the medium during the time interval Δt. When one may introduce the radius vector $\boldsymbol{r}$ in the space (and in Euclidean space this is always possible), the quantity $\Delta \boldsymbol{r}$ is, obviously, the differential of the radius vector of the material point under consideration.

The limit of the ratios of the two infinitesimal quantities $\Delta \boldsymbol{r}$ and Δt for $\Delta t \to 0$ in the case of non-Euclidean space or the partial derivative of the radius vector of the point of the medium with respect to time $\partial \boldsymbol{r}/\partial t$ in the case of Euclidean space is referred to as the velocity of the point of the medium. The symbol $\boldsymbol{v}$ will be used to denote the velocity.

The radius vector $\boldsymbol{r}$ depends in the general case on the three parameters ξ^1, ξ^2, ξ^3, individualizing the point of the medium, and on the time t. The velocity is evaluated for individual points of the medium, i.e., for fixed ξ_1, ξ_2, ξ_3, therefore it is equal to the partial derivative of $\boldsymbol{r}$ with respect to t

$$\boldsymbol{v} = \frac{\partial \boldsymbol{r}}{\partial t}.$$

Velocity is evaluated with respect to a frame of reference. Obviously, the medium is at rest with respect to the concomitant coordinate system, and hence the velocity with respect to the concomitant system always vanishes.

Base vectors. Through each point of space, there passes three coordinate lines, and at each space point $M(x^1, x^2, x^3)$ one can consider elementary rectilinear directions $\Delta \boldsymbol{r}_1, \Delta \boldsymbol{r}_2, \Delta \boldsymbol{r}_3$, which radiate from M and join it to the points $M_1(x^1+\Delta x^1, x^2, x^3)$, $M_2(x^1, x^2+\Delta x^2, x^3)$ and $M_3(x^1, x^2, x^3+\Delta x^3)$, respectively. Thus, one can introduce at every point of space the limits of the ratios $\Delta \boldsymbol{r}_i/\Delta x_i$ or $\Delta \boldsymbol{r}_i/\Delta \xi^i$ (as $\Delta x^i \to 0$ or $\Delta \xi^i \to 0$), i.e., vectors which obviously will be directed along the tangents to the corresponding coordinate lines through M. In Euclidean space, these limits will be partial derivatives of $\boldsymbol{r}$ with respect to the corresponding coordinates. If Δx^i and $\Delta \xi^i$ denote the arc-lengths along the corresponding coordinate lines, then the magnitudes of $\partial \boldsymbol{r}/\partial x^i$ and $\partial \boldsymbol{r}/\partial \xi^i$ are equal to unity.

Introduce the notation

$$\frac{\partial \boldsymbol{r}}{\partial x^i} = \mathfrak{z}_i \quad \text{and} \quad \frac{\partial \boldsymbol{r}}{\partial \xi^i} = \hat{\mathfrak{z}}_i \tag{1.4}$$

and call $\mathfrak{z}_i$ and $\hat{\mathfrak{z}}_i$ the base vectors of the reference and concomitant systems, respectively. If the coordinate system x^1, x^2, x^3 is Cartesian, then one may use the notation

$$\mathfrak{z}_1 = \boldsymbol{i}, \quad \mathfrak{z}_2 = \boldsymbol{j}, \quad \mathfrak{z}_3 = \boldsymbol{k},$$

where $\boldsymbol{i}, \boldsymbol{j}, \boldsymbol{k}$ are the unit vectors along the coordinate axes x, y, z, respectively. If the coordinate systems x^1, x^2, x^3 or ξ^1, ξ^2, ξ^3 are curvilinear, then, generally speaking, $\mathfrak{z}_i$ and $\hat{\mathfrak{z}}_i$ vary from point to point in space and form at every point of space a non-orthogonal trihedron.

Velocity components. An infinitesimal displacement $\overline{MM'} = \Delta \boldsymbol{r}$ of a point of the continuum may be expanded in terms of the base vectors $\mathfrak{z}_i$ at the point M:

$$\Delta \boldsymbol{r} = \Delta x^1 \mathfrak{z}_1 + \Delta x^2 \mathfrak{z}_2 + \Delta x^3 \mathfrak{z}_3, \tag{1.5'}$$

where Δx^i are the components of the displacement $\Delta \boldsymbol{r}$. The expression (1.5′) may be rewritten, more briefly,

$$\Delta \boldsymbol{r} = \sum_{i=1}^{3} \Delta x^i \mathfrak{z}_i = \Delta x^i \mathfrak{z}_i, \tag{1.5}$$

where the sum sign $\Sigma_{i=1}^3$ has been omitted in the last expression. In the sequel, the summation sign will often be omitted on the understanding that in expressions of the type (1.5) summations are carried out over indices which appear simultaneously as subscripts and superscripts.

Divide (1.5) by the time increment Δt corresponding to the displacement of a point of the continuum from M to M' in the reference space and pass to the limit $\Delta t \to 0$; one then obtains, by definition, the velocity of a point of the continuum

$$\boldsymbol{v} = \frac{\partial \boldsymbol{r}}{\partial t} = \frac{\partial x^i}{\partial t} \mathbf{\mathfrak{z}}_i = v^i \mathbf{\mathfrak{z}}_i = v^1 \mathbf{\mathfrak{z}}_1 + v^2 \mathbf{\mathfrak{z}}_2 + v^3 \mathbf{\mathfrak{z}}_3 , \tag{1.6}$$

whence

$$v^1 = \left(\frac{\partial x^1}{\partial t}\right)_{\xi^i}, \quad v^2 = \left(\frac{\partial x^2}{\partial t}\right)_{\xi^i}, \quad v^3 = \left(\frac{\partial x^3}{\partial t}\right)_{\xi^i},$$

where the subscripts ξ^i indicate that the derivatives are taken for constant parameters ξ^1, specifying a point of the medium. The quantities v^i are called the components of the velocity vector $\boldsymbol{v}$ referred to the base vectors $\mathbf{\mathfrak{z}}_1, \mathbf{\mathfrak{z}}_2, \mathbf{\mathfrak{z}}_3$. The velocity and its components depend, generally speaking, on the ξ^i and t:

$$v^1 = v^1 (\xi^1, \xi^2, \xi^3, t) ,$$
$$v^2 = v^2 (\xi^1, \xi^2, \xi^3, t) ,$$
$$v^3 = v^3 (\xi^1, \xi^2, \xi^3, t) .$$

In the sequel, the following notation will be used: the letter v with superscripts 1, 2, 3 will denote the components of the velocity vector $\boldsymbol{v}$ in any coordinate system (including Cartesian systems), the letters u, v, w exclusively the components of the velocity vector in Cartesian coordinate systems. Thus, u, v, w are the projections of $\boldsymbol{v}$ on to the x, y, z-axes, respectively. In a Cartesian coordinate system, the position of a point of the medium is characterised by the radius-vector

$$\boldsymbol{r} = x\boldsymbol{i} + y\boldsymbol{j} + z\boldsymbol{k} ,$$

and its velocity by

$$\boldsymbol{v} = \left(\frac{\partial \boldsymbol{r}}{\partial t}\right)_{\xi^i} = \left(\frac{\partial x}{\partial t}\right)_{\xi^i} \boldsymbol{i} + \left(\frac{\partial y}{\partial t}\right)_{\xi^i} \boldsymbol{j} + \left(\frac{\partial z}{\partial t}\right)_{\xi^i} \boldsymbol{k} ,$$

i.e.

$$u = \left(\frac{\partial x}{\partial t}\right)_{\xi^i}, \quad v = \left(\frac{\partial y}{\partial t}\right)_{\xi^i}, \quad w = \left(\frac{\partial z}{\partial t}\right)_{\xi^i}.$$

The concept of a vector. Certain vectors have already been considered, for instance, the velocity vector $\boldsymbol{v}$, the radius vector $\boldsymbol{r}$, the displacement vector $d\boldsymbol{r}$. What is a vector? It is not a scalar; however, at the same time, like a scalar, it is invariant, being independent of the choice of coordinate system. In defining vectors, it is often said that they represent triplets of numbers, the so-called components of the vector, which transform in a definite manner during the transition from one coordinate system to another. However, this definition is insufficient, since a vector is always related to a definite base and, in stating its components, one must always specify the base to which these components refer.

In a Cartesian coordinate system, the components of a vector are related to $\boldsymbol{i}, \boldsymbol{j}, \boldsymbol{k}$, in an arbitrary curvilinear coordinate system, to the base vectors $\mathfrak{z}_i$ which vary in space from point to point. Hence, in contrast to the components of a vector in a Cartesian system, the components of a vector in a curvilinear coordinate system are essentially linked to the point at which they are considered.

For instance, in the case of the velocity vector $\boldsymbol{v}$ at every point of space, one must consider the numbers v^1, v^2, v^3 together with the base vectors $\mathfrak{z}_1$, $\mathfrak{z}_2$, $\mathfrak{z}_3$ and define the vector $\boldsymbol{v}$ by (1.6) through the base vectors one may represent by analogous methods every vector in a given coordinate system.

Acceleration. Beside the velocity $\boldsymbol{v}$, one must still consider the acceleration vector $\boldsymbol{a}$ of a point of the continuum

$$\boldsymbol{a} = \left(\frac{\partial \boldsymbol{v}}{\partial t}\right)_{\xi^i} = a^i \mathfrak{z}_i ,$$

with the components $a^i = a^i(\xi^1, \xi^2, \xi^3, t)$. The acceleration $\boldsymbol{a}$, as the velocity $\boldsymbol{v}$, is evaluated for individual points of the continuum. This definition of the acceleration depends on the observer's choice of the coordinate system x^i which is used in the description of the law of motion (1.2) and which may move.

Note that the relations

$$a^1 = \frac{\partial v^1}{\partial t}, \quad a^2 = \frac{\partial v^2}{\partial t}, \quad a^3 = \frac{\partial v^3}{\partial t}$$

are valid only in Cartesian and not in curvilinear coordinate systems. In fact, the acceleration vector is defined as the time derivative of the velocity vector: $\boldsymbol{a} = (\partial \boldsymbol{v}/\partial t)_{\xi^i}$, and in the evaluation of its components it must be kept in mind that the point of the medium moves in space with time and that the base vector $\mathbf{3}_i$ of a curvilinear system vary in space from point to point.

In Cartesian coordinate systems, one has also

$$a^1 = \frac{\partial^2 x}{\partial t^2}, \quad a^2 = \frac{\partial^2 y}{\partial t^2}, \quad a^3 = \frac{\partial^2 z}{\partial t^2}.$$

In many studies of motions of continua, the basic problem of finding the laws of motion may be replaced by the problem of the determination of the functional dependence of the components of the velocity v^i or acceleration a^i on ξ^1, ξ^2, ξ^3 and t.

Note especially that the Lagrangian approach to the study of motion of continua is fundamental for physical laws, since they are concerned with motions of individual material points.

§2. Eulerian approach to the study of motion of continua

The essence of the Eulerian approach. Assume now that not the history of the motion of individual points of a continuum is of interest, but what happens at different times at a given geometrical point of space which is linked to the reference system of an observer. Concentrate attention on a given point of space at which arrive at different times different particles of the medium. In essence, this is the Eulerian approach to the study of the motion of continua. For instance, the motion of water in a river can be studied either by following the motion of each particle from the upper reaches to the mouth of the river (the Lagrangian approach), or by observation of the changes of the flow of water at definite locations in the river without tracing the motion of individual water particles along the entire river (the Eulerian approach).

Eulerian variables. An Eulerian approach is often employed in applications. The geometric space coordinates x^1, x^2, x^3 and the time t are referred to as Eulerian variables. In the Eulerian approach, a motion is considered to be known, if its velocity, acceleration, temperature and other quantities of interest are given as functions of x^1, x^2, x^3 and t. The functions $\boldsymbol{v} = \boldsymbol{v}(x^1, x^2, x^3, t)$, $\boldsymbol{a} = \boldsymbol{a}(x^1, x^2, x^3, t)$, $T = T(x^1, x^2, x^3, t)$, etc., for fixed x^1, x^2, x^3 and variable t determine the time variations of the velocity, acceleration, temperature, etc., at a given space point for the different particles arriving at this point. For fixed t and variable x^1, x^2, x^3, these functions yield space distributions of the characteristics of the motion at a given instant of time; for variables x^1, x^2, x^3 and t-space distribution of the motion at different instants of time.

Differences between the Lagrangian and Eulerian approaches in the study of motions of continua. Thus, from a Lagrangian point of view, the laws of change of the velocity, acceleration, temperature, etc., of a given particle of a continuum are of interest. In the Eulerian approach, one studies the velocity, acceleration, temperature, etc. at a given location, i.e., one focusses attention on some region of space and wishes to know all the information on the particles which pass through it.

Clearly, the only distinction between the two points of view lies in the fact that the variables in the Eulerian approach are the space coordinates x^1, x^2, x^3 and the time t, and in the Lagrangian approach the parameters ξ^1, ξ^2, ξ^3 of an individualized point of the continuum and the time t.

The transition from Lagrangian to Eulerian variables. The law of motion of a continuum has the form

$$x^i = x^i(\xi^1, \xi^2, \xi^3, t)\,, \tag{2.1}$$

where the independent variables ξ^1, ξ^2, ξ^3 and t are Lagrangian variables. Solving these equations for ξ^1, ξ^2, ξ^3, one obtains

$$\xi^i = \xi^i(x^1, x^2, x^3, t)\,, \tag{2.2}$$

i.e., one arrives at Eulerian variables. For fixed x^1, x^2, x^3, Equation (2.2) selects those points (ξ^1, ξ^2, ξ^3) of the continuum which at different instants of time pass through a given point of space. If the velocity

$$\boldsymbol{v} = \boldsymbol{v}(\xi^1, \xi^2, \xi^3, t)\,,$$

the acceleration

$$\boldsymbol{a} = \boldsymbol{a}(\xi^1, \xi^2, \xi^3, t)\,,$$

the temperature

$$T = T(\xi^1, \xi^2, \xi^3, t)$$

and other quantities are given from a Lagrangian point of view, i.e., as functions of ξ^1, ξ^2, ξ^3 and t, then (2.2) permits to determine the velocity, acceleration, temperature, etc., as functions of the Eulerian variables x^1, x^2, x^3 and t. Thus, if a motion is known from the Lagrangian and to be described from the Eulerian point of view, one must only solve the law of motion (2.1) with respect to the ξ^1, ξ^2, ξ^3, i.e., rewrite it in the form (2.2); the transition from a motion, given in Lagrangian terms, to a description in Eulerian terms is reduced to the determination of implicit functions.

The transition from Eulerian to Lagrangian variables. Conversely, let the space distribution of the velocity be given from the Eulerian point of view. How does one proceed to find the Lagrangian laws of motion? Choose a Cartesian coordinate system x, y, z and let there be given the velocity components

$$u = u(x, y, z, t), \quad v = v(x, y, z, t), \quad w = w(x, y, z, t).$$

The velocity components u, v, w are the derivatives with respect to the time t of the corresponding coordinates x, y, z for fixed parameters, ξ^1, ξ^2, ξ^3, individualizing a point of the continuum. Hence, if u, v, w are known as functions of the Eulerian variables x, y, z and t, then

$$\frac{dx}{dt} = u(x, y, z, t),$$

$$\frac{dy}{dt} = v(x, y, z, t),$$

$$\frac{dz}{dt} = w(x, y, z, t) \tag{2.3}$$

can be interpreted as three ordinary differential equations in the dependent variables x, y, z. Solving this system, one finds x, y, z as functions of t and of three arbitrary constants C_i which are determined by the values of x, y, z at a given instant of time t_0, and, consequently, they are parameters, individualizing a point of the medium, i.e., they are Lagrangian variables. Thus, by solution of the above system of differential equations, one arrives at the law of motion (2.1) with the aid of which one may proceed from the Eulerian to Lagrangian variables in all formulae for $\boldsymbol{a}$, T, etc. Consequently, the transition from Eulerian to Lagrangian variables for a given velocity field is, in general, connected with an integration of ordinary differential equations.

Clearly, specifications of a motion of a continuum from Lagrangian and Eulerian points of view are in a mechanical sense equivalent.

§3. Scalar and vector fields and their characteristics

Definition of scalar and vector fields. In the study of the motion of a continuum, one must introduce into the consideration scalar and vector quantities: the temperature T, the velocity $\boldsymbol{v}$, etc. In general, these can be considered in different coordinate systems: In an observer's coordinate system and in a system concomitant with the medium. They may be functions of x^1, x^2, x^3 or functions of ξ^1, ξ^2, ξ^3. In each of these coordinate systems, one can select a finite or infinite region and associate with every point of these regions a number, for instance, a temperature T, or a vector, for instance, a velocity $\boldsymbol{v}$, or, as will be seen later, some other more complicated characteristics.

The set of values of this or that quantity, which is given at every point of a region under consideration, is called the field of this quantity. If the quantity in question is a scalar, i.e., a number the value of which at a given point is independent of the choice of the coordinate system, the field is said to be scalar. Examples of scalar fields are temperature, density, etc. If a quantity in question is a vector, for instance, velocity, acceleration, etc., the field is called a vector field. The velocity in every coordinate system x^i has three components v^i and, consequently, at a given point and in a given coordinate system it is determined by three numbers. Hence, the velocity field and any vector field is equivalent to the three fields of the projections of the vector under consideration. However, although a vector itself does not depend on the coordinate system, its projections depend on the system of coordinates.

Next, certain general characteristics of scalar and vector fields will be studied using the temperature field T and the velocity field $\boldsymbol{v}$ as examples.

Individual and local derivatives with respect to time. A distribution of temperature may be given from a Lagrangian point of view by $T(\xi^i, t)$ and from an Eulerian point of view by $T(x^i, t)$. If the distribution T is given in Lagrangian variables, then the change of the temperature T in a particle of the continuum with time is readily computed. It is given by the derivative

$$\left(\frac{\partial T}{\partial t}\right)_{\xi^i}.$$

How does one evaluate this quantity, if the distribution of the temperature is given in terms of Eulerian variables by $T(x^1, x^2, x^3, t)$? Obviously, for this purpose, one must step over from Eulerian to Lagrangian variables and apply the chain rule. Then

$$T(x^1,x^2,x^3,t) = T[x^1(\xi^1,\xi^2,\xi^3,t),\ x^2(\xi^1,\xi^2,\xi^3,t),\ x^3(\xi^1,\xi^2,\xi^3,t),t]$$

and

$$\left(\frac{\partial T}{\partial t}\right)_{\xi^i} = \left(\frac{\partial T}{\partial t}\right)_{x^i} + \frac{\partial T}{\partial x^1}\left(\frac{\partial x^1}{\partial t}\right)_{\xi^i} + \frac{\partial T}{\partial x^2}\left(\frac{\partial x^2}{\partial t}\right)_{\xi^i} + \frac{\partial T}{\partial x^3}\left(\frac{\partial x^3}{\partial t}\right)_{\xi^i},$$

where the derivatives $(\partial x^1/\partial t)$, $(\partial x^2/\partial t)$, $(\partial x^3/\partial t)$ are evaluated for fixed ξ^1, ξ^2, ξ^3, and, consequently, they are the corresponding velocity components v^1, v^2, v^3. Hence

$$\left(\frac{\partial T}{\partial t}\right)_{\xi^i} = \left(\frac{\partial T}{\partial t}\right)_{x^i} + v^i \frac{\partial T}{\partial x^i}.$$

Note that for a given $T(x^1, x^2, x^3, t)$ for the evaluation of $(\partial T/\partial t)_{\xi^i}$ one need not know completely the law of motion, but only the velocity field $\boldsymbol{v}$.

The derivative $(\partial T/\partial t)_{\xi^i}$ characterizes the time variation of the temperature at a given point of the medium; it is referred to as the individual or total derivative of the temperature T with respect to the time t and often denoted by $\mathrm{d}T/\mathrm{d}t$. The derivative $(\partial T/\partial t)_{x^i}$ characterizes the rate of change of the temperature T at a given point of space x^i. It is referred to as the local derivative and denoted by $\partial T/\partial t$. In the general case, the total derivative $\mathrm{d}T/\mathrm{d}t$ is not equal to the local derivative $\partial T/\partial t$; it differs from it by the quantity, which depends on the motion of the particles and is called the convective derivative. Thus,

$$\frac{\mathrm{d}T}{\mathrm{d}t} = \frac{\partial T}{\partial t} + v^i \frac{\partial T}{\partial x^i}.$$

The significance of the definition of the convective derivative will be discussed in detail below, and now, using again the temperature field as an example, concepts will be presented which can be introduced for any scalar field.

Level surfaces. If the temperature T is given as a function of Eulerian variables, then, at any given instant of time, one can consider surfaces

$$T(x, y, z, t) = \text{const.},$$

which are referred to as level or equipotential surfaces. In the case of a temperature field, they are called isothermal surfaces.

Directional derivative. Selecting some point M on the level surface $T = \text{const.}$, one can investigate the variation of the temperature as a function of the direction in which this point moves. Denote this direction by $\boldsymbol{s}$.

The limit

$$\lim_{\Delta s \to 0} \frac{\Delta T}{\Delta s} = \frac{\partial T}{\partial s}$$

is said to be the derivative of T in the direction s. Obviously, if $\boldsymbol{s} = \boldsymbol{s}_1$, i.e., the direction $\boldsymbol{s}$ lies in the tangent plane to the level surface $T = \text{const.}$ at point M, then

$$\frac{\partial T}{\partial s_1} = 0.$$

Since ΔT is equal to $T_2 - T_1$, where $T_1 = \text{const.}$ and $T_2 = \text{const.}$ are the equations of adjoining level surfaces and $\Delta n = \Delta s \cos \alpha$ for a given ΔT (fig. 4), one has

$$\frac{\partial T}{\partial s} = \frac{\partial T}{\partial n} \cos \alpha, \tag{3.1}$$

where $\partial T/\partial n$ is the derivative in the direction of the normal $\boldsymbol{n}$ to the level surface $T = \text{const.}$ and α the angle between $\boldsymbol{n}$ and s. Clearly, the maximum value of the derivative $\partial T/\partial s$ is attained in the direction of the normal $\boldsymbol{n}$ (for $\alpha = 0$).

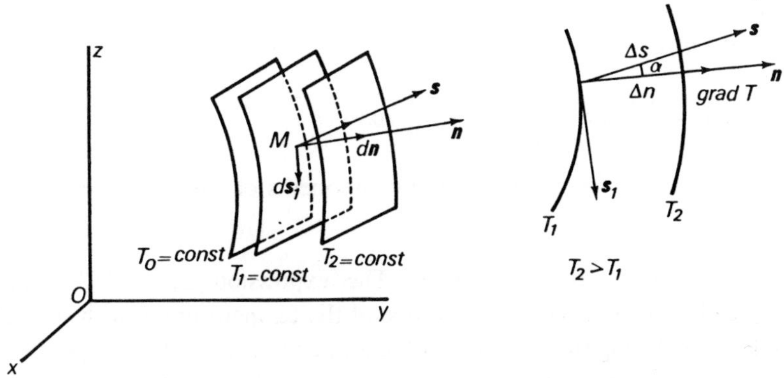

Fig. 4. Surfaces of equal level and the gradient vector of the temperature.

Gradient vector. Consider the vector which is parallel to the normal $\boldsymbol{n}$, points in the direction of increasing T and has the magnitude $\partial T/\partial n$. This vector is called the gradient vector of the scalar function under consideration, in the present case, of the temperature T; it is denoted by grad T, i.e.,

$$\operatorname{grad} T = \frac{\partial T}{\partial n}\boldsymbol{n}^0 ,$$

where $\boldsymbol{n}^0$ is the unit vector of the normal $\boldsymbol{n}$ directed towards increasing T. Obviously, the absolute value of grad T is greater whenever the level surfaces $T = \text{const.}$ are more crowded. By (3.1), the projection of the gradient vector of the temperature on to any direction $\boldsymbol{s}$ is the derivative of T in this direction:

$$(\operatorname{grad} T)_s = \frac{\partial T}{\partial s} .$$

In particular, the projections of the vector gradient on the coordinate axes x^i are

$$(\operatorname{grad} T)_{x^i} = \frac{\partial T}{\partial x^i}$$

and, in a Cartesian coordinate system,

$$\operatorname{grad} T = \frac{\partial T}{\partial x}\boldsymbol{i} + \frac{\partial T}{\partial y}\boldsymbol{j} + \frac{\partial T}{\partial z}\boldsymbol{k} .$$

Note that $\partial T/\partial x$, $\partial T/\partial y$, $\partial T/\partial z$ can be interpreted as the projections of a vector, since

$$\mathrm{d}T = \frac{\partial T}{\partial x}\,\mathrm{d}x + \frac{\partial T}{\partial y}\,\mathrm{d}y + \frac{\partial T}{\partial z}\,\mathrm{d}z$$

is invariant and $\mathrm{d}x$, $\mathrm{d}y$, $\mathrm{d}z$ are the components of the vector $\mathrm{d}\boldsymbol{r}$.

The convective derivative. The expression $\Sigma_{i=1}^{3} v^i \partial T/\partial x^i$ has been called the convective derivative of the temperature T with respect to the time. Using the concepts of the gradient vector of the temperature and of the scalar product, this expression may be rewritten

$$v^i \frac{\partial T}{\partial x^i} = \boldsymbol{v}\cdot \mathrm{grad}\, T\,.$$

The word derivative is always interpreted as the limit of the ratio of the increment of a function to the increment of an argument as the latter tends to zero. However, what increment of a function is taken in the case of the definition of the convective derivative? Rewrite the convective derivative of the temperature in the form

$$\frac{\boldsymbol{v}\,\Delta t\cdot \mathrm{grad}\, T}{\Delta t}\,.$$

Clearly, the quantity $\boldsymbol{v}\Delta t$ is equal to the displacement $\Delta \boldsymbol{s}$ (fig. 5) and

$$v^i \frac{\partial T}{\partial x^i} = \lim_{\Delta t\to 0} \frac{\Delta \boldsymbol{s}\cdot \mathrm{grad}\, T}{\Delta t} = \lim_{\Delta t\to 0} \frac{(\mathrm{grad}\, T)_s \Delta s}{\Delta t} =$$

$$= \frac{\partial T}{\partial s} \lim_{\Delta t\to 0} \frac{\Delta s}{\Delta t} = \lim_{\Delta t\to 0} \frac{\Delta T}{\Delta t}\,,$$

where the increment ΔT of the temperature occurs as a particle of the continuum is displaced in the direction $\boldsymbol{s}$ with velocity $\boldsymbol{v}$ at time Δt from one point of space to another (the points A and B of fig. 5).

In the general case, the convective derivative differs from zero, since the values of the temperature at the point A and B differ. It can be equal to zero, for example, in the absence of motion, or of a temperature gradient, i.e., when the temperature at a given instant of time does not vary in space

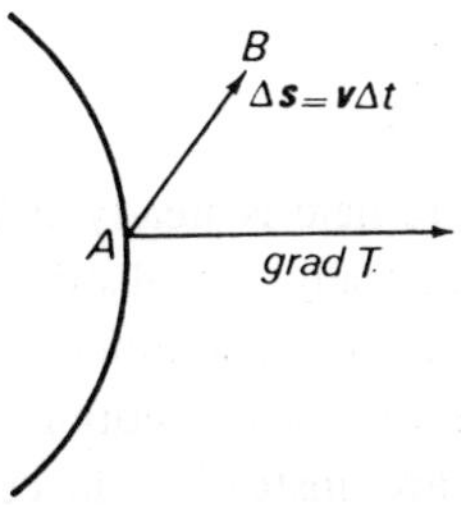

Fig. 5. On the concept of the convective derivative.

from point to point (such a field is said to be homogeneous) or for motion along a level surface $T(x, y, z, t) = \text{const.}$ at $t = \text{const.}$

Formulae for the determination of acceleration components in a Cartesian system. The formulae for the determination of the acceleration components in the case when the velocity is given in Eulerian variables are closely related to the concepts of individual local and convective derivatives. Let the velocity components v^1, v^2, v^3 be given in a Cartesian coordinate system as functions of Eulerian variables.

How are the acceleration components determined? The acceleration is defined for particles of a continuum; hence the components of the acceleration will be determined as the individual derivatives with respect to time of the corresponding velocity components, i.e.,

$$a_x = \left(\frac{\partial u}{\partial t}\right)_{\xi^i} = \frac{du}{dt} = \left(\frac{\partial u}{\partial t}\right)_{x^i} + u\frac{\partial u}{\partial x} + v\frac{\partial u}{\partial y} + w\frac{\partial u}{\partial z},$$

$$a_y = \left(\frac{\partial v}{\partial t}\right)_{\xi^i} = \frac{dv}{dt} = \left(\frac{\partial v}{\partial t}\right)_{x^i} + u\frac{\partial v}{\partial x} + v\frac{\partial v}{\partial y} + w\frac{\partial v}{\partial z},$$

$$a_z = \left(\frac{\partial w}{\partial t}\right)_{\xi^i} = \frac{dw}{dt} = \left(\frac{\partial w}{\partial t}\right)_{x^i} + u\frac{\partial w}{\partial x} + v\frac{\partial w}{\partial y} + w\frac{\partial w}{\partial z}.$$

Note the fact that these formulae are valid only in Cartesian coordinate systems (where all $\mathbf{3}_i$ are independent of coordinates and time).

Steady and unsteady motions. Different processes and motions are said to be steady or stationary, if all their characteristics in terms of Eulerian variables depend only on x^1, x^2, x^3 and not on the time t. Thus, for steady processes and motions, the local derivatives of all characteristics vanish, i.e.,

$$\frac{\partial T}{\partial t} = \frac{\partial v^1}{\partial t} = \frac{\partial v^2}{\partial t} = \frac{\partial v^3}{\partial t} = \ldots = 0 .$$

In particular, a temperature field is steady, if $T = T(x^i)$; in other words, if $T = T(x^i, t)$, the field is unsteady. The distribution in the space x^1, x^2, x^3 of the temperature T, the velocity $\boldsymbol{v}$ and other quantities at different instants of time coincide with each other when motions are steady, and differ from each other when motions are unsteady. The concept of steady motion is very important for applications. Firstly, many flows encountered in practice are stationary and, secondly, the study of such motions is simplified by the fact that the number of independent variables is reduced by one (as the time ceases to be a variable).

Note that one and the same motion can simultaneously be steady and unsteady. This property depends on the choice of the coordinate system with respect to which it is being examined. For instance, the wave motion of water behind a ship moving with a constant speed is steady for an observer moving with the ship and unsteady for an observer on the shore. The concept of stationary motion is relative.

Note yet that both the observers above study the motion in their own coordinate systems which are defined with respect to one and the same coordinate system: either absolute motion with respect to shore or relative motion with respect to the ship.

Vector lines; stream lines. Next, consider the concept of vector lines which may be introduced for any vector field, for instance, for the velocity field $\boldsymbol{v}$, the acceleration field $\boldsymbol{a}$, the temperature gradient grad T, etc. For the sake of definiteness, this concept will be explained by means of the example of the vector lines of the velocity field, which are referred to as stream lines.

As has been shown above, the field $\boldsymbol{v}$ will be defined, if at each point of the space x^1, x^2, x^3 and at each instant of time t one is given the vector

$$\boldsymbol{v} = v^i \boldsymbol{\mathfrak{z}}_i = v^1 \boldsymbol{\mathfrak{z}}_1 + v^2 \boldsymbol{\mathfrak{z}}_2 + v^3 \boldsymbol{\mathfrak{z}}_3 ,$$

where $v^i(x^1, x^2, x^3, t)$ are the velocity components referred to the base $\boldsymbol{\mathfrak{z}}_i$. The requirement regarding the specification of the velocity is very strong and sometimes it can be weakened; for example, one may only demand to know the velocity at some instant of time t_0 instead of at all time t. One may even go further and require to know only the direction of the velocity vector for every point of the space x^1, x^2, x^3 at the instant t_0, and not its magnitude. Obviously, this information permits the construction of a family of lines the tangents of which at every space point will coincide at time t with the directions of the velocity vector $\boldsymbol{v}$ there. In case of a velocity field, these lines are referred to as stream lines, in the case of an arbitrary vector field, as vector lines.

For every velocity field $\boldsymbol{v}$, the family of stream lines may be constructed; as soon as this has been done, the direction of the velocity vector $\boldsymbol{v}$ will be exactly known at every point. In practice, it is often essential to know the stream lines which may also be determined by experiment. For example, for the experimental determination of the stream lines one may photograph with short exposure times the flow of a fluid mixed with special particles, with bubbles of air, etc. Short streaks on the photographs will then recreate the flow pattern. The vector lines of a magnetic field are easily visualized by pouring iron powder on a piece of paper and placing a magnet underneath. The stream lines in the flow around the wings of an aircraft can also be visualized, by glueing thin silk threads to the surface of the wing and photographing the flow in an aerodynamic wind-tunnel or directly in flight.

Equations of the stream lines. How does one find analytical expressions for the stream lines? For this purpose, one must formulate the mathematical problem the solution of which determines the stream lines. Consider the condition that the element

$$d\boldsymbol{r} = dx^i \boldsymbol{\mathfrak{z}}_i = dx^1 \boldsymbol{\mathfrak{z}}_1 + dx^2 \boldsymbol{\mathfrak{z}}_2 + dx^3 \boldsymbol{\mathfrak{z}}_3 ,$$

taken along a stream line, and the velocity vector

$$\boldsymbol{v} = v^i \boldsymbol{\mathfrak{z}}_i = v^1 \boldsymbol{\mathfrak{z}}_1 + v^2 \boldsymbol{\mathfrak{z}}_2 + v^3 \boldsymbol{\mathfrak{z}}_3$$

are to be parallel to each other:

$$d\boldsymbol{r} = d\lambda \cdot \boldsymbol{v},$$

where $d\lambda$ is a scalar parameter, which one can consider on each stream line $\mathscr{L}$ as differential of some function $\lambda(s, \mathscr{L})$, where s-length of arc in the direction of the stream lines. For the components, one finds

$$\frac{dx^1}{v^1} = \frac{dx^2}{v^2} = \frac{dx^3}{v^3} = d\lambda ,$$

or

$$\frac{dx^i}{d\lambda} = v^i(x^1, x^2, x^3, t), \qquad i = 1, 2, 3, \tag{3.2}$$

i.e., the set of differential equations for the stream lines. They differ from the differential equations, defining the law of motion or the trajectories of the motion of the particles of a continuous medium, which, obviously, have the form

$$\frac{dx^i}{dt} = v^i(x^1, x^2, x^3, t). \tag{3.3}$$

The time enters on the right-hand as well as on the left-hand side of Equations (3.3). In Equations (3.2), the derivatives are taken with respect to λ and the right-hand side depends on t. In the integration of (3.2), the variable t must be considered to be a constant parameter, whereas in Equation (3.3) it is a variable.

Thus, generally speaking, the stream lines do not coincide with the particle trajectories. The family of stream lines $x^i = x^i(c^1, c^2, c^3, \lambda, t)$ depends on the time and varies with it. However, the parameter t appears on the right-hand side of (3.2) and (3.3) only in the case of unsteady motions. In the case of steady motion, the distinction between Equations (3.2) and (3.3) disappears; it reduces to different notation for the differentiation parameter, a fact which has no significance. Therefore stream lines and trajectories coincide for steady motions.

Examples of stream lines and trajectories. Consider some examples. A motion of a rigid body is said to be translatory, if any straight segment, selected inside the body, moves parallel to itself. All points of a rigid body in translation have at a given instant of time the same velocity in magnitude and direction. Consequently, the stream lines are in this case always straight lines. What about the trajectories? Translation of a rigid body can proceed along any trajectories and hence also along a circle (fig. 6). Therefore, in the general case of translation of a rigid body, the stream lines and trajectories do not coincide.

In the case of arbitrary motion of a rigid body, the stream lines are spirals and the trajectories may be arbitrary.

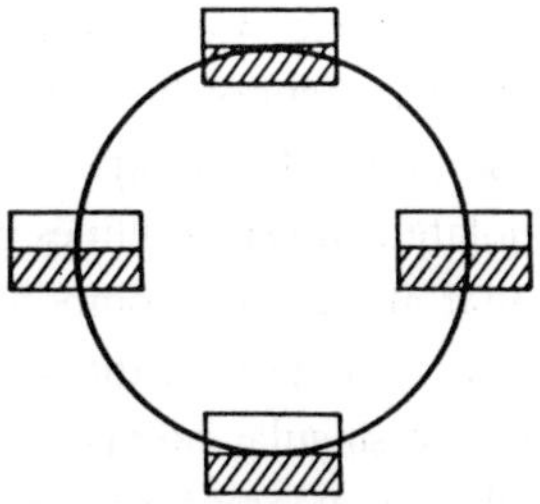

Fig. 6. Translation of a rigid body along a circle.

Do there exist unsteady motions for which the stream lines coincide with the trajectories? For instance, consider rectilinear translation of a rigid body with variable velocity. In this case, stream lines and trajectories are straight lines, and the motion itself is, of course, unsteady. Analogously, stream lines and trajectories will coincide in the case of the revolution of a rigid body with variable angular velocity about a fixed axis. In general, the stream lines and trajectories will coincide with each other during those unsteady motions in which the magnitude of the velocity varies at a given space point in time, but not its direction.

Consequently, stream lines and trajectories coincide for fields $\boldsymbol{v}(x^1, x^2, x^3)$ and $\boldsymbol{v}_1 = f(x^1, x^2, x^3, t)\boldsymbol{v}(x^1, x^2, x^3)$, where $f(x^1, x^2, x^3, t)$ is a scalar function of its arguments.

The existence of stream lines. The differential equations (3.2) of the stream lines may be rewritten in the form

$$\frac{dx^2}{dx^1} = \frac{v^2}{v^1}, \quad \frac{dx^3}{dx^1} = \frac{v^3}{v^1}, \tag{3.4}$$

and the Cauchy problem for this system posed, i.e., the problem of finding, with t being a constant parameter those solutions $x^2(x^1)$, $x^3(x^1)$ which for given $x^1 = x^1_0$ become given functions x^2_0 and x^3_0. In other words, one should draw the integral curve of the system of Equations (3.4) (a stream line) through a given point x^1_0, x^2_0, x^3_0. As is well-known from the general theory of ordinary differential equations, Cauchy problem has always a unique solution in every case when the right-hand sides in (3.4) and their derivatives, with respect to x^i, are continuous and single-valued. (This theorem is valid under some more general assumptions.) Thus, under very general limitations, through each point of a given volume, occupied by a continuous medium in motion, one may draw a unique stream line.

Singular and critical points. Conditions, which provide a unique solution of the Cauchy problem cannot be satisfied in the same points of the velocity field. At these points the uniqueness of the solution of the Cauchy problem may be violated and stream lines may intersect or bifurcate. However, it can happen that all velocity components v^i vanish or become infinite at some point x^i. At these points, the right-hand sides of (3.4) are undetermined. Such points are singular points of the differential equations of the stream lines. There the uniqueness theorem, may be violated. Singular points can be of the type of a centre, focus, saddle, node or more complicated patterns. The singular points of the differential equation (3.4) of the stream lines are called critical points or lines. The important examples of such problems are critical points or lines of the flow, when the value of the velocity is equal to zero. In fig. 7 the stream lines in meridional section of the flow, generated by collision of two axisymmetrical streams, moving in opposite direction. At some point A in the centre of the region of the interaction of the streams appears a critical point, in which the velocity vanishes and the stream lines are intersecting and branching.

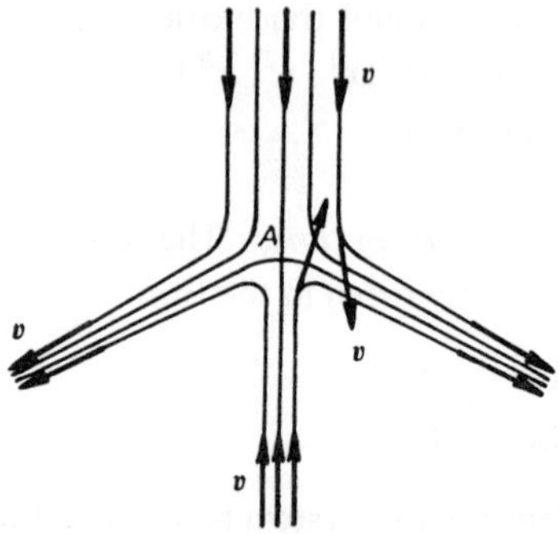

Fig. 7. At critical point A the velocity of fluid is equal zero and one can observe that the stream lines bifurcate.

Stream surfaces, vector surfaces. One may draw a stream line through every point of an arbitrary curve C. Then, if C is not a stream line (and not tangent to a stream line), a surface is formed, at every point of which the velocity $\boldsymbol{v}$ lies in the tangent plane. This surface is called a stream surface. In the case of an arbitrary vector field, an analogously constructed surface is called a vector surface. How does one find the stream surface $f(x^i)=\text{const.}$? Obviously, grad f, being perpendicular to the normal to $f(x^i) = \text{const.}$, will be perpendicular to the vector $\boldsymbol{v}$ and, consequently,

$$\operatorname{grad} f \cdot \boldsymbol{v} = 0 ,$$

i.e.,

$$u \frac{\partial f}{\partial x} + v \frac{\partial f}{\partial y} + w \frac{\partial f}{\partial z} = 0 \tag{3.5}$$

which is a partial differential equation for the determination of the function $f(x, y, z)$.

The method for the construction of stream surfaces, described above, suggests a method for the integration of partial differential equations of the form (3.5). This integration reduces to the determination of the family of the stream lines passing through the contour C, i.e., to the integration of a system of ordinary differential equations (in the general case, the solutions of these ordinary equations are called characteristics of partial differential equations (3.5)). It is obvious that one may construct in this way a unique solution only in the case when the contour C itself is not a stream line (i.e., a characteristic).

Stream tube, vector tube. If the curve C is closed, then the set of all stream lines passing through it form a stream tube. In the case of an arbitrary vector field, analogously constructed tubes are called vector tubes.

Potential vector field, potential flow. Above, consideration has been given to the gradient vector of the temperature T. There arises now the following question. Is it possible to represent the velocity vector in the form of the gradient of some scalar function $\varphi(x, y, z, t)$? If there exists a function $\varphi(x, y, z, t)$ such that

$$u = \partial\varphi/\partial x, \quad v = \partial\varphi/\partial y, \quad w = \partial\varphi/\partial z ,$$

then the velocity field $\boldsymbol{v}$ is said to be a potential field and the function φ a velocity potential. Analogously, an arbitrary vector field $\boldsymbol{A}(x, y, z, t)$ is a potential field, if there exists a function $\Phi(x, y, z, t)$ such that

$$\boldsymbol{A} = \operatorname{grad} \Phi(x, y, z, t) .$$

In accordance with the properties of the gradient vector, the velocity $\boldsymbol{v}$ in the case of potential flow is orthogonal to the equipotential surface

$\varphi=\text{const.}$, and it is larger, wherever the equipotential surfaces are more closely spaced. The projection of the velocity $\boldsymbol{v}$ on to an arbitrary direction $\boldsymbol{s}$ is the derivative of the potential φ in this direction: $v_s=\partial\varphi/\partial s$.

Necessary and sufficient conditions for the existence of a potential. Consider the expression

$$u\,dx+v\,dy+w\,dz\,. \tag{3.6}$$

If the flow is potential, then (3.6) is the total differential of the function φ (with respect to the coordinates x, y, z). In fact,

$$u\,dx+v\,dy+w\,dz=\frac{\partial\varphi}{\partial x}dx+\frac{\partial\varphi}{\partial y}dy+\frac{\partial\varphi}{\partial z}dz=d\varphi\,.$$

The converse is also true: If the expression (2.6) is a total differential, then the flow is potential. It is well-known that it is necessary and sufficient, for (3.6) to be a total differential, that

$$\frac{\partial u}{\partial y}=\frac{\partial v}{\partial x}\,,\quad \frac{\partial u}{\partial z}=\frac{\partial w}{\partial x}\,,\quad \frac{\partial v}{\partial z}=\frac{\partial w}{\partial y}\,,$$

which are the necessary and sufficient conditions for a flow to be potential.

It will be shown later that potential flow plays an important role; certain examples of potential flow will now be studied.

Translatory flow. Translatory flow with constant velocity u_0 along the x-axis is an example of potential flow. In this case, $u=u_0$, $v=w=0$; since

$$\frac{\partial\varphi}{\partial x}=u_0\,,\quad \frac{\partial\varphi}{\partial y}=\frac{\partial\varphi}{\partial z}=0\,,$$

the potential is given by $\varphi=u_0x+\text{const.}$ This example easily leads to two obvious conclusions. Firstly, a potential is always determined exactly, apart from an additive constant, and, secondly, any translatory flow is always potential. In fact, in the general case of translatory flow: $u=u_0$, $v=v_0$, $w=w_0$ and $\varphi=u_0x+v_0y+w_0z+C$, where u_0, v_0, w_0 and C may depend on t.

Note that it is simpler to study flows with potentials than those without potentials, since the first are determined by single functions $\varphi(x, y, z, t)$

and the latter by three functions $v^1(x, y, z, t)$, $v^2(x, y, z, t)$, $v^3(x, y, z, t)$.

Source and sink in space. Consider another example of potential flow which will be important later on. Let

$$\varphi = -\frac{Q}{4\pi r}, \tag{3.7}$$

where $r = \sqrt{(x^2+y^2+z^2)}$ and $Q = \text{const.}$ or $Q = Q(t)$. Obviously, the equipotential surfaces $\varphi = \text{const.}$ are in this case given by $r = \text{const.}$, i.e., they are concentric spheres with centre at the origin of coordinates. The velocity $\boldsymbol{v} = \text{grad}\ \varphi$ is orthogonal to these spheres, i.e., it is directed along the radii, emanating from the origin. The stream lines are rays along these radii. Let $Q > 0$; then, since grad φ points in the direction of increasing φ,

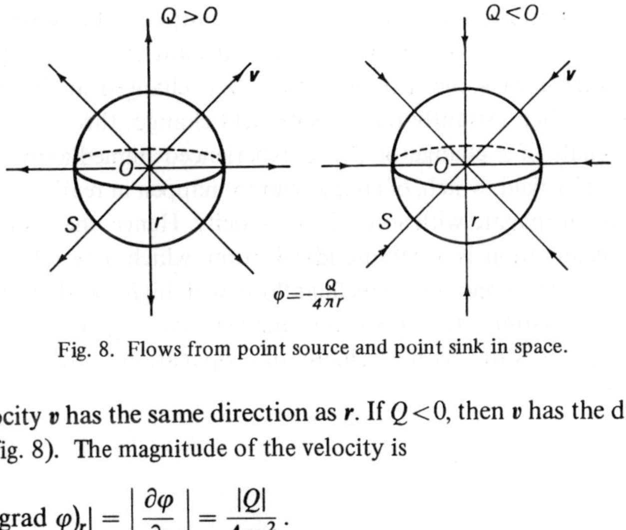

Fig. 8. Flows from point source and point sink in space.

the velocity $\boldsymbol{v}$ has the same direction as $\boldsymbol{r}$. If $Q < 0$, then $\boldsymbol{v}$ has the direction of $-\boldsymbol{r}$ (fig. 8). The magnitude of the velocity is

$$|(\text{grad}\ \varphi)_r| = \left|\frac{\partial\varphi}{\partial r}\right| = \frac{|Q|}{4\pi r^2}.$$

It tends to zero for $r \to \infty$ and becomes infinite for $r \to 0$. Thus, the origin and the point at infinity are critical points. For $Q > 0$, the liquid flows out of the origin of coordinates in all directions, and the flow is said to be due to a point source. For $Q < 0$, the liquid flows into the origin, i.e., one has a sink. In the first case, the point at infinity is a sink, in the second case it is a source.

Next, compute the volume of liquid passing in unit time through the surface of a sphere S with some radius r with centre at the origin. The fluid volume passing through the surface element $d\sigma$ of the sphere in unit time is $v\,d\sigma$, so that the volume of fluid crossing the entire spherical surface is given by

$$\int_S v\,d\sigma = v\int_S d\sigma = 4\pi r^2 v = Q$$

(where v can be taken out from underneath the integral sign, because $v=$const. on the surface of the sphere). Note that the first two equations are always valid when $v=v(r)$ and $\boldsymbol{v}$ is orthogonal to the surface of the sphere S. The resulting fluid volume does not depend on the radius r. Thus, although on different spheres of different radii with centre at the origin the velocities differ, the constant Q in the potential (3.7) is the volume of liquid which passes in unit time through each sphere. The quantity Q is referred to as the strength or output of the source (sink).

If $Q=$ const., the source or sink has constant strength, if $Q=Q(t)$, it is variable. If at some instant of time Q is changed at the origin of coordinates, then instantly the velocity field changes through the entire space. Signals of a change of Q are experienced immediately over the entire velocity field, which, of course, cannot happen in reality. Perturbations must propagate with some finite velocity. Hence, the velocity field under consideration is a definite idealization, which may reflect reality only in cases when one considers fluid flows with high speeds of propagation of perturbations. In many cases, one may assume, for example, that such a liquid is water with a velocity of propagation of disturbances of 1450 m/sec.

§4. Elements of tensor calculus

Many characteristics of the motion of continua have a tensor character; therefore the elements of tensor calculus will now be considered. Note that scalars and vectors are also tensors, namely those of the simplest type; vector and scalar quantities by themselves are insufficient for the description of the motion of continua.

A coordinate system establishes a correspondence between numbers and points of space. Three coordinate lines pass through each point of space. These may be coordinate lines of the concomitant coordinate system ξ^1, ξ^2, ξ^3 or of an observer's reference system x^1, x^2, x^3 or of any other coordinate system. Therefore, in this section, the letters $\zeta^1, \zeta^2, \zeta^3$ or η^1, η^2, η^3 will be used to denote coordinate systems.

A coordinate system is introduced into a study by the observer and its choice depends on him and not on the phenomena under consideration. The laws of motion may involve coordinates, but may not depend on the choice of coordinate systems. They must be invariant with respect to the choice of the coordinate system which imposes known limitations on the form of the mathematical description of these laws.

Transformation of coordinates. A start will be made with necessary expositions from the theory of transformations of coordinates. Let there be given, together with the system $\zeta^1, \zeta^2, \zeta^3$, a coordinate system η^1, η^2, η^3 and consider the laws of motion referred to both systems. Assume that there exists a relationship

$$\zeta^i = \zeta^i(\eta^1, \eta^2, \eta^3) \tag{4.1}$$

between these two systems, a so-called coordinate transformation. Only continuous single-valued coordinate transformations will be studied. They constitute a group. Consider those relations which are invariant with respect to the group of continuous single-valued transformations. By (4.1), one has

$$\begin{aligned}
\mathrm{d}\zeta^1 &= \frac{\partial \zeta^1}{\partial \eta^1}\,\mathrm{d}\eta^1 + \frac{\partial \zeta^1}{\mathrm{d}\eta^2}\,\mathrm{d}\eta^2 + \frac{\partial \zeta^1}{\partial \eta^3}\,\mathrm{d}\eta^3\,,\\
\mathrm{d}\zeta^2 &= \frac{\partial \zeta^2}{\partial \eta^1}\,\mathrm{d}\eta^1 + \frac{\partial \zeta^2}{\partial \eta^2}\,\mathrm{d}\eta^2 + \frac{\partial \zeta^2}{\partial \eta^3}\,\mathrm{d}\eta^3\,,\\
\mathrm{d}\zeta^3 &= \frac{\partial \zeta^3}{\partial \eta^1}\,\mathrm{d}\eta^1 + \frac{\partial \zeta^3}{\partial \eta^2}\,\mathrm{d}\eta^2 + \frac{\partial \zeta^3}{\partial \eta^3}\,\mathrm{d}\eta^3\,,
\end{aligned} \tag{4.2}$$

or

$$\mathrm{d}\zeta^i = \frac{\partial \zeta^i}{\partial \eta^j}\,\mathrm{d}\eta^j\,, \tag{4.2$'$}$$

where the summation extends over 1 to 3 and i assumes the values 1, 2, 3, a fact which will not be specifically mentioned in the sequel, but it will always be implied.

Thus, near any given point, there exists a relationship between the coordinate increments $d\zeta^i$ and $d\eta^i$. The derivatives $\partial\zeta^i/\partial\eta^j$ are functions of a point; however, in (4.2), at a given point, they are constants and one has there linear relations between the coordinate increments $d\zeta^i$ and $d\eta^i$. Let

$$\frac{\partial\zeta^i}{\partial\eta^j} = a^{i}_{\cdot j} .$$

(In the sequel, it will be shown that the arrangement of the upper and lower indices and their order of writing is very important). The quantities $a^{i}_{\cdot j}$ form the matrix

$$\|a^{i}_{\cdot j}\| = \mathbf{A},$$

where the first index i corresponds to its rows and the second j to its columns. It follows from the mutual single-valuedness that the Jacobian of the transformation, equal to the determinant of the matrix $\|a^{i}_{\cdot j}\|$, does not vanish, i.e., that $\Delta = |a^{i}_{\cdot j}| \neq 0$. Since $\Delta \neq 0$, the linear relations (4.2) can be solved for the $d\eta^i$ and one can write down together with (4.2) the formulae

$$d\eta^i = \frac{\partial\eta^i}{d\zeta^j}\, d\zeta^j . \tag{4.3}$$

Introduce the matrix

$$\mathbf{B} = \|b^{i}_{\cdot j}\| ,$$

where

$$b^{i}_{\cdot j} = \frac{\partial\eta^i}{\partial\zeta^j} .$$

The matrices **A** and **B**, introduced for the transformation and its inverse, are mutually inverse, i.e., their product is equal to the unit matrix. In fact,

$$\mathbf{A}\cdot\mathbf{B} = \|a^{i}_{\cdot j}\| \cdot \|b^{j}_{\cdot k}\| = \|a^{i}_{\cdot j} b^{j}_{\cdot k}\| ;$$

however,

$$a^{i}_{\cdot j} b^{j}_{\cdot k} = \frac{\partial\zeta^i}{\partial\eta^j}\frac{\partial\eta^j}{\partial\zeta^k} = \frac{\partial\zeta^i}{\partial\zeta^k} = \begin{cases} 1 \text{ for } i=k , \\ 0 \text{ for } i\neq k , \end{cases}$$

since $\zeta^1, \zeta^2, \zeta^3$ and η^1, η^2, η^3 are independent coordinates. In what follows, use will be made of Kronecker's delta defined by

$$\delta^{i\cdot}_{\cdot k} = \begin{cases} 1 \text{ for } i=k \\ 0 \text{ for } i \neq k\,. \end{cases}$$

Then

$$\mathbf{A} \cdot \mathbf{B} = \|\delta^{i\cdot}_{\cdot k}\| = \begin{Vmatrix} 1 & 0 & 0 \\ 0 & 1 & 0 \\ 0 & 0 & 1 \end{Vmatrix} = \mathbf{E}\,,$$

where **E** is the unit matrix. Obviously, for the determinant of the matrix **B**,

$$|b^{i\cdot}_{\cdot j}| = \frac{1}{\Delta}\,.$$

The reasoning above has been carried out for three-dimensional space; however, the same arguments apply for any n-dimensional space, including the one-dimensional, two-dimensional and four-dimensional spaces encountered in continuum mechanics.

Note that these considerations do not require the introduction of a metric of the space. The space ζ^i can be a space without metric or a space with a very complicated metric.

Base vectors. Now, for the sake of completeness and to underline the points of view adopted, recall the question of the introduction of the base vectors $\mathbf{э}_1, \mathbf{э}_2, \mathbf{э}_3$. In the coordinate system $\zeta^1, \zeta^2, \zeta^3$, consider a point M with coordinates $\zeta^1, \zeta^2, \zeta^3$ and infinitesimally close to it a point M' with coordinates $\zeta^1 + \mathrm{d}\zeta^1$, $\zeta^2 + \mathrm{d}\zeta^2$, $\zeta^3 + \mathrm{d}\zeta^3$ (fig. 9). Introduce into the consideration the new object

$$\mathrm{d}\boldsymbol{r} = \overline{MM'}\,,$$

i.e., a pair of infinitesimally close points M and M', taken in a definite order (an ordered pair of points) and mark it in the figure by an arrow; $\overline{MM'}$ is defined only by the coordinates of the points M and M'. Side by side with $\mathrm{d}\boldsymbol{r}$ introduce another object

$$k\,\mathrm{d}\boldsymbol{r}\,, \tag{4.4}$$

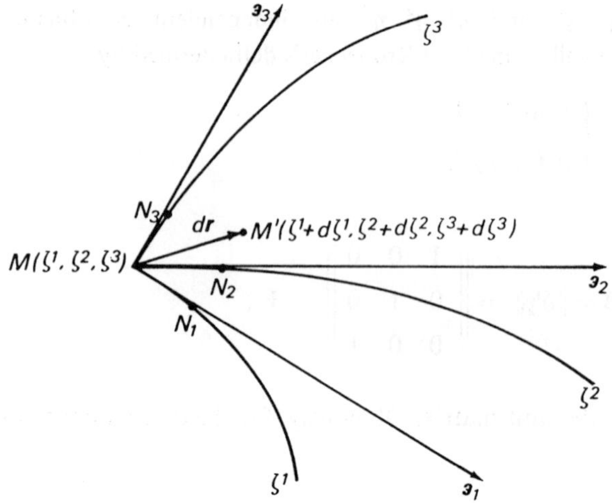

Fig. 9. Base vectors $\mathbf{э}_i$.

where k is a certain number; the object $k\mathrm{d}\boldsymbol{r}$ has the direction of $\mathrm{d}\boldsymbol{r}$, if $k>0$, and the opposite direction of $\mathrm{d}\boldsymbol{r}$ if $k<0$. Draw coordinate lines from the point M and consider on them points N_1, N_2, N_3, each of which is determined by an increment $\mathrm{d}\zeta^1, \mathrm{d}\zeta^2, \mathrm{d}\zeta^3$ along only one of the coordinates $\zeta^1, \zeta^2, \zeta^3$. Analogously to $\mathrm{d}\boldsymbol{r}$, introduce the objects MN_1, MN_2, MN_3 or, by (4.4), setting $k_i = 1/\mathrm{d}\zeta^i$, the objects

$$\frac{\partial \boldsymbol{r}}{\partial \zeta^i} = \mathbf{э}_i ,$$

the so-called base vectors which are directed along the tangents to the coordinate lines.

In the general case, $\mathrm{d}\boldsymbol{r}$ is directed arbitrarily and it may, by definition, be written in the form

$$\mathrm{d}\boldsymbol{r} = \mathrm{d}\zeta^1 \mathbf{э}_1 + \mathrm{d}\zeta^2 \mathbf{э}_2 + \mathrm{d}\zeta^3 \mathbf{э}_3 ,$$

where the $\mathrm{d}\zeta^1, \mathrm{d}\zeta^2, \mathrm{d}\zeta^3$ are called the components of $\mathrm{d}\boldsymbol{r}$. Obviously, the base vectors $\mathbf{э}_1, \mathbf{э}_2, \mathbf{э}_3$ of the coordinate system $\zeta^1, \zeta^2, \zeta^3$ always have the components $(1, 0, 0)$, $(0, 1, 0)$, $(0, 0, 1)$ in the $\zeta^1, \zeta^2, \zeta^3$ system of coordinates respectively. The base vectors may be introduced into the coordinate system $\zeta^1, \zeta^2, \zeta^3$ as well as into the coordinate system η^1, η^2, η^3. In different coordinate systems, at one and the same point, the base vectors will differ. Denote the base vectors in the coordinate system η^1, η^2, η^3 by $\mathbf{з}_1, \mathbf{з}_2, \mathbf{з}_3$. In the system η^1, η^2, η^3, one has

$$\mathrm{d}\boldsymbol{r} = \mathrm{d}\eta^j \mathbf{з}'_j .$$

Obviously, the components of $\mathrm{d}\boldsymbol{r}$ and the base vectors $\mathbf{з}_i$ depend on the choice of the coordinate system.

Transformation of the base vectors and the components of $\mathrm{d}\boldsymbol{r}$ *for transformation of coordinates.* Now, formulae will be derived which express the base vectors $\mathbf{з}'_i$ in the new system η^1, η^2, η^3 in terms of the base vectors $\mathbf{з}_i$ in the old system $\zeta^1, \zeta^2, \zeta^3$. We have

$$\mathrm{d}\boldsymbol{r} = \mathrm{d}\eta^i \mathbf{з}'_j = \mathrm{d}\zeta^i \mathbf{з}_i = \frac{\partial \zeta^i}{\partial \zeta^j} \mathrm{d}\eta^j \mathbf{з}_i .$$

Then

$$\mathbf{з}'_j = \mathbf{з}_i \frac{\partial \zeta^i}{\partial \eta^j} = \mathbf{з}_i a^{i\cdot}_{\cdot j} . \tag{4.5}$$

By (4.3), the components of $\mathrm{d}\boldsymbol{r}$ are interrelated by the formula

$$\mathrm{d}\eta^i = b^{i\cdot}_{\cdot j} \mathrm{d}\zeta^j . \tag{4.6}$$

Note that the base vectors $\mathbf{з}_i$ transform by (4.5) with the aid of the matrix $\mathbf{A}$, and the components of $\mathrm{d}\boldsymbol{r}$, by (4.6), with the aid of the matrix $\mathbf{B}$ which is the inverse of $\mathbf{A}$ (it is important to focus attention on the arrangement of the indices in (4.5) and (4.6)).

Invariance of $\mathrm{d}\boldsymbol{r}$ *with respect to coordinate transformations.* The object $\mathrm{d}\boldsymbol{r}$ is invariant with respect to coordinate transformations. In fact,

$$\mathrm{d}\boldsymbol{r} = \mathrm{d}\eta^j \mathbf{з}'_j = b^{j\cdot}_{\cdot i} \mathrm{d}\zeta^i a^{s\cdot}_{\cdot j} \mathbf{з}_s = \mathrm{d}\zeta^i \mathbf{з}_i ,$$

since

$$b^{j\cdot}_{\cdot i} a^{s\cdot}_{\cdot j} = \begin{Bmatrix} 1, & i = s \\ 0, & i \neq s \end{Bmatrix} = \delta^{s\cdot}_{\cdot i} .$$

Hence, the expression for $\mathrm{d}\boldsymbol{r}$ in terms of the components and base vectors of the corresponding coordinate systems does not change during transition from one coordinate system to another; it is invariant with respect to transformations of coordinates.

Covariant and contravariant quantities. Quantities which transform like the base vectors $\mathbf{з}_i$, by (4.5), are said to be covariant. Quantities which transform like the components of d$\boldsymbol{r}$, by (4.6), are called contravariant. It must be emphasized that the transformations, generating covariant and contravariant quantities, are mutually inverse.

Definition of a vector. Now, using the example of d$\boldsymbol{r}$, one may introduce a new object $\boldsymbol{A}$ which can be represented in the following manner in terms of the base vectors:

$$\boldsymbol{A} = A^i \mathbf{з}_i \ ;$$

its components A^i transform like those of d$\boldsymbol{r}$:

$$A'^j = b'^j_{.i} A^i \ .$$

Object $\boldsymbol{A}$ is called a vector. Like d$\boldsymbol{r}$, it is invariant under coordinate transformations

$$\boldsymbol{A} = A^j \mathbf{з}_j = A'^i \mathbf{з}'_i \ . \tag{4.7}$$

The invariance of the vector $\boldsymbol{A}$ is ensured by the mutually inverse transformation of the vector components A^i and of the base vectors $\mathbf{з}_i$. The base vectors are the "supports" of each vector; their coefficients in (4.7) are, in the general case, numerical functions of the point M.

The vector $\boldsymbol{A}$ may have any geometrical or physical character; however, it is always expressed in terms of the base vectors by the expansion (4.7), where the numbers (functions) A^i depend on the coordinate system. The base vectors $\mathbf{з}_i$ govern the numbers A^i and create a new object, namely, the vector $\boldsymbol{A}$.

Polyadic products of the base vectors. There arises the following question: is it possible to introduce, besides $\mathbf{з}_i$, still some other base objects, which, in an analogous manner to $\mathbf{з}_i$, governing numbers, permit to propose concepts which are yet more complex than vectors, invariant with respect to coordinate transformations? Such objects can indeed be introduced; in particular, one can take as such objects

$$E_1 = \mathfrak{z}_1\mathfrak{z}_1, \quad E_2 = \mathfrak{z}_1\mathfrak{z}_2, \quad E_3 = \mathfrak{z}_1\mathfrak{z}_3, \quad E_4 = \mathfrak{z}_2\mathfrak{z}_1,$$
$$E_5 = \mathfrak{z}_2\mathfrak{z}_2, \quad E_6 = \mathfrak{z}_2\mathfrak{z}_3, \quad E_7 = \mathfrak{z}_3\mathfrak{z}_1, \quad E_8 = \mathfrak{z}_3\mathfrak{z}_2, \quad E_9 = \mathfrak{z}_3\mathfrak{z}_3$$

and consider

$$T = T^i E_i, \tag{4.8}$$

where the T^i are numbers, called components of T to the base E_i ($i = 1, 2, \ldots, 9$).

The base objects E_i are referred to as polyadic products of the base vectors $\mathfrak{z}_i$ (in this particular case, they may be called dyadics, since each product consists of two vectors; however, one can introduce products of many vectors of the form $\mathscr{E}_s = \mathfrak{z}_i\mathfrak{z}_j\mathfrak{z}_k\mathfrak{z}_l$, in three-dimensional space $s = 1, 2, \ldots, 81$). By definition, polyadic products of base vectors are assumed to be linearly independent, i.e., the equality $T = 0$ is possible only if the nine numbers T^i are equal to zero. Instead of the new notation E_i, it is convenient to use directly the symbols $\mathfrak{z}_k\ \mathfrak{z}_j$ and to rewrite (4.8) in the form

$$T = T^{ij}\mathfrak{z}_i\mathfrak{z}_j.$$

Polyadic multiplication of vectors represents a certain operation on vectors which leads to new objects (neither vectors nor scalars). For the definition of this operation, one need only state its properties. In particular, the order of the vectors in the product is essential: $\mathfrak{z}_1\mathfrak{z}_2 \neq \mathfrak{z}_2\mathfrak{z}_1$. By definition, the operation of polyadic multiplication is linear (it has the property of distributivity, i.e., the position of numerical factors in the product is inessential). For example, one has the equality

$$\mathfrak{z}_i(a\mathfrak{z}_j + b\mathfrak{z}_k) = a\mathfrak{z}_i\mathfrak{z}_j + b\mathfrak{z}_i\mathfrak{z}_k, \tag{4.9}$$

where a and b are numbers.

Polyadic products of the base vectors $\mathfrak{z}_i\mathfrak{z}_j$, just as the base vectors themselves, depend on the coordinate system. The transformation formulae for the quantities $\mathfrak{z}_i\mathfrak{z}_j$ are readily obtained, knowing the transformation formulae of the $\mathfrak{z}_i$ and using the property of linearity of the polyadic product. These formulae have the form

$$\mathfrak{z}'_i\mathfrak{z}'_j = a^{p\cdot}_{\cdot i}\, a^{q\cdot}_{\cdot j}\, \mathfrak{z}_p\mathfrak{z}_q. \tag{4.10}$$

The components of the polyadic (dyadic) products $\mathfrak{z}_i\mathfrak{z}_j$ in the corresponding coordinate system may be rewritten in the form of matrices consisting of a single one and zeros. For example, the components of $\mathbf{3}_1\mathbf{3}_2$ form the matrix

$$\begin{Vmatrix} 0 & 1 & 0 \\ 0 & 0 & 0 \\ 0 & 0 & 0 \end{Vmatrix} .$$

With the aid of polyadic products, one may introduce objects referred to as tensors.

Let it be required that $T^{ij}\,\mathbf{3}_i\mathbf{3}_j$ be invariant with respect to coordinate transformations, i.e., that

$$T^{ij}\mathbf{3}_i\mathbf{3}_j = T'^{ij}\mathbf{3}'_i\mathbf{3}'_j\,, \tag{4.11}$$

where T^{ij} and T'^{ij} refer to different coordinate systems. It is then clear from the transformation rule (4.10) for polyadic products that T^{ij} must transform in the case of change of coordinates in the following manner:

$$T'^{ij} = b^{i.}_{.p}\, b^{j.}_{.q}\, T^{pq}\,. \tag{4.12}$$

Definition of a tensor. The invariant object $T = T^{ij}\mathbf{3}_i\mathbf{3}_j$ is called a tensor of the second order or of the second rank. The number of indices attached to its components is called its rank or order. Obviously, a vector is a tensor of order unity.

As in the case of a vector A, the invariance of a tensor T, i.e., of (4.11), is ensured by the mutually complementary character of the transformation formulae for polyadic products (4.10) and the components of a tensor (4.12).

In an analogous manner to the second order tensor, one may introduce tensors of arbitrary order, for instance, the fifth order tensor

$$T = T^{ijklm}\mathbf{3}_i\mathbf{3}_j\mathbf{3}_k\mathbf{3}_l\mathbf{3}_m = T'^{ijklm}\mathbf{3}'_i\mathbf{3}'_j\mathbf{3}'_k\mathbf{3}'_l\mathbf{3}'_m\,, \tag{4.13}$$

where the objects, governed by the numbers T^{ijklm}, are the polyadic products $\mathbf{3}_i\,\mathbf{3}_j\,\mathbf{3}_k\,\mathbf{3}_l\,\mathbf{3}_m$, which transform like (4.10), and the components of the tensor T transform like (4.12).

Note that vectors and tensors are defined as objects which do not depend on coordinate transformations, and not simply as a set of components which transform by a given law.[1)]

Symmetric and antisymmetric tensors. The tensor components T^{ij} and T^{ijklm}, introduced by (4.11) and (4.13), transform in a contravariant manner and are called contravariant components of the tensor.

In the general case, all components of a tensor T are different. However, if the values of the components of a tensor do not alter when an arbitrary pair of indices are interchanged, i.e., $T^{ijklm}=T^{jiklm}$, then the tensor T is said to be symmetric with respect to these indices. It is clear from the transformation rules for tensor components (4.12) that the property of symmetry is invariant with respect to coordinate transformations.

If the components of a tensor T change sign for permutations of some pair of indices, i.e., $T^{ijklm}=-T^{jiklm}$, the tensor is said to be antisymmetric with respect to these indices. The property of antisymmetry is also invariant under coordinate transformations.

Let $T=T^{ij}\mathbf{3}_i\mathbf{3}_j$ be a tensor, then the object $T^*=T^{*ij}\mathbf{3}_i\mathbf{3}_j$, where $T^{*ij}=T^{ji}$, is also a tensor, where $T=T^*$ only if the tensor is symmetric.

Addition of tensors and multiplication by a number. Consider two tensors $A=A^{ijk}\mathbf{3}_i\mathbf{3}_j\mathbf{3}_k$ and $B=B^{ijk}\mathbf{3}_i\mathbf{3}_j\mathbf{3}_k$ and form the combination $A+B=(A^{ijk}+B^{ijk})\mathbf{3}_i\mathbf{3}_j\mathbf{3}_k$ which, obviously, will also be a tensor. This new tensor $A+B$ is called the sum of the tensors A and B. Thus, it is possible, with the aid of this rule, to form from given tensors new tensors which are their sum and difference. One may add and subtract only tensors of equal order.

[1)] Invariant objects, vectors and tensors have been defined, the base objects and components of which transform under coordinate transformations $\eta^i=\eta^i(\xi^j)$ with the aid of the invertible matrices $a^{i}_{\cdot j}=\partial\xi^i/\partial\eta^j$ and $b^{i}_{\cdot j}=\partial\eta^i/\partial\xi^j$. Analogously, one may introduce other base objects e_i the transformations of which are determined by other (linked to coordinate transformations by a different method) matrices A^j_i and B^i_j and construct on them corresponding invariant objects $Q=Q^ie_i=Q'^je'_j$, $P=P^{ij}e_ie_j=P'^{kl}e'_ke'_l$, etc. in such a manner that

$$e'_i=A^j_ie_i,\quad Q'^j=B^j_iQ^i,\quad P'^{kl}=B^k_iB^l_jP^{ij},\quad A^i_jB^j_k=\delta^i_k\,.$$

For example, when dealing with orthogonal transformations, besides vectors and tensors, one introduces spinors and spin-tensors, the base objects and components of which transform with the aid of certain matrices A^i_j and B^i_j which are with respect to each other (not coinciding with $a^{i}_{\cdot j}$ and $b^{i}_{\cdot j}$) matrix representations of the group of orthogonal transformations of space.

Obviously, if A is a tensor, then the object $C=k\cdot A$, where k is an arbitrary number which does not depend on the coordinate system (a scalar), is also a tensor.

The operation of symmetrization and alternation. Using the rules of addition of tensors and multiplication by a number, any second order tensor $T=T^{ij}\mathbf{э}_i\mathbf{э}_j$ can yield a symmetric tensor

$$T_0=\tfrac{1}{2}(T^{ij}+T^{ji})\mathbf{э}_i\mathbf{э}_j$$

and an antisymmetric tensor

$$T_1=\tfrac{1}{2}(T^{ij}-T^{ji})\mathbf{э}_i\mathbf{э}_j\,.$$

The operations of generating the tensors T_0 and T_1 are called symmetrization and alternation, respectively. If the tensor T is symmetric, then $T_0=T$ and $T_1=0$; if it is antisymmetric, then $T_0=0$, and $T_1=T$.

Note that, by definition, a tensor is equal to zero, if all its components are zero.

Transformation formulae of contravariant base vectors. The base vectors $\mathbf{э}_i$, transforming by (4.5), are called covariant base vectors. Consider some second order tensor $\varkappa=\varkappa^{ij}\mathbf{э}_i\mathbf{э}_j$ and in some coordinate system ξ^1, ξ^2, ξ^3 introduce

$$\mathbf{э}^i=\varkappa^{ij}\mathbf{э}_j\,, \tag{4.14}$$

where, for example, $\varkappa^{1j}\mathbf{э}_j=\varkappa^{11}\mathbf{э}_1+\varkappa^{12}\mathbf{э}_2+\varkappa^{13}\mathbf{э}_3$ is the sum of the three base vectors $\mathbf{э}_i$, multiplied by the numbers $\varkappa^{1i}$. Similarly, in another coordinate system η^1, η^2, η^3, one may introduce

$$\mathbf{э}'^p=\varkappa'^{pq}\mathbf{э}'_q\,.$$

The formulae (4.12) for the transformation of the $\varkappa^{pq}$ and (4.5) for the $\mathbf{э}_q$ are known; with their aid, one obtains the formulae for the transformation of

$$\mathbf{э}'^p=\varkappa'^{pq}\mathbf{э}'_q=b^{p}_{\cdot i}\,b^{q}_{\cdot j}\varkappa^{ij}a^{k}_{\cdot q}\mathbf{э}_k=b^{p}_{\cdot i}\varkappa^{ij}\mathbf{э}_j=b^{p}_{\cdot i}\mathbf{э}^i\,, \tag{4.15}$$

since $b^{q}_{\cdot j}a^{k}_{\cdot q}=\delta^{k}_{\cdot j}$. Evidently, the $\mathbf{э}^i$ transform in a contravariant manner. They are called contravariant base vectors.

Thus, with the aid of an arbitrary second order tensor $\varkappa$, one may introduce contravariant base vectors $\mathbf{э}^i$. Note that, if the covariant base vectors $\mathbf{э}_i$ depend only on the coordinate system, the contravariant base vectors $\mathbf{э}^i$ also depend on the coordinate system, and on the tensor $\varkappa$, used in their formation.

Covariant components of the tensor $\varkappa$. If the contravariant base vectors $\mathbf{э}^i$ are known, one may find covariant base vectors $\mathbf{э}_i$, i.e., one may solve (4.14) with respect to $\mathbf{э}_i$. For this purpose, introduce the matrix $\|\varkappa_{ij}\|$, the inverse of the matrix $\|\varkappa^{ij}\|$, provided Det $\|\varkappa^{ij}\| \neq 0$. It is known from elementary algebra that

$$\varkappa_{ij} = \frac{k_{ji}}{\Delta}, \tag{4.16}$$

where the k_{ij} are the cofactors of the matrix $\|\varkappa^{ij}\|$ and $\Delta = \text{Det}\,\|\varkappa^{ij}\|$. Thus, knowing the matrix $\|\varkappa^{ij}\|$ with non-zero determinant, one may form by (4.16) the matrix $\|\varkappa_{ij}\|$ and solve (4.14) with respect to $\mathbf{э}_i$. In some coordinate system $\zeta^1, \zeta^2, \zeta^3$, one will have

$$\mathbf{э}_j = \varkappa_{ji}\,\mathbf{э}^i . \tag{4.17}$$

Similarly, in another coordinate system η^1, η^2, η^3,

$$\mathbf{э}'_j = \varkappa'_{ji}\,\mathbf{э}'^i .$$

Using the transformation formulae (4.5) and (4.15) for $\mathbf{э}_i$ and $\mathbf{э}^i$, respectively, one obtains the tranformation formulae for the $\varkappa_{ij}$. In fact,

$$\mathbf{э}'_j = \varkappa'_{ji}\mathbf{э}'^i = a^{i\cdot}_{\cdot j}\mathbf{э}_i = a^{i\cdot}_{\cdot j}\varkappa_{ik}\mathbf{э}^k = a^{i\cdot}_{\cdot j}a^{k\cdot}_{\cdot l}\varkappa_{ik}\mathbf{э}'^l ,$$

whence

$$\varkappa'_{ji} = a^{p\cdot}_{\cdot j}a^{q\cdot}_{\cdot i}\varkappa_{pq} . \tag{4.18}$$

It is seen that the expression $\varkappa_{ij}\mathbf{э}^i\mathbf{э}^j$, where $\mathbf{э}^i\mathbf{э}^j$ are the polyadic products of the contravariant base vectors $\mathbf{э}^i$ which transform by the formulae

$$\mathbf{э}'^i\mathbf{э}'^j = b^{i\cdot}_{\cdot k}b^{j\cdot}_{\cdot l}\mathbf{э}^k\mathbf{э}^l ,$$

will represent itself an object which does not depend on the choice of the coordinate system, because the $\varkappa_{ij}$ transform covariantly and the polyadic products $\mathbf{э}^i\,\mathbf{э}^j$ contravariantly. Besides, by (4.14) and bearing in mind that the matrix $\varkappa_{ij}$ is the inverse of the matrix $\varkappa^{ij}$, we have

$$\varkappa_{ij}\mathbf{э}^i\mathbf{э}^j = \varkappa^{ip}\varkappa^{jq}\varkappa_{ij}\mathbf{э}_p\mathbf{э}_q = \varkappa^{pq}\mathbf{э}_p\mathbf{э}_q .$$

Thus, it is seen that the $\varkappa_{ij}$ may be called covariant components of the same second order tensor $\varkappa$ on the contravariant base $\mathbf{3}^i$. For the sake of simplicity, it will be assumed below that $\varkappa$ is a symmetric tensor, i.e., that $\varkappa^{ij}=\varkappa^{ji}$, and hence also that $\varkappa_{ij}=\varkappa_{ji}$.

Covariant components of an arbitrary vector. Obviously, for any vector A,

$$A = A^j\mathbf{3}_j = A^j\varkappa_{ij}\mathbf{3}^i = A_i\mathbf{3}^i ,$$

if

$$A_i = \varkappa_{ij}A^j . \tag{4.19}$$

It is seen that the index of the contravariant components A^j of the vector A and of the contravariant base vectors $\mathbf{3}^i$ is lowered with the aid of the covariant components (4.19) and (4.17) of the tensor $\varkappa$. Consequently, the A_i transform in the same way as $\mathbf{3}_i$, i.e., in a contravariant manner:

$$A'_i = a^{k\cdot}_{\cdot i}\, A_k .$$

The A_i are called the covariant components of the vector A on the contravariant base $\mathbf{3}^i$. Consequently, for each vector A, one can introduce components A^i, which transform with the aid of the matrix $\mathbf{B}$, i.e., so called contravariant components, and components A_i, which transform with the aid of the matrix $\mathbf{A}$, i.e., so called covariant components. In the general case, the covariant and contravariant components of a vector differ from each other: $A^j \neq A_j$.

Covariant and mixed tensor components. The argument, used in the case of a vector, may be applied to tensors of any order, for example, to the fourth order tensor

$$\begin{aligned} T &= T^{ijkl}\mathbf{3}_i\mathbf{3}_j\mathbf{3}_k\mathbf{3}_l = T^{ijkl}\varkappa_{ip}\varkappa_{jq}\varkappa_{km}\varkappa_{ln}\mathbf{3}^p\mathbf{3}^q\mathbf{3}^m\mathbf{3}^n = T_{pqmn}\mathbf{3}^p\mathbf{3}^q\mathbf{3}^m\mathbf{3}^n \\ &= T^{ijkl}\varkappa_{jp}\varkappa_{kq}\mathbf{3}_i\mathbf{3}^p\mathbf{3}^q\mathbf{3}_l = T^{i\,..\,l}_{\cdot pq\cdot}\mathbf{3}_i\mathbf{3}^p\mathbf{3}^q\mathbf{3}_l . \end{aligned} \tag{4.20}$$

The components T_{pqmn} are said to be covariant and the components $T^{i\,..\,l}_{\cdot pq\cdot}$ mixed (covariant in the indices p, q and contravariant in the indices i, l) components of the tensor T. The transformation formulae for mixed components have the form

$$T'^{m\,..\,s}_{\cdot nr\cdot} = T^{i\,..\,l}_{\cdot pq\cdot}\, b^{m\cdot}_{\cdot i}\, a^{p\cdot}_{\cdot n}\, a^{q\cdot}_{\cdot r}\, b^{s\cdot}_{\cdot l} ,$$

i.e., the transformation is covariant with respect to the lower indices n, r, and contravariant with respect to the upper indices m, s.

Manipulation of indices. It has been seen that one can lower and raise the indices of the components of any tensor with the aid of the tensor $\varkappa$. This operation is referred to as manipulation of the indices. Consider, for example,

$$T = T_{ij}\mathbf{3}^i\mathbf{3}^j = T_{ij}\varkappa^{ik}\mathbf{3}_k\mathbf{3}^j = T^{k.}_{.j}\mathbf{3}_k\mathbf{3}^j \; ; \tag{4.21}$$

instead of expressing the tensor T in terms of the covariant components T_{ij}, it has been expressed in terms of its mixed components $T^{k.}_{.j}$. Clearly, the lowering of the indices (4.20) takes place with the aid of the $\varkappa_{ij}$, the raising (4.21) with the aid of the $\varkappa^{ij}$.

Note that one may add and subtract only components of tensors with the same index structure. The properties of symmetry and anti-symmetry of tensors likewise are defined with respect to identically placed indices.

Length of a vector. The reasoning above referred to one arbitrary, but fixed point of space. The metric of the space will now be introduced, i.e., a method will be described for the determination of length in space. For the determination of the length of a vector, it is sufficient to define the scalar products of the base vectors

$$\mathbf{3}_i \cdot \mathbf{3}_j = g_{ij} \, ,$$

which, generally speaking, at a given point may be arbitrary numbers. The square of the length of $d\mathbf{r}$, by definition, is given by

$$|d\mathbf{r}|^2 = ds^2 = d\mathbf{r} \cdot d\mathbf{r} = d\zeta^i d\zeta^j \mathbf{3}_i \cdot \mathbf{3}_j = d\zeta^i d\zeta^j g_{ij} \, , \tag{4.22}$$

and the square of the length of any vector A by

$$|A|^2 = A^i A^j g_{ij} \, .$$

The length of any vector is expressible in terms of its components and scalar products of the base vectors g_{ij}.

The condition of invariance of the length of $|d\mathbf{r}|$ with respect to the choice of coordinate system has the form

$$|d\mathbf{r}|^2 = g'_{pq} d\eta^p d\eta^q = g_{ij} d\zeta^i d\zeta^j = g_{ij} a^{i.}_{.p} a^{j.}_{.q} d\eta^p d\eta^q \, .$$

Fundamental metric tensor. The last formula yields the transformation formula for the tensor g_{ij}:

$$g'_{pq} = a^{i}_{\cdot p} a^{j}_{\cdot q} g_{ij} .$$

Thus, since the length $|d\boldsymbol{r}|$ is invariant, one must consider the g_{ij} as covariant components of the tensor $g = g_{ij} \mathbf{Э}^i \mathbf{Э}^j$, which is referred to as the fundamental metric tensor.

By the definition of the scalar product, the tensor g is symmetric, i.e.,

$$g_{ij} = g_{ji} .$$

The quadratic form (4.22) in the coordinate increments $d\zeta^i$ is called the fundamental quadratic form which yields the metric of the distances between closeby points in space.

It is known from algebra that every quadratic form with constant coefficients can be reduced to its canonical form, i.e., that at every given point one can find coordinates x^1, x^2, x^3 such that quadratic form (4.22) can be rewritten as the sum of squares

$$ds^2 = (dx^1)^2 + (dx^2)^2 + (dx^3)^2 , \tag{4.23}$$

and the matrix of the tensor g is reduced to the form

$$\begin{Vmatrix} 1 & 0 & 0 \\ 0 & 1 & 0 \\ 0 & 0 & 1 \end{Vmatrix} .$$

Note that, generally speaking, it is impossible to realize such a type of transformation immediately for an entire space, i.e., to find a coordinate system x^1, x^2, x^3 such that (4.22) reduces to (4.23) throughout a space. If such a coordinate system exists, then a space is said to be Euclidean, otherwise non-Euclidean. If (4.22) can be reduced in n-dimensional space by means of a real transformation to the form $ds^2 = \alpha_i (dx^i)^2$ where $\alpha_i = \pm 1$, $i = 1, 2, \ldots, n$ and at least one of the α_i differs in sign from the others, then the space is said to be pseudo-Euclidean.

Reciprocity of covariant and contravariant bases if g is used in place of $\varkappa$. One can introduce the matrix g^{ij}, the inverse of the matrix g_{ij} in the same way as it was done earlier for $\varkappa^{ij}$. The transformation of the components of the matrix g^{ij} is contravariant. For the introduction of g^{ij} one requires only that Det $\|g_{ij}\| \neq 0$. With the aid of the g^{ij}, one can introduce the contravariant base vectors $\mathbf{з}^j$

$$\mathbf{з}^j = g^{ij}\mathbf{з}_i \tag{4.24}$$

and manipulate indices not with the aid of the arbitrary vector $\varkappa$, but with the aid of the fundamental metric tensor g. In what follows we will use $\mathbf{з}^i$, introduced with the aid of g^{ij}.

Consider a property of the scalar products $\mathbf{з}^j \cdot \mathbf{з}_p$.

From (4.24), one has

$$\mathbf{з}^j \cdot \mathbf{з}_p = g^{ij}\mathbf{з}_i \cdot \mathbf{з}_p = g^{ij} g_{ip} = \delta^j_p\,, \tag{4.25}$$

i.e.,

$$\mathbf{з}^1 \cdot \mathbf{з}_1 = 1, \quad \mathbf{з}^1 \cdot \mathbf{з}_2 = 0, \quad \mathbf{з}^1 \cdot \mathbf{з}_3 = 0\,, \text{ etc.,}$$

whence it follows that the base vector $\mathbf{з}^1$ is orthogonal to the plane formed by the vectors $\mathbf{з}_2, \mathbf{з}_3$, etc. It is not difficult to verify that for contravariant base vectors

$$\mathbf{з}^1 = \frac{\mathbf{з}_2 \times \mathbf{з}_3}{\mathbf{з}_1 \cdot (\mathbf{з}_2 \times \mathbf{з}_3)}, \quad \mathbf{з}^2 = \frac{\mathbf{з}_3 \times \mathbf{з}_1}{\mathbf{з}_1 \cdot (\mathbf{з}_2 \times \mathbf{з}_3)}, \quad \mathbf{з}^3 = \frac{\mathbf{з}_1 \times \mathbf{з}_2}{\mathbf{з}_1 \cdot (\mathbf{з}_2 \times \mathbf{з}_3)}, \tag{4.26}$$

and for covariant base vectors

$$\mathbf{з}_1 = \frac{\mathbf{з}^2 \times \mathbf{з}^3}{\mathbf{з}^1 \cdot (\mathbf{з}^2 \times \mathbf{з}^3)}, \quad \mathbf{з}_2 = \frac{\mathbf{з}^3 \times \mathbf{з}^1}{\mathbf{з}^1 \cdot (\mathbf{з}^2 \times \mathbf{з}^3)}, \quad \mathbf{з}_3 = \frac{\mathbf{з}^1 \times \mathbf{з}^2}{\mathbf{з}^1 \cdot (\mathbf{з}^2 \times \mathbf{з}^3)}, \tag{4.27}$$

where the sign $\times$ denotes the ordinary vector product. One says that covariant and contravariant base vectors are dual. It is obvious that in an orthogonal Cartesian coordinate system $\mathbf{з}^j = \mathbf{з}_j$, and hence in such a coordinate system there is no distinction between the covariant and contravariant components of a vector and tensor; in such a coordinate system, the use of superscripts and subscripts has no significance.

Mixed components of the metric tensor. It follows from (4.25) that the mixed components $g^{i\cdot}_{\cdot j}$ of the fundamental metric tensor g in any coordinate system form the unit matrix:

$$\|g^{i\cdot}_{\cdot j}\| = \begin{Vmatrix} g^{1\cdot}_{\cdot 1} & g^{1\cdot}_{\cdot 2} & g^{1\cdot}_{\cdot 3} \\ g^{2\cdot}_{\cdot 1} & g^{2\cdot}_{\cdot 2} & g^{2\cdot}_{\cdot 3} \\ g^{3\cdot}_{\cdot 1} & g^{3\cdot}_{\cdot 2} & g^{3\cdot}_{\cdot 3} \end{Vmatrix} = \begin{Vmatrix} 1 & 0 & 0 \\ 0 & 1 & 0 \\ 0 & 0 & 1 \end{Vmatrix} = \|\delta^{i\cdot}_{\cdot j}\|\,.$$

Direct tensor multiplication. Certain tensor operations have already been encountered. Consider next the operation of multiplication of tensors. Let there be given a vector $A = A^i \mathfrak{z}_i$ and the tensor $T = T^{k}_{.j} \mathfrak{z}_k \mathfrak{z}^j$; formally, construct

$$B = A^i T^{k}_{.j} \mathfrak{z}_i \mathfrak{z}_k \mathfrak{z}^j$$

and

$$B^* = A^i T^{k}_{.j} \mathfrak{z}_k \mathfrak{z}^j \mathfrak{z}_i = T^{ik} A_j \mathfrak{z}_i \mathfrak{z}_k \mathfrak{z}^j \, .$$

Obviously, B and B^* are tensors, but $B \neq B^*$. This operation, which generates tensors of order higher than that of the original tensors, is called direct tensor multiplication. Its result depends on the order of the factors. With the aid of direct multiplication of vectors, one may form tensors of any order: $A^i A^j A^k \ldots \mathfrak{z}_i \mathfrak{z}_j \mathfrak{z}_k \ldots$; however, not every tensor can be represented as a product of vectors. By multiplication, one may introduce tensors of the form

$$\mathscr{D} = g^{ij} g^{pq} g^{rs} \ldots \mathfrak{z}_i \mathfrak{z}_j \mathfrak{z}_p \mathfrak{z}_q \mathfrak{z}_r \mathfrak{z}_s \ldots =$$
$$= g_{ij} g_{pq} g_{rs} \ldots \mathfrak{z}^i \mathfrak{z}^j \mathfrak{z}^p \mathfrak{z}^q \mathfrak{z}^r \mathfrak{z}^s \ldots = \delta^{i.}_{.j} \delta^{p.}_{.q} \delta^{r.}_{.s} \ldots \mathfrak{z}_i \mathfrak{z}^j \mathfrak{z}_p \mathfrak{z}^q \mathfrak{z}_r \mathfrak{z}^s \ldots$$

Obviously, the mixed components of tensors $\mathscr{D}$ in any coordinate system are the same, i.e., they are invariant under coordinate transformations. These components are equal to zero or unity.

Number of components of a tensor. A scalar k may be considered as a tensor of zero order; it is characterized by one number ($3^0 = 1$). A vector is a tensor of first order; it has, in three-dimensional space, three components ($3^1 = 3$). A second order tensor has $3^2 = 9$ components; a tensor of order p has, in three-dimensional space, 3^p components, a tensor of order r in n-dimensional space has n^r components. Sometimes, for example, in the presence of symmetry, the number of independent components of a tensor can be smaller. In particular, the symmetric second order tensor ($T_{ij} = T_{ji}$) has only 6 independent components, the anti-symmetric second order tensor ($T_{ij} = -T_{ji}$) has only three independent components. The concept of symmetry of a tensor implies invariance of its components with respect to some group of transformations. For instance, the mixed components of the tensor $\mathscr{D}$ are invariant with respect to the group of all continuous transformations. The components of a tensor with any index structure are invariant with respect to the group of orthogonal transformations, defined by the condition of invariance of the components of the fundamental tensor g_{ij}.

Scalar invariants of a tensor. In general, the components of a tensor depend on the choice of the coordinate system; however, one may pose the problem of finding functions of tensor components $\Phi(T^{i\cdot}_{\cdot j})$ which will be invariant with respect to choice of coordinate system, i.e., for which

$$\Phi(T^{i\cdot}_{\cdot j}) = \Phi(T'^{i\cdot}_{\cdot j}) .$$

Such functions of the components of a tensor are called invariants of the tensor. They are numbers or functions of points in space; namely, such functions of tensor and vector components must, together with other invariant objects, enter the mathematical description of physical laws which must be invariant with respect to methods of description of physical effects, and, in particular, cannot depend on a coordinate system. Analogous methods may determine invariant functions of the components of some tensors. Such functions are called scalars. Simple rules for the formation of tensor and vector invariants will now be given. Consider the vector

$$A = A^i \mathbf{э}_i = A_j \mathbf{э}^j = A^i g_{ij} \mathbf{э}^j$$

and form the scalar product

$$A \cdot A = A^i A^j \mathbf{э}_i \cdot \mathbf{э}_j = A^i A^j g_{ij} = A^i A_j .$$

The expression obtained is an invariant (the square of the length of the vector A), since the transformations of the dual components of a vector are dual. A vector has only one independent invariant—its length; all other invariants are functions of its length.

Next, consider any second order tensor

$$T = T^{ij} \mathbf{э}_i \mathbf{э}_j$$

and contract it with the metric tensor with respect to both indices, i.e., form $T^{ij} g_{ij}$ (contraction is the operation of summation over an upper and a lower index) to obtain a number which does not depend on the coordinate system, since transformations of components with upper and lower indices are dual. One may write

$$T^{ij} g_{ij} = T^{i\cdot}_{\cdot i} = T^{1\cdot}_{\cdot 1} + T^{2\cdot}_{\cdot 2} + T^{3\cdot}_{\cdot 3} . \tag{4.28}$$

Similarly, the expressions

$$T^{i\cdot}_{\cdot j} T^{j\cdot}_{\cdot i} , \quad T^{i\cdot}_{\cdot j} T^{j\cdot}_{\cdot p} T^{p\cdot}_{\cdot i}$$

are also invariants. Thus, for a second order tensor, three invariants have been obtained which are linear, quadratic and cubic with respect to the components. It will be shown below that in the case of symmetric order tensors, which are especially important from a point of view of applications, all remaining scalar invariants are functions of these three invariants.

Tensor surface. Select an arbitrary point O and near it a point M. Construct at O coordinate lines $\zeta^1, \zeta^2, \zeta^3$ and consider the vector

$$\overline{OM} = d\mathbf{r} = d\zeta^i \mathbf{з}_i$$

and the symmetric tensor

$$T = T^{ij}\mathbf{з}_i\mathbf{з}_j = T_{ij}\mathbf{з}^i\mathbf{з}^j .$$

Obviously, the quantity $T_{ij}d\zeta^i d\zeta^j$ is an invariant and one may write

$$T_{ij}d\zeta^i d\zeta^j = T'_{ij}d\eta^i d\eta^j = c\,, \tag{4.29}$$

where c is some number. In a small neighbourhood of O, Equation (4.29) determines, for fixed c and the values of T_{ij} at O, a second order surface, which is referred to as a tensor surface. The differentials $d\zeta^i$ or $d\eta^i$ are considered as coordinates of points of this tensor surface. For every symmetric second order tensor T, one can construct at every point the corresponding second order surface (4.29).

Principal axes and principal tensor components. As is well known, the equation of this second order surface can be reduced, with the aid of a coordinate transformation, to its canonical form, i.e., one may select at the point O a coordinate system x^1, x^2, x^3 such that (4.29) assumes the form

$$T_{11}(dx^1)^2 + T_{22}(dx^2)^2 + T_{33}(dx^3)^2 = c\,.$$

Then, at the point O, the system of coordinates x^1, x^2, x^3 will be orthogonal. Consequently, one can introduce at every point of space coordinate axes such that only the three components T_{11}, T_{22}, T_{33} of a second order symmetric tensor are non-zero. Such axes are called the principal axes of the tensor, and the orthogonal Cartesian coordinate ($g_{ii} = 1$, $g_{ij} = 0$ for $i \neq j$) with axes parallel to the principle axes are said to form the principal coordinate system of the tensor. Obviously, the difference between covariant and contravariant components disappears in the principal coordinate system:

$$T^{ii} = T_{ii} = T^i_i = T_i .$$

(Summation with respect to i does not occur here). The three, in general different, non-zero components of the tensor in the principal system are referred to as principal components.

The method of finding the principal axes and the principal components of the symmetric tensor is described on pages 170–172.

The question regarding the number of independent invariants of a second order symmetric tensor is now readily answered. All of them must in the principal coordinate system be functions of only three components and, consequently, their number cannot exceed three; from the listing of invariants (4.28) in the principal system it is clear that all three invariants, found earlier, are independent. In the sequel, a number of deductions of tensor analysis will be required which will be stated as required.

§5. Theory of deformation

Let an absolutely rigid body move with respect to the observer's coordinate system x^1, x^2, x^3 (fig. 10). Select two of its positions, one at the initial instant t_0, the other at an arbitrary instant t.

On the dependence of the base vectors of concomitant systems on the time. A concomitant coordinate system ξ^1, ξ^2, ξ^3 may be associated with every point M of the body. The concomitant system will move with the body, and the base vectors of the concomitant system at times t_0 and t will differ. Denote those at time t_0 by $\mathring{\mathfrak{z}}_i$, those at time t by $\mathfrak{z}_i$. Clearly, the base vectors of the concomitant system depend, generally speaking, on the point M of the body and, besides, change with time. Obviously, since the system ξ^1, ξ^2, ξ^3 is frozen in the medium and the medium moves as an absolutely rigid body, then the trihedron $\mathfrak{z}_i$ can be obtained from the trihedron $\mathring{\mathfrak{z}}_i$ by means of a translation and a rotation

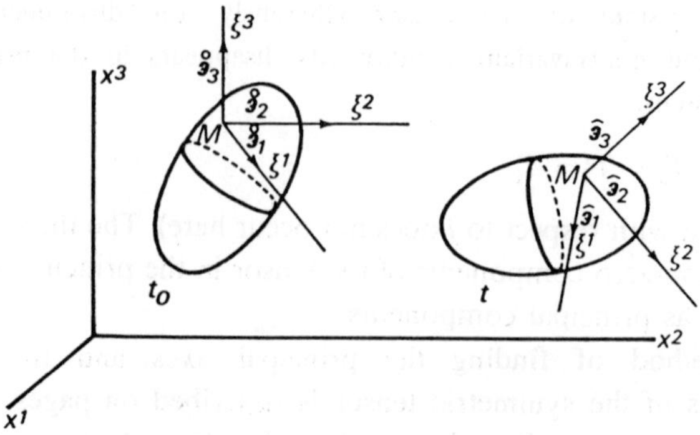

Fig. 10. Motion of an absolutely rigid body.

$$|\mathring{\mathfrak{z}}_i| = |\hat{\mathfrak{z}}_i| \quad \text{and} \quad \angle \mathring{\mathfrak{z}}_i \mathring{\mathfrak{z}}_j = \angle \hat{\mathfrak{z}}_i \hat{\mathfrak{z}}_j, \text{ i.e., } \mathring{\mathfrak{z}}_i \cdot \mathring{\mathfrak{z}}_j = \hat{\mathfrak{z}}_i \cdot \hat{\mathfrak{z}}_j \,.$$

In the case of the motion of a deformable body, the circumstances will be more complicated. In fact, for the motion of a deformable body, distances between its points M and M' change. The coordinate lines of the concomitant coordinate system deform and the base vectors $\hat{\mathfrak{z}}_i$ change in time, so that their magnitudes as well as the angles between them vary.

The effects of changes of the distances between the points of a continuum during the time of motion are very important. In particular, note that the forces of interaction between the particles depend on the changes in the distances between them.

Consider two arbitrary positions of a deformable body and, in particular, its points M and M' at arbitrary instants of time t and t' (fig. 11). Denote the base vectors at M at time t' by $\mathfrak{z}'_i$, those at time t by $\hat{\mathfrak{z}}_i$. Obviously, in the concomitant coordinate system,

$$d\boldsymbol{r} = d\xi^i \hat{\mathfrak{z}}_i \quad \text{and} \quad d\boldsymbol{r}' = d\xi^i \mathfrak{z}'_i \,.$$

One wishes to introduce into the consideration the characteristics of changes of distances; therefore one must introduce the metric tensors of the concomitant coordinate systems at times t and t'.

However, prior to the introduction of the metric, note that any infinitesimal segment of a straight line through M during the motion of the continuum goes over into an infinitesimal segment of a straight line through the point corresponding to M.

In fact, side by side with the infinitesimal element $d\boldsymbol{r}$ of the continuum at time t, to which corresponds $d\boldsymbol{r}'$at time t', one may introduce at time t

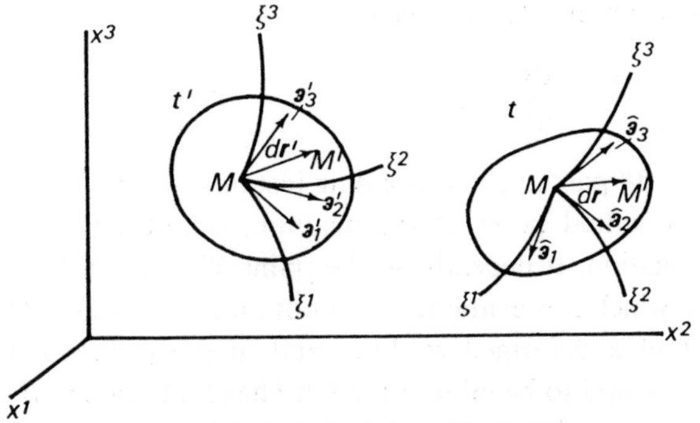

Fig. 11. Motion of a deformable medium.

an element $k\mathrm{d}\boldsymbol{r}$ of the continuum, where k is a numerical factor. In the space ξ^1, ξ^2, ξ^3 at time t', there will correspond to this element the element $k\,\mathrm{d}\boldsymbol{r}'$, since in this space, as a result of the conservation of the Lagrangian coordinates of all points of a continuum, the expansion in terms of the base vectors $\mathbf{з}'_i$ must hold true, i.e.,

$$k\,\mathrm{d}\xi^i \mathbf{з}'_i = k\,\mathrm{d}\boldsymbol{r}' .$$

For different finite k and given $\mathrm{d}\boldsymbol{r}$, the elements $k\mathrm{d}\boldsymbol{r}$ determine at time t a small segment of a straight line starting from M to which in the space ξ^1, ξ^2, ξ^3 at time t' will correspond the small segment $k\,\mathrm{d}\boldsymbol{r}'$ of a straight line.

Now introduce the metrics of the spaces of the concomitant coordinate systems at times t and t'. Let at time t

$$|\mathrm{d}\boldsymbol{r}| = \mathrm{d}s, \quad \mathrm{d}s^2 = \hat{g}_{ij}\mathrm{d}\xi^i \mathrm{d}\xi^j , \tag{5.1}$$

where

$$\hat{g}_{ij} = \hat{\mathbf{з}}_i \cdot \hat{\mathbf{з}}_j ,$$

and at time t'

$$|\mathrm{d}\boldsymbol{r}'| = \mathrm{d}s', \quad \mathrm{d}s'^2 = g'_{ij}\mathrm{d}\xi^i \mathrm{d}\xi^j , \tag{5.2}$$

where

$$g'_{ij} = \mathbf{з}'_i \cdot \mathbf{з}'_j .$$

Note that the coordinates of the points M and M' at times t and t' are the same in the concomitant coordinate systems, while the components of $\hat{g}_{ij}$ and g'_{ij} differ.

Coefficient of relative elongation. The ratio

$$l = \frac{ds - ds'}{ds'} = \frac{ds}{ds'} - 1 , \tag{5.3}$$

where ds and ds' pass at corresponding instants of time through one and the same individual point of the medium, is called the coefficient of relative elongation. It depends on the point M and the direction of the element, for which it is evaluated, and not on d$\boldsymbol{r}$. If l is infinitely small at every point of a deformed medium and in every direction, then the deformation is said to be infinitesimal. If l has finite values, the deformation is said to be finite. By definition, all coefficients l vanish for an absolutely rigid body.

Note that the deformations and coefficients of relative elongation l may be introduced, by considering two quite arbitrary positions of a continuum, and that l can be computed for any d$\boldsymbol{r}$, if one knows $\hat{g}_{ij}$, g'_{ij} and the direction of d$\boldsymbol{r}$.

Strain tensor. Introduce the notation

$$\varepsilon_{ij} = \tfrac{1}{2}(\hat{g}_{ij} - g'_{ij}) ; \tag{5.4}$$

by (5.1) and (5.2),

$$ds^2 - ds'^2 = 2\varepsilon_{ij} d\xi^i d\xi^j .$$

It is seen from (5.4) that the ε_{ij} can be considered as covariant tensor components. As is known, one may, with the aid of any second order tensor $\varkappa$, transform its contravariant components with respect to the covariant components of some tensor. In metric space, it was agreed to use in the capacity of $\varkappa$ the fundamental tensor g. In the present case, one may raise indices either with the aid of g'^{ij} or with that of $\hat{g}_{ij}$; therefore one may transform with respect to the covariant components ε_{ij} two different sets of contravariant components: $\hat{\varepsilon}^{ij}$ (the indices are raised by means of $\hat{g}^{ij}$) and ε'^{ij} (the indices are raised by means of g'^{ij}). This means that one may form the two different tensors

$$\hat{\mathscr{E}} = \varepsilon_{ij} \hat{\mathbf{3}}^i \hat{\mathbf{3}}^j \quad \text{and} \quad \mathscr{E}' = \varepsilon_{ij} \mathbf{3}'^i \mathbf{3}'^j ,$$

which have the same covariant components (5.4), but referred to different bases $\hat{\mathbf{3}}^i$ and $\mathbf{3}'^i$. These two tensors are called strain tensors. The contravariant and mixed components of the tensors $\hat{\mathscr{E}}$ and $\mathscr{E}'$ differ, and one has for them the notations $\hat{\varepsilon}^{ij}$, ε'^{ij} and $\hat{\varepsilon}^i{}_j$, $\varepsilon'^i{}_j$, respectively; obviously, $\varepsilon'^i_j \neq \hat{\varepsilon}^i{}_j$, since $\varepsilon'^i_j = \varepsilon_{pj} g'^{pi}$ and $\hat{\varepsilon}^i{}_j = \varepsilon_{pj} \hat{g}^{pi}$ and $g'^{pi} \neq \hat{g}^{pi}$.

Strain tensors are fundamental characteristics which arise in the deformation of bodies, and their components enter into the basic equations which describe the motion of continua.

Initial states and "initial states". It is clear that, at the moment of interest, deformations do not only depend on the state of a body considered, but also on the state with respect to which these strains were computed. How does one select this state, if one wishes to obtain definite physical characteristics of a deformation? Obviously, it may be quite arbitrary, but it must be determined by real physical considerations. Note that it may be defined differently in the theory of deformation; this method of definition will not be fixed and the state which is chosen by any method for the purpose of comparison with a given state of a medium is called initial. It should be pointed out, however, that the following conditions may be encountered. This initial state need not really occur. For example, one may select as initial state some imagined state in which the structure of each element of a continuum is in equilibrium and the element is left to itself, i.e., it is not acted upon by any forces. Let $\mathring{g}_{ij}$ denote the metric in this imagined state and $\mathring{\mathbf{э}}_i$ the base vectors of the concomitant system in the initial state. Obviously, the metric so introduced may be non-Euclidean. However, the real motion of a continuum takes place in an Euclidean system and, consequently, in the general case there may not exist a real transition of a continuum from the initial to a given state. Ideal, imagined "initial" states can be employed to estimate changes of metric and for the introduction of a strain tensor.

These observations will now be explained by the example of motion in two-dimensional Euclidean space, i.e., in the plane. Consider the motion of some film in a plane and select an initial state such that no forces act in the film. Let the film be stretched along the edges and remain plane only due to this stretching. However, if one were to release the film from the extending forces, it would warp, curl up and, remaining two-dimensional, it would already no longer be plane. The problem is now to establish the interrelationship between the points of the plane film at a given instant and when it is warped and possibly wrinkled (after removal of all load); in general, one must go for this purpose to three-dimensional space. Therefore the inextended warped state of a film with respect to motions in two-dimensional space can only be considered as an "initial state" (i.e., as a state in inverted commas). Thus, if an initial state, introduced for the sake of physical reasoning as a reference state, may be realized mentally or actually with the aid of some motion, this initial state may be used to define an initial state without inverted commas. However, if an imagined reference state cannot be obtained by continuous motion of the medium, in the same space, it will be referred to as "initial state".

In the general case, the components $\mathring{g}_{ij}$ may depend on the ξ^1, ξ^2, ξ^3 and t; once an imagined "initial state" is fixed, then the $\mathring{g}_{ij}$ may depend only on the ξ^1, ξ^2, ξ^3.

Geometric significance of covariant strain tensor components. The mechanical meaning of the covariant components of the strain tensors $\hat{\mathscr{E}}$ and $\mathring{\mathscr{E}}$ will now be explained. Rewrite the components of the metric tensors as

$$\hat{g}_{ij} = \hat{\mathfrak{z}}_i \cdot \hat{\mathfrak{z}}_j = |\hat{\mathfrak{z}}_i| \cdot |\hat{\mathfrak{z}}_j| \cos \psi_{ij}, \tag{5.5}$$

where ψ_{ij} are the angles between the vectors $\hat{\mathfrak{z}}_i$ and $\hat{\mathfrak{z}}_j$, and

$$\mathring{g}_{ij} = \mathring{\mathfrak{z}}_i \cdot \mathring{\mathfrak{z}}_j = |\mathring{\mathfrak{z}}_i| \cdot |\mathring{\mathfrak{z}}_j| \cos \mathring{\psi}_{ij} \tag{5.6}$$

with $\mathring{\psi}_{ij}$ the angles between the vectors $\mathring{\mathfrak{z}}_i$ and $\mathring{\mathfrak{z}}_j$. Form the ratio

$$\frac{|\hat{\mathfrak{z}}_i|}{|\mathring{\mathfrak{z}}_i|} = \frac{|\partial \boldsymbol{r}/\partial \xi^i|}{|\partial \boldsymbol{r}_0/\partial \xi^i|} = \frac{|\mathrm{d}\boldsymbol{r}_i|}{|\mathrm{d}\boldsymbol{r}_{0i}|} = \frac{\mathrm{d}s_i}{\mathrm{d}s_{0i}} = l_i + 1, \tag{5.7}$$

where $\mathrm{d}s_i$ and $\mathrm{d}s_{0i}$ are the elements of arc of the coordinate lines ξ^i and l_i are the coefficients of the relative elongations in the directions ξ^i. Now, by (5.7) and (5.5),

$$\hat{g}_{ij} = |\mathring{\mathfrak{z}}_i| \cdot |\mathring{\mathfrak{z}}_j| (1 + l_i)(1 + l_j) \cos \psi_{ij} \tag{5.8}$$

and, by (5.6), (5.8) and (5.4), selecting as state of the medium at time t' the initial state or "initial state" $\mathring{g}_{ij}$, one finds the formula

$$2\varepsilon_{ij} = [(1 + l_i)(1 + l_j) \cos \psi_{ij} - \cos \mathring{\psi}_{ij}] \, |\mathring{\mathfrak{z}}_i| \cdot |\mathring{\mathfrak{z}}_j|, \tag{5.9}$$

which is convenient for geometrical interpretations of ε_{ij}.

Consider first the geometrical interpretation of ε_{ij} when $i = j$. It follows from (5.9) that

$$2\varepsilon_{ii} = [(1 + l_i)^2 - 1] \mathring{g}_{ii}, \tag{5.10}$$

whence

$$l_i = \sqrt{\left(1 + \frac{2\varepsilon_{ii}}{\mathring{g}_{ii}}\right)} - 1 . \tag{5.11}$$

If the deformations are small, then the ε_{ij} are small; expanding (5.11) in a power series, one finds

$$l_i \simeq \frac{\varepsilon_{ii}}{\mathring{g}_{ii}} . \tag{5.12}$$

Besides, if one selects as concomitant system in the "initial state" a Cartesian coordinate system, then $\mathring{g}_{ii}=1$ and, consequently,

$$l_i \simeq \varepsilon_{ii} , \tag{5.13}$$

i.e., the covariant strain tensor components with a single index in the case of infinitesimal deformations coincide with the coefficients of relative extension along the Cartesian coordinate axes of the initial state.

Next, consider the question of the geometrical interpretation of the components ε_{ij} when $i \neq j$. For the sake of simplicity, select in the "initial state" at a given point a coordinate system such that the $\mathring{\mathfrak{z}}_i$ are orthogonal to each other, i.e.,

$$\mathring{\psi}_{ij} = \frac{\pi}{2} .$$

Then, setting

$$\psi_{ij} = \frac{\pi}{2} - \chi_{ij} .$$

one finds from (5.5), (5.6) and (5.4)

$$2\varepsilon_{ij} = |\hat{\mathfrak{z}}_i| \cdot |\hat{\mathfrak{z}}_j| \sin \chi_{ij} ,$$

or

$$\sin \chi_{ij} = \frac{2\varepsilon_{ij}}{\sqrt{(\hat{g}_{ii})}\sqrt{(\hat{g}_{jj})}} . \tag{5.14}$$

Hence it is clear that in the general case the angles, which are right angles in the "initial state", after deformation cease to be right angles, and that the covariant tensor components ε_{ij}, $i \neq j$, characterize a shearing of the right angles between the original coordinate axes. If the deformation is infinitesimal and the coordinate system in the "initial state" is Cartesian, then $\mathring{g}_{ii}=1$ and $\hat{g}_{ii}=1+O(\varepsilon)$ (where ε is an infinitesimal quantity). Expanding in a power series, one readily obtains

$$\sin \chi_{ij} \simeq 2\varepsilon_{ij}\,, \tag{5.15}$$

or

$$\chi_{ij} \simeq 2\varepsilon_{ij}\,. \tag{5.16}$$

Principal axes of the strain tensor. All symmetric tensors, including strain tensors, may be related to a quadratic form $\varepsilon_{ij}\, d\xi^i d\xi^j$. It has been shown in §4, that at every point one may find an orthogonal coordinate system η^1, η^2, η^3 at which this quadratic form reduces to

$$\varepsilon_{ij} d\xi^i d\xi^j = \varepsilon_{11}(d\eta^1)^2 + \varepsilon_{22}(d\eta^2)^2 + \varepsilon_{33}(d\eta^3)^2\,. \tag{5.17}$$

The transformation from ξ^1, ξ^2, ξ^3 to η^1, η^2, η^3 depends on the components ε_{ij}; therefore the corresponding orthogonal trihedron η^i in the presence of motion will, in general, change with time. Such axes η^1, η^2, η^3 will now be selected in the space $\mathring{g}_{ij}$ and it will be shown that, as a consequence of the motion, they will transfer into the space $\hat{g}_{ij}$ (for concomitant system) with such directions η^1, η^2, η^3 for which $\hat{\mathfrak{z}}_i$ will again be orthogonal. In fact, for such axes η^1, η^2, η^3, the components $\varepsilon_{ij}, i \neq j$, vanish, and consequently, by (5.14), $\chi_{ij} = 0$, i.e., the axes η^1, η^2, η^3 remain orthogonal. Thus, it is obvious that the coordinate trihedron for variables η^1, η^2, η^3 in the spaces $\mathring{g}_{ij}$ and $\hat{g}_{ij}$ coincide with the principal axes of $\hat{\mathscr{E}}$ and $\mathring{\mathscr{E}}$ tensor.

Referred to these axes, the matrices

$$\|\mathring{g}_{ij}\|,\ \|\hat{g}_{ij}\|,\ \ \|\mathring{g}^{ij}\|,\ \|\hat{g}^{ij}\|,\ \|\varepsilon_{ij}\|,\ \|\hat{\varepsilon}^{i}_{\cdot j}\|,\ \|\hat{\varepsilon}^{ij}\|,\ \|\mathring{\varepsilon}^{i}_{\cdot j}\|,\ \|\mathring{\varepsilon}^{ij}\|\ \cdot$$

become diagonal. The orthogonal trihedron, formed by the principal axes for a given displacement, remain orthogonal: the angles between the principal axes do not shear; however, the orthogonal trihedron of the principal axes may move as a rigid body, i.e., it may translate and rotate. Thus, at each point of a deformed medium, one may construct an orthogonal trihedron of principal axes which for a given displacement behaves like a rigid body. Note that elements $d\boldsymbol{r}$ along the principal axes, at the time of motion, may compress or stretch. Note that the concept of the principal axes of a strain tensor has been introduced also for the case of finite deformations. The principal axes of the strain tensors $\hat{\mathscr{E}}$ and $\mathring{\mathscr{E}}$ in corresponding spaces pass through one and the same individual point of the medium.

Principal components of strain tensors. Along the principal axes η^i of a strain tensor one has at time t

$$ds_i^2 = \hat{g}_{ii}(d\eta^i)^2 \tag{5.18}$$

(no summation with respect to i), and in the principal axes the square of the length of an arbitrary directed element $d\boldsymbol{r}$ may be represented in the form

$$ds^2 = ds_1^2 + ds_2^2 + ds_3^2 . \tag{5.19}$$

Similarly, in an "initial state",

$$ds_{0i}^2 = \mathring{g}_{ii}(d\eta^i)^2 \tag{5.20}$$

(no summation with respect to i) and

$$ds_0^2 = ds_{01}^2 + ds_{02}^2 + ds_{03}^2 . \tag{5.21}$$

The elementary segments ds_i and ds_{0i}, constructed in this manner along the principal axes, may be conceived as ordinary Cartesian coordinates in the neighbourhood of a given point in a given state and in an "initial state", respectively (where the scales of the coordinates ds_i as well as those of the coordinates ds_{0i} along the different principal axes are the same). In the space of the observer, the systems ds_{0i} and ds_i do not, in general, coincide.

Using (5.18) and (5.20) and the definition of the covariant components of the strain tensor (5.4), one readily finds

$$ds^2 - ds_0^2 = 2\sum_i \frac{\varepsilon'_{ii}}{\hat{g}_{ii}} ds_i^2 = 2\sum_i \frac{\varepsilon'_{ii}}{\mathring{g}_{ii}} ds_{0i}^2 , \tag{5.22}$$

where the stroke on ε'_{ii} indicates that the covariant strain tensor components are referred to principal axes. The matrices $\|\mathring{g}_{ij}\|$ and $\|\hat{g}_{ij}\|$ are diagonal, when referred to principal axes; therefore their inverse matrices $\|\mathring{g}^{ij}\|$ and $\|\hat{g}^{ij}\|$, referred to principal axes, also are diagonal and $\mathring{g}^{ii} = 1/\mathring{g}_{ii}$, $\hat{g}^{ii} = 1/\hat{g}_{ii}$. Thus, the ratios $\varepsilon'_{ii}/\mathring{g}_{ii}$ and $\varepsilon'_{ii}/\hat{g}_{ii}$ in (5.22) will be equal to $\varepsilon'_{ii}\mathring{g}^{ii} = \mathring{\varepsilon}^i{}_i = \mathring{\varepsilon}_i$ and $\varepsilon'_{ii}\hat{g}^{ii} = \hat{\varepsilon}^i{}_i = \hat{\varepsilon}_i$, respectively (the last two expressions not being summed with respect to i); consequently, they are mixed components of the strain tensor referred to the corresponding principal axes. The expression (5.22) may now be rewritten in the form

$$\begin{aligned} ds^2 - ds_0^2 &= 2(\hat{\varepsilon}_1 ds_1^2 + \hat{\varepsilon}_2 ds_2^2 + \hat{\varepsilon}_3 ds_3^2) \\ &= 2(\mathring{\varepsilon}_1 ds_{01}^2 + \mathring{\varepsilon}_2 ds_{02}^2 + \mathring{\varepsilon}_3 ds_{03}^2) . \end{aligned} \tag{5.23}$$

Hence, at every point of a moving medium, one may construct an ordinary, orthogonal Cartesian coordinate system s_{01}, s_{02}, s_{03} directed along the principal axes of the strain tensor, which in the process of motion will also go over into an ordinary orthogonal Cartesian coordinate system s_1, s_2, s_3. The arrangement of the subscripts and superscripts in these systems is not essential, since the systems are orthogonal and Cartesian. The corresponding components of the strain tensors $\hat{\mathscr{E}}_i$ and $\mathring{\mathscr{E}}_i$ in these systems are principal components.

The link between the principal components of the strain tensors $\hat{\mathscr{E}}$ and $\mathring{\mathscr{E}}$. The strain tensors $\hat{\mathscr{E}}$ and $\mathring{\mathscr{E}}$ have different principal components, i.e., $\hat{\mathscr{E}}_i \neq \mathring{\mathscr{E}}_i$; but there exists a relationship between them. It follows from (5.22) for the directions of dr_i, taken along the i-th principal axis, that

$$ds_i^2 - ds_{0i}^2 = 2\hat{\mathscr{E}}_i ds_i^2 , \tag{5.24}$$

whence

$$2\hat{\mathscr{E}}_i = 1 - \frac{ds_{0i}^2}{ds_i^2} . \tag{5.25}$$

Analogously,

$$ds_i^2 - ds_{0i}^2 = 2\mathring{\mathscr{E}}_i ds_{0i}^2 \tag{5.26}$$

and

$$2\mathring{\mathscr{E}}_i = \frac{ds_i^2}{ds_{0i}^2} - 1 . \tag{5.27}$$

It is seen, in particular, from (5.25) and (5.27) that $\mathring{\mathscr{E}}_i \neq \hat{\mathscr{E}}_i$, one readily derives from these formulae

$$2\hat{\mathscr{E}}_i = 1 - \frac{1}{1+2\mathring{\mathscr{E}}_i} = \frac{2\mathring{\mathscr{E}}_i}{1+2\mathring{\mathscr{E}}_i}, \tag{5.28}$$

i.e., the unknown relationship between the components of the tensors $\hat{\mathscr{E}}$ and $\mathring{\mathscr{E}}$.

Next, consider a relationship which exists between the coefficients of relative extension in the directions of the principal axes:

$$l_i = \frac{ds_i - ds_{0i}}{ds_{0i}}$$

and the principal components of the strain tensor. It follows from (5.25) that

$$l_i = \sqrt{\left(\frac{1}{1-2\hat{\varepsilon}_i}\right)} - 1 \tag{5.29}$$

and, analogously, from (5.27), that

$$l_i = \sqrt{(1+2\mathring{\varepsilon}_i)} - 1 \, . \tag{5.30}$$

Formulae (5.29) and (5.30) are true for finite deformations. However, if the deformation is infinitesimal, the components of the strain tensors $\hat{\mathscr{E}}$ and $\mathring{\mathscr{E}}$ are small and one obtains from (5.29) and (5.30), after expansion in power series,

$$l_i = \hat{\varepsilon}_i = \mathring{\varepsilon}_i \, ,$$

i.e., the coefficients of relative extension along the principal axes in the case of infinitesimal deformation coincide with the principal components of the strain tensor $\hat{\mathscr{E}}$ in the real space as well as with the principal components of the strain tensor $\mathring{\mathscr{E}}$ in the "initial" space.

The method of determination of the principal components of a tensor. A method for the evaluation of the principal components of the strain tensor will now be recalled. For the sake of brevity, consider the matrix

$$C = \|c^i_j\| = \|\lambda\delta^i_j - \varepsilon^i_j\| \, ,$$

where λ is some numerical parameter; let it represent the matrix $\|\lambda\delta^i{}_j - \mathring{\varepsilon}^i{}_j\|$ as well as the matrix $\|\lambda\delta^i{}_j - \hat{\varepsilon}^i{}_j\|$. Referred to principal axes, the matrix C has the form

$$C^* = \begin{Vmatrix} \lambda-\varepsilon_1 & 0 & 0 \\ 0 & \lambda-\varepsilon_2 & 0 \\ 0 & 0 & \lambda-\varepsilon_3 \end{Vmatrix} .$$

If one forms the determinant of C^* and sets it equal to zero, one obviously obtains, in terms of λ, the cubic equation

$$(\lambda-\varepsilon_1)(\lambda-\varepsilon_2)(\lambda-\varepsilon_3) = 0 \, ,$$

or, in expanded form, (5.31)

$$\lambda^3 - I_1\lambda^2 + I_2\lambda - I_3 = 0 \, .$$

The roots $\lambda_1, \lambda_2, \lambda_3$ of this equation will be the principal components $\varepsilon_1, \varepsilon_2, \varepsilon_3$ of the corresponding strain tensor, provided (5.31) is referred to the principal coordinate system η^1, η^2, η^3.

Next, select an arbitrary coordinate system ξ^1, ξ^2, ξ^3, which is not principal, and form again the matrix C. Consider the transformation from the η^1, η^2, η^3 to the ξ^i. The components of the matrix C, as differences of the components of the two tensors, are components of a tensor; hence one finds from the formulae for the transformation of mixed tensor components

$$C^* = \| c^p{}_q b^i{}_p a^q{}_j \| = \mathbf{BCB}^{-1} ,$$

whence it is seen that $\operatorname{Det} C^* = \operatorname{Det} C$ and, consequently, Equation (5.31) or

$$|\lambda\delta^i{}_j - \varepsilon^i{}_j| = 0 \tag{5.32}$$

are invariant with respect to the choice of coordinate systems; their roots always determine the principal components of strain tensors. If one replaces in (5.32) the quantities $\varepsilon^i{}_j$ by $\hat{\varepsilon}^i{}_j$, and $\varepsilon^i{}_j$ by $\mathring{\varepsilon}^i{}_j$, one obtains the roots $\hat{\varepsilon}_i$ and $\mathring{\varepsilon}_i$, respectively. Equation (5.32) is referred to as the characteristic or secular equation; it is known that it has always three real roots for a symmetric tensor. The coefficients of the secular equation (5.31) are invariant with respect to coordinate transformations, since they are determined completely by the roots, i.e., by the principal components of the strain tensor. Expanding (5.31) and (5.32), one finds

$$
\begin{aligned}
I_1 &= \varepsilon_1 + \varepsilon_2 + \varepsilon_3 = \varepsilon^\alpha{}_\alpha ,\\
I_2 &= \varepsilon_1 \varepsilon_2 + \varepsilon_2 \varepsilon_3 + \varepsilon_3 \varepsilon_1 = \tfrac{1}{2}\left[(\varepsilon^\alpha{}_\alpha)^2 - \varepsilon^\alpha{}_\beta \varepsilon^\beta{}_\alpha \right] ,\\
I_3 &= \varepsilon_1 \varepsilon_2 \varepsilon_3 = \operatorname{Det} \|\varepsilon^i{}_j\| .
\end{aligned}
\tag{5.33}
$$

Thus, for the determination of the principal components of strain tensors, one must form in a given coordinate system ξ^1, ξ^2, ξ^3 the secular equation (5.32) with the coefficients (5.33) and find its roots.

The invariants I_1, I_2, I_3 of the tensor $\mathring{\mathscr{E}}$ will be denoted by $\mathring{I}_1, \mathring{I}_2, \mathring{I}_3$, those of the tensor $\hat{\mathscr{E}}$ by $\hat{I}_1, \hat{I}_2, \hat{I}_3$, obviously, the invariants $\mathring{I}_1, \mathring{I}_2, \mathring{I}_3$ and $\hat{I}_1, \hat{I}_2, \hat{I}_3$ are expressed in terms of the $\mathring{\varepsilon}_1, \mathring{\varepsilon}_2, \mathring{\varepsilon}_3$ and $\hat{\varepsilon}_1, \hat{\varepsilon}_2, \hat{\varepsilon}_3$, respectively, and since $\mathring{\varepsilon}_j \neq \hat{\varepsilon}_j$, $\mathring{I}_i \neq \hat{I}_i$. The principal components $\mathring{\varepsilon}_i$ and $\hat{\varepsilon}_i$ are interrelated, and hence so are the invariants $\mathring{I}_i$ and $\hat{I}_i$. In the case of infinitesimal deformation, $\mathring{\varepsilon}_i = \hat{\varepsilon}_i$ and $\mathring{I}_i = \hat{I}_i$. In the case of finite deformation, by (5.28) and (5.33), one finds readily the following relations between the invariants $\mathring{I}_i$ and $\hat{I}_i$.

$$\hat{I}_1 = \hat{\varepsilon}_1 + \hat{\varepsilon}_2 + \hat{\varepsilon}_3 = \frac{\mathring{I}_1 + 4\mathring{I}_2 + 12\mathring{I}_3}{1 + 2\mathring{I}_1 + 4\mathring{I}_2 + 8\mathring{I}_3},$$

$$\hat{I}_2 = \hat{\varepsilon}_1\hat{\varepsilon}_2 + \hat{\varepsilon}_2\hat{\varepsilon}_3 + \hat{\varepsilon}_3\hat{\varepsilon}_1 = \frac{\mathring{I}_2 + 6\mathring{I}_3}{1 + 2\mathring{I}_1 + 4\mathring{I}_2 + 8\mathring{I}_3}, \tag{5.34}$$

$$\hat{I}_3 = \hat{\varepsilon}_1\hat{\varepsilon}_2\hat{\varepsilon}_3 = \frac{\mathring{I}_3}{1 + 2\mathring{I}_1 + 4\mathring{I}_2 + 8\mathring{I}_3}.$$

Coefficient of cubical dilatation. By a study of the correspondence of the linear elements $\mathrm{d}s$ and $\mathrm{d}s_0$ in the real and "initial states", one finds the correspondence of elementary volumes in these states. Select in the initial state, referred to principal axes of the strain tensor, an elementary parallelepiped with sides $\mathrm{d}s_{0i}$ and volume $\mathrm{d}V_0 = \mathrm{d}s_{01}\,\mathrm{d}s_{02}\,\mathrm{d}s_{03}$. After deformation, one obtains the corresponding parallelepiped with sides $\mathrm{d}s_1, \mathrm{d}s_2, \mathrm{d}s_3$ and volume $\mathrm{d}V = \mathrm{d}s_1\,\mathrm{d}s_2\,\mathrm{d}s_3$. The coefficient of cubical dilatation θ is defined as the relative change of volume

$$\theta = \frac{\mathrm{d}V - \mathrm{d}V_0}{\mathrm{d}V_0}. \tag{5.35}$$

By (5.26), the expression (5.35) assumes the form

$$\theta = \sqrt{\{(1+2\mathring{\varepsilon}_1)(1+2\mathring{\varepsilon}_2)(1+2\mathring{\varepsilon}_3)\}} - 1, \tag{5.36}$$

and, by (5.33), the form

$$\theta = \sqrt{\{1 + 2\mathring{I}_1 + 4\mathring{I}_2 + 8\mathring{I}_3\}} - 1. \tag{5.37}$$

The quantity θ is defined as an invariant geometric characteristic. Formula (5.37) gives an expression for θ which is convenient for use in any coordinate system.

In an analogous manner, one may introduce θ for elementary parallelepipeds in any curvilinear coordinate system. It will be shown below that the coefficient of cubical dilatation defined by (5.35) does not depend on the form of the original volume $\mathrm{d}V_0$. It is the same for any change of any small volume near a given point in the case of finite deformation.

In the case of infinitesimal deformations, it follows from (5.36) or (5.37) that

$$\theta \simeq I_1 = \hat{\varepsilon}^i{}_i \simeq \mathring{\varepsilon}^i{}_i.$$

Thus, the first invariant of the strain tensor in the case of infinitesimal deformation may be interpreted as coefficient of cubical dilatation.

Computation of strain tensor components for a law of motion. Next, consider the problem of the determination of the covariant components of the strain tensor $\hat{\varepsilon}_{ij}$ for the given laws of motion

$$x^i = x^i(\xi^1, \xi^2, \xi^3, t), \quad \xi^i = \xi^i(x^1, x^2, x^3, t), \tag{5.38}$$

$$x_0^i = x^i(\xi^1, \xi^2, \xi^3, t_0), \quad \xi^i = \xi^i(x_0^1, x_0^2, x_0^3, t_0), \tag{5.39}$$

and known metric g_{ij} of the observer's space x^1, x^2, x^3. Note that the time t is considered as a parameter in the transformation of the coordinates from the concomitant to the observer's system. If the "initial state" corresponds to the position of the medium at time t_0, then the coordinate transformation from the observer's to a Lagrangian system for the initial state is given by (5.39). The covariant components of the strain tensor in the concomitant coordinate system are determined by

$$\hat{\varepsilon}_{ij} = \tfrac{1}{2}\{\hat{g}_{ij} - \mathring{g}_{ij}\},$$

where $\hat{g}_{ij}$ is the metric of the real space in the concomitant system. Since

$$\hat{g}_{ij} d\xi^i d\xi^j = g_{pq} dx^p dx^q,$$

one has

$$\hat{g}_{ij} = g_{pq} \frac{\partial x^p}{\partial \xi^i} \frac{\partial x^q}{\partial \xi^j}$$

and, consequently, in the concomitant coordinate system,

$$\hat{\varepsilon}_{ij} = \tfrac{1}{2}\left(g_{pq} \frac{\partial x^p}{\partial \xi^i} \frac{\partial x^q}{\partial \xi^j} - \mathring{g}_{ij}\right), \tag{5.40}$$

where the derivatives $\partial x^p / \partial \xi^i$ are determined by (5.38). Note that in the general case it is impossible to say anything about the metric $\mathring{g}_{ij}$ of the space of the "initial state", since it may be introduced in different cases with the aid of different physical reasoning. However, one may influence the metric $\mathring{g}_{ij}$ only by choice of the system ξ^1, ξ^2, ξ^3 in the "initial state".

If the transition from the observer's system to the intial state is determined by (5.39), one has in the Lagrangian system of the initial state

$$\mathring{\varepsilon}_{ij} = \tfrac{1}{2}\left(\hat{g}_{ij} - g_{pq} \frac{\partial x^{p_0}}{\partial \xi^i} \frac{\partial x^{q_0}}{\partial \xi^j}\right),$$

where $\partial x_0^m / \partial \xi^n$ is given by (5.39).

On the basis of the transformation formulae for the components of the tensor $\mathscr{E}$ from the concomitant to the observer's system, one has, side by side with (5.40) in the concomitant system, in the observer's system

$$\varepsilon_{ij}^{(x^i)} = \tfrac{1}{2}\left(g_{ij} - \mathring{g}_{pq}\frac{\partial \xi^p}{\partial x^i}\frac{\partial \xi^q}{\partial x^j}\right),$$

where the derivatives $\partial\xi^m/\partial x^n$ are given by (5.38).

Displacement vector. Consider the case when the initial state may really exist and its metric $\mathring{g}_{ij}$, as well as the metric $\hat{g}_{ij}$, are Euclidean. Then one may introduce the displacement vector $\boldsymbol{w}$ (fig. 12):

$$\boldsymbol{r} = \boldsymbol{r}_0 + \boldsymbol{w}\,, \tag{5.41}$$

where $\boldsymbol{r}_0$ and $\boldsymbol{r}$ are the radius vectors, referred to the reference system x^i, of the same point M of the continuum at the initial instant t_0 and at time t, respectively.

From (5.41), one may readily establish the relationship between

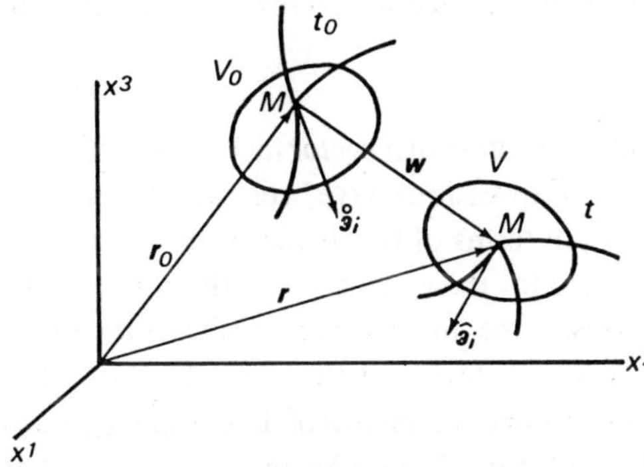

Fig. 12. Displacement vector.

the base vectors $\mathfrak{z}_i$ and $\mathring{\mathfrak{z}}_i$, and write down the formulae for the strain tensor component ε_{ij}. Differentiating (5.41) with respect to ξ^i, one finds

$$\frac{\partial \boldsymbol{w}}{\partial \xi^i} = \frac{\partial \boldsymbol{r}}{\partial \xi^i} - \frac{\partial \boldsymbol{r}_0}{\partial \xi^i} = \mathfrak{z}_i - \mathring{\mathfrak{z}}_i\,,$$

whence

$$\hat{\mathfrak{z}}_i = \mathring{\mathfrak{z}}_i + \frac{\partial \boldsymbol{w}}{\partial \xi^i}, \quad \text{or} \quad \mathring{\mathfrak{z}}_i = \hat{\mathfrak{z}}_i - \frac{\partial \boldsymbol{w}}{\partial \xi^i}; \tag{5.42}$$

therefore

$$\hat{g}_{ij} = \hat{\mathfrak{z}}_i \cdot \hat{\mathfrak{z}}_j = \mathring{\mathfrak{z}}_i \cdot \mathring{\mathfrak{z}}_j + \mathring{\mathfrak{z}}_i \cdot \frac{\partial \boldsymbol{w}}{\partial \xi^j} + \mathring{\mathfrak{z}}_j \cdot \frac{\partial \boldsymbol{w}}{\partial \xi^i} + \frac{\partial \boldsymbol{w}}{\partial \xi^i} \cdot \frac{\partial \boldsymbol{w}}{\partial \xi^j}$$

and

$$\mathring{g}_{ij} = \mathring{\mathfrak{z}}_i \cdot \mathring{\mathfrak{z}}_j = \hat{\mathfrak{z}}_i \cdot \hat{\mathfrak{z}}_j - \hat{\mathfrak{z}}_i \cdot \frac{\partial \boldsymbol{w}}{\partial \xi^j} - \hat{\mathfrak{z}}_j \cdot \frac{\partial \boldsymbol{w}}{\partial \xi^i} + \frac{\partial \boldsymbol{w}}{\partial \xi^i} \cdot \frac{\partial \boldsymbol{w}}{\partial \xi^j}.$$

Consequently,

$$\mathring{\varepsilon}_{ij} = \hat{\varepsilon}_{ij} = \tfrac{1}{2}(\hat{g}_{ij} - \mathring{g}_{ij}) = \tfrac{1}{2}\left[\mathring{\mathfrak{z}}_i \cdot \frac{\partial \boldsymbol{w}}{\partial \xi^j} + \mathring{\mathfrak{z}}_j \cdot \frac{\partial \boldsymbol{w}}{\partial \xi^i} + \frac{\partial \boldsymbol{w}}{\partial \xi^i} \cdot \frac{\partial \boldsymbol{w}}{\partial \xi^j}\right]$$

$$= \tfrac{1}{2}\left[\frac{\partial \boldsymbol{w}}{\partial \xi^i} \cdot \hat{\mathfrak{z}}_j + \frac{\partial \boldsymbol{w}}{\partial \xi^j} \cdot \hat{\mathfrak{z}}_i - \frac{\partial \boldsymbol{w}}{\partial \xi^i} \cdot \frac{\partial \boldsymbol{w}}{\partial \xi^j}\right]. \qquad ((5.43)$$

These formulae are true for any choice of general curvilinear Lagrangian coordinates ξ^1, ξ^2, ξ^3. Note that the expression (5.43) for ε_{ij} contains only the first derivatives of the displacement vector $\boldsymbol{w}$ with respect to the coordinates ξ^1, ξ^2, ξ^3 which characterize the relative displacements of the points of the continuum.

On the differentiation of a vector and of its components with respect to the coordinates. Expressions have been derived for the components of the strain tensor ε_{ij} in terms of the displacement vector $\boldsymbol{w}$. Now, expressions will be obtained for the ε_{ij} in terms of the components of the vector $\boldsymbol{w}$. For this purpose, one must establish the rules for expressing the derivative of a vector in terms of the derivatives of its components.

Obviously, the ordinary derivatives of the components do not determine the change of the vector itself, since for the transition from space point to space point, generally speaking, also the base vectors change. In fact, select as an example a plane polar coordinate system and consider the field of a vector $\boldsymbol{A}$ which is constant in magnitude as well as in direction at all points of the plane. For transition from point to point of the plane, the vector $\boldsymbol{A}$ does not change, and, obviously, its derivative must vanish. The radius r and the angle φ will represent the coordinates ξ^1 and ξ^2; the base vectors will be directed in the following manner: Along the ray from the origin of coordinates, $\mathfrak{z}_1$, tangential to the circle $r=$const., $\mathfrak{z}_2$. At different points of the plane, the vectors $\mathfrak{z}_1$ and $\mathfrak{z}_2$ will have different directions, and the projections of the constant vector $\boldsymbol{A}$ on the directions of $\mathfrak{z}_1$ and $\mathfrak{z}_2$ at different points of the plane will differ (e.g., the points B and

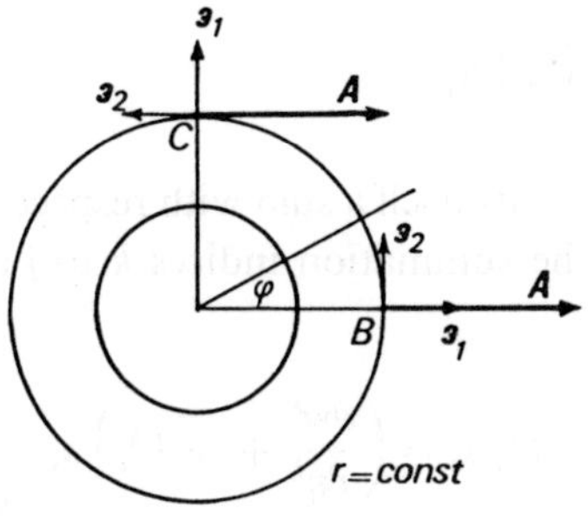

Fig. 13. Plane polar coordinate system.

C of fig. 13), i.e., the derivatives of the components of the constant vector will not vanish.

In a Cartesian coordinate system

$$\frac{\partial \boldsymbol{w}}{\partial x^i} = \frac{\partial}{\partial x^i}(w^k \mathfrak{z}_k) = \frac{\partial w^k}{\partial x^i}\mathfrak{z}_k\,,$$

since the base vectors $\mathfrak{z}_1 = \boldsymbol{i}$, $\mathfrak{z}_2 = \boldsymbol{j}$, $\mathfrak{z}_3 = \boldsymbol{k}$ do not change from point to point.

Covariant differentiation of tensor and vector components and its properties. In an arbitrary curvilinear coordinate system η^1, η^2, η^3, the base vectors $\mathfrak{z}_i$ vary, and therefore one must write

$$\frac{\partial \boldsymbol{w}}{\partial \eta^i} = \frac{\partial w^k}{\partial \eta^i}\mathfrak{z}_k + w^k \frac{\partial \mathfrak{z}_k}{\partial \eta^i}\,. \tag{5.44}$$

Obviously, by definition, one may assume that the derivatives $\partial \mathfrak{z}_k/\partial \eta^i$ also form a vector, characterizing a property of the curvilinear coordinate system. Decomposing this vector with respect to the base $\mathfrak{z}_j$ and denoting the components of this expansion by Γ^j_{ki}, one finds

$$\frac{\partial \mathfrak{z}_k}{\partial \eta^i} = \Gamma^j_{ki}\mathfrak{z}_j\,. \tag{5.45}$$

The quantities Γ^j_{ki} are functions of the coordinates η^1, η^2, η^3; they are called

Christoffel symbols.[1)] The Γ^i_{ki} will be studied in detail below. On the basis of (5.45), Equality (5.44) assumes the form

$$\frac{\partial \boldsymbol{w}}{\partial \eta^i} = \frac{\partial w^k}{\partial \eta^i} \boldsymbol{\mathfrak{z}}_k + w^k \Gamma^j_{ki} \boldsymbol{\mathfrak{z}}_j .$$

The second term represents itself a sum with respect to k and j. Changing in it the notation of the summation indices k to j and j to k, one may write

$$\frac{\partial \boldsymbol{w}}{\partial \eta^i} = \frac{\partial w^k}{\partial \eta^i} \boldsymbol{\mathfrak{z}}_k + w^j \Gamma^k_{ji} \boldsymbol{\mathfrak{z}}_k = \left(\frac{\partial w^k}{\partial \eta^i} + w^j \Gamma^k_{ji} \right) \boldsymbol{\mathfrak{z}}_k . \tag{5.46}$$

The coefficients of $\boldsymbol{\mathfrak{z}}_k$, i.e., $\partial w^k/\partial \eta_i + w^j \Gamma^k_{ji}$, with two indices, are specially denoted by $\nabla_i w^k$; they are called the covariant derivatives of the contravariant components of the vector $\boldsymbol{w}$:

$$\nabla_i w^k = \frac{\partial w^k}{\partial \eta^i} + w^j \Gamma^k_{ji} . \tag{5.47}$$

Properties of $\nabla_i w^k$ will now be established.

In a Cartesian coordinate system ($\eta^i = x^i$), since $\partial \boldsymbol{\mathfrak{z}}_k/\partial x^i = 0$, i.e., $\Gamma^k_{ji} = 0$, one has

$$\nabla_i w^k = \frac{\partial w^k}{\partial \eta^i} = \frac{\partial w^k}{\partial x^i} ,$$

[1)] These symbols are closely linked to calculations of geometrical properties introduced by the mathematical models of physical or phase spaces. In the general case of geometrical spaces, the symbols can be given by different formulae. Below, one considers Euclidean space, pseudo-Euclidean space and the more general case of Riemannian space, in which, after calculations, Γ^i_{kj} are given by the same formulae in terms of components of the matrix tensor g_{mn} and their derivatives with respect to the coordinates only.

The specification of Γ^i_{kj} allows the transition from vector algebra at each point of space to tensor analysis, in which it is necessary to translate vectors and tensors from the given point into another arbitrary point of space, in such a way that one can compare vectors and tensors in neighbouring points, which is necessary in construction the adjoining derivatives of vectors and tensors of any order, with respect to coordinates x^k.

The following conclusions and formulae are valid for the manipulation of tensors in the spaces of arbitrary number of indices and in particular, in the models of metric four-dimensional physical spaces, used in the classical theory of relativity.

the covariant derivative coincides with the ordinary derivatives of the component of a vector with respect to a coordinate.

The covariant derivatives form the components of a tensor. In fact, let $\zeta^1, \zeta^2, \zeta^3$ be the new, η^1, η^2, η^3 the old coordinate system. Then

$$\frac{\partial w}{\partial \zeta^k} = \frac{\partial w}{\partial \eta^i}\frac{\partial \eta^i}{\partial \zeta^k}$$

and, obviously, since w is invariant, the derivatives $\partial w/\partial \eta^i$ transform like the covariant components of a vector. Therefore

$$T = \frac{\partial \mathbf{w}}{\partial \eta^i}\mathbf{3}^i$$

represents itself an invariant object; however, by (5.46) and (5.47), one has

$$T = \nabla_i w^k \mathbf{3}_k \mathbf{3}^i ,$$

i.e., T is a second order tensor the mixed components of which are the covariant derivatives $\nabla_i w^k$.

Note that the derivatives $\partial w^k/\partial \eta^i$ are not components of a tensor. In fact, if one replaces under the differentiation $\partial/\partial \eta^i$ the w^k by their expressions in the new coordinate system

$$w^k = w'^j \frac{\partial \eta^k}{\partial \zeta^j},$$

one must differentiate with respect to η^i also $\partial \eta^k/\partial \zeta^j$, and one does not obtain the tensor transformation law for $\partial w^k/\partial \eta^i$.

It is obvious from the definition of the covariant derivative that the covariant derivatives of a scalar coincide with the ordinary derivatives, *viz.*,

$$\nabla_i \varphi = \frac{\partial \varphi}{\partial \eta^i},$$

and that they define a vector which is the vector-gradient of the scalar field φ. This vector will be considered in detail below as the characteristic of the field φ.

Next, determine the covariant derivative of the contravariant components of a tensor. For the sake of definiteness, select the second order tensor $H = H^{jk}\mathbf{3}_j\mathbf{3}_k$ and compute in the following manner

$$\frac{\partial H}{\partial \eta^i} = \frac{\partial H^{jk}}{\partial \eta^i} \mathfrak{z}_j \mathfrak{z}_k + H^{jk} \frac{\partial \mathfrak{z}_j}{\partial \eta^i} \mathfrak{z}_k + H^{jk} \mathfrak{z}_j \frac{\partial \mathfrak{z}_k}{\partial \eta^i}$$

$$= \frac{\partial H^{jk}}{\partial \eta^i} \mathfrak{z}_j \mathfrak{z}_k + H^{jk} \Gamma^l_{ji} \mathfrak{z}_l \mathfrak{z}_k + H^{jk} \mathfrak{z}_j \Gamma^l_{ki} \mathfrak{z}_l .$$

In the second sum, replace the summation index l by j, and in the third l by k, to obtain

$$\frac{\partial H}{\partial \eta^i} = \left(\frac{\partial H^{jk}}{\partial \eta^i} + H^{lk} \Gamma^j_{li} + H^{jl} \Gamma^k_{li} \right) \mathfrak{z}_j \mathfrak{z}_k = \nabla_i H^{jk} \mathfrak{z}_j \mathfrak{z}_k ,$$

where, by definition,

$$\nabla_i H^{jk} = \frac{\partial H^{jk}}{\partial \eta^i} + H^{lk} \Gamma^j_{li} + H^{jl} \Gamma^k_{li}$$

is called the covariant derivative of the contravariant component of the second order tensor H. It is readily seen that, in relation to the second order tensor H, one may introduce the third order tensors

$$T_1 = \frac{\partial H}{\partial \eta^i} \mathfrak{z}^i = \nabla_i H^{jk} \mathfrak{z}_j \mathfrak{z}_k \mathfrak{z}^i ,$$

or

$$T_2 = \nabla_i H^{jk} \mathfrak{z}^i \mathfrak{z}_j \mathfrak{z}_k ,$$

or

$$T_3 = \nabla_i H^{jk} \mathfrak{z}_j \mathfrak{z}^i \mathfrak{z}_k .$$

Obviously, in general, the tensors T_i will differ.

In an analogous manner, one may construct the covariant derivative of the contravariant components of tensors of any order.

It is clear from the definition of the covariant derivative (its linearity with respect to the components of a vector) that the covariant derivative of the sum of the contravariant components is equal to the sum of the covariant derivatives:

$$\nabla_i (v^k + w^k) = \nabla_i v^k + \nabla_i w^k .$$

It will now be shown that the differentiation rules for products in the covariant and ordinary senses coincide. Let it be required to evaluate $\nabla_i (v^j w^k)$. For this purpose, one must use the rule for the covariant differentiation of the contravariant components of a tensor, since the products $v^j w^k$, as is known from Sec. 4, Chapter II, are the components of a second order tensor. Thus

$$\nabla_i(v^j w^k) = \frac{\partial(v^j w^k)}{\partial\eta^i} + v^l w^k \Gamma^j_{li} + v^j w^l \Gamma^k_{li}$$

$$= \left(\frac{\partial v^j}{\partial\eta^i} + v^l \Gamma^j_{li}\right) w^k + v^j \left(\frac{\partial w^k}{\partial\eta^i} + w^l \Gamma^k_{li}\right)$$

$$= w^k \nabla_i v^j + v^j \nabla_i w^k ,$$

proving the assertion. In quite an analogous manner, products of any number of terms will be differentiated covariantly.

Consider the problem of the covariant differentiation in the case when a vector is given in terms of its covariant and not of its contravariant components. Let

$$\mathbf{w} = w_j \mathfrak{z}^j$$

and compute $\partial\mathbf{w}/\partial\eta^i$. Then

$$\frac{\partial \mathbf{w}}{\partial\eta^i} = \frac{\partial w_j}{\partial\eta^i}\mathfrak{z}^j + w_j \frac{\partial \mathfrak{z}^j}{\partial\eta^i} . \tag{5.48}$$

Obviously, the derivative $\partial\mathfrak{z}^j/\partial\eta^i$, just as $\partial\mathfrak{z}_j/\partial\eta^i$, will be a vector; expand it in terms of the $\mathfrak{z}^k$. In the case of Euclidean space and in the more general case of Riemannian space, one has

$$\frac{\partial \mathfrak{z}^j}{\partial\eta^i} = -\Gamma^j_{ki}\mathfrak{z}^k , \tag{5.49}$$

where Γ^j_{ki} are the Christoffel symbols, introduced earlier. In order to prove (5.49), consider the scalar product

$$\mathfrak{z}^j \cdot \mathfrak{z}_k = \delta^j{}_k$$

and differentiate this equation, which is true at all points of space, with respect to η^i:

$$\frac{\partial \mathfrak{z}^j}{\partial\eta^i}\cdot \mathfrak{z}_k + \mathfrak{z}^j \cdot (\Gamma^l_{ki}\mathfrak{z}_l) = 0 .$$

In the last sum, only those terms will be non-zero for which $l{=}j$, i.e., one finds

$$\frac{\partial \mathbf{3}^j}{\partial \eta^i} \cdot \mathbf{3}_k = -\Gamma^j_{ki} .$$

Obviously, this formula is the same as (5.49). By (5.49), Formula (5.48) assumes the form

$$\frac{\partial \boldsymbol{w}}{\partial \eta^i} = \frac{\partial w_j}{\partial \eta^i} \mathbf{3}^j - w_j \Gamma^j_{ki} \mathbf{3}^k .$$

Replacing in the last sum the summation indices j and k by k and j, respectively, one finds

$$\frac{\partial \boldsymbol{w}}{\partial \eta^i} = \left(\frac{\partial w_j}{\partial \eta^i} - w_k \Gamma^k_{ji} \right) \mathbf{3}^j = \nabla_i w_j \mathbf{3}^j .$$

The expression $\partial w_j / \partial \eta^i - w_k \Gamma^k_{ji}$ determines the covariant derivative of the covariant components of a vector:

$$\nabla_i w_j = \frac{\partial w_j}{\partial \eta^i} - w_k \Gamma^k_{ji} .$$

In an analogous manner, one may introduce the covariant derivative of the covariant components of tensors of any order.

Note that the $\nabla_i w_j$ are the covariant, and the $\nabla_i w^j$ the mixed components of one and the same second order tensor:

$$T = \frac{\partial \boldsymbol{w}}{\partial \eta^i} \mathbf{3}^i = \nabla_i w_j \mathbf{3}^j \mathbf{3}^i = \nabla_i w^j \mathbf{3}_j \mathbf{3}^i .$$

It follows from the preceding result that the components of the metric tensors g_{ij} and g^{ij}, in spite of the fact that they depend on the η^i, must behave with respect to covariant differentiation like constant quantities. In other words, without changing the result, one may take them under or bring them outside the ∇_i sign. In fact, there exists between the expressions $\nabla_i w^j$ and $\nabla_i w_k$, as between the different components of one and the same tensor, the relationship

$$\nabla_i w^j = g^{jk} \nabla_i w_k ; \tag{5.50}$$

however,

$$w^j = g^{jk} w_k , \tag{5.51}$$

whence

$$\nabla_i (g^{jk} w_k) = g^{jk} \nabla_i w_k \,,$$

i.e., $\nabla_i g^{jk} = 0$. In an analogous manner, one obtains

$$\nabla_i g_{jk} = 0 \,,$$

if (5.50) is replaced by $\nabla_i w_k = g_{kj} \nabla_i w^j$ and (5.51) by $w_k = g_{kj} w^j$.

Properties of Christoffel symbols. Next, the question of the calculation of the Christoffel symbols in metric Euclidean space will be discussed and properties of the symbols explained. Note that there exist spaces, more complicated than Euclidean or Riemannian spaces, in which the Christoffel symbols are not computed, but are given and the method of their specification enters into the definition of the space.

Christoffel symbols are not components of any tensor. For example, this fact can be verified by noting that for one and the same space they may vanish for a Cartesian and be non-zero in a curvilinear coordinate system. Obviously, the components of a tensor cannot have such a property.

In Euclidean space, the Christoffel symbols are symmetric with respect to their subscripts

$$\Gamma^i_{kj} = \Gamma^i_{jk} \,.$$

This fact may be proved as follows. In Euclidean space, there always exists a radius vector $\boldsymbol{r}(\eta^1, \eta^2, \eta^3)$, and $\mathfrak{Z}_j = \partial r / \partial \eta^j$ and

$$\frac{\partial \mathfrak{Z}_j}{\partial \eta^k} = \frac{\partial^2 \boldsymbol{r}}{\partial \eta^k \partial \eta^j} = \frac{\partial^2 \boldsymbol{r}}{\partial \eta^j \partial \eta^k} = \frac{\partial \mathfrak{Z}_k}{\partial \eta^j} \,, \tag{5.52}$$

whence

$$\Gamma^i_{jk} \mathfrak{Z}_i = \Gamma^i_{kj} \mathfrak{Z}_i \,.$$

Consider the formulae for the evaluation of the Christoffel symbols in terms of the components of the metric tensor $\boldsymbol{g}$.

In Riemannian space, the symmetry of the Christoffel symbols with respect to their subscripts is taken by definition. Then, the formula (5.53) for Γ^i_{kj} is also valid in Riemannian space. Select

$$\frac{\partial g_{js}}{\partial \eta^k} = \frac{\partial \mathfrak{Z}_j}{\partial \eta^k} \cdot \mathfrak{Z}_s + \frac{\partial \mathfrak{Z}_s}{\partial \eta^k} \cdot \mathfrak{Z}_j$$

and obtain from this relation

$$\frac{\partial g_{js}}{\partial \eta^k} - \frac{\partial \mathfrak{z}_s}{\partial \eta^k} \cdot \mathfrak{z}_j = \Gamma'_{jk} \mathfrak{z}_l \cdot \mathfrak{z}_s = \Gamma^l_{jk} g_{ls}$$

and, analogously,

$$\frac{\partial g_{ks}}{\partial \eta^j} - \frac{\partial \mathfrak{z}_s}{\partial \eta^j} \cdot \mathfrak{z}_k = \Gamma^l_{kj} \mathfrak{z}_l \cdot \mathfrak{z}_s = \Gamma^l_{kj} g_{ls} \, .$$

Adding these two equations and using the symmetry of the Christoffel symbols with respect to their subscripts, Equation (5.52) and the fact that

$$\frac{\partial \mathfrak{z}_k}{\partial \eta^s} \cdot \mathfrak{z}_j + \frac{\partial \mathfrak{z}_j}{\partial \eta^s} \cdot \mathfrak{z}_k = \frac{\partial g_{jk}}{\partial \eta^s} \, ,$$

one obtains

$$\frac{\partial g_{js}}{\partial \eta^k} + \frac{\partial g_{ks}}{\partial \eta^j} - \frac{\partial g_{jk}}{\partial \eta^s} = 2\Gamma'_{jk} g_{ls} \, .$$

Contracting the last equation with $g^{is}/2$, one finds the required result

$$\Gamma^i_{jk} = \tfrac{1}{2} g^{is} \left(\frac{\partial g_{js}}{\partial \eta^k} + \frac{\partial g_{ks}}{\partial \eta^j} - \frac{\partial g_{jk}}{\partial \eta^s} \right) . \tag{5.53}$$

Expression for the strain tensor in terms of the components of the displacement vector. Consider (5.43) and derive from it formulae expressing the strain tensor components in terms of the components of the displacement vector. The displacement vector $\boldsymbol{w}$ may be expanded in terms of the actual $\hat{\mathfrak{z}}_i$ as well as in terms of the initial base vector $\mathring{\mathfrak{z}}_i$; correspondingly, two kinds of components of one and the same vector $\boldsymbol{w}$ may be introduced, i.e., $\hat{w}^k$ and $\mathring{w}^k$:

$$\boldsymbol{w} = \mathring{w}^k \mathring{\mathfrak{z}}_k = \hat{w}^k \hat{\mathfrak{z}}_k \, .$$

Further, one may also introduce two types of covariant derivatives:

$$\frac{\partial \boldsymbol{w}}{\partial \xi^i} = \mathring{\nabla}_i \mathring{w}^k \mathring{\mathfrak{z}}_k \tag{5.54}$$

and

$$\frac{\partial \boldsymbol{w}}{\partial \xi^i} = \hat{\nabla}_i \hat{w}^k \hat{\mathfrak{z}}_k \,. \tag{5.55}$$

The first of the covariant derivatives is computed in the initial state, where the Christoffel symbols are calculated in terms of the $\mathring{g}_{ij}$, the second in real space, where the Christoffel symbols are expressed in terms of the $\hat{g}_{ij}$. Substituting (4.54) into the first equation (5.43), one obtains

$$\mathring{\varepsilon}_{ij} = \tfrac{1}{2}[(\mathring{\nabla}_i \mathring{w}^k)\mathring{g}_{kj} + (\mathring{\nabla}_j \mathring{w}^k)\mathring{g}_{ki} + (\mathring{\nabla}_i \mathring{w}^k \cdot \mathring{\nabla}_j \mathring{w}^l)\mathring{g}_{kl}]\,.$$

Using the fact that the components of the metric tensor may be introduced, without changing the result, under the covariant differentiation sign, one finds

$$\mathring{\varepsilon}_{ij} = \tfrac{1}{2}[\mathring{\nabla}_i \mathring{w}_j + \mathring{\nabla}_j \mathring{w}_i + \mathring{\nabla}_i \mathring{w}_k \mathring{\nabla}_j \mathring{w}^k]\,. \tag{5.56}$$

Analogously, by (5.55) and the second equation (5.43),

$$\hat{\varepsilon}_{ij} = \tfrac{1}{2}[\hat{\nabla}_i \hat{w}_j + \hat{\nabla}_j \hat{w}_i - \hat{\nabla}_i \hat{w}_k \hat{\nabla}_j \hat{w}^k]\,. \tag{5.57}$$

In the case of infinitesimal relative displacements, one obtains, after neglecting squares in $|\, \boldsymbol{w} \,|$,

$$\varepsilon_{ij} = \tfrac{1}{2}[\mathring{\nabla}_i \mathring{w}_j + \mathring{\nabla}_j \mathring{w}_i) = \tfrac{1}{2}[\hat{\nabla}_i \hat{w}_j + \hat{\nabla}_j \hat{w}_i)\,. \tag{5.58}$$

Obviously, the ε_{ij} coincide with the components of the symmetrized tensor $\nabla_i w_j \mathfrak{z}^i \mathfrak{z}^j$.

In a Cartesian coordinate system,

$$\varepsilon_{ij} = \tfrac{1}{2}\left(\frac{\partial w_j}{\partial x^i} + \frac{\partial w_i}{\partial x^j}\right). \tag{5.59}$$

Note that the formulae (5.43) and (5.56) for the strain tensor components are true only when the displacement vector $\boldsymbol{w}$ may be introduced for all points of a moving medium, since then the strain tensor and its components are determined by the metrics ds^2 and ds_0^2 from (5.4) and (5.40) independently of an assumption regarding the existence of the displacement vector.

On the existence of the equations of compatibility. The strain tensor has nine components only six of which differ as a consequence of the symmetry of the ε_{ij}. In the presence of the displacement $\boldsymbol{w}$, these six components ε_{ij} are expressed, by (5.56), at each given point in terms of the nine derivatives $\partial w_i/\partial \xi^j$; consequently, they may assume at a point of space arbitrary values. However, the ε_{ij} cannot be arbitrary functions of the points ξ^1, ξ^2, ξ^3 of space, since, by the same formula (5.56), the six functions ε_{ij} of ξ^1, ξ^2, ξ^3 are expressed in terms of the derivatives of only three functions w_i of the ξ^1, ξ^2, ξ^3. Therefore the ε_{ij} must satisfy definite equations which are referred to as equations of compatibility of strains.

The equations of compatibility must exist only when the displacement vector $\boldsymbol{w}$ exists, i.e., when the actual as well as the initial state of the continuous medium belong to Euclidean space. Therefore, as a first step, consideration will be given to the conditions under which a space is Euclidean.

The transformation formulae for the Christoffel symbols. The Christoffel symbols Γ^k_{ij} are known not to be the components of any tensor; in three-dimensional space, they form a set of 27 quantities. The Christoffel symbols are linked to the components of the metric tensor by the formulae

$$\Gamma^k_{ij} = \tfrac{1}{2} g^{ks}\left(\frac{\partial g_{is}}{\partial x^j} + \frac{\partial g_{js}}{\partial x^i} - \frac{\partial g_{ij}}{\partial x^s}\right), \tag{5.53a}$$

which have been obtained above for Euclidean space and, by definition, are true for Riemannian space.

Denote the Christoffel symbols in the coordinate system η^i by Γ'^k_{ij} and in the system ξ^i by Γ^k_{ij} and establish their transformation formulae for the transition from the system ξ^i to the system η^i. Obviously,

$$\mathbf{3}'_i = \mathbf{3}_\alpha \frac{\partial \xi^\alpha}{\partial \eta^i};$$

differentiating this equality with respect to η^j and taking into consideration that

$$\frac{\partial \mathbf{3}'_i}{\partial \eta^j} = \Gamma'^\alpha_{ij}\mathbf{3}'_\alpha, \quad \frac{\partial \mathbf{3}_\alpha}{\partial \xi^\beta} = \Gamma^\omega_{\alpha\beta}\mathbf{3}_\omega = \Gamma^\omega_{\alpha\beta}\frac{\partial \eta^\gamma}{\partial \xi^\omega}\mathbf{3}'_\gamma,$$

since $\mathbf{3}_\omega = (\partial \eta^\gamma/\partial \xi^\omega)\mathbf{3}'_\gamma$, one finds

$$\Gamma'^\alpha_{ij}\mathbf{3}'_\alpha = \left[\Gamma^\omega_{\alpha\beta}\frac{\partial \eta^\gamma}{\partial \xi^\omega}\frac{\partial \xi^\alpha}{\partial \eta^i}\frac{\partial \xi^\beta}{\partial \eta^j} + \frac{\partial^2 \xi^\omega}{\partial \eta^i \partial \eta^j}\frac{\partial \eta^\gamma}{\partial \xi^\omega}\right]\mathbf{3}'_\gamma .$$

Multiplying both sides of this equality scalarly by $\mathfrak{z}'^{\gamma}$, one arrives at the known formula[1)]

$$\Gamma'^{\gamma}_{ij} = \left(\Gamma^{\omega}_{\alpha\beta}\frac{\partial\xi^{\alpha}}{\partial\eta^{i}}\frac{\partial\xi^{\beta}}{\partial\eta^{j}} + \frac{\partial^{2}\xi^{\omega}}{\partial\eta^{i}\partial\eta^{j}}\right)\frac{\partial\eta^{\gamma}}{\partial\xi^{\omega}}.$$

The equation defining the coordinate system in which $\Gamma^{k}_{ij}=0$. Is it possible to find a coordinate system η^{i} such that all Γ'^{γ}_{ij} vanish? In Euclidean space, one may introduce a Cartesian coordinate system, unique for the entire space, in which $g_{ik}=$const. and, consequently, all $\Gamma'^{\gamma}_{ij}=0$ at all points of the space. In Riemannian spaces, this is not possible. The equations which determine the coordinate system in which $\Gamma^{\omega}_{\alpha\beta}=0$ will now be written down. Since Det $\|\partial\eta^{\gamma}/\partial\xi^{\omega}\|\neq 0$, all Γ'^{γ}_{ij} may vanish only when

$$\Gamma^{\omega}_{\alpha\beta}\frac{\partial\xi^{\alpha}}{\partial\eta^{i}}\frac{\partial\xi^{\beta}}{\partial\eta^{j}} + \frac{\partial^{2}\xi^{\omega}}{\partial\eta^{i}\partial\eta^{j}} = 0 \quad (\omega, i, j=1, 2, 3). \tag{5.60}$$

In Euclidean as well as in Riemannian space, these equations may always be satisfied at a given point, i.e., one may always introduce new coordinates η^{i} such that at a given point η^{i}_{0}, corresponding to the point ξ^{i}_{0}, all $\Gamma'^{\gamma}_{ij}=0$. Obviously, it is sufficient for this purpose to set

$$\xi^{\omega}-\xi^{\omega}_{0} = \delta^{\omega}_{s}(\eta^{s}-\eta^{s}_{0}) - \tfrac{1}{2}\Gamma^{\omega}_{0\alpha\beta}(\eta^{\alpha}-\eta^{\alpha}_{0})(\eta^{\beta}-\eta^{\beta}_{0}) + \dots\ .$$

It is possible to construct a coordinate system such that $\Gamma^{\gamma}_{ij} = 0$ at all points of a given curve (Fermi's coordinates). To prove such an interesting and useful assumption it is enough to show the method of construction of a coordinate system ξ^{i}, in which at all points of an arbitrary given curve C the following equalities are satisfied

$$\Gamma^{\gamma}_{ij} = 0\ . \tag{A}$$

1) The established formula of transition and the condition (5.60) are independent of the definition (5.53a) and they appear to be the result of (4.5) and (5.45). Next, for the sake of brevity, we assume that $\Gamma^{\omega}_{\alpha\beta} = \Gamma^{\omega}_{\beta\alpha}$, hence the formula (5.53a) is valid for the metric space.

If the space is Euclidean, then, introducing a Cartesian coordinate system, one obtains that the equality (A) is satisfied not only on a curve C, but at all points of the space. Consequently, verification of equality (A) along the curve C should be made in the case of Riemannian space only, where not all components of $R_{\beta s \alpha}^{\cdot\cdot\cdot\omega}$ (5.61) vanish.

In the case of n-dimensional Riemannian space it is impossible to satisfy the equality (A) exactly, neither at all points of an infinitesimal volume, nor on arbitrary elements of the surface of a dimension bigger than one. Let us assume, that the equation of the curve C in the initial coordinate system has the form:

$$x^{\alpha} = \varphi^{\alpha}(x^1), \quad \alpha = 2, 3, \ldots n, \quad \xi^1 = x^1 \quad .$$

Let us also assume that functions $\varphi^{\alpha}(x^1)$ are defined at all x^1 in intervals selected by the appropriate method. Performing transformation of coordinates

$$\begin{aligned} \xi^2 &= x^2 - \varphi^2(x^1) \\ &\ldots\ldots\ldots\ldots \\ \xi^n &= x^n - \varphi^n(x^i) \end{aligned}$$

one obtains that the curve C in the coordinate system ξ^i corresponds to

$$\xi^2 = \xi^3 = \ldots = \xi^n = 0$$

and a variable ξ^1. It is obvious that in the system ξ^i the curve C coincides with the coordinates of a curve ξ^1.

The transformation of the Christoffel symbols can be written in the following form:

$$\frac{\partial \xi^k}{\partial \eta^{\gamma}} \Gamma_{ij}^{\prime\gamma} = \Gamma_{pq}^{k} \frac{\partial \xi^p}{\partial \eta^i} \frac{\partial \xi^q}{\partial \eta^j} + \frac{\partial^2 \xi^k}{\partial \eta^i \partial \eta^j} \quad .$$

Since Det $\| d\xi^k/d\eta^{\nu} \| \neq 0$, then, if for each $k = 1, 2, \ldots n$ and for fixed indices i and j, the equalities (5.60) are valid, likewise it is true that

$$\Gamma_{ij}^{\prime k} = 0 \qquad \text{for } 1, 2, \ldots n \quad .$$

At the beginning let us consider the transformation from a system ξ^i to a system η^i, in which $\Gamma_{ij}^{\cdot\cdot k} = 0$ for all k and all i,j, except when $i = j = 1$,

$$\xi^1 = \eta^1 + b^1_\alpha \eta^\alpha - \frac{1}{2}[b^\nu_\varkappa b^\mu_\chi \Gamma^1_{\nu\mu} + (b^1_\varkappa b^\mu_\chi + b^1_\chi b^\mu_\varkappa)\Gamma^1_{\mu 1} + b^1_\varkappa b^1_\chi \Gamma^1_{11}]\,\eta^\varkappa\eta^\chi + F^1 \,, \tag{B}$$

$$\xi^\alpha = b^\alpha_\varkappa \eta^\varkappa - \frac{1}{2}[b^\nu_\varkappa b^\mu_\chi \Gamma^\alpha_{\nu\mu} + (b^1_\varkappa b^\mu_\chi + b^1_\chi b^\mu_\varkappa)\Gamma^\alpha_{\mu 1} + b^1_\varkappa b^1_\chi \Gamma^\alpha_{11}]\,\eta^\varkappa\eta^\chi + F^\alpha \,,$$

where $\varkappa, \mu, \chi, \nu$ and α are taking values $2, 3, \ldots n$; $b^\beta_1(\eta^1)$, $b^\alpha_\beta(\eta^1)$ ($\mathrm{Det}\,\| \beta^\alpha_\beta \| \neq 0$) — functions of η^1; Γ^1 and Γ^α — arbitrary functions of η^1 and η^α, tends to zero as infinitesimally small quantities of an order higher than two with respect to η^α, when η^α tends to zero. It is obvious that along the curve C

$$\eta^2 = \eta^3 = \ldots = \eta^n \,, \quad \xi^1 = \eta^1 \,.$$

In the formula (B) quantities $\Gamma^{\cdot\cdot k}_{ij}(\xi)$ can be identified with the values of $\Gamma^{\cdot\cdot k}_{ij}$ on the curve C. Substituting $\xi^i(\eta^k)$ from (B) into the conditions (5.60), one obtains the relations for the functions $b^1_\beta(\eta^1)$ and $b^\alpha_\beta(\eta^1)$.

Let us rewrite the conditions (5.60) in the form

$$\frac{\partial^2 \xi^k}{\partial\eta^i\,\partial\eta^j} = -\Gamma^k_{\alpha\beta}\frac{\partial\xi^\alpha}{\partial\eta^i}\frac{\partial\xi^\beta}{\partial\eta^j} - \Gamma^k_{1\mu}\frac{\partial\xi^1}{\partial\eta^i}\frac{\partial\xi^\mu}{\partial\eta^j} - \Gamma^k_{\mu 1}\frac{\partial\xi^\mu}{\partial\eta^i}\frac{\partial\xi^1}{\partial\eta^j} - \Gamma^k_{11}\frac{\partial\xi^1}{\partial\eta^i}\frac{\partial\xi^1}{\partial\eta^j} \,.$$

From (B) it results, that at the curve C the following equalities are valid

$$\frac{\partial \xi^1}{\partial \eta^1} = 1\ ; \quad \frac{\partial \xi^1}{\partial \eta^\varkappa} = b^1_\varkappa\ ; \quad \frac{\partial^2 \xi^1}{(\partial \eta^1)^2} = 0\ ; \quad \frac{\partial^2 \xi^1}{\partial \eta^1\, \partial \eta^\varkappa} = \frac{db^1_\varkappa}{d\eta^1}\ ;$$

$$\frac{\partial^2 \xi^1}{\partial \eta^\varkappa\, \partial \eta^\chi} = -\,[b^\nu_\varkappa b^\mu_\chi \Gamma^1_{\nu\mu} + (b^1_\varkappa b^\mu_\chi + b^1_\chi b^\mu_\varkappa)\Gamma^1_{\mu 1} + b^1_\varkappa b^1_\chi \Gamma^1_{11}]\ ;$$

$$\frac{\partial \xi^\alpha}{\partial \eta^1} = 0\ ; \quad \frac{\partial \xi^\alpha}{\partial \eta^\varkappa} = b^\alpha_\varkappa\ ; \quad \frac{\partial^2 \xi^\alpha}{(\partial \eta^1)^2} = 0\ ; \quad \frac{\partial^2 \xi^\alpha}{\partial \eta^\varkappa\, \partial \eta^1} = \frac{db^\alpha_\varkappa}{d\eta^1}\ ;$$

$$\frac{\partial^2 \xi^\alpha}{\partial \eta^\varkappa\, \partial \eta^\chi} = -\,[b^\nu_\varkappa b^\mu_\chi \Gamma^\alpha_\mu + (b^1_\varkappa b^\mu_\chi + b^1_\chi b^\mu_\varkappa)\Gamma^\alpha_{\mu 1} + b^1_\varkappa b^1_\chi \Gamma^\alpha_{11}]\ .$$

Substituting those values of the derivatives at C into (5.60), one can conclude that for $k = \alpha$, $i = \varkappa$, $j = \chi$ the relations (5.60) are satisfied identically. For $k = i = 1$ and $j = \varkappa$ or $k = j = 1$ and $i = \varkappa$ one has

$$\frac{db^1_\varkappa}{d\eta^1} = -\,\Gamma^1_{1\mu} b^\mu_\varkappa\ .$$

For $k = \alpha$, $i = 1$, $j = \varkappa$ or $k = \alpha$, $i = \varkappa$, $j = 1$ one has

$$\frac{db^\alpha_\varkappa}{\partial \eta^1} = -\,\Gamma^\alpha_{1\mu} b^\mu_\varkappa\ .$$

Integrating the above differential equations, one can determine $b^1_\varkappa$ and $b^\alpha_\varkappa$. With the aid of (B), one obtains a multiplicity of transformations into the variables η^i, in which all symbols Γ'^k_{ij} on the curve C, may be except the symbols Γ'^k_{11}, $k = 1, 2 \ldots n$, vanish. For the transformation from the system η^i to the system ξ^i in which all symbols $\widetilde{\Gamma}^k_{ij}$ vanish on the curve C, the Equations (5.60) on the curve C have the form

$$\Gamma'^k_{11} \frac{\partial \eta^1}{\partial \zeta^i} \frac{\partial \eta^1}{\partial \zeta^j} + \frac{\partial^2 \eta^k}{\partial \zeta^i\, \partial \zeta^j} = 0\ . \tag{5.60'}$$

To obtain the transformation which satisfies the above equations one can put

$$\eta^1 = f(\zeta^1)\,, \quad \eta^\alpha = g^\alpha(\zeta^1) + \zeta^\alpha\ .$$

If $i \neq 1$ or $j \neq 1$ the Equations (5.60′) are satisfied identically. If $i = j = 1$ one obtains

$$\Gamma'^{1}_{11} f'^{2} + f'' = 0 \quad \text{and} \quad \Gamma'^{\alpha}_{11} f'^{2} + g''^{\alpha} = 0 \ .$$

Replacing in $\Gamma'^{k}_{11}(\eta^1)$ argument η^1 by $f(\zeta^1)$ one obtains after integration, the transformation which solves the problem.

Note, that obtained results are not linked to the formulas (5.53a), nor to the initial metric of the space. Only the symmetry $\Gamma^k_{ij} = \Gamma^k_{ji}$ and the formulas of transformations of the Christoffel symbols Γ^k_{ij} are essential. In the coordinate system ζ^i all first covariant derivatives of each tensor along a curve obviously coincides with ordinary derivatives. The condition for a tensor to be constant along the curve C reduces to the constancy of its components in the coordinate system ζ^i.

If the curve C is closed or passes the points of intersection, then the values of the coordinates ζ^k and the base vectors $\mathbf{з}_k(\zeta^i)$ (after making the round on C and coming back to the initial point or to the point of intersection) in general, differ from the initial values. Obviously, in this case the constant vectors and tensors which have the same components in the system ξ on the closed curve C, will not coincide, after making a round with the initial vectors and tensors. Consequently, two tensors equal at the point M will not be equal to each other in another point N, provided that the values of tensors at that point are reached on the different curves. It can be shown, that these results are essentially linked to the non-Euclidean properties of the space.

In Riemannian space the components of a metric tensor always satisfies the equalities $\nabla_k g_{ij} = 0$. In the coordinate system ζ^k along C one has $\nabla_k g_{ij} = \partial g_{ij}/\partial \zeta^k = 0$ and consequently, on C in the coordinates ζ^k one has $g_{ij}(\zeta^k) = \mathbf{з}_i \cdot \mathbf{з}_j =$ const. It results that the base vectors $\mathbf{з}_i(\zeta^k)$ and $\mathbf{з}^i(\zeta^k)$ at the points of C describe the invariant system which, when translated along C may rotate as a rigid body only.

Starting from general formulas for the transformation of the symbols $\hat{\Gamma}^k_{ij}$, it is easy to notice that each linear transformation

$$\zeta'^i = a^i_k \, \zeta^k$$

where a^i_k — constant coefficients, also reduces the symbols $\approx\!\!\Gamma^k_{ij}$ to zero on a curve C in coordinates ζ'^i. By the appropriate choice of the coefficients one can satisfy, on the curve C in the system ζ'^i, the following equalities: $g_{ij} = 0$ for $i \neq j$ and $g_{ii} = \pm 1$ (The signs are determined by the signature of a metric). The orientation of the corresponding system of the base vectors can be arbitrarily given, in general case, only at an arbitrary given point of the curve C.

The condition that a space is Euclidean. However, if it is required that Equality (5.60) is to hold throughout space, i.e., that the space is Euclidean (or pseudo-Euclidean, or generally, when one can introduce the coordinates x^s in such a manner, that g_{ik} changes into a constant value globally), then these equalities will themselves represent a system of differential equations for the determination of the transformation of the given system ξ^i into a Cartesian system η^i in the entire space. In the general case, this system of equations cannot be integrated. The condition for a space to be Euclidean or pseudo-Euclidean coincides with the conditions of integrability of the system of differential equations (5.60), to be written down next. For this purpose, differentiate (5.60) with respect to η^k and eliminate from the equation obtained the second derivatives with the aid of (5.60) to obtain

$$\left(\frac{\partial \Gamma^{\omega}_{\alpha\beta}}{\partial \xi^s} - \Gamma^{\omega}_{\lambda\beta}\Gamma^{\lambda}_{\alpha s} - \Gamma^{\omega}_{\alpha\lambda}\Gamma^{\lambda}_{\beta s}\right)\frac{\partial \xi^s}{\partial \eta^k}\frac{\partial \xi^\alpha}{\partial \eta^i}\frac{\partial \xi^\beta}{\partial \eta^j} + \frac{\partial^3 \xi^\omega}{\partial \eta^i \partial \eta^j \partial \eta^k} = 0 .$$

Interchange the summation indices s and β and the indices k and j and use the symmetry of the Christoffel symbols with respect to their subscripts, to find another analogous equality

$$\left(\frac{\partial \Gamma^{\omega}_{\alpha s}}{\partial \xi^\beta} - \Gamma^{\omega}_{\lambda s}\Gamma^{\lambda}_{\alpha\beta} - \Gamma^{\omega}_{\alpha\lambda}\Gamma^{\lambda}_{\beta s}\right)\frac{\partial \xi^\beta}{\partial \eta^j}\frac{\partial \xi^\alpha}{\partial \eta^i}\frac{\partial \xi^s}{\partial \eta^k} + \frac{\partial^3 \xi^\omega}{\partial \eta^i \partial \eta^k \partial \eta^j} = 0 .$$

By subtraction of these equations, eliminate the third derivatives. Further, using the fact that the determinant of the transformation from ξ^i to η^i

must be non-zero, one arrives at the necessary and sufficient[1)] conditions for the integrability of the system (5.60) in the form[2)]

$$R_{\beta s\alpha.}^{\cdot\cdot\cdot\omega} = \frac{\partial \Gamma^{\omega}_{\alpha\beta}}{\partial \xi^s} - \frac{\partial \Gamma^{\omega}_{\alpha s}}{\partial \xi^\beta} + \Gamma^{\omega}_{\lambda s}\Gamma^{\lambda}_{\alpha\beta} - \Gamma^{\omega}_{\lambda\beta}\Gamma^{\lambda}_{\alpha s} = 0\,. \tag{5.61}$$

If the space is Euclidean, the equalities must be fulfilled in any coordinate system. If a space is non-Euclidean, Equation (5.61) is not satisfied.

The Riemann–Christoffel tensor. In the general case of Riemannian space, the quantities $R_{\beta s\mu}^{\cdot\cdot\cdot\alpha}$, introduced in the above manner, may be considered as the components of a fourth order tensor. In order to prove this statement, select some differentiable vector $\boldsymbol{a}$ and consider the two tensors

$$T = \nabla_j \nabla_i a^\alpha \mathbf{3}_\alpha \mathbf{3}^i \mathbf{3}^j \quad \text{and} \quad T^* = \nabla_i \nabla_j a^\alpha \mathbf{3}_\alpha \mathbf{3}^i \mathbf{3}^j\,.$$

Obviously, in general, $T \neq T^*$. Evaluating directly the difference $T - T^*$, one finds

$$T - T^* = R_{ij\mu.}^{\cdot\cdot\cdot\alpha} a^\mu \mathbf{3}_\alpha \mathbf{3}^i \mathbf{3}^j\,.$$

Since $T - T^*$ is a tensor, and $\boldsymbol{a}$ an arbitrary vector, then the $R_{ij\mu}^{\cdot\cdot\cdot\alpha}$ must transform like the components of a fourth order tensor. This tensor is referred to as Riemann–Christoffel tensor.

In Euclidean space, the Riemann–Christoffel tensor vanishes identically and *the result of a repeated covariant differentiation in Euclidean space*

---

1) Necessity is obvious. Sufficiency of Equality (5.60) can be proved in the following manner. Let us write (5.60) in the form $\partial^2 \xi^\omega / \partial \eta^j \partial \eta^i = f_{ij}^\omega = f_{ji}^\omega$ (just as $\Gamma^\omega_{\alpha\beta} = \Gamma^\omega_{\beta\alpha}$). It follows from (5.61), that

$$f_{ij}^\omega \, \mathrm{d}\eta^j = \mathrm{d}Q_i^\omega \quad \text{and} \quad f_{ji}^\omega \, \mathrm{d}\eta^i = \mathrm{d}Q_j^\omega\,, \tag{$*$}$$

where $\mathrm{d}Q_i^\omega$ represents the total differential of the function Q_i^ω, determined by integration of (∗). It follows from Formula (∗) that the differential forms $Q_i^\omega \, \mathrm{d}\eta^i$ are a total differentials, just as $\partial Q_i^\omega / \partial \eta^j = \partial Q_j^\omega / \partial \eta^i = f_{ij}^\omega$. On the basis of (5.60) one can assume that $Q_i^\omega = \partial \xi^\omega / \partial \eta^i$. Thus, the functions $\xi^\omega(\eta^k)$ which fulfill (5.60), can be obtained by integration of equalities $\mathrm{d}\xi^\omega = (\partial \xi^\omega / \partial \eta^i)\, \mathrm{d}\eta^i = Q_i^\omega \, \mathrm{d}\eta^i$.

2) The arrangement of indexes corresponds to the sign of the components of $R_{\beta s\alpha}^{\cdot\cdot\cdot\omega}$, used by H. Weyl in his book, *Space-Time Matter,* (1922) and afterwards by the majority of authors in the scientific literature.

does not depend on the order of its execution.[3)]

Expression for the components of the Riemann-Christoffel tensor in terms of the components of the metric tensor. Purely covariant components of the Riemann–Christoffel tensor have the form

$$R_{ij\mu\nu} = g_{\alpha\nu} R_{ij\mu\cdot}^{\cdot\cdot\cdot\alpha} = \frac{\partial \Gamma_{\nu\mu i}}{\partial \xi^j} - \frac{\partial \Gamma_{\nu\mu j}}{\partial \xi^i} + g^{\alpha\omega}\left[\Gamma_{\omega\mu j}\Gamma_{\alpha\nu i} - \Gamma_{\omega\mu i}\Gamma_{\alpha\nu j}\right], \tag{5.61'}$$

where, by (5.53a)

$$\Gamma_{\nu\alpha j} = \frac{1}{2}\left[\frac{\partial g_{\alpha\nu}}{\partial \xi^j} + \frac{\partial g_{j\nu}}{\partial \xi^\alpha} - \frac{\partial g_{\alpha j}}{\partial \xi^\nu}\right].$$

At any given point, one can select a coordinate system such that $\Gamma_{\nu\alpha j}=0$; however, derivatives of $\Gamma_{\nu\alpha j}$, if the space is not Euclidean, are non-zero, and therefore, in connection with the fact that $\Gamma_{\nu\alpha j}=\Gamma_{\alpha\nu j}$, one may always write down for the components of the Riemann–Christoffel tensor in such a coordinate system x^i the formula

$$R_{ij\mu\nu} = \frac{1}{2}\left[\frac{\partial^2 g_{\nu i}}{\partial x^j \partial x^\mu} + \frac{\partial^2 g_{\mu j}}{\partial x^i \partial x^\nu} - \frac{\partial^2 g_{\mu i}}{\partial x^j \partial x^\nu} - \frac{\partial^2 g_{\nu j}}{\partial x^i \partial x^\mu}\right].$$

Symmetry properties of the components of the Riemann–Christoffel tensor. This formula leads directly to the following symmetry properties which, as a consequence of properties of tensor transformations, are fulfilled in any coordinate system and at any point of space:

$$R_{ij\mu\nu} = -R_{ji\mu\nu}, \quad R_{ii\mu\nu} = 0,$$

$$R_{ij\mu\nu} = -R_{ij\nu\mu}, \quad R_{ij\nu\nu} = 0,$$

$$R_{ij\mu\nu} = R_{\mu\nu ij}, \quad R_{ij\mu\nu} + R_{\mu ij\nu} + R_{j\mu i\nu} = 0.$$

[3)] For better exposure of existing matter it is useful to note the following equalities

$$T = \frac{\partial}{\partial x^j}\left(\frac{\partial a}{\partial x^i}\mathbf{э}^i\right)\mathbf{э}^j = \frac{\partial}{\partial x^i}\left(\frac{\partial a}{\partial x^j}\mathbf{э}^j\right)\mathbf{э}^i = \nabla_j \nabla_i a^\alpha \mathbf{э}_\alpha \mathbf{э}^i \mathbf{э}^j ;$$

but for Riemannian space

$$T \neq T^* = \nabla_j \nabla_i a^\alpha \mathbf{э}_\alpha \mathbf{э}^j \mathbf{э}^i = \nabla_i \nabla_j a^\alpha \mathbf{э}_\alpha \mathbf{э}^i \mathbf{э}^j .$$

Note that not all the symmetry properties given here are independent of each other.

The number of independent components of the Riemann–Christoffel tensor for $n=3$. In three-dimensional space ($n=3$), the Riemann–Christoffel tensor has only six independent components which in the general case of a Riemann space may differ from zero. For example, one has the components

$$\begin{aligned} &R_{1212}, \quad R_{1313}, \quad R_{2323} \\ &R_{1213}, \quad R_{2123}, \quad R_{3132}\,. \end{aligned} \tag{5.62}$$

For the listing of these components, the following reasoning may be used. According to the symmetry properties stated above, one has $R_{iiii}=0$. If only two of the indices differ, then for two such fixed subscripts all components can be expressed in terms of one. For example, for $n=3$, with two indices different, there are only three independent components, say, those given in the first row of (5.62). If there are three different subscripts, then, for $n=3$, among the four indices two are always the same. These equal subscripts of the non-zero components must belong to different pairs. One may assume that they occupy the first and third positions. For fixed equal subscripts, only one component R_{ijik} is independent. One has altogether three such independent components which are listed in the second row of (5.62).

The condition for a space to be Euclidean consists of the vanishing of the Riemann–Christoffel tensor. The condition that the Riemann–Christoffel tensor vanishes, and consequently that a space is Euclidean or pseudo-Euclidean, is equivalent to six equations which are obtained by setting, for example, the six components (5.62) equal to zero.

On Ricci and Weyl tensors. It is easy to observe, that together with the fourth order tensors, $R_{ij\mu\nu}$ one can introduce the second order tensor with components $R_{i\nu}$, which is obtained by raising the index j, equating j to the index μ and contracting with respect to μ, i.e.

$$R_{i\nu} = P^{\mu\cdot\cdot}_{i\cdot\mu\nu}\,.$$

It is obvious from the property of symmetry that the same tensor can be obtained by performing the above operation on the edge indices. Such a constructed tensor is referred to as the Ricci tensor.

After contraction with respect to ν of the Ricci tensor $R^{\nu\cdot}_{\cdot\nu}=R$ one obtains a scalar which is referred to as the curvature of Riemanian space.

It is obvious that for Euclidean and pseudo-Euclidean space $R_{ijkl}=0$, $R_{ij}=0$ and $R=0$, therefore Euclidean and pseudo-Euclidean spaces are referred to as plane spaces.

It is easy to observe, that for the k-dimensional Riemannian space one can introduce the fourth order tensor with components W_{ijkl} by the relation

$$W_{ijkl}=R_{ijkl}-\frac{1}{k-2}\,(R_{ik}\,g_{jl}+R_{jl}\,g_{ik}-R_{il}\,g_{kj}-R_{kj}\,g_{il})$$

$$+\frac{R}{(k-1)(k-2)}\,(g_{ik}\,g_{jl}-g_{il}\,g_{jk})\,.$$

The tensor

$$W=W_{ijkl}\,\mathfrak{z}^i\mathfrak{z}^j\mathfrak{z}^k\mathfrak{z}^l$$

is referred to as the Weyl tensor. The components W_{ijkl} have the same properties of symmetry as the components R_{ijkl}, and, in addition, the following contractions with respect to the index k vanish: $W_{i\cdot kl}^{k}=0$ and $W_{kil\cdot}^{k}=0$. The Weyl tensor, generally speaking, differs from zero only for $k\geqslant 4$. Taking into account the conditions of symmetry of tensors, one obtains that in two-dimensional and three-dimensional cases, the Weyl tensor is identically equal to zero. In the general theory of relativity, in four-dimensional Riemannian space, it is obtained that generally $W_{ijkl}\neq 0$ and the Weyl tensor differs from zero.

The mathematical properties of the fourth order matrix W_{AB}, where $A=ij$ and $B=kl$, and, between them, its canonical form, represent themselves the important characteristic of Riemannian space, in particular for $R_{ij}=0$.

The equations of compatibility of strain. The components of the strain tensor are defined by

$$2\varepsilon_{\alpha\beta}=\hat{g}_{\alpha\beta}-\mathring{g}_{\alpha\beta}\,.$$

When the displacement vector $\boldsymbol{w}$ exists, both the quadratic forms

$$ds^2=\hat{g}_{\alpha\beta}\,d\xi^\alpha\,d\xi^\beta \quad\text{and}\quad ds_0^2=\mathring{g}_{\alpha\beta}\,d\xi^\alpha\,d\xi^\beta$$

determine the square of the element of length in Euclidean space. Therefore the Riemann–Christoffel tensor, formed for the fundamental tensors $\mathring{g}_{\alpha\beta}$ and $\hat{g}_{\alpha\beta}$, must vanish. This condition leads to the equations

$$\mathring{R}_{ij\mu\nu}=0 \quad \text{and} \quad \hat{R}_{ij\mu\nu}=0\,. \tag{5.63}$$

One of the coordinate base vectors $\mathring{\mathfrak{z}}_i$ and $\hat{\mathfrak{z}}_i$ may be chosen arbitrarily; the second is completely determined by the deformations. Consequently, one of the equations (5.63) is satisfied automatically as a result of the choice of the coordinate base in Euclidean space; the second may be considered as equations for the components of the strain tensor. These equations, referred to as equations of compatibility of strain, are readily written down explicitly with the aid of (5.61′). In particular, if one chooses in the actual deformed state a rectilinear Cartesian (in general, not orthogonal) coordinate system, then $\partial\hat{g}_{\alpha\beta}/\partial\xi^j=0$, and therefore the compatibility equations $\mathring{R}_{ij\mu\nu}=0$ may be written in the form

$$\frac{\partial^2 \varepsilon_{\nu i}}{\partial\xi^j\partial\xi^\mu}+\frac{\partial^2 \varepsilon_{\mu j}}{\partial\xi^i\partial\xi^\nu}-\frac{\partial^2 \varepsilon_{\mu i}}{\partial\xi^j\partial\xi^\nu}-\frac{\partial^2 \varepsilon_{\nu j}}{\partial\xi^i\partial\xi^\mu}+\mathring{g}^{\alpha\omega}\left[G_{\omega\mu j}G_{\alpha\nu i}-G_{\omega\mu i}G_{\alpha\nu j}\right]=0$$

where (5.63a)

$$G_{\nu\alpha j}=\frac{\partial\varepsilon_{\alpha\nu}}{\partial\xi^j}+\frac{\partial\varepsilon_{j\nu}}{\partial\xi^\alpha}-\frac{\partial\varepsilon_{\alpha j}}{\partial\xi^\nu}\,,$$

and the components $\mathring{g}^{\alpha\omega}$ are determined as the elements of a matrix which is the inverse of the matrix with the components $\hat{g}_{\alpha\omega}-2\varepsilon_{\alpha\omega}$:

$$\|\mathring{g}^{\alpha\omega}\|=\|\hat{g}_{\alpha\omega}-2\varepsilon_{\alpha\omega}\|^{-1}\,.$$

In an analogous manner, the compatibility equation $\hat{R}_{ij\mu\nu}=0$ may be rewritten, provided the Lagrangian system in the initial state is rectilinear and $\mathring{g}_{\alpha\beta}=\text{const}$.

Equations (5.63a) represent second order partial differential equations for the six functions $\varepsilon_{\alpha\beta}(\xi^i)$ which are linear in the second derivatives and non-linear in the first derivatives.

For all possible values of the indices i, j, μ, ν taken from 1,2,3, the system (5.63a) comprises altogether six independent equations. Clearly, Formulae (5.57), expressing ε_{ij} in terms of w_α, are general integrals of the system of compatibility equations (5.63a).

The equation of compatibility in the case of infinitesimal deformations. In the case of infinitesimal deformations, the compatibility equations (5.63a) assume the form

$$\frac{\partial^2 \varepsilon_{\nu i}}{\partial \xi^j \partial \xi^\mu} + \frac{\partial^2 \varepsilon_{\mu j}}{\partial \xi^i \partial \xi^\nu} - \frac{\partial^2 \varepsilon_{\mu i}}{\partial \xi^j \partial \xi^\nu} - \frac{\partial^2 \varepsilon_{\nu j}}{\partial \xi^i \partial \xi^\mu} = 0 \tag{5.63b}$$

and are referred to as the compatibility equations of Saint-Venant. By direct substitution of (5.58) into (5.63b), it may be verified that Formulae (5.58) yield for the three arbitrary functions w_α a general integral of the system (5.63b).

The compatibility equations (5.63b) in the case of infinitesimal strain are six independent linear second order partial differential equations in ε_{ij}.

Thus, in the case when between the initial state and the state under consideration of a continuum one may introduce a displacement vector $\boldsymbol{w}$, the compatibility equations must be fulfilled and the expressions for ε_{ij} in terms of the components of $\boldsymbol{w}$ may be conceived as general solutions of these equations.

Next, more detailed consideration will be given to the geometrical pattern of the deformations of a continuous medium in the presence of displacements.

Transformation for the displacement of an absolutely rigid body. First investigate the displacements of an absolutely rigid body. Select any two of its positions I and II (fig. 14). Let M and M' denote two positions of one and the same point of the body. If the points M' and M correspond, i.e., if one excludes from the consideration translation, then the displacement of an absolutely rigid body must be a simple rotation with respect to the axis, passing through the point M.

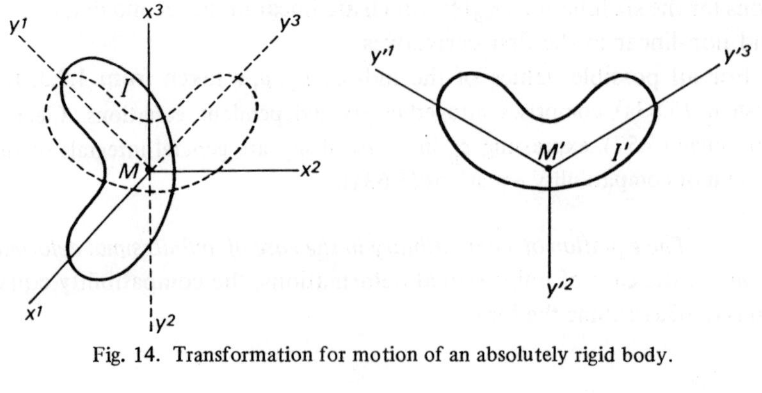

Fig. 14. Transformation for motion of an absolutely rigid body.

In Position I, select "frozen" coordinate axes x^1, x^2, x^3 with origin at M; in Position II, these axes will go over into y'^1, y'^2, y'^3 with origin at M' (fig. 14). Denote by y^1, y^2, y^3 the axes which are obtained for the translation of y'^1, y'^2, y'^3 from M' to M. The transformation of rotation for transition from x^i to y^i may be rewritten in the form

$$y^i = c^i{}_j x^j ,$$

where $\|c^i{}_j\|$ is one and the same orthogonal matrix for all points M of the rigid body. Thus, an arbitrary displacement of a rigid body (excluding translations) is simply an orthogonal transformation.

Now, let the body deform. The corresponding transformation has a very general form; it will only be assumed that it satisfies the properties of mutual single-valuedness, continuity and differentiability with respect to the coordinates.

Transformation of infinitesimal particles of a continuum. If one considers an infinitesimal neighbourhood of a point of a continuum, one may assume the transformation to be affine exactly up to small first order quantities, as will now be shown. At instant t_0, denote the base vectors of the Lagrangian coordinate system at the point M by $\mathring{\mathfrak{z}}_i$; the positions of all points of the neighbourhood of M are completely defined by $\mathrm{d}\boldsymbol{r}_0$, where

$$\mathrm{d}\boldsymbol{r}_0 = \mathrm{d}\xi^i \mathring{\mathfrak{z}}_i .$$

The position of all points of the neighbourhood of M', to which M has moved at the instant t under consideration, is determined by the vector $\mathrm{d}\boldsymbol{r}$ the components of which to the base $\mathfrak{z}_i$ likewise are equal to $\mathrm{d}\xi^i$, so that

$$\mathrm{d}\boldsymbol{r} = \mathrm{d}\xi^i \mathfrak{z}_i .$$

If one combines M and M' and takes the decomposition of $\mathrm{d}\boldsymbol{r}$ with respect to the base vectors $\mathring{\mathfrak{z}}_i$, the components of this decomposition will differ from $\mathrm{d}\xi^i$; denote them by $\mathrm{d}\eta^i$, so that

$$\mathrm{d}\boldsymbol{r} = \mathrm{d}\eta^i \mathring{\mathfrak{z}}_i .$$

The connection between the $d\eta^i$ and $d\xi^i$ determines the transformation of the particles of the continuum.[1)] This transformation is found from

$$d\boldsymbol{r} = d\xi^i \mathfrak{z}_i = d\eta^i \mathfrak{z}_i . \tag{5.64}$$

On the basis of (5.42), linking $\mathfrak{z}_i$, $\mathring{\mathfrak{z}}_i$ and $\boldsymbol{w}$,

$$\mathfrak{z}_i = \mathring{\mathfrak{z}}_i + \frac{\partial \boldsymbol{w}}{\partial \xi^i} = \mathring{\mathfrak{z}}_i + \mathring{\nabla}_i \mathring{w}^k \mathring{\mathfrak{z}}_k = (\delta_i^k + \mathring{\nabla}_i \mathring{w}^k)\mathring{\mathfrak{z}}_k = c^k_{\ i}\, \mathring{\mathfrak{z}}_k ,$$

where

$$c^k_{\ i} = \delta^k_i + \mathring{\nabla}_i \mathring{w}^k , \tag{5.65}$$

one may write

$$d\boldsymbol{r} = d\xi^i \mathfrak{z}_i = d\xi^i c^k_{\ i}\, \mathring{\mathfrak{z}}_k = d\eta^k \mathring{\mathfrak{z}}_k ,$$

whence

$$d\eta^k = c^k_{\ i}\, d\xi^i . \tag{5.66}$$

The transformation from $d\xi^i$ to $d\eta^i$ is homogeneous and linear with the matrix $\|c^k_{\ i}\|$ which does not depend on the differentials $d\xi^i$, i.e., the approximate coordinates of the points of the neighbourhood; the $c^k_{\ i}$ can only depend on the coordinates of the point M. Consequently, one may assume the coefficients $c^k_{\ i}$ to be constant for small particles, i.e., the transformation (5.66) to be affine.

Properties of affine transformations. Next, the properties of affine transformations, which follow directly from the linearity of (5.66), will be written down.

Under affine transformations, straight lines go over into straight lines, planes into planes; in particular, parallel straight lines and planes become parallel straight lines and planes, parallelogrammes become parallelogrammes. It follows from this result that all equal, equally directed segments stretch (or contract) in the same manner.

The ratio of the length of any segment before and after deformation (by strength of the fact that it is a ratio of homogeneous first order functions) does not depend on the original length of the segment, it

[1)] For the infinitesimal particles one can consider $d\eta^i$ and $d\xi^i$ as the Cartesian coordinates in one and the same curvilinear system of coordinates with the base $\mathring{\mathfrak{z}}_i$.

depends only on its direction. It follows from this result that the coefficient of relative extension of any segment also does not depend on the length of the segment, but only on its direction.

Segments always become segments, where the ratio in which a point divides a segment remains unchanged.

Algebraic curves or surfaces become algebraic curves or surfaces of equal degree. For example, a second degree surface becomes a second degree surface: a sphere becomes an ellipsoid or sphere, where conjugate diameters of the sphere become conjugate diameters of the ellipsoid. In a sphere, all conjugate diameters are orthogonal; in an ellipsoid, in the general case, there exists a unique set of three orthogonal conjugate diameters. Consequently, there exists always at least one orthogonal triplet of axes which goes over into an orthogonal triplet of axes, i.e., there always exist principal directions.

In general, under affine tranformations, volumes change, but the magnitude of the relative change of volume $\theta = (V - V_0)/V_0$ does not depend on the original form or the dimensions of the volume.

Therefore, in calculating the magnitude of the relative change of volume in terms of the components of the strain tensor, using for this purpose elementary parallelepipeds, one obtains results which are true for any small volumes.

Geometrical pattern of the transformation of small particles of a continuum. Every infinitesimal sphere, separated from a continuum, transforms as a result of deformation into an ellipsoid. If during this process the transformations of the principal directions do not change their orientation in space, one speaks of pure deformation which reduces to extension or compression along three orthogonal principal axes.

If a sphere is transformed into an ellipsoid in such a manner that the principal directions change their orientations in space, one is faced with the general case of an affine transformation which reduces to pure deformation (extension along the three principal axes) and rotation in space. Note that in the case of pure deformation any segment in a particle, not directed along principal axes, changes, generally speaking, its direction in space.

For motion of a particle as an absolutely rigid body, a sphere becomes a sphere of the same radius, where all mutually perpendicular trihedra may be considered to be principal; all of them rotate about one and the same axis by the same angle. One says in this case that pure rotation takes place.

By (5.66), the matrix of the affine transformation $\|c^k_{\cdot i}\|$ is determined by the nine derivatives of the components of the displacement vector $\boldsymbol{w}$ with respect to the coordinates ξ^i; in the general case, this matrix is formed at a point by nine arbitrary numbers. Pure deformation is characterized by the three principal components of the strain tensor and three parameters, determining the directions of the principal axes in space (or six components of the strain tensor); the rotation of the principal axes in space is characterized by the three remaining parameters. In the case of pure rotation, the matrix is orthogonal and depends only on three independent parameters (the direction of the axis of rotation and the angle of rotation).

Thus, an arbitrary displacement of an infinitesimal particle of a concontinuum reduces to a translation in space, a rotation and pure deformation (compression or extension along three mutually perpendicular principal axes).

The geometrical characteristics of deformation are important for solids. In fluids and gases, the characteristics of deformation themselves play a much smaller role. For example, as a liquid (which is homogeneous) is poured from one vessel into another, it remains the same liquid, although during the pouring one may subject it to very complicated and big deformations. In fluids and gases, the property of deformation manifests itself, in essence, only through volume changes. Fluids and gases exhibit resistance to pressure; compressed and uncompressed fluids and gases differ.

The strain tensor plays a basic and decisive role in the theory of deformation of solids. In the theory of flow of liquids and gases, i.e., in hydrodynamics (and in the theory of deformation of certain solids) a big role is played by another characteristic, the strain rate tensor. It may happen that a deformation itself is inessential, but that great importance attaches to the rate at which it takes place.

§6. Strain rate tensor

Definition of strain rate tensor. The strain tensor

$$\varepsilon_{ij} = \tfrac{1}{2}(\hat{g}_{ij} - \mathring{g}_{ij}) \tag{6.1}$$

has been introduced in connection with two states of a continuum: a given state, $\hat{g}_{ij}$, and, generally speaking, an "initial state" $\mathring{g}_{ij}$. If the initial state $\mathring{g}_{ij}$ is actually realized, there exists a displacement vector $\boldsymbol{w}$ from all points of the continuum in the initial state, attained at the instant t_0, to the state under consideration at time t; the strain tensor is given by (6.56) and (5.57). Besides these two states of a continuum, consider yet its state at time $t+\Delta t$, close to that under consideration. Denote the components of the metric tensor at time $t+\Delta t$ by $\hat{g}'_{ij}$. Obviously, the components of the strain tensor may be introduced with respect to the states of the medium at times t and $t+\Delta t$. Denote these components by $\Delta\varepsilon_{ij}$, so that

$$\Delta\varepsilon_{ij}=\tfrac{1}{2}(\hat{g}'_{ij}-\hat{g}_{ij})=\tfrac{1}{2}(\nabla_i w_j+\nabla_j w_i+\nabla_i w^p\nabla_j w_p), \tag{6.2}$$

where $\boldsymbol{w}=w_i\hat{\mathfrak{Z}}^i$ and the covariant derivatives are computed in this case in the initial space $\hat{g}_{ij}$. The formula (6.2) is valid, because there exists the displacement $\boldsymbol{w}$ from the state at time t to the state at time $t+\Delta t$. Obviously,

$$\boldsymbol{w}=\boldsymbol{v}\,\Delta t=v_i\hat{\mathfrak{Z}}^i\,\Delta t\,,$$

i.e., $w_i=v_i\Delta t$ is of order Δt and it is an infinitesimal displacement if Δt is infinitesimal. Therefore

$$\lim_{\Delta t\to 0}\frac{\Delta\varepsilon_{ij}}{\Delta t}=\tfrac{1}{2}(\nabla_i v_j+\nabla_j v_i)=e_{ij}\,. \tag{6.3}$$

The quantities e_{ij} are the components of a symmetric tensor which is referred to as the strain rate tensor. If the velocity field $\boldsymbol{v}$ is known, then the components e_{ij} may be calculated from (5.3). Obviously, the formula $e_{ij}=\frac{1}{2}(\nabla_i v_j+\nabla_j v_i)$ conserves its form in any moving curvilinear coordinate system, in view of the fact that the vector $\boldsymbol{v}$ is determined with the aid of the observer's and the concomitant systems, which form the basis of the study of the motion of a continuum.

Link between the components of the strain and strain rate tensors. It follows directly from (6.2) that the components of the strain rate tensor in the concomitant coordinate system are given by

$$\hat{e}_{ij}=\frac{1}{2}\,\frac{d\hat{g}_{ij}}{dt}\,. \tag{6.4}$$

With the aid of (5.4), one obtains readily from (6.3), if the "initial" state $\mathring{g}_{ij}$ does not depend on the time t, formulae relating the components of the strain and strain rate tensors in the concomitant coordinate system:

$$\hat{e}_{ij} = \frac{d\hat{\varepsilon}_{ij}}{dt}. \tag{6.5}$$

(6.5) is only true when $\mathring{g}_{ij} \neq$ const. in time, while (6.4) is, by definition of $\hat{e}_{ij}$, always true.

The strain and strain rate tensors are different, but the quantities $e_{ij}\Delta t$ are components of the tensor of infinitesimal deformation, corresponding to the displacement in time Δt, i.e.,

$$e_{ij}\Delta t = \varepsilon_{ij}. \tag{6.6}$$

Note that the strain tensor $\mathscr{E}$ has been introduced as a result of the comparison of two states of a medium and that the strain rate tensor is a characteristic of a given state at a given time.

Compatibility equations for the components of the strain rate tensor. Clearly, the components of the strain tensor (6.6) must satisfy the compatability equations. Substituting (6.6) into (5.63a) and going to the limit $\Delta t \to 0$, one arrives at the equations of compatibility for the components of the strain rate tensor (in the Cartesian coordinates)

$$\frac{\partial^2 e_{\nu i}}{\partial \xi^j \partial \xi^\mu} + \frac{\partial^2 e_{\mu j}}{\partial \xi^i \partial \xi^\nu} - \frac{\partial^2 e_{\mu i}}{\partial \xi^j \partial \xi^\nu} - \frac{\partial^2 e_{\nu j}}{\partial \xi^i \partial \xi^\mu} = 0. \tag{6.7}$$

Just as the system of equations (5.63b), the system (6.7) contains six independent linear partial differential equations of the second order. The corresponding independent equations may be obtained for the sets of indices (5.62). Formulae (6.3) for arbitrary three functions v_i yield general integrals of the system of equations (6.7).

Analogously to the strain rate tensor, one may introduce other tensors the components of which are derivatives of ε_{ij} with respect to t. One may likewise consider tensors the components of which are derivatives of ε_{ij} with respect to the space coordinates, for example, the tensor $\nabla_k \varepsilon_{ij} \mathbf{э}^k \mathbf{э}^j \mathbf{э}^i$.

§ 7. Distribution of velocities in an infinitesimal particle of a continuum

Infinitesimal affine transformation of a small particle of a continuum in time Δt. Select an infinitesimal particle of a continuum and study the question of the distribution of velocities in it. By an infinitesimal particle will be understood the set of points of a medium with coordinates $\xi^i + d\xi^i = \xi^i + \rho^i$, which lie an infinitesimal distance ρ from a given point O with coordinates ξ^1, ξ^2, ξ^3, the so-called centre of the particles. Let it be assumed that the velocity field v is continuous and has derivatives at least of first order.

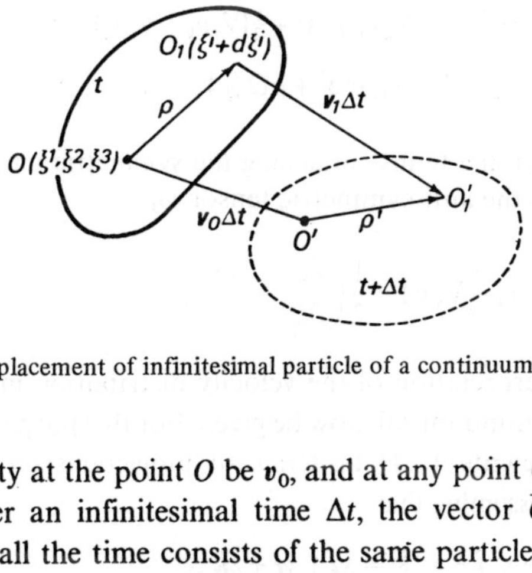

Fig. 15. Displacement of infinitesimal particle of a continuum in time Δt.

Let the velocity at the point O be $\boldsymbol{v}_0$, and at any point O_1 of the small particle $\boldsymbol{v}_1$. After an infinitesimal time Δt, the vector $\overline{OO_1} = \boldsymbol{\rho}$ in the medium, which all the time consists of the same particles, becomes the vector $\overline{O'O'_1} = \boldsymbol{\rho}'$ and, obviously, (fig. 15)

$$\boldsymbol{\rho}' = \boldsymbol{\rho} + (\boldsymbol{v}_1 - \boldsymbol{v}_0)\Delta t \,. \tag{7.1}$$

Select a decomposition of $\boldsymbol{v}$ near O which is exact to first order in ρ

$$\boldsymbol{v}_1 = \boldsymbol{v}_0 + \left(\frac{\partial \boldsymbol{v}}{\partial \xi^i}\right)_0 \rho^i + \rho O(\rho) \;, \qquad \lim_{\rho \to 0} O(\rho) = 0 \tag{7.2}$$

and substitute (7.2) into (7.1) to obtain

$$\boldsymbol{\rho}' = \boldsymbol{\rho} + \left(\frac{\partial \boldsymbol{v}}{\partial \xi^i}\right) \rho^i \Delta t + \boldsymbol{\rho} O(\rho)\Delta t \,. \tag{7.3}$$

It is seen that, exactly to $\rho O(\rho)\Delta t$, an infinitesimal particle of the continuum in the infinitesimal time Δt undergoes an infinitesimal affine transformation (the derivatives of $\boldsymbol{v}$ with respect to ξ^i being taken at the centre O of the particle).

Distribution of velocities in an infinitesimal element of a deforming medium. Rewrite (7.2), expressing the velocity $\boldsymbol{v}_1$ of any point O_1 of an infinitesimal particle of a continuous medium in terms of the velocity of its centre $\boldsymbol{v}_0$, the derivative of $\boldsymbol{v}$ at the centre and the coordinates of the point under consideration, in the form

$$\begin{aligned} \boldsymbol{v} &= \boldsymbol{v}_0 + \nabla_i v_k \rho^i \mathbf{3}^k + \rho O(\rho) \\ &= \boldsymbol{v}_0 + \tfrac{1}{2}(\nabla_i v_k + \nabla_k v_i)\rho^i \mathbf{3}^k + \tfrac{1}{2}(\nabla_i v_k - \nabla_k v_i)\,\rho^i \mathbf{3}^k + \rho O(\rho) \\ &= \boldsymbol{v}_0 + e_{ki}\rho^i \mathbf{3}^k + \omega_{ki}\rho^i \mathbf{3}^k + \rho O(\rho)\,. \end{aligned} \tag{7.4}$$

In formula (7.4), the terms containing the symmetric tensor e_{ki} have been separated from the anti-symmetric tensor ω_{ki}:

$$\omega_{ki} = \tfrac{1}{2}(\nabla_i v_k - \nabla_k v_i) = \frac{1}{2}\left(\frac{\partial v_k}{\partial \xi^i} - \frac{\partial v_i}{\partial \xi^k}\right). \tag{7.5}$$

A physical interpretation of the velocity distribution in an infinitesimal element of a continuum will now be given. For this purpose, for the sake of convenience, rewrite (7.4) in terms of projections on to Cartesian coordinate axes, assuming that

$$\boldsymbol{\rho} = x^1 \boldsymbol{i} + x^2 \boldsymbol{j} + x^3 \boldsymbol{k} = x\boldsymbol{i} + y\boldsymbol{j} + z\boldsymbol{k}\,.$$

One finds

$$\left.\begin{aligned} u_1 &= u_0 + e_{1i}x^i + \omega_{1i}x^i, \\ v_1 &= v_0 + e_{2i}x^i + \omega_{2i}x^i, \\ w_1 &= w_0 + e_{3i}x^i + \omega_{3i}x^i. \end{aligned}\right\} \tag{7.6}$$

The components e_{ji} and ω_{ji} in these formulae do not depend on x^i, and terms of higher order in x^i than the first have been omitted. Introduce the quadratic form

$$\Phi = \tfrac{1}{2} e_{pq} x^p x^q\ ; \tag{7.7}$$

obviously,

$$e_{ki}x^i = \frac{\partial \Phi}{\partial x^k}\,. \tag{7.8}$$

Formulae (7.6) may be rewritten in the form

$$\left.\begin{aligned} u_1 &= u_0 + \frac{\partial \Phi}{\partial x^1} + \omega_{1i}x^i\,, \\ v_1 &= v_0 + \frac{\partial \Phi}{\partial x^2} + \omega_{2i}x^i\,, \\ w_1 &= w_0 + \frac{\partial \Phi}{\partial x^3} + \omega_{3i}x^i\,. \end{aligned}\right\} \tag{7.9}$$

Thus, the velocity of a point of an infinitesimal particle of a continuum has been decomposed into three parts: the first, $\boldsymbol{v}_0(u_0, v_0, w_0)$, does not depend on the coordinates x^i and, consequently, represents itself the velocity of translation of the entire particle (which coincides with the velocity of the centre of the particle), the second, $\partial\Phi/\partial x^i$, has the potential Φ. For a more detailed study of the third, $(\omega_{1i}x^i, \omega_{2i}x^i, \omega_{3i}x^i)$, introduce in the Cartesian coordinate system the anti-symmetric matrix

$$\|\omega_{ji}\| = \left\|\begin{matrix} 0 & \omega_{12} & \omega_{13} \\ \omega_{21} & 0 & \omega_{23} \\ \omega_{31} & \omega_{32} & 0 \end{matrix}\right\| = \left\|\begin{matrix} 0 & \omega_{12} & \omega_{13} \\ -\omega_{12} & 0 & \omega_{23} \\ -\omega_{13} & -\omega_{23} & 0 \end{matrix}\right\|$$

$$= \left\|\begin{matrix} 0 & -\omega_3 & \omega_2 \\ \omega_3 & 0 & -\omega_1 \\ -\omega_2 & \omega_1 & 0 \end{matrix}\right\|,$$

i.e., introduce the notation

$$\omega_1 = \omega_{32}, \quad \omega_2 = \omega_{13}, \quad \omega_3 = \omega_{21}\,. \tag{7.10}$$

By (7.5) and (7.10), one has, in a Cartesian coordinate system,

$$\left.\begin{aligned} \omega_1 &= \frac{1}{2}\left(\frac{\partial w}{\partial y} - \frac{\partial v}{\partial z}\right), \\ \omega_2 &= \frac{1}{2}\left(\frac{\partial u}{\partial z} - \frac{\partial w}{\partial x}\right), \\ \omega_3 &= \frac{1}{2}\left(\frac{\partial v}{\partial x} - \frac{\partial u}{\partial y}\right), \end{aligned}\right\} \tag{7.11}$$

which, as is readily verified, can be represented symbolically by

$$\boldsymbol{\omega} = \omega_1 \boldsymbol{i} + \omega_2 \boldsymbol{j} + \omega_3 \boldsymbol{k} = \frac{1}{2}\begin{vmatrix} \boldsymbol{i} & \boldsymbol{j} & \boldsymbol{k} \\ \dfrac{\partial}{\partial x} & \dfrac{\partial}{\partial y} & \dfrac{\partial}{\partial z} \\ u & v & w \end{vmatrix}. \tag{7.12}$$

After introduction of the notation $\omega_1, \omega_2, \omega_3$, from (7.10), Equation (7.9) is rewritten in the form

$$\begin{aligned} u_1 &= u_0 + \frac{\partial \Phi}{\partial x} + \omega_2 z - \omega_3 y, \\ v_1 &= v_0 + \frac{\partial \Phi}{\partial y} + \omega_3 x - \omega_1 z, \\ w_1 &= w_0 + \frac{\partial \Phi}{dz} + \omega_1 y - \omega_2 x \end{aligned} \tag{7.13}$$

or, by (7.12), in the form

$$\left.\begin{aligned} u_1 &= u_0 + \frac{\partial \Phi}{\partial x} + (\boldsymbol{\omega} \times \boldsymbol{\rho})_x, \\ v_1 &= v_0 + \frac{\partial \Phi}{\partial y} + (\boldsymbol{\omega} \times \boldsymbol{\rho})_y, \\ w_1 &= w_0 + \frac{\partial \Phi}{dz} + (\boldsymbol{\omega} \times \boldsymbol{\rho})_z. \end{aligned}\right\} \tag{7.14}$$

Finally, one has in vector form

$$\boldsymbol{v}_1 = \boldsymbol{v}_0 + \operatorname{grad} \Phi + \boldsymbol{\omega} \times \rho + \boldsymbol{\rho} O(\rho), \tag{7.15}$$

which replaces (7.4) or (7.14).

Comparison of (7.15) *with Euler's formula for the velocity distribution in an absolutely rigid body.* The velocity distribution in an absolutely rigid body is known to be given by Euler's formula

$$\boldsymbol{v}_1 = \boldsymbol{v}_0 + \boldsymbol{\Omega} \times \boldsymbol{\rho}\,;$$

where v_0 is the velocity of a definite point of an absolutely rigid body, $\boldsymbol{v}$ is the velocity of an arbitrary point, Ω is the vector of the instantaneous angular velocity of the body and $\boldsymbol{\rho}$ is the radius vector OO_1. Formula (7.15) for the velocity distribution in an infinitesimal element of a continuum differs in form from Euler's formula by the presence of the terms grad Φ and $\rho O(\rho)$, the last of which is infinitely small compared with ρ and may be neglected in a first approximation.

Rate of relative elongation. The role of the term grad Φ will now be explained. As a result of the motion of the continuum, the vector $\boldsymbol{\rho}$ goes over into the vector $\boldsymbol{\rho}'$. The change of the vector ρ, i.e., $\rho' - \rho = \Delta\boldsymbol{\rho}$ may arise only when different points of the infinitesimal particle move with different velocities; in fact, it follows from (7.1), in the limit $\Delta t \to 0$, that

$$\frac{\mathrm{d}\boldsymbol{\rho}}{\mathrm{d}t} = \boldsymbol{v}_1 - \boldsymbol{v}_0\,. \tag{7.16}$$

Let us compute the rate of relative elongation of a segment of the medium in the direction of $\boldsymbol{\rho}$:

$$e_\rho = \frac{1}{|\boldsymbol{\rho}|}\frac{\mathrm{d}|\boldsymbol{\rho}|}{\mathrm{d}t} = \frac{1}{\rho}\frac{\mathrm{d}\rho}{\mathrm{d}t} = \frac{1}{2}\frac{1}{\rho^2}\frac{\mathrm{d}\rho^2}{\mathrm{d}t}$$

$$= \frac{1}{2}\frac{1}{\rho^2}\frac{\mathrm{d}(\boldsymbol{\rho}\cdot\boldsymbol{\rho})}{\mathrm{d}t} = \frac{1}{\rho^2}\left(\boldsymbol{\rho}\cdot\frac{\mathrm{d}\boldsymbol{\rho}}{\mathrm{d}t}\right).$$

By (7.16) and (7.15), since $\boldsymbol{\rho}\cdot(\boldsymbol{\omega}\times\boldsymbol{\rho}) = 0$; one obtains

$$e_\rho = \frac{1}{\rho^2}\left(\boldsymbol{\rho}\cdot\frac{\mathrm{d}\boldsymbol{\rho}}{\mathrm{d}t}\right) = \frac{1}{\rho^2}(\boldsymbol{\rho}\cdot\operatorname{grad}\Phi) = \frac{1}{\rho^2}\left(\frac{\partial\Phi}{\partial x}x + \frac{\partial\Phi}{\partial y}y + \frac{\partial\Phi}{\partial z}z\right)$$

$$= \frac{2\Phi}{\rho^2} = e_{ij}\frac{x^i}{\rho}\frac{x^j}{\rho} = e_{ij}\alpha^i\alpha^j\,, \tag{7.17}$$

where

$$\alpha^i = \frac{x^i}{\rho} = \cos(\rho, x^i).$$

Thus, if the components of the strain rate tensor e_{ij} and the direction of ρ are known, then the rate of relative elongation e_ρ in this direction may be determined.

Kinematic interpretation of the components of the strain rate tensor. From (7.17) follows directly a mechanical interpretation of the components of the strain rate tensor with equal indices. Let ρ be directed along x^i, then all terms but one on the right-hand side of (7.17) vanish and one obtains

$$e_{x^i} = e_{ii}\,.$$

Thus

$$e_x = e_{11}, \quad e_y = e_{22}, \quad e_z = e_{33}\,,$$

i.e., the diagonal components of the strain rate tensor are the rates of relative elongation of segments of the medium which originally are parallel to the Cartesian orthogonal coordinate axes. The same interpretation may be obtained by a different reasoning. In time Δt, by (7.1), an infinitesimal element of the medium undergoes infinitesimal deformation with respect to the state of the continuum at time t. One can introduce the strain tensor $\mathscr{E}$ with respect to the states of the medium at times t and $t+\Delta t$ and obtain

$$\varepsilon_{ij} = e_{ij}\Delta t\,.$$

From this result follows a kinematic interpretation of the components e_{ij} which, apart from the multiplier Δt, coincide with the components ε_{ij} of the infinitesimal strain tensor. The quantities ε_{ij}, $i \neq j$, characterize shearing of originally right angles between segments of the medium which were originally aligned with the coordinate axes x, y, z. The components e_{ij}, $i \neq j$, are equal to half the rate of shearing of the originally right angles, formed by segments of the medium which at a given instant of time were parallel to the corresponding coordinate axes. It is seen from (7.17) that the term grad Φ in (7.15), characterizing the velocity distribution in an infinitesimal element of the medium, corresponds to the deformation of the particle.

Introduce the notation

$$\boldsymbol{v}^* = \text{grad } \Phi \tag{7.18}$$

and call it the rate of pure deformation. If $\boldsymbol{v}^* = 0$, then all $e_\rho = 0$ and no deformation occurs, i.e., the length of a segment $\boldsymbol{\rho}$ in any direction does not change. In contrast, if there is no deformation, then all $e_\rho = 0$ and, by (7.17),

$$\text{grad } \Phi = \boldsymbol{v}^* = 0\,.$$

Principal axes and principal components of the strain rate tensor. As for every second order symmetric tensor, one can introduce for the strain rate tensor principal axes; in a Cartesian coordinate system, aligned with the principal axes, the matrix of the components e_{ij} of the strain rate tensor has the form

$$\begin{Vmatrix} e_1 & 0 & 0 \\ 0 & e_2 & 0 \\ 0 & 0 & e_3 \end{Vmatrix}.$$

One may specify the principal axes of the strain rate tensor at any given instant of time t and at any given point O in the medium. The quantities e_i are called the principal components of the strain rate tensor. Obviously, $e_i > 0$ corresponds to extension, $e_i < 0$ to compression along the i-th axis.

As in the case of every symmetric tensor, one may link a tensor surface to the strain rate tensor. It will be an ellipsoid if all e_i have the same sign, a hyperboloid if the e_i have different signs. In general, the principal axes of the strain rate tensor and the rates of deformation differ.

The vector $\boldsymbol{\omega}$; *the link between anti-symmetric tensors and vectors in three-dimensional space.* Consider the third term in (7.15), viz., $\boldsymbol{\omega} \times \boldsymbol{\rho}$. First of all, it will be shown that the quantity $\boldsymbol{\omega}$, introduced above in a Cartesian coordinate system, is a vector. In fact, Formula (7.15) represents itself a vector equation and, consequently, $\boldsymbol{\omega} \times \boldsymbol{\rho}$ is a vector; the scalar product $(\boldsymbol{\omega} \times \boldsymbol{\rho}) \cdot \boldsymbol{c}$, where $\boldsymbol{c}$ is an arbitrary vector, is invariant. Obviously, one may transpose the terms in the invariant quantity $(\boldsymbol{\omega} \times \boldsymbol{\rho}) \cdot \boldsymbol{c}$ and rewrite it in the form

$$\boldsymbol{\omega} \cdot (\boldsymbol{\rho} \times \boldsymbol{c}) = (\boldsymbol{\omega}, \boldsymbol{b}),$$

where $\boldsymbol{\rho}$, $\boldsymbol{c}$ and $\boldsymbol{\rho}\times\boldsymbol{c}=\boldsymbol{b}$ are arbitrary vectors. It follows now from the fact that the scalar product of $\boldsymbol{\omega}$ with an arbitrary vector $\boldsymbol{b}$ is invariant that $\boldsymbol{\omega}$ is a vector; using the general transformation rule and (7.11), the components ω_i may be written down in any coordinate system.

It follows from the above reasoning that the velocity field $\boldsymbol{v}$ may always be placed in correspondence with the tensor $e_{ij}\mathbf{Э}^i\mathbf{Э}^j$ and the vector $\boldsymbol{\omega}$.

The method of introduction of $\boldsymbol{\omega}$ leads to the general result that in three-dimensional space any second order anti-symmetric tensor

$$\Omega = \omega_{ik}\mathbf{Э}^i\mathbf{Э}^k$$

may always be linked to a vector $\boldsymbol{\omega}$ in such a way[1)] that in a Cartesian coordinate system the components of $\boldsymbol{\Omega}$ and $\boldsymbol{\omega}$ are interrelated by (7.10).

On the commutativity of infinitesimal affine transformations. As has been shown above, infinitesimal particles of a medium during continuous motion over an infinitesimal time dt undergo infinitesimal affine transformation which may now be represented in the form

$$\boldsymbol{\rho}' = \boldsymbol{\rho} + \operatorname{grad}\Phi\,\mathrm{d}t + (\boldsymbol{\omega}+\boldsymbol{\rho})\mathrm{d}t + \boldsymbol{\rho}O(\rho)\mathrm{d}t\,, \tag{7.19}$$

where the second and third terms have the orders of magnitude $\rho\,\mathrm{d}t$. Equation (7.19) thus may be rewritten

$$x'^i = (\delta^i{}_j + c^i{}_j)x^j = x^i + c^i{}_j x^j\,, \tag{7.20}$$

where x'^i and x^i are the components of $\boldsymbol{\rho}'$ and $\boldsymbol{\rho}$, respectively, and $c^i{}_j$ are of order $\mathrm{d}t$.

Recall that in the case of finite deformation an infinitesimal particle of the medium experiences the affine, but finite transformation with the matrix (5.65). Let there be given two successive affine transformations

$$x'^i = (\delta^i{}_j + a^i{}_j)x^{*j} \tag{a}$$

and

$$x^{*j} = (\delta^j{}_p + b^j{}_p)x^p\,, \tag{b}$$

and construct the resulting transformation

$$x'^i = (\delta^i{}_p + a^i{}_p + b^i{}_p + a^i_j b^j{}_p)x^p\,, \tag{7.21}$$

[1)] Note that ω (just as $\mathbf{b}$, equal to the vector product of the polar vectors) does not behave as an ordinary polar vector for all coordinate transformations. *Cf.*, pp. 202–207.

corresponding to (b) followed by (a). If one were to take (a) first and then (b), one would obtain

$$x'^i = (\delta^i{}_p + b^i{}_p + a^i{}_p + b^i{}_j a^j{}_p) x^p .$$

However, since, in general, $a^i{}_j b^j{}_p \neq b^i{}_j a^j{}_p$, it follows that the affine transformation in the general case is not commutative. On the other hand, if the affine transformation is infinitesimal, then the terms of the matrices $\|a^i{}_j b^j{}_p\|$ and $\|b^i{}_j a^j{}_p\|$ are small of second order; infinitesimal transformations are commutative exactly to the order of these terms.

Decomposition of the transformation of an infinitesimal particle of a medium during time Δt *into a sum of simple transformations.* Consider now (7.19) which describes the transformation of an infinitesimal particle of a medium. This transformation may be decomposed, without concern about the sequence of the factors, into two transformations, namely, one which determines the strain rate tensor:

$$\rho^* = \rho + \text{grad}\, \Phi \, dt \tag{7.22}$$

and another which determines the vector ω:

$$\rho' = \rho^* + (\omega \times \rho^*) dt . \tag{7.23}$$

The quadratic form $\Phi = \frac{1}{2} e_{ij} x^i x^j$ may be reduced to the canonical form

$$\Phi = \tfrac{1}{2}(e_1 x^2 + e_2 y^2 + e_3 z^2)$$

and the transformation (7.22), referred to principal axes, becomes

$$\left.\begin{aligned} x^* &= (1 + e_1 dt) x , \\ y^* &= (1 + e_2 dt) y , \\ z^* &= (1 + e_3 dt) z , \end{aligned}\right\} \tag{7.24}$$

where

$$e_1 \simeq \frac{x^* - x}{x dt}, \quad e_2 \simeq \frac{y^* - y}{y dt}, \quad e_3 \simeq \frac{z^* - z}{z dt}$$

are the principal rates of elongation ($e_i > 0$) or compression ($e_i < 0$). Obviously, the transformation (7.22) may be replaced by three transformations of the form

$$\left.\begin{aligned} x^{**} &= (1 + e_1 dt) x , \\ y^{**} &= y , \\ z^{**} &= z , \end{aligned}\right\} \tag{7.25}$$

each of which represents pure extension or compression of one of the principal axes.

Thus, any infinitesimal transformation of an infinitesimal particle of a continuum may be decomposed into four transformations, one of which, namely (7.23), is determined by the vector $\boldsymbol{\omega}$, while three of the form (7.25) represent pure elongations along three mutually perpendicular principal axes. In this context, in contrast to the case of finite deformation of infinitesimal particles of a medium, the order of the execution of these transformations is inessential.

Note that all the reasoning was applied with reference to the vector $\boldsymbol{\rho}$ from the centre of the particle O. However, on the basis of the properties of affine transformations, the change of the length of all parallel segments is the same, and therefore an arbitrary small vector (which does not start from O) experiences the same transformation. All vectors, parallel to the x-axis, elongate by $e_1 \mathrm{d}t$, parallel to the y-axis by $e_2 \mathrm{d}t$, and parallel to the z-axis by $e_3 \mathrm{d}t$. An arbitrary vector $\boldsymbol{\rho}$ elongates by $e_\rho \mathrm{d}t$ for every unit of length.

The vorticity vector and its kinematic interpretation. Next consider the kinematic interpretation of the vector $\boldsymbol{\omega}$. Take the transformation (7.23), caused by the vector $\boldsymbol{\omega}$, and construct the change of $\boldsymbol{\rho}$ due to this transformation:

$$\boldsymbol{\rho}' - \boldsymbol{\rho}^* = \mathrm{d}\boldsymbol{\rho}^* .$$

If one forms the scalar product $\boldsymbol{\rho}^* \cdot \mathrm{d}\boldsymbol{\rho}^*$, then, by (7.23), it vanishes, i.e., the change of the vector $\boldsymbol{\rho}$ is orthogonal to $\boldsymbol{\rho}^*$. Consequently, all $e_{\rho^*} = 0$. Thus, under the transformation (7.23), an infinitesimal particle of the medium behaves like an absolutely rigid body and $(\boldsymbol{\omega} \times \boldsymbol{\rho}^*)\mathrm{d}t$ may be interpreted as a rotational displacement with instantaneous angular velocity $\boldsymbol{\omega}$ of an infinitesimal particle of the continuum, instantaneously rigidified before or after the onset of deformation. Thus, the vector $\boldsymbol{\omega}$ must be interpreted as an instantaneous angular velocity of rotation of a body, linked to an infinitesimal particle of the medium, which over the time $\mathrm{d}t$ remains rigid, i.e., of the trihedron of the principal axes of the strain rate tensor. Thus, the vector $\boldsymbol{\omega}$, called the vorticity vector, is the instantaneous angular velocity of rotation of the principal axes of the strain rate tensor.

In the case of finite deformation of an infinitesimal particle of a medium, the motion likewise reduces to rotation and pure deformation. One may find the rotation, knowing the components of the matrix of the affine transformation $\|c^i{}_j\|$; however, this problem is complicated. In the case of motion of an infinitesimal particle of a continuum in time dt, when the corresponding transformation is an infinitesimal affine transformation, the rotation vector is given by $\boldsymbol{\omega}\,dt$.

The theorem of Cauchy–Helmholtz on the decomposition of the velocities of points of an infinitesimal particle of a medium. In conclusion, the results of all the preceding reasoning will be summarized and the theorem of Cauchy–Helmholtz on the decomposition of the velocities of an infinitesimal particle of a continuum formulated. The velocity $\boldsymbol{v}_1$ of any point O_1 of an infinitesimal particle of a continuum with centre at O is, by (7.15),

$$\boldsymbol{v}_1 = \boldsymbol{v}_0 + \boldsymbol{\omega} \times \boldsymbol{\rho} + \operatorname{grad} \Phi \tag{7.26}$$

and consists of the velocities of translation $\boldsymbol{v}_0$ and rotation $\boldsymbol{\omega} \times \boldsymbol{\rho}$ of the motion of the particle as an absolutely rigid body and of the velocity $\boldsymbol{v}^* = \operatorname{grad} \Phi$ of pure deformation:

$$\boldsymbol{v}_1 = \boldsymbol{v}_0 + \boldsymbol{v}_{\text{rot}} + \boldsymbol{v}^* . \tag{7.27}$$

On the divergence of the velocity. The concept of the divergence of the velocity vector $\boldsymbol{v}$ will now be introduced. Consider at time t in the medium an infinitesimal sphere

$$x^2 + y^2 + z^2 = R^2$$

formed from the points of a medium. During the time interval Δt, it becomes an ellipsoid the equation of which, referred to principal axes, has the form

$$\frac{x^{*2}}{(1+e_1\,\Delta t)^2} + \frac{y^{*2}}{(1+e_2\,\Delta t)^2} + \frac{z^{*2}}{(1+e_3\,\Delta t)^3} = R^2$$

(where the sphere invariably becomes an ellipsoid or, in a particular case, a sphere, since $e_i dt$ is small compared with unity).

Consider the change experienced by the volume of the infinitesimal sphere during the time interval dt. Obviously, at time t, it has the volume $V_0 = \frac{4}{3}\pi R^3$ and, at time $t + \Delta t$, the same particles of the medium will occupy the volume of the ellipsoid

$$V = \tfrac{4}{3}\pi R^3(1+e_1\Delta t)(1+e_2\Delta t)(1+e_3\Delta t).$$

Compute the limit of the relative change of the infinitesimal volume of the medium as $\Delta t \to 0$ and $V_0 \to 0$:

$$\lim_{\substack{\Delta t\to 0\\ V_0\to 0}} \frac{V-V_0}{V_0\Delta t} = e_1+e_2+e_3\,. \tag{7.28}$$

Obviously, the sum $e_1 + e_2 + e_3$ is invariant quantity; in fact, it is the first invariant of the strain rate tensor. It is well-known that this invariant may be written in terms of the components of the strain rate tensor in any arbitrary coordinate system in the form

$$e_1+e_2+e_3 = e^{\alpha}_{\alpha} = g^{\alpha\beta}e_{\alpha\beta}\,.$$

It follows from the definition (6.3) of the e_{ij} that

$$e^{\alpha}_{\alpha} = \nabla_{\alpha}v^{\alpha}\,.$$

By definition, the invariant quantity $\nabla_{\alpha} v^{\alpha}$ is called the divergence of the velocity vector with the notation

$$\operatorname{div}\boldsymbol{v} = \nabla_{\alpha}v^{\alpha}\,. \tag{7.29}$$

In Cartesian coordinates, it obviously has the form

$$\operatorname{div}\boldsymbol{v} = \frac{\partial u}{\partial x} + \frac{\partial v}{\partial y} + \frac{\partial w}{\partial z}\,.$$

Equality (7.28) is established for the relative change experienced by the volume V_0 of the infinitesimal sphere. From the general properties of the affine transformation stated above, it follows that this change is independent of the shape of the volume V_0 of the infinitesimal particle.

§ 8. Theorems of Stokes and Gauss-Ostrogradskii and certain related properties of vector fields

Curl and divergence of a vector. Let there be given a continuous field of some vector $\boldsymbol{A}$ and let the vector have first derivatives with respect to the coordinates. All the reasoning, applied to the vector field $\boldsymbol{v}$, may be applied to the field $\boldsymbol{A}$ to obtain

$$\boldsymbol{A}_1 = \boldsymbol{A} + \text{grad}\, \Psi + \tfrac{1}{2}\boldsymbol{\Omega} \times \boldsymbol{\rho} + \rho O(\rho),$$

where

$$\Psi = \tfrac{1}{2} a_{ij} x^i x^j, \quad a_{ij} = \tfrac{1}{2}(\nabla_i A_j + \nabla_j A_i)$$

and, in Cartesian coordinates,

$$\boldsymbol{\Omega} = \begin{vmatrix} \boldsymbol{i} & \boldsymbol{j} & \boldsymbol{k} \\ \dfrac{\partial}{\partial x} & \dfrac{\partial}{\partial y} & \dfrac{\partial}{\partial z} \\ A_1 & A_2 & A_3 \end{vmatrix}. \tag{8.1}$$

This symbolic determinant can also be written down when one takes instead of the vector components A_i any three differentiable functions P, Q, R of x, y, z (in a fixed coordinate system, any three numbers may be interpreted as vector components).

If A is a vector, then the vector Ω, introduced by definition by (8.1), is called the curl of A and denoted by

$$\boldsymbol{\Omega} = \text{curl}\, \boldsymbol{A}.$$

Analogously to (7.29), one may define the divergence of the vector $\boldsymbol{A}$ by

$$\text{div}\, \boldsymbol{A} = \nabla_\alpha A^\alpha$$

or, in a Cartesian system of coordinates, by

$$\text{div}\, \boldsymbol{A} = \frac{\partial A_1}{\partial x} + \frac{\partial A_2}{\partial y} + \frac{\partial A_3}{\partial z}.$$

Thus, one may write

$$\boldsymbol{\omega} = \tfrac{1}{2}\, \text{curl}\, \boldsymbol{v},$$

i.e., the vorticity vector is equal to half the curl of the velocity vector.

Circulation of a vector. Select in the region of definition of a vector field $\boldsymbol{A}$ some open line $\mathscr{L}$ or closed contour C. Construct the scalar product $\boldsymbol{A}\cdot d\boldsymbol{s}$, where $d\boldsymbol{s}$ is a directed element of $\mathscr{L}$ or C. Obviously, this scalar product is invariant. Form the integral

$$\int_{AB} (\boldsymbol{A}\cdot d\boldsymbol{s}) = \Gamma \, .$$

The scalar Γ, introduced in this manner, is referred to as the circulation of the vector $\boldsymbol{A}$ along $\mathscr{L}$. The direction of motion along the contour must

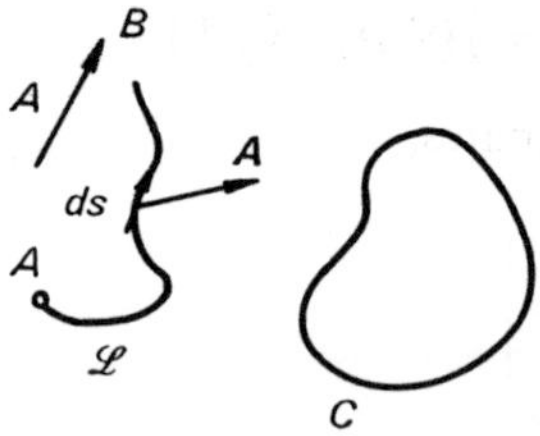

Fig. 16. For the definition of circulation.

be stated. The circulation depends, in the general case, on the line $\mathscr{L}$ along which it is calculated. Obviously,

$$\Gamma_{AB} = -\Gamma_{BA} \, .$$

If the vector $\boldsymbol{A}$ is the velocity $\boldsymbol{v}$ of a point of a continuum, then

$$\Gamma = \int_{AB} (\boldsymbol{v}\cdot d\boldsymbol{s}) = \int_{AB} (u\,dx + v\,dy + w\,dz)$$

is called the circulation of the velocity.

Let the velocity vector $\boldsymbol{v}$ have a potential

$$\boldsymbol{v} = \operatorname{grad} \varphi \; ;$$

then

$$\Gamma = \int_{AB} (\boldsymbol{v}\cdot d\boldsymbol{s}) = \int_{AB} \frac{\partial \varphi}{\partial s} \, ds = \varphi_B - \varphi_A \, .$$

Hence it is seen that in the case of potential flow the circulation of the velocity depends on the coordinates of the points A and B; however, the value of Γ does not depend on the shape of the contour $\mathscr{L}$, if the potential

φ is a single-valued function of the coordinates. For example, for $\varphi = Q/4\pi r$, where Q is a constant, the circulation Γ does not depend on $\mathscr{L}$; it follows from this result that the circulation along a closed contour C vanishes: $\Gamma_C = 0$. However, if, for example,

$$\varphi = k\theta = k \arctan \frac{y}{x} \tag{8.2}$$

(k is a constant), then

$$\varphi_B - \varphi_A = k(\theta_B - \theta_A),$$

and it follows from this formula that there exists a closed contour C, surrounding the origin of coordinates, for which the circulation differs from zero (fig. 17).

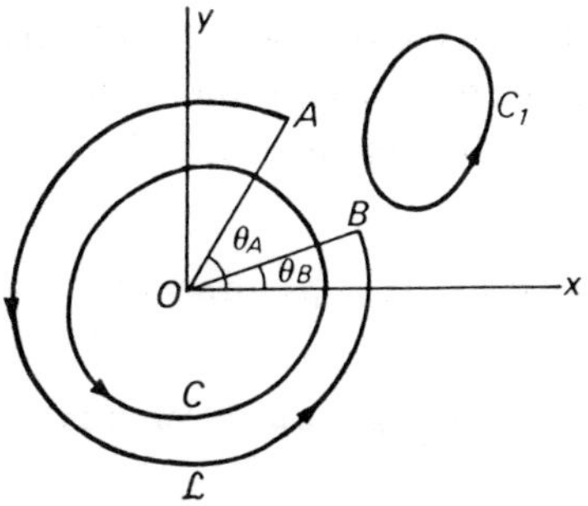

Fig. 17. Circulation in the case (8.2): $\Gamma_C = 2\pi k$, $\Gamma_{C_1} = 0$, $\Gamma_C = 2\pi kn$ for n circuits around O.

Stokes' theorem. Next, let the velocity $\boldsymbol{v}$ be without potential. Consider a closed contour C and assume that one may stretch over it a surface Σ on which the field of $\boldsymbol{v}$ is continuous and differentiable, i.e., assume that the contour C may be shrunk into a point, remaining in the region of continuity and differentiability of $\boldsymbol{v}$. Subdividing the surface Σ by the contours C_k in the way shown in fig. 18, one will have

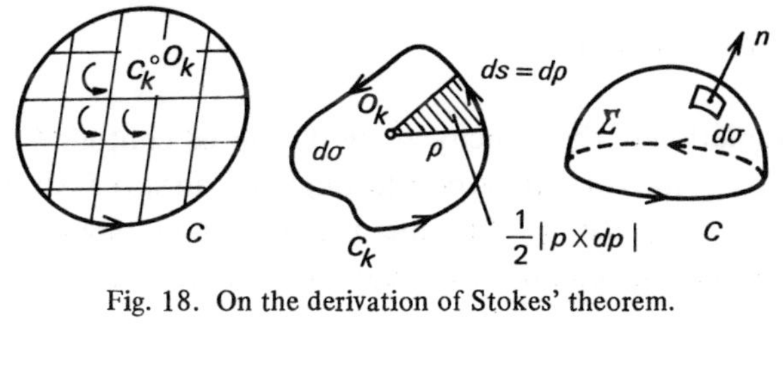

Fig. 18. On the derivation of Stokes' theorem.

$$\Gamma = \int_C (\boldsymbol{v}\cdot \mathrm{d}\boldsymbol{s}) = \sum_k \int_{C_k} \boldsymbol{v}\cdot \mathrm{d}\boldsymbol{s}\,. \tag{8.3}$$

This equality is obvious, since the integrals, taken along common sides of the contours C_k, cancel on summation (fig. 18).

The contours C_k may be taken as small as one pleases, and for the evaluation of $\Gamma_{C_k} = \int_{C_k} \boldsymbol{v}\cdot \mathrm{d}\boldsymbol{s}$ one may assume that the velocity $\boldsymbol{v}$ on C_k is determined by the Cauchy–Helmholtz theorem on the velocity distribution of points of infinitesimal particles of a continuum with centre at some point O_k on the surface Σ inside the contour,

$$\boldsymbol{v}_{C_k} = \boldsymbol{v}_{O_k} + \boldsymbol{\omega}\times\boldsymbol{\rho} + \operatorname{grad}\Phi + \rho O(\rho)\,. \tag{8.4}$$

In the calculation of the Γ_{C_k}, the terms with $\boldsymbol{v}_{O_k}$ and grad Φ render no contribution, since potential vectors and their potentials are single-valued; the contribution from the term $\rho O(\rho)$ will be small of higher order than the term $\int_{C_k}[(\boldsymbol{\omega}\times\boldsymbol{\rho})\cdot \mathrm{d}\boldsymbol{s}]$. It is readily verified (fig. 18) that

$$\begin{aligned}\int_{C_k} (\boldsymbol{\omega}\times\boldsymbol{\rho})\cdot \mathrm{d}\boldsymbol{s} &= \int_{C_k} (\boldsymbol{\omega}\times\boldsymbol{\rho})\cdot \mathrm{d}\boldsymbol{\rho} = \int_{C_k} \boldsymbol{\omega}\cdot(\boldsymbol{\rho}\times \mathrm{d}\boldsymbol{\rho})\\ &= \boldsymbol{\omega}\cdot\int_{C_k} \boldsymbol{\rho}\times \mathrm{d}\boldsymbol{\rho} = 2\boldsymbol{\omega}\cdot\boldsymbol{n}\,\mathrm{d}\sigma = 2\omega_n \mathrm{d}\sigma\,,\end{aligned} \tag{8.5}$$

since the vector $\boldsymbol{\omega}$, as the contour C_k shrinks, remains constant, because it depends only on O_k. The magnitude of the integral $\int_{C_k} \boldsymbol{\rho}\times \mathrm{d}\boldsymbol{\rho}$ is $2\mathrm{d}\sigma$ and it is directed along the normal $\boldsymbol{n}$ to $\mathrm{d}\sigma$ towards the side with which $\boldsymbol{\rho}$ turned to $\mathrm{d}\boldsymbol{\rho}$ is counter-clockwise; the region of the surface Σ, stretched over the infinitesimal contour C_k, may always be assumed to be plane with $\boldsymbol{n}$ the unit normal vector.

Now, by (8.3) and (8.5), one obtains in the limit $k\to\infty$ and with C_k shrinking into a point Stokes' theorem

$$\int_C v_s \mathrm{d}s = 2\int_\Sigma \omega_n \mathrm{d}\sigma\,, \tag{8.6}$$

i.e., the circulation of the velocity along the closed contour C is equal to twice the flux of the vorticity vector through the surface Σ, stretched over this contour. Note that in (8.6) the direction of the normal $\boldsymbol{n}$ must be chosen such that the circuit of the contour C is counter-clockwise with respect to it.

Obviously, Stokes' Theorem is not only true for the velocity vector $\boldsymbol{v}$ of a continuum, but also for any other vector $\boldsymbol{A}=A_i\boldsymbol{3}^i$, satisfying the necessary conditions of continuity and differentiability. Stokes' theorem for the vector $\boldsymbol{A}$ will now be written down in different forms:

$$\int_C A_s \mathrm{d}s = \int_C A_i \mathrm{d}x^i = \int_\Sigma (\mathrm{curl}\, \boldsymbol{A})_n \mathrm{d}\sigma$$

$$= \int_\Sigma \left[\left(\frac{\partial A_3}{\partial y} - \frac{\partial A_2}{\partial z} \right) \cos(\boldsymbol{n}, x) + \left(\frac{\partial A_1}{\partial z} - \frac{\partial A_3}{\partial x} \right) \cos(\boldsymbol{n}, y) + \left(\frac{\partial A_2}{\partial x} - \frac{\partial A_1}{\partial y} \right) \cos(\boldsymbol{n}, z) \right] \mathrm{d}\sigma .$$

Potential and irrotational motions. A motion of a continuum is said to be irrotational in some region, if $\boldsymbol{\omega}=0$ at all points of this region, and rotational, if $\boldsymbol{\omega}\neq 0$ there. In the case of irrotational flow, a filament lying along the principal axes of the strain rate tensor at a given time preserves during an infinitesimal time interval its direction in space.

It is readily verified formally that, if $\boldsymbol{v} = \mathrm{grad}\, \varphi$, then $\boldsymbol{\omega}=0$, and, consequently, that the circulation around any closed contour, satisfying the conditions of Stokes' theorem, vanishes. Thus, if a motion is potential, it is irrotational.

Conversely, one may show that, if a flow is irrotational, i.e., $\boldsymbol{\omega}=0$, then it has a potential, i.e., there exists a function φ such that $\boldsymbol{v}=\mathrm{grad}\, \varphi$.

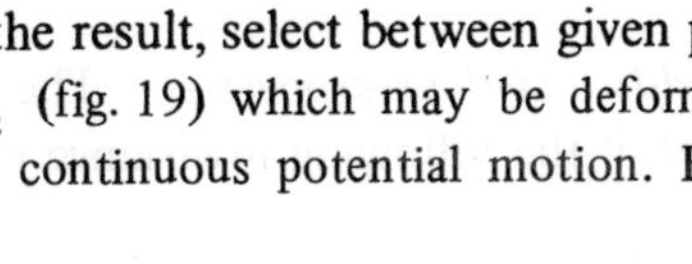

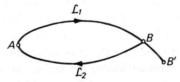

Fig. 19. On the equivalence of irrotational and potential flows.

In order to prove the result, select between given points A and B two contours $\mathscr{L}_1$ and $\mathscr{L}_2$ (fig. 19) which may be deformed into each other within the region of continuous potential motion. By Stokes' theorem, one has

$$\int_{\mathscr{L}_1+\mathscr{L}_2} (u\mathrm{d}x + v\mathrm{d}y + w\mathrm{d}z) = 0 \,,$$

whence

$$\Gamma_{AB}=\int_{\mathscr{L}_1}(u\,dx+v\,dy+w\,dz)=\int_{-\mathscr{L}_2}(u\,dx+v\,dy+w\,dz)\,.$$

Since the contours $\mathscr{L}_1$ and $\mathscr{L}_2$ are arbitrary, it follows that

$$\int_{AB}(u\,dx+v\,dy+w\,dz)=\varphi(x,y,z)\,,$$

i.e., the circulation between the points A and B does not depend on the path of integration; it depends only on the coordinates of the end point B, if the starting point A is fixed. The increase of Γ for any infinitesimal segment BB' will obviously be given by

$$u\,dx+v\,dy+w\,dz=d\varphi\,.$$

Consequently, since dx, dy and dz are arbitrary, one has

$$u=\frac{\partial\varphi}{\partial x}\,,\qquad v=\frac{\partial\varphi}{\partial y}\,,\qquad w=\frac{\partial\varphi}{\partial z}\,.$$

Thus, the concepts of potential and irrotational flows are equivalent.

Multi-valuedness of a potential and multiply connected flow regions. As is known, a region is called simply connected, if any closed contour in it can be shrunk into a point without leaving the region; otherwise it is said to be multiply connected. Obviously, if a region of continuous potential motion is simply connected, the potential φ is a single-valued function of the coordinates; if this region is multiply connected, φ may be a multi-valued function of the coordinates. In a multiply connected region, the circulation Γ along a contour, which cannot be shrunk into a point, may be non-zero and have the same value along contours which may intersect each other without leaving the region. In the example (8.2), the region of continuity of the potential flow $\varphi=k\theta$ is not simply connected, since the z-axis is a singular line.

Solenoidal fields and their properties. The field of a vector $\boldsymbol{B}$ is said to be solenoidal, if the following invariance exists:

$$\operatorname{div}\boldsymbol{B}=\nabla_\alpha B^\alpha=0\,.$$

Using the definitions of $\operatorname{div}\boldsymbol{B}$ and $\operatorname{curl}\boldsymbol{A}$, one verifies readily directly that the field of rotation of any vector $\boldsymbol{A}$ is always solenoidal, i.e., if

$$\boldsymbol{B} = \operatorname{curl} \boldsymbol{A}\,, \quad \text{then} \quad \operatorname{div} \boldsymbol{B} = 0\,.$$

In particular,

$$\boldsymbol{\omega} = \tfrac{1}{2} \operatorname{curl} \boldsymbol{v}\;;$$

hence, for the motion of any continuum, the vortex field satisfies the equation

$$\operatorname{div} \boldsymbol{\omega} = 0$$

or, in Cartesian coordinates,

$$\frac{\partial \omega_1}{\partial x} + \frac{\partial \omega_2}{\partial y} + \frac{\partial \omega_3}{\partial z} = 0\,.$$

Thus, the vortex field is always solenoidal.

If a medium is incompressible, i.e., if its volume does not change during motion, then, by (7.28).

$$\operatorname{div} \boldsymbol{v} = 0\,,$$

i.e., the velocity field of an incompressible medium is solenoidal.

As is known from physics, in vacuum the magnetic field vector $\boldsymbol{H}$ likewise is always solenoidal:

$$\operatorname{div} \boldsymbol{H} = 0\,.$$

Any solenoidal vector $\boldsymbol{B}$ may be represented in the form

$$\boldsymbol{B} = \operatorname{curl} \boldsymbol{A}\,. \tag{8.7}$$

In fact, the particular representation of the vector $\boldsymbol{B}$ by means of (8.7) in terms of a vector $\boldsymbol{A}_1$ can be constructed in the following manner. Select a Cartesian coordinate system and set $A_{1z}=0$. Then the equality $\boldsymbol{B}=\operatorname{curl} A_1$ leads to the following system of equations for the determination of A_{1x} and A_{1y}:

$$-\frac{1}{2}\frac{\partial A_{1y}}{\partial z} = B_x\,,$$

$$\frac{1}{2}\frac{\partial A_{1x}}{\partial z} = B_y\,, \tag{8.7'}$$

$$\frac{1}{2}\left(\frac{\partial A_{1y}}{\partial x} - \frac{\partial A_{1x}}{\partial y}\right) = B_z\,.$$

Under the condition div $\boldsymbol{B}=0$, this system will be satisfied if one sets

$$A_{1y} = -2\int_{z_0}^{z} B_x \mathrm{d}z + 2\int_{x_0}^{x} B_z(x, y, z_0)\mathrm{d}x ,$$

$$A_{1x} = 2\int_{z_0}^{z} B_y \mathrm{d}z .$$

Actually, it is seen directly that the first two equations (8.7′) are then satisfied. The third equation (8.7′) is also fulfilled, since, by the strength of the relation

$$\frac{\partial B_x}{\partial x} + \frac{\partial B_y}{\partial y} = -\frac{\partial B_z}{\partial z} ,$$

one obtains

$$\frac{1}{2}\left(\frac{\partial A_{1y}}{\partial x} - \frac{\partial A_{1x}}{\partial y}\right) = -\int_{z_0}^{z}\left(\frac{\partial B_x}{\partial x} + \frac{\partial B_y}{\partial y}\right)\mathrm{d}z + B_z(x, y, z_0)$$

$$= \int_{z_0}^{z} \frac{\partial B_z}{\partial z}\mathrm{d}z + B_z(x, y, z_0) = B_z(x, y, z) .$$

Obviously, all vectors $\boldsymbol{A}$ which satisfy (8.7) may be represented in the form

$$\boldsymbol{A} = \boldsymbol{A}_1 + \operatorname{grad} \Psi ,$$

where Ψ is an arbitrary scalar function. In fact, the difference $\boldsymbol{A} - \boldsymbol{A}_1$ must satisfy the equation

$$\operatorname{curl}(\boldsymbol{A} - \boldsymbol{A}_1) = \operatorname{curl} \boldsymbol{A} - \operatorname{curl} \boldsymbol{A}_1 = 0 ,$$

i.e., it must be possible to represent it in the form of the gradient of some function Ψ.

As an example of a vorticity field $\boldsymbol{\omega}$, consider the general properties of a solenoidal field. As for every vector field, one may introduce for a vorticity field (*cf.*, Sec. 3) the concepts of vector lines, surfaces and tubes, i.e., the concepts of vortex lines, surfaces and tubes. Vortex lines are lines the tangents to every point of which coincide with the directions of the vorticity vector $\boldsymbol{\omega}$. The differential equations of the vortex lines have the form

$$\frac{\mathrm{d}x}{\omega_1} = \frac{\mathrm{d}y}{\omega_2} = \frac{\mathrm{d}z}{\omega_3} . \tag{8.8}$$

A vortex surface $f(x, y, z) = \text{const.}$ consists entirely of vortex lines and its equation has the form

$$\omega_1 \frac{\partial f}{\partial x} + \omega_2 \frac{\partial f}{\partial y} + \omega_3 \frac{\partial f}{\partial z} = 0 \,. \tag{8.9}$$

A vortex tube is formed, if through all points of a closed contour C (which is not a vortex line) there passes a vortex line. The side surface of a vortex tube is a vortex surface and on it $\omega_n = 0$.

Consider the properties of vortex tubes. On the side surface of a vortex tube, select two contours C_1 and C_2, as shown in fig. 20. Join these

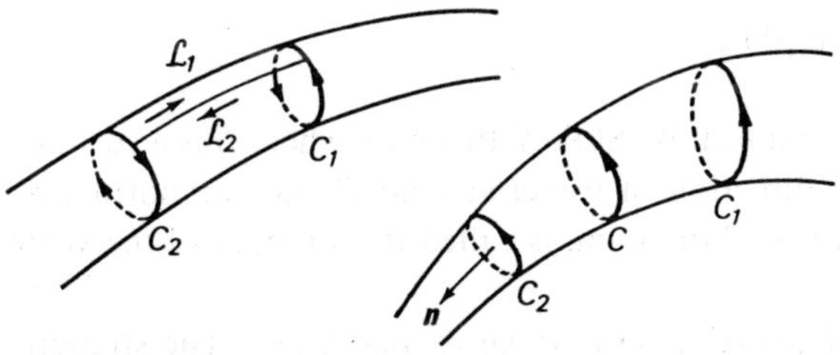

Fig. 20. On properties of vortex tubes.

contours by the cuts $\mathscr{L}_1$ and $\mathscr{L}_2$. The surface Σ which is formed in this way lies entirely on the side surface of the vortex tube; apply Stokes' theorem to obtain

$$\int_{\text{along boundary of } \Sigma} v \cdot ds = 0 \,.$$

The direction of the circuit of the boundary of the surface Σ is shown in fig. 20; the two sides $\mathscr{L}_1$ and $\mathscr{L}_1$ of the cut are covered during the integration in different directions; therefore the corresponding integrals cancel each other. The contours C_1 and C_2 likewise are travelled in opposite directions and, consequently, replacing the circuit of one of them by the other, one finds

$$\int_{C_1} (\boldsymbol{v} \cdot d\boldsymbol{s}) = \int_{C_2} (\boldsymbol{v} \cdot d\boldsymbol{s}) \,,$$

or

$$\Gamma_{C_1} = \Gamma_{C_2} \,.$$

The contours C_1 and C_2 may obviously be arbitrary in this context, as long as they encircle the given vortex tube once. Thus

$$\Gamma_C = \text{const.},$$

where C is an arbitrary contour enveloping the given vortex tube once.

The circulation

$$\Gamma_C = \int_C \boldsymbol{v} \cdot d\boldsymbol{s}$$

or the quantity

$$2\int_\Sigma \omega_n d\sigma,$$

which is equal to it by Stokes' theorem, where Σ is a surface bounded by C and the direction of travel around C and the normal $\boldsymbol{n}$ to Σ are as stated in Stokes' theorem, is called the strength of the vortex tube.

Helmholtz' kinematic vortex theorems. The strength of a vortex tube is the same along the tube and is a characteristic of a given tube. This statement is referred to as Helmholtz' first kinematic vortex theorem.

Helmholtz' second kinematic theorem says that vortex tubes may not begin or end inside a medium. This result follows directly from the condition of continuity of the field $\boldsymbol{\omega}$ and conservation of the strength of vortex tubes. Thus, vortex tubes either may be closed or may end and begin on boundaries of moving media or, if a medium is unbounded, they may extend to infinity.

Examples of vortex flows. Intuition suggests that the flow of a liquid is always rotational, i.e., $\boldsymbol{\omega} \neq 0$, if it contains closed flow lines. In fact, if one observes in a flow a velocity distribution which is similar to that shown in fig. 21a, then the circulation along the stream line drawn is non-zero:

$$\Gamma_C = \int_C v_s ds \neq 0,$$

since the integrand during the entire integration does not change sign; by Stokes' theorem, there must exist on the surface stretched over the contour C points with $\omega_n \neq 0$; consequently, the flow is rotational. But this result is only true for conditions under which Stokes' theorem is applicable, i.e., when one may stretch over C a surface Σ on which $\boldsymbol{v}$ together with its partial derivatives is continuous. For example, it is impossible to arrive at the result that a flow in the presence of the velocity distribution of fig. 21a is rotational, if one has inside C a rigid cylindrical body with generators parallel to the z-axis (fig. 21b). It is also impossible to arrive at the result that the fields $\boldsymbol{v}$ or $\boldsymbol{\omega}$ have singularities inside C.

In this context, consider in greater detail the example of the flow

$$\varphi = k\theta = k \arctan \frac{y}{x} \; . \tag{8.10}$$

This flow is potential, i.e., $\boldsymbol{v} = \operatorname{grad}\varphi$, and its flow lines are orthogonal

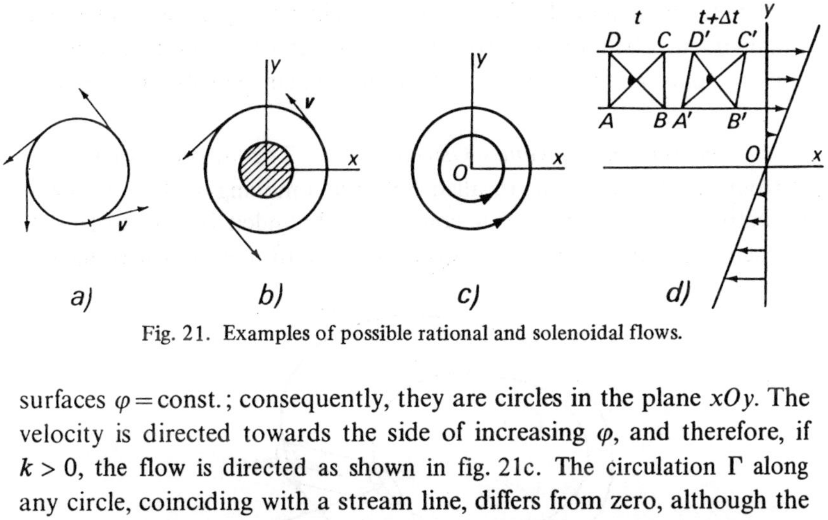

Fig. 21. Examples of possible rational and solenoidal flows.

surfaces $\varphi = \text{const.}$; consequently, they are circles in the plane xOy. The velocity is directed towards the side of increasing φ, and therefore, if $k > 0$, the flow is directed as shown in fig. 21c. The circulation Γ along any circle, coinciding with a stream line, differs from zero, although the flow is potential everywhere except at the origin of coordinates where the potential is not defined. If one computes the vorticity vector for this flow, it is seen that it vanishes everywhere, except along the z-axis; on the z-axis, the quantity $\boldsymbol{\omega}$ must become infinite. Thus, along the z-axis, the fields $\boldsymbol{v}$ and $\boldsymbol{\omega}$ have singularities. Along the z-axis, one has an isolated vortex thread of finite intensity $\Gamma = 2\pi k$. This flow is referred to as the flow of an isolated vortex filament.

It must not be thought that vortex flows are, as a rule, connected with the presence of closed stream lines. Consider the flow (fig. 21d)

$$u = ay\,, \quad v = w = 0\,,$$

where a is a positive constant. The trajectories, coinciding with the stream lines in this case, are straight, parallel to the x-axis. The velocity distribution along any straight line $x = \text{const.}$ is linear. Direct evaluation of the components of $\boldsymbol{\omega}$ in Cartesian coordinates yields

$$\omega_1 = \omega_2 = 0\,, \quad \omega_3 = -\frac{a}{2}\,,$$

i.e., the flow is rotational, $\boldsymbol{\omega}$ is directed along the negative z-axis and does not change from point to point. Obviously, an infinitesimal fluid particle, taken at time t in the form of a square $ABCD$, becomes at time $t+\Delta t$ a rhombus $A'B'C'D'$. It may be shown that the principal axes of the strain rate tensor at time t coincide with the diagonals of the square and at time $t+\Delta t$ with the diagonals of the rhombus. The principal axes, obviously, remain during the motion perpendicular to each other, but their orientations in space change. They rotate with angular velocity $\boldsymbol{\omega} = -\frac{1}{2}a\boldsymbol{k}$.

The theorem of Gauss–Ostrogradskii. The theorem of Gauss–Ostrogradskii will now be recalled. Select in a moving medium at time t an individual volume V of a continuum, bounded by the surface Σ. At each point of the surface Σ, erect the outward normal $\boldsymbol{n}$ with respect

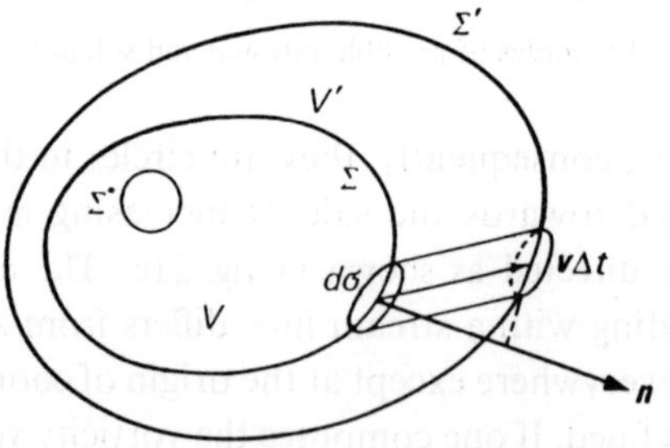

Fig. 22. On the theorem of Gauss-Ostrogradskii.

to Σ, At time $t + \Delta t$, the volume V becomes the volume V', the surface Σ becomes the surface Σ' (fig. 22), bounding V'. The change of volume $V' - V$ is, obviously, given by

$$V' - V = \int_\Sigma v_n \Delta t \, d\sigma\,.$$

An increase of V' with respect to V is taken into account in this formula, since the normal is always outward with respect to V. The rate of change of the volume is given by

$$\lim_{\Delta t\to 0}\frac{V'-V}{\Delta t}=\int_{\Sigma} v_n \mathrm{d}\sigma\ .$$

Analogously, for an infinitesimal volume V^* bounded by the surface Σ^*, one has

$$\lim_{\Delta t\to 0}\frac{V'^*-V^*}{\Delta t}=\int_{\Sigma^*} v_n \mathrm{d}\sigma\ .$$

Using the definition of the divergence of a velocity vector (7.28) and recalling its mechanical interpretation, one obtains in Cartesian coordinates

$$\int_{\Sigma^*} v_n \mathrm{d}\sigma=\int_{\Sigma^*}[u\cos(\boldsymbol{n},x)+v\cos(\boldsymbol{n},y)+w\cos(\boldsymbol{n},z)]\mathrm{d}\sigma$$

$$=V^*\operatorname{div}\boldsymbol{v}+V^*\varepsilon=\left(\frac{\partial u}{\partial x}+\frac{\partial v}{\partial y}+\frac{\partial w}{\partial z}\right)V^*+V^*\varepsilon\ , \qquad (8.11)$$

where ε is an infinitesimal quantity.

A finite volume V may always be decomposed into infinitesimal volumes V^*, and for each of these one may write down (8.11), if v is continuous and differentiable inside V. Summing (8.11) over all subvolumes V^* and taking the limit as the number of volumes V^* tends to infinity and $V^*\to 0$, one finds

$$\int_{\Sigma}[u\cos(\boldsymbol{n},x)+v\cos(\boldsymbol{n},y)+w\cos(\boldsymbol{n},z)]\mathrm{d}\sigma$$

$$=\int_{V}\left(\frac{\partial u}{\partial x}+\frac{\partial v}{\partial y}+\frac{\partial w}{\partial z}\right)\mathrm{d}\tau\ , \qquad (8.12)$$

since on the left-hand side the integrals, taken over continuous surfaces Σ^*, as a result of the fact that the normals to them point in opposite directions, cancel, and in the limit there remains only the integral over the external surface Σ.

The equality (8.12) represents the Gauss–Ostrogradskii theorem on the transformation of an integral, taken over a closed surface Σ, into an integral taken through the volume V bounded by Σ. It may be rewritten in a form which does not depend on the choice of coordinate systems:

$$\int_{\Sigma} v_n \mathrm{d}\sigma = \int_V \operatorname{div} \boldsymbol{v} \mathrm{d}\tau . \tag{8.13}$$

Obviously, any continuous vector A which has continuous first derivatives inside V and on the surface Σ may be treated as a velocity $\boldsymbol{v}$; one may obtain for it the Gauss–Ostrogradskii formula

$$\int_{\Sigma} \boldsymbol{A}\cdot\boldsymbol{n}\, \mathrm{d}\sigma = \int_V \operatorname{div} \boldsymbol{A}\, \mathrm{d}\tau .$$

Moreover, since in a given coordinate system any three quantities P, Q, R may be treated as components of a vector, the theorem of Gauss–Ostrogradskii may be written down for any three continuous and differentiable functions P, Q, R of x, y, z in the form

$$\int_{\Sigma} [P\cos(\boldsymbol{n}, x) + Q\cos(\boldsymbol{n}, y) + R\cos(\boldsymbol{n}, z)]\, \mathrm{d}\sigma$$

$$= \int_V \left(\frac{\partial P}{\partial x} + \frac{\partial Q}{\partial y} + \frac{\partial R}{\partial z}\right) \mathrm{d}\tau .$$

In the Gauss–Ostrogradskii formula, one has under the integrals on the right-hand and left-hand sides invariants which do not depend on the choice of the coordinate system. If they are known in Cartesian coordinates, one may easily compute them in any other system. In fact, let in a system η^1, η^2, η^3

$$\boldsymbol{A} = A^k \boldsymbol{\mathfrak{z}}_k , \qquad \boldsymbol{n} = n_i \boldsymbol{\mathfrak{z}}^i ;$$

then

$$\boldsymbol{A}\cdot\boldsymbol{n} = A^k n_k$$

and

$$\operatorname{div} \boldsymbol{A} = \nabla_k A^k = \frac{\partial A^k}{\partial \eta^k} + A^i \Gamma^k_{ki} ,$$

where the Christoffel symbols Γ^i_{kj} are evaluated, using the formulae above, in terms of g_{ij} in the space η^1, η^2, η^3 (the g_{ij} may be computed from the formulae for the transition from a Cartesian coordinate system into the given system η^1, η^2, η^3).

Now, the theorem of Gauss–Ostrogradskii may be rewritten in the form

$$\int_{\Sigma} A^k n_k \mathrm{d}\sigma = \int_V \nabla_k A^k \mathrm{d}\tau \,, \tag{8.14}$$

which is true in an arbitrary curvilinear coordinate system. Note that the number of dimensions of the space in the derivation of the theorem may be arbitrary. In mechanics and physics, this theorem is often applied in two, three or four dimensions.

The formula for the differentiation with respect to time of an integral taken over a moving volume. Finally, a formula from vector analysis will be derived which will be useful later on. Let there be given an arbitrary function (which may be a tensor) which depends on space coordinates and the time t. Consider the integral

$$\int_V f(x, y, z, t)\; \mathrm{d}\tau$$

over the moving volume V. Compute the derivative

$$\frac{\mathrm{d}}{\mathrm{d}t}\int_V f(x, y, z, t)\mathrm{d}\tau \,,$$

where not only the integrand, but also the region of integration V depends on t. By definition, the derivative (*cf.*, fig. 22) may be written in the form

$$\frac{\mathrm{d}}{\mathrm{d}t}\int_{V(t)} f(x, y, z, t)\mathrm{d}\tau$$

$$= \lim_{\Delta t \to 0} \frac{\int_{V'} f(x, y, z, t+\Delta t)\mathrm{d}\tau - \int_V f(x, y, z, t)\mathrm{d}\tau}{\Delta t}$$

$$= \lim_{\Delta t \to 0} \frac{\int_V [f(x, y, z, t+\Delta t) - f(x, y, z, t)]\,\mathrm{d}\tau + \int_{V'-V} f(x, y, z, t+\Delta t)\mathrm{d}\tau}{\Delta t}$$

$$= \int_V \frac{\partial f(x, y, z, t)}{\partial t}\,\mathrm{d}\tau + \int_{\Sigma} f v_n \mathrm{d}\sigma \;; \tag{8.15}$$

since the volume $V' - V$ consists of elementary cylinders (fig. 22),

$$\mathrm{d}\tau = v_n \mathrm{d}\sigma\, \Delta t$$

and, for $\Delta t \to 0$, the surface Σ' becomes the surface Σ and

$$f(x, y, z, t+\Delta t) \to f(x, y, z, t) .$$

Applying to the last integral of (8.15) the formula of Gauss–Ostrogradskii, one finds

$$\frac{\mathrm{d}}{\mathrm{d}t}\int_V f(x, y, z, t)\mathrm{d}\tau = \int_V \left[\frac{\partial f}{\partial t} + \nabla_i(fv^i)\right] \mathrm{d}\tau . \tag{8.16}$$

The region of integration V moves, and it is natural that the result of the differentiation should depend on the velocity field $\boldsymbol{v}$ with which the points of the volume V move.

Obviously, one has always the kinematic relation

$$\frac{\partial f}{\partial t} + \nabla_i(fv^i) = \frac{\partial f}{\partial t} + v^i\nabla_i f + f\nabla_i v^i = \frac{\mathrm{d}f}{\mathrm{d}t} + f\nabla_i v^i , \tag{8.17}$$

where

$$\frac{\mathrm{d}f}{\mathrm{d}t} = \frac{\partial f}{\partial t} + v^i\nabla_i f \tag{8.18}$$

is the expression for the total derivative of f with respect to the time t in an arbitrary coordinate system.

Therefore (8.16) assumes the form

$$\frac{\mathrm{d}}{\mathrm{d}t}\int_V f(x, y, z, t)\mathrm{d}\tau = \int_V \left[\frac{\mathrm{d}f}{\mathrm{d}t} + f\nabla_i v^i\right]\mathrm{d}\tau . \tag{8.19}$$

Formula (8.16) will now be applied to a particular case.

Let

$$f = \frac{1}{V} ,$$

where V is the volume of a continuum. Obviously, in this case, the function f depends on the variable volume V, namely the region of integration of (8.15), i.e., it depends only on t and not on the coordinates. Clearly, one has always the kinematic identity

$$\int_V \frac{\mathrm{d}\tau}{V(t)} = 1$$

and, by (8.16),

$$\frac{\mathrm{d}}{\mathrm{d}t}\int_V \frac{\mathrm{d}r}{V(t)} \int_V \left[\frac{\partial \frac{1}{V}}{\partial t} + \nabla_i\left(\frac{1}{V} v^i\right)\right] \mathrm{d}\tau = 0 \tag{8.20}$$

or

$$\int_V \left[\frac{\mathrm{d}\frac{1}{V}}{\mathrm{d}t} + \frac{1}{V}\operatorname{div} \boldsymbol{v}\right] \mathrm{d}\tau = 0\,. \tag{8.21}$$

This identity may be written down for the entire volume V of a moving medium as well as for any part of it.

Applying (8.21) to an infinitesimal volume $\mathrm{d}V$, one obtains

$$\frac{\mathrm{d}\left(\frac{1}{\Delta V}\right)}{\mathrm{d}t} + \frac{1}{\Delta V}\operatorname{div} \boldsymbol{v} = 0\,, \tag{8.22}$$

where div $\boldsymbol{v}$ is taken at the point into which $\mathrm{d}V$ shrinks. Note that this equality is true for any medium and is not at all connected with properties of the moving medium. In particular, it is true also for non-material media, for example, for phase space.

Chapter III

Dynamic concepts and dynamic equations of continuum mechanics

§1. Equation of continuity

A beginning will now be made with the study of motions of physical objects, i.e., of material bodies and fields. Here, and in some later sections, the basic concern will be the laws of motion of material bodies only. Material bodies are those which possess the property of inertia. The property of inertia is characterized by the mass of a body. One may introduce the mass m of a body as a whole, as well as the mass m_i of any of its parts. By definition, in Newtonian mechanics the mass m of an entire body is equal to the sum of the masses m_i of all its component parts.

Law of conservation of mass; density; equation of continuity in Eulerian variables. A fundamental law of Newtonian mechanics is the law of conservation of the mass of any volume element, i.e., of a volume consisting of one and the same particles of a medium. This law may be regarded as an experimentally established natural law which is true to a definite approximation.

One of the fundamental equations of continuum mechanics states that for any volume element

$$m = \text{const.}$$

This equation may be written in the alternative form, namely

$$\frac{dm}{dt} = 0. \tag{1.1}$$

Introduce the mean density

$$\rho_{\text{mean}} = \frac{\Delta m}{\Delta V},$$

where ΔV is the volume occupied by the mass Δm; the true density is defined by the limit

$$\rho = \lim_{\Delta V \to 0} \frac{\Delta m}{\Delta V}.$$

In continuum mechanics, one considers almost always instead of the mass m the density ρ. For a small volume, one has

$$\Delta m \simeq \rho \Delta V ;$$

for a finite volume,

$$m = \int_V \rho \, \mathrm{d}\tau ,$$

where the integral is taken over a moving individual volume. Thus, if ρ is known, the mass m may be computed.

The density ρ of individual particles may or may not be conserved, since the volume of a particle may change during a motion.

Obviously, the law of conservation of mass for an individual volume of a continuous medium may now be written in the form

$$\frac{\mathrm{d}}{\mathrm{d}t} \int_V \rho \, \mathrm{d}\tau = 0 . \tag{1.2}$$

Applying the rule of differentiation (8.16) for an integral, taken over a moving volume, under the conditions of the law of conservation of mass, one obtains

$$0 = \frac{\mathrm{d}m}{\mathrm{d}t} = \int_V \left(\frac{\partial \rho}{\partial t} + \operatorname{div} \rho \boldsymbol{v} \right) \mathrm{d}\tau = \int_V \left(\frac{\mathrm{d}\rho}{\mathrm{d}t} + \rho \operatorname{div} \boldsymbol{v} \right) \mathrm{d}\tau$$

or, since this equality holds for any given volume, the first fundamental differential equation of continuum mechanics assumes the form

$$\frac{\mathrm{d}\rho}{\mathrm{d}t} + \rho \operatorname{div} \boldsymbol{v} = 0 , \tag{1.3}$$

which is referred to as the equation of continuity in Eulerian variables.

Conditions on quantities which conserve their value in an individual volume. Obviously, this equation may be derived directly from (8.22), since the mass of any individual volume is conserved.

Besides the mass m, there are other physical characteristics which remain constant during the motion of any given volume of a continuous medium. For example, let N be the number of molecules or atoms in an arbitrary individual volume. This quantity N is constant in the individual volume. Introducing the number of molecules or atoms per unit volume $n = \lim_{V\to 0} N/V$, one can obtain, on the basis of the assumed constancy of N, from (8.22) a differential equation for n which is analogous to (1.3):

$$\frac{dn}{dt} + n \operatorname{div} \boldsymbol{v} = 0 . \tag{1.4}$$

If chemical reactions occur in a continuous medium, then (1.3) applies but not (1.4).

There exist also other scalar, vector or tensor quantities which conserve their values within any individual volume element. Such conserved quantities will be denoted by Φ; their density will be defined by

$$f = \lim_{\Delta V \to 0} \frac{\Delta \Phi}{\Delta V} .$$

Clearly, the quantities Φ and f satisfy the conditions

$$\frac{d\Phi}{dt} = 0 , \quad \Phi = \int_V f d\tau , \quad \frac{df}{dt} + f \operatorname{div} \boldsymbol{v} = 0 .$$

In physics, in many cases, the density of the electric charge e satisfies this last condition. One of the basic problems of physics is concerned with the establishment of the characteristics which conserve their quantities in a volume element.

Continuity equations for multi-component mixtures. Consider a mixture with n components, for example, a mixture of hydrogen, oxygen and water vapour ($N = 3$), an alloy of tin and copper, a saline solution, a plasma, i.e., a mixture of free electrons and ions, etc. Any such multi-component mixture may be regarded as a set of n continua filling one and the same volume occupied by the mixture. For each of these continua, introduce their densities and velocities and denote them by $\rho_1, \rho_2, \ldots, \rho_n$ and $\boldsymbol{v}_1, \boldsymbol{v}_2, \ldots, \boldsymbol{v}_n$. There will be N densities ρ_i and N velocities $\boldsymbol{v}_i$ at each point of the volume, occupied by the mixture, each referring to its continuum.

Thus, in this case, the mechanics of a mixture becomes the mechanics of continuum filling one and the same volume.

Consider first the case when no chemical reactions or ionizations occur in the mixture. Here, the law of conservation of mass should be satisfied for each of the N components, and one obtains the N equations

$$\frac{dm_i}{dt} = 0$$

or

$$\frac{\partial \rho_i}{\partial t} + \operatorname{div} \rho_i \boldsymbol{v}_i = 0 . \tag{1.5}$$

However, if chemical reactions or ionizations occur (a case of interest from the point of view of applications), then the masses m_i of the components may change. Let $\varkappa_i$ be the change of the mass m_i of the i-th component per unit time in a unit volume, as a result of chemical reactions or ionizations. The quantities $\varkappa_i$ are defined in chemistry. Then the continuity equations for the components of the mixture may be rewritten in the form

$$\frac{dm_i}{dt} = \int_V \varkappa_i d\tau$$

or

$$\frac{\partial \rho_i}{\partial t} + \operatorname{div} \rho_i \boldsymbol{v}_i = \varkappa_i . \tag{1.6}$$

A basic law of chemical reactions consists of the fact that the total mass of a mixture remains constant, and hence

$$\sum_{i=1}^{n} \varkappa_i = 0 . \tag{1.7}$$

In addition to the N densities and N velocities of the components of a mixture, one may introduce the single density ρ and single velocity $\boldsymbol{v}$ of the mixture as a whole. By definition, the mass m of a given volume of a mixture is equal to the sum of the masses of the components of that volume:

$$m = \sum_{i=1}^{N} m_i ,$$

and the density ρ of a mixture is defined by

$$\lim_{\Delta V \to 0} \frac{\Delta m}{\Delta V}.$$

The density of a component of a mixture is

$$\rho_i = \lim_{\Delta V \to 0} \frac{\Delta m_i}{\Delta V},$$

and therefore,

$$\rho = \sum_{i=1}^{N} \rho_i .$$

Summing (1.5) and using (1.7), one may write

$$\frac{\partial \rho}{\partial t} + \operatorname{div} \sum_{i=1}^{N} \rho_i \boldsymbol{v}_i = 0 .$$

This equation will have the usual form of the continuity equation (1.3), if the velocity $\boldsymbol{v}$ of the mixture as a whole is defined in the following manner:

$$\rho \boldsymbol{v} = \sum_{i=1}^{N} \rho_i \boldsymbol{v}_i , \qquad m\boldsymbol{v} = \sum_{i=1}^{N} m_i \boldsymbol{v}_i ,$$

i.e.,

$$\boldsymbol{v} = \frac{\sum_{i=1}^{N} m_i \boldsymbol{v}_i}{m} = \frac{\sum_{i=1}^{N} \rho_i \boldsymbol{v}_i}{\rho} . \tag{1.8}$$

Note that the defined velocity $\boldsymbol{v}$ is itself the velocity of the common centre of mass of the N individual volumes, corresponding to the N components of the mixture.

Continuity equation in the case of processes with diffusion. It may happen that all the components of a mixture move with the same velocity, which in this case coincides with the velocity of the mixture as a whole:

$$\boldsymbol{v}_1 = \boldsymbol{v}_2 = \dots = \boldsymbol{v}_N = \boldsymbol{v} .$$

This type of process is said to be a process without diffusion.

If the velocities $\boldsymbol{v}_i$ of the components are different, one is dealing with diffusion; in this case, components of a mixture move with respect to each other. An electric current is an example of such a process. If an electric current passes through a stationary conductor, one has $\boldsymbol{v}=0$ and $\boldsymbol{v}_i \neq 0$, because the motion of the electrons and ions in the conductor forms the electric current.

In processes with diffusion, the continuity equations (1.5) or (1.6) may be modified by substituting the velocity of the mixture as a whole into the continuity equation of each component. In general, in the presence of chemical interactions and diffusion, (1.6) may be written in the form

$$\frac{\partial \rho_i}{\partial t} + \operatorname{div} \rho_i \boldsymbol{v} = \varkappa_i - \operatorname{div} \boldsymbol{I}_i \,, \tag{1.9}$$

where

$$\boldsymbol{I}_i = \rho_i(\boldsymbol{v}_i - \boldsymbol{v}) \,.$$

The difference $\boldsymbol{v}_i - \boldsymbol{v}$ is clearly the velocity of the i-th component relative to the medium as a whole. The term div $\boldsymbol{I}_i$ in (1.9) characterizes the change in mass of the i-th component in a volume moving with velocity v, even though, if $\boldsymbol{v}_i \neq \boldsymbol{v}$, this volume is not an individual volume element for the i-th component.

Particles, comprising the i-th component, enter the volume and leave it. The vectors $\boldsymbol{I}_i$ are called diffusion flow vectors.

In order to evaluate the diffusion flow vectors $\boldsymbol{I}_i$, one must apply laws of physics. Although the laws of diffusion may differ, in particular cases, it follows from (1.8) that in any case

$$\sum_{i=1}^{N} \boldsymbol{I}_i = 0 \,.$$

Instead of N equations of continuity (1.9) for the components of a mixture, one may use $N-1$ independent equations (1.9) for the components of the mixture together with the equation of continuity of the mixture as a whole

$$\frac{\partial \rho}{\partial t} + \operatorname{div} \rho \boldsymbol{v} = 0 \,.$$

Thus, when studying the motion of a multi-component mixture, one does not need to introduce explicitly N continua, filling the same volume and moving with different velocities v_i. Instead, one may consider only the vectors of diffusion flow I_i, and regard Equations (1.9) as equations for the densities ρ_i of the components of the mixture.

It is clear that in the study of the motion of a reacting multi-component mixture, one must combine laws of mechanics with laws of physics and chemistry for the quantities $\varkappa_i$ and $\boldsymbol{I}_i$.

Equation of continuity and the property of flow tubes in the case of incompressible media. The equation of continuity (1.3) has been derived for arbitrary continuous media. A medium is said to be incompressible, if the magnitude of any of its individual volumes remains constant during the time of motion. Hence, the density in a particle of an incompressible medium remains constant. The equation of continuity assumes the form

$$\operatorname{div} \boldsymbol{v} = 0 \,. \tag{1.10}$$

A medium is said to be homogeneous, if the density ρ is the same for all particles of the medium, i.e., if ρ does not depend on the space co-ordinates x, y, z. It is non-homogeneous, if ρ differs for different particles of the medium: $\rho = \rho(x, y, z)$. Obviously, *the equation of continuity* (1.10) *is valid both for homogeneous and for non-homogeneous incompressible media.*

The velocity field of an incompressible fluid is always solenoidal and, consequently, its vector tubes, i.e., its flows tubes, have the properties, studied in Sec. 8, Chapter II. For example, the strength of the flow tube

$$\int_{\Sigma} v_n \mathrm{d}\sigma = Q \,,$$

(where Σ is the cross-section of the tube and $\boldsymbol{n}$ its normal) which is called the transport of the tube, remains constant along the flow tube.

In the case of continuous motion, a flow tube cannot begin and end inside a volume of an incompressible medium.

Equation of continuity in Lagrangian variables. Next, the continuity equation will be derived in a different form, namely in terms of Lagrangian variables. For this purpose, construct at an arbitrary point M of a continuum at a given time t the elementary infinitesimal oblique parallelepiped with edges $\mathfrak{z}^1 \, \mathrm{d}\xi^1$, $\mathfrak{z}^2 \, \mathrm{d}\xi^2$, $\mathfrak{z}^3 \, \mathrm{d}\xi^3$, directed along the axes of the concomitant system ξ^1, ξ^2, ξ^3. Its volume will be

$$V = \left| \mathfrak{z}_1 \cdot (\mathfrak{z}_2 \times \mathfrak{z}_3) \mathrm{d}\xi^1 \mathrm{d}\xi^2 \mathrm{d}\xi^3 \right| .$$

At another arbitrary time t_0, this parallelepiped corresponds to an elementary oblique parallelepiped, constructed on the vectors $\mathring{\mathfrak{z}}^1 \, \mathrm{d}\xi^1$, $\mathring{\mathfrak{z}}^2 \, \mathrm{d}\xi^2$, $\mathring{\mathfrak{z}}^3 \, \mathrm{d}\xi^3$ taken at the same point M. The volume of this parallelepiped is

$$V_0 = \left|\mathring{\mathbf{э}}_1 \cdot (\mathring{\mathbf{э}}_2 \times \mathring{\mathbf{э}}_3) \mathrm{d}\xi^1 \mathrm{d}\xi^2 \mathrm{d}\xi^3\right| .$$

Denote the densities of the medium at times t and t_0 by ρ and ρ_0, respectively. By the law of conservation of mass, one has

$$\rho_0 V_0 = \rho V ,$$

or

$$\rho = \rho_0 \frac{V_0}{V} = \rho_0 \left| \frac{\mathring{\mathbf{э}}_1 \cdot (\mathring{\mathbf{э}}_2 \times \mathring{\mathbf{э}}_3)}{\hat{\mathbf{э}}_1 \cdot (\hat{\mathbf{э}}_2 \times \hat{\mathbf{э}}_3)} \right| . \tag{1.11}$$

In order to evaluate the triple scalar products of the base vectors, introduce a Cartesian rectilinear reference system x^1, x^2, x^3 with the base vectors $\mathbf{э}_1 = \boldsymbol{i}$, $\mathbf{э}_2 = \boldsymbol{j}$ and $\mathbf{э}_3 = \boldsymbol{k}$, with respect to which the medium is moving. In this system, denote the coordinates of a point of the medium at time t by x^1, x^2, x^3 and at time t_0 by x_0^1, x_0^2, x_0^3. Obviously,

$$x_0^i = x^i(\xi^1, \xi^2, \xi^3, t_0), \quad x^i = x^i(\xi^1, \xi^2, \xi^3, t),$$

i.e., x_0^i and x^i are values of the functions given by the law of motion, taken for various values of the independent variable t. Since in this system the radius vector of the point M is

$$\boldsymbol{r} = x^k \mathbf{э}_k \quad \text{and} \quad \hat{\mathbf{э}}_i = \frac{\partial \boldsymbol{r}}{\partial \xi^i},$$

then

$$\hat{\mathbf{э}}_i = \frac{\partial x^k}{\partial \xi^i} \mathbf{э}_k ,$$

and the triple scalar product $\hat{\mathbf{э}}_1 \cdot (\hat{\mathbf{э}}_2 \times \hat{\mathbf{э}}_3)$ may be written in the form of the determinant

$$\hat{\mathbf{э}}_1 \cdot (\hat{\mathbf{э}}_2 \times \hat{\mathbf{э}}_3) = \begin{vmatrix} \dfrac{\partial x^1}{\partial \xi^1} & \dfrac{\partial x^2}{\partial \xi^1} & \dfrac{\partial x^3}{\partial \xi^1} \\[2ex] \dfrac{\partial x^1}{\partial \xi^2} & \dfrac{\partial x^2}{\partial \xi^2} & \dfrac{\partial x^3}{\partial \xi^2} \\[2ex] \dfrac{\partial x^1}{\partial \xi^3} & \dfrac{\partial x^2}{\partial \xi^3} & \dfrac{\partial x^3}{\partial \xi^3} \end{vmatrix} = \hat{\Delta} ,$$

where $\hat{\Delta}$ is the Jacobian of the transformation from the variables ξ^1, ξ^2, ξ^3 to the variables x^1, x^2, x^3. Analogously, one has

$$\mathring{\mathfrak{z}}_1 \cdot (\mathring{\mathfrak{z}}_2 \times \mathring{\mathfrak{z}}_3) = \begin{vmatrix} \dfrac{\partial x_0^1}{\partial \xi^1} & \dfrac{\partial x_0^2}{\partial \xi^1} & \dfrac{\partial x_0^3}{\partial \xi^1} \\ \dfrac{\partial x_0^1}{\partial \xi^2} & \dfrac{\partial x_0^2}{\partial \xi^2} & \dfrac{\partial x_0^3}{\partial \xi^2} \\ \dfrac{\partial x_0^1}{\partial \xi^3} & \dfrac{\partial x_0^2}{\partial \xi^3} & \dfrac{\partial x_0^3}{\partial \xi^3} \end{vmatrix} = \mathring{\Delta},$$

where $\mathring{\Delta}$ is the Jacobian of the transformation from the variables ξ^1, ξ^2, ξ^3 to the variables x_0^1, x_0^2, x_0^3. Using a property of Jacobians, Equation (1.11) may now be expressed in the form

$$\rho = \rho_0 \frac{\mathring{\Delta}}{\hat{\Delta}} = \rho_0 \operatorname{Det} \left\| \frac{\partial x_0^i}{\partial x^k} \right\|. \tag{1.12}$$

Equation (1.12) will still be transformed further. For the sake of clarity, denote the components $\partial x^i/\partial \xi^j$ of the vectors $\hat{\mathfrak{z}}_j$ in the x, y, z system by $\hat{\mathfrak{z}}_{jx}, \hat{\mathfrak{z}}_{jy}, \hat{\mathfrak{z}}_{jz}$. Then, obviously,

$$[\hat{\mathfrak{z}}_1 \cdot (\hat{\mathfrak{z}}_2 \times \hat{\mathfrak{z}}_3)]^2 = \hat{\Delta}^2 = \begin{vmatrix} \hat{\mathfrak{z}}_{1x} & \hat{\mathfrak{z}}_{1y} & \hat{\mathfrak{z}}_{1z} \\ \hat{\mathfrak{z}}_{2x} & \hat{\mathfrak{z}}_{2y} & \hat{\mathfrak{z}}_{2z} \\ \hat{\mathfrak{z}}_{3x} & \hat{\mathfrak{z}}_{3y} & \hat{\mathfrak{z}}_{3z} \end{vmatrix}^2 =$$

$$= \begin{vmatrix} \hat{\mathfrak{z}}_{1x} & \hat{\mathfrak{z}}_{1y} & \hat{\mathfrak{z}}_{1z} \\ \hat{\mathfrak{z}}_{2x} & \hat{\mathfrak{z}}_{2y} & \hat{\mathfrak{z}}_{2z} \\ \hat{\mathfrak{z}}_{3x} & \hat{\mathfrak{z}}_{3y} & \hat{\mathfrak{z}}_{3z} \end{vmatrix} \times \begin{vmatrix} \hat{\mathfrak{z}}_{1x} & \hat{\mathfrak{z}}_{2x} & \hat{\mathfrak{z}}_{3x} \\ \hat{\mathfrak{z}}_{1y} & \hat{\mathfrak{z}}_{2y} & \hat{\mathfrak{z}}_{3y} \\ \hat{\mathfrak{z}}_{1z} & \hat{\mathfrak{z}}_{2z} & \hat{\mathfrak{z}}_{3z} \end{vmatrix} = \operatorname{Det} \|\hat{g}_{ik}\| = \hat{g},$$

because $\hat{g}_{ik} = \hat{\mathfrak{z}}_i \cdot \hat{\mathfrak{z}}_k$.

Analogously, one finds

$$[\mathring{\mathfrak{z}}_1 \cdot (\mathring{\mathfrak{z}}_2 \times \mathring{\mathfrak{z}}_3)]^2 = \operatorname{Det} \|\mathring{g}_{ik}\| = \mathring{g}$$

and, consequently, (1.11) may be expressed in the form

$$\rho = \rho_0 (\mathring{g}/\hat{g})^{\frac{1}{2}}. \tag{1.13}$$

Equations (1.11), (1.12) and (1.13) are different forms of the equation of continuity in Lagrangian variables.

Note that, in general, the density f of any quantity Φ, which conserves its value in a given volume of a continuous medium, satisfies the equation

$$f = f_0 (\mathring{g}/\hat{g})^{\frac{1}{2}} = f_0 \Delta \, , \tag{1.14}$$

where Δ is the determinant of the matrix of the transformation from the variables x^i to the variables x_0^i.

The equation of continuity has an extremely general character. It is satisfied for motions of any material medium, independently of the properties of that medium. It is the same for all media: water, air, metal, etc. In the equation of continuity in the case of a compressible medium (1.3), there appear four unknown functions: the density ρ and three velocity components; for an incompressible medium (1.10), there are only three unknown functions: the velocity components. Clearly, the single equation of continuity is not sufficient for the solution of problems of continuum mechanics. Next, further equations will be derived which are satisfied for motions of any continuous medium.

§2. The equations of motion of a continuous medium

Examine the motion of a material continuum in relation to the causes of that motion. For this purpose, introduce forces into the consideration. Forces are vector quantities.

A basic classification of the forces which occur in continuum mechanics will now be given. The concept of force in continuum mechanics is more complicated than in the mechanics of a single point mass or of a rigid unchanging system.

Concentrated and distributed forces. Fundamentally, theoretical mechanics deals with concentrated forces, i.e., with finite forces acting at a point. In continuum mechanics, one encounters, basically, distributed forces, i.e., forces acting on every part of a volume V or on every element of a surface Σ of a continuous medium, where, when an infinitesimal element of volume or surface tends to zero, the principal vector of the forces acting on it tends also to zero.

Only in exceptional cases does one find point forces in continuum mechanics. By Newton's second law,

$$\mathbf{F} = \Delta m \boldsymbol{a} ,$$

where Δm is the mass of an infinitesimal element of the continuous medium, and $\boldsymbol{a}$ is its acceleration; it is apparent that concentrated forces may exist only at points where $\boldsymbol{a}$ (or ρ) become infinite.

Volume or mass forces. Forces, distributed over a volume V, are called volume or mass forces. Denote by $\mathbf{F}$ the principal vector of the mass forces, acting on an element of mass Δm. Then the density $\boldsymbol{F}$ of the mass force at a given point is

$$\boldsymbol{F} = \lim_{\Delta m \to 0} \frac{\mathbf{F}}{\Delta m} .$$

For a small particle,

$$\mathbf{F} \simeq \boldsymbol{F} \Delta m .$$

Sometimes the force $\boldsymbol{\Phi}$ per unit volume is considered rather than that per unit mass. It is obvious that

$$\boldsymbol{\Phi} = \lim_{\Delta V \to 0} \frac{\mathbf{F}}{\Delta V} ,$$

i.e.,

$$\boldsymbol{\Phi} = \rho \boldsymbol{F} .$$

The dimensions of $\boldsymbol{\Phi}\, \mathrm{d}V$ and $\boldsymbol{F}\, \mathrm{d}m$ are force, that of $\boldsymbol{F}$ is acceleration, and that of $\boldsymbol{\Phi}$ acceleration times density.

The number of different types of body forces is small. They comprise the force of gravity (weight), $\boldsymbol{F} = \boldsymbol{g}$, $\boldsymbol{\Phi} = \rho \boldsymbol{g}$, and general gravitational forces, satisfying Newton's law of universal attraction; electro-magnetic forces; inertial forces, which one must introduce when studying motion in non-inertial frames of reference and which are perfectly ordinary real external mass forces from the point of view of a self-propelling body. Occasionally, in the study of actual motions of continuous media, mass forces are introduced artificially. For example, when studying the motion of a wing in a fluid, it may be assumed that the region occupied by the wing is also filled with fluid. However, in order that the artificially introduced fluid continues to move like the wing, one must subject it to distributed mass forces.

In the mechanics of a rigid body, the effect of any system of forces is equivalent to the effect of its resultant vector and moment. In the mechanics of deformable media, the character of the distributions of the forces over a body is essential.

Surface forces. In continuum mechanics, surface forces, i.e., forces which are distributed over the surface of a continuous medium play a more basic role than mass forces. For instance, consider water in a vessel; obviously, there will occur an interaction of forces over the surface S of contact between the water and the walls of the container. Select an element $d\sigma$ of the surface S and introduce the surface force $d\boldsymbol{P}=\boldsymbol{p}\,d\sigma$, where $\boldsymbol{p}=\lim_{\Delta\sigma\to 0}\Delta\boldsymbol{P}/\Delta\sigma$ is the density of the surface forces acting on $d\sigma$. The density $\boldsymbol{p}$ of the surface force may be introduced at each point of the surface S. In general, it will vary from point to point.

Internal and external forces. One may distinguish between internal and external forces. Forces are said to be internal, if they are caused by objects belonging to the system the motion of which is under consideration; they are called external, if they are caused by objects which are outside with respect to the system under study.

The concept of external and internal forces is relative. For example, when dealing with the motion of the air in the atmosphere and Earth together, the force of the weight of the air alone is an internal force, however, when dealing with the motion of the air only, the force of the weight is external. If one looks at the motion of a material body and an electromagnetic field, then electromagnetic forces are internal; however, if one deals with the motion of a material body by itself, the field is an external agent and the electromagnetic forces are external.

The forces of internal stress. Imagine an arbitrary volume V, separated from a continuous medium, and cut it into two parts V_1 and V_2,

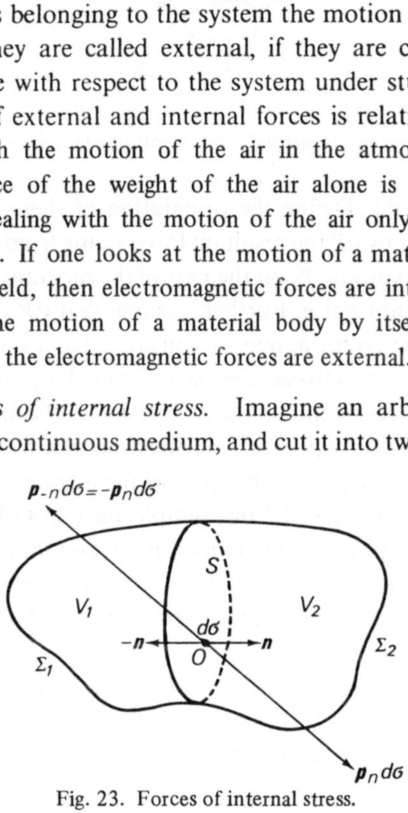

Fig. 23. Forces of internal stress.

by a cross-section S (fig. 23). If one considers the motion of one part of V, say of V_1, one must replace the effect of V_2 on it by mass forces distributed over V_1 and surface forces spread over S. The forces introduced in this way are external with respect to V_1. On the other hand, if one considers the motion of the volume V as a whole, then these forces will be internal. The cross-section S could have been placed elsewhere. Clearly, the surface forces acting on S will be different for different cross-sections S.

Select some point M inside a body and consider at this point different

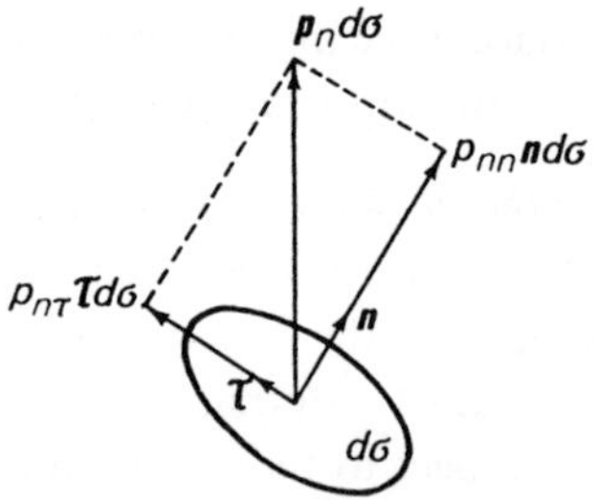

Fig. 24. Normal and tangential components of the force of internal stress.

elements of area $d\sigma$. Define the orientation of these elements by their normals $\boldsymbol{n}$; denote by $d\boldsymbol{P}$ the resultant force acting from the side of the part of the medium in volume V_2 on the part of the medium in volume V_1 over the area $d\sigma$ with normal $\boldsymbol{n}$. Further, assume that $d\boldsymbol{P}=\boldsymbol{p}_n d\sigma$, where $\boldsymbol{p}_n$ is a finite vector. The vector $\boldsymbol{p}_n$ may be considered as a surface density of the forces of interaction between the parts divided by the area $d\sigma$. In the general case, $\boldsymbol{p}_n$ may depend on the orientation of the area element $d\sigma$ and its other geometrical properties. The direction of $\boldsymbol{n}$ will always be chosen to be external with respect to that part of the medium which is subjected to the force $\boldsymbol{p}_n d\sigma$ introduced. For example, the effect of the volume V_2 on V_1 will be replaced by distributed forces $\boldsymbol{p}_n d\sigma$, and the effect of V_1 on V_2 by distributed forces $\boldsymbol{p}_{-n} d\sigma$ (fig. 23). Such surface forces may be introduced at any point of a continuous medium; they are called the forces of internal stress.

The force of internal stress $\boldsymbol{p}_n d\sigma$ at any point of a continuous medium may be decomposed into two components, one along the normal $\boldsymbol{n}$ and one along the tangent $\boldsymbol{\tau}$ to the area $d\sigma$ (fig. 24).

$$\boldsymbol{p}_n \, \mathrm{d}\sigma = p_{nn} \boldsymbol{n} \, \mathrm{d}\sigma + p_{n\tau} \boldsymbol{\tau} \, \mathrm{d}\sigma \, ,$$

where $p_{nn} \, \mathrm{d}\sigma$ is the normal component of the force of internal stress and $p_{n\tau} \, \mathrm{d}\sigma$ is its tangential component, which is also called the tangential force or, in the case of fluids, the internal friction force.

Clearly, the surface forces $\boldsymbol{p}_n \, \mathrm{d}\sigma$ may also be external forces, i.e., forces which act on the outer surface bounding a continuous medium.

At each point M of a continuous medium, there exist infinitely many vectors $\boldsymbol{p}_n$, corresponding to the infinitely many choices of the areas $\mathrm{d}\sigma$, passing through M. However, among them there exists a general relationship which does not depend on the particular properties of a moving medium. This relationship will be established below.

The basic dynamic equation of motion of a point mass is Newton's second law:

$$\mathbf{F} = m\boldsymbol{a} \, .$$

The more complex law for the motion of a continuous material body will now be formulated; it is a direct generalization of Newton's second law.

Equation of momentum for a material point and a system of points. Consider the motion of a point mass m relative to a system of coordinates x, y, z. Since the mass m is constant, one has

$$m \frac{\mathrm{d}\boldsymbol{v}}{\mathrm{d}t} = \frac{\mathrm{d}m\boldsymbol{v}}{\mathrm{d}t} = \mathbf{F} \, . \tag{2.1}$$

The product of mass and velocity $m\boldsymbol{v}$ is called the momentum of the point, and one obtains the equation of momentum for the point mass: the derivative with respect to time of the momentum of a point mass is equal to the sum of all forces acting on that point mass.

Using the fundamental equation of momentum (2.1), one can solve two typical problems: given the forces, find the law of motion of a point, or, given the law of motion of a point, find the forces acting on it.

If one has a system of n material points, each with mass m_i, each moving with velocity $\boldsymbol{v}_i$, as the result of the action of the force $\mathbf{F}_i$, then one may write down the equation of momentum (2.1) for each point

$$\frac{\mathrm{d}m_i \boldsymbol{v}_i}{\mathrm{d}t} = \mathbf{F}_i \, ,$$

where $\mathbf{F}_i$ comprises all forces acting on the i-th point, both internal and external with respect to the system of points. Summing the momentum equation (2.1) over all n points, one obtains

$$\sum_{i=1}^{n} \frac{\mathrm{d}m_i \boldsymbol{v}_i}{\mathrm{d}t} = \frac{\mathrm{d}}{\mathrm{d}t}\left(\sum_{i=1}^{n} m_i \boldsymbol{v}_i\right) = \sum_{i=1}^{n} \mathbf{F}_i^{(e)},$$

where the sum on the right-hand side consists only of forces external to the system of n points, since, by Newton's third law, the internal forces of interaction occur in pairs and therefore cancel in the sum.

The sum

$$Q = \sum_{i=1}^{n} m_i \boldsymbol{v}_i = m\boldsymbol{v}^*,$$

where $m = \Sigma_{i=1}^{n} m_i$ is the total mass of the system and $\boldsymbol{v}^*$, the velocity of the centre of mass of the system, is given by

$$\boldsymbol{v}^* = \frac{\sum_{i=1}^{n} m_i \boldsymbol{v}_i}{m},$$

is called the momentum of the system. Thus, one arrives at the equation of momentum for a system of n point masses

$$\frac{\mathrm{d}Q}{\mathrm{d}t} = \sum_{i=1}^{n} \mathbf{F}_i^{(e)} \quad \text{or} \quad m\frac{\mathrm{d}\boldsymbol{v}^*}{\mathrm{d}t} = \sum_{i=1}^{n} \mathbf{F}_i^{(e)}.$$

The time derivative of the momentum of a system of point masses is equal to the sum of all external forces acting on it. In other words, the mass multiplied by the acceleration of the centre of mass of the system is equal to the sum of all external forces acting on the system.

Thus, the motion of a system of mass points is equivalent to the motion of a single point mass, namely of the centre of mass of the system. For distant observers, the motion of a very small system of points reduces, in many problems, to the motion of the centre of mass of the system.

The equation of motion of a point mass has universal significance. It may be applied to all possible mechanical systems—to galaxies, stars, planets, to any flying devices, to people, birds, insects, etc.

Equation of momentum for a finite volume of a continuous medium. The equation of momentum will now be generalized to the case of a finite individual volume V of a continuum, bounded by a surface Σ. Since

$$\int_V \frac{\mathrm{d}}{\mathrm{d}t}(\boldsymbol{v}\rho\,\mathrm{d}\tau) = \int_M \frac{\mathrm{d}\boldsymbol{v}}{\mathrm{d}t}\,\mathrm{d}m = \frac{\mathrm{d}}{\mathrm{d}t}\int_M \boldsymbol{v}\,\mathrm{d}m = \frac{\mathrm{d}}{\mathrm{d}t}\int_V \boldsymbol{v}\rho\,\mathrm{d}\tau\,,$$

this equation will be written in the form

$$\frac{\mathrm{d}\boldsymbol{Q}}{\mathrm{d}t} = \int_V \boldsymbol{F}\rho\,\mathrm{d}\tau + \int_\Sigma \boldsymbol{p}_n\,\mathrm{d}\sigma\,,$$

where

$$\boldsymbol{Q} = \int_V \boldsymbol{v}\rho\,\mathrm{d}\tau$$

is, by definition, the momentum of the continuous medium, occupying the volume V, and

$$\int_V \boldsymbol{F}\rho\,\mathrm{d}\tau \quad \text{and} \quad \int_\Sigma \boldsymbol{p}_n\,\mathrm{d}\sigma$$

are the sums of the external mass forces and the surface forces, respectively, acting on the medium in the volume V.

Thus, for any individual volume V of a continuous medium, the equation of momentum with respect to the invariant system of reference may be written in the form

$$\frac{\mathrm{d}}{\mathrm{d}t}\int_V \boldsymbol{v}\rho\,\mathrm{d}\tau = \int_V \boldsymbol{F}\rho\,\mathrm{d}\tau + \int_\Sigma \boldsymbol{p}_n\,\mathrm{d}\sigma\,, \tag{2.2}$$

i.e., the time derivative of the momentum of a volume V of a continuous medium is equal to the sum of all external mass and surface forces acting on it. The mentally selected volume V may be any substantial, moving, deformable volume consisting, by definition, of the same particles of the medium.

If, in addition to external distributed forces, there act on the mass in the volume V forces concentrated at a point or along some line or along some surface inside V then their sum must be added to the right-hand side of (2.2).

Equation (2.2) is a basic postulated dynamic relationship of continuum mechanics. Just as Newton's second law is the starting equation in point mechanics, Equation (2.2) is the foundation of continuum mechanics. All the preceding reasoning must be understood as leading heuristic arguments, interrelating Equation (2.2) for continuous media and Newton's equation for point masses.

The following reasoning may also be used as basis for Equation (2.2). One may introduce the velocity $\boldsymbol{v}^*$ of the centre of mass of a volume V of a continuous medium by means of the formula

$$m\boldsymbol{v}^* = \int_V \boldsymbol{v}\rho \, \mathrm{d}t$$

and formulate the momentum equation (2.2) as the equation of motion of the centre of mass of an individual volume V of a continuous medium

$$m \frac{\mathrm{d}\boldsymbol{v}^*}{\mathrm{d}t} = \int_V \boldsymbol{F}\rho \, \mathrm{d}\tau + \int_\Sigma \boldsymbol{p}_n \, \mathrm{d}\sigma .$$

Note that (2.2) is often called the impulse equation, as it can be rewritten in the form

$$\mathrm{d} \int_V \boldsymbol{v}\rho \, \mathrm{d}\tau = \int_V \boldsymbol{F}\rho \, \mathrm{d}\tau \, \mathrm{d}t + \int_\Sigma \boldsymbol{p}_n \, \mathrm{d}t \, \mathrm{d}\sigma .$$

Relation (2.2) may also be regarded as an equation for the definition of force. Indeed, all known laws for forces may be derived from this equation, i.e., from Newton's generalized second law.[1)] These laws may be obtained on the basis of observations during experiments, with the aid of different hypotheses or with the help of "mental experiments", formulated as a generalization of practical data. Having determined the laws for the forces with the aid of (2.2) during prior studies, one may in other cases when the forces are known find from (2.2) the motions which correspond to these forces.

The momentum equation (2.2) *is the fundamental equation for any motion of continuous media.* It applies even if the motion is discontinuous, i.e., when the characteristics of the motion and the state of the continuum are not continuous functions of the coordinates everywhere in the volume V, and for processes involving shocks, when the charac-

[1)] This question is treated in more detail in L. I. Sedov, *Similarity and dimensional methods in mechanics.* 7th ed. Izd-vo "Nauka", Moscow, Chapter I, Sec. 5. An English translation of the fourth edition of this book, edited by M. Holt, has been published by Infosearch, London, 1959.

teristics of the motion and the state of the medium in the volume under consideration are discontinuous functions of the time.

In particular, in regions of continuous motion, the integral momentum theorem (2.2) is equivalent to the differential equations of motion of a continuous medium which will be derived below. It also follows from (2.2) that the surface stresses $\boldsymbol{p}_n$ for any media and any motions must satisfy some universal relations to be derived next.

A fundamental property of internal stresses. Consider now the limitations which the momentum equation (2.2) imposes on the possible form of the dependence of the stresses $\boldsymbol{p}_n$ on the orientations of the corresponding planes at a given point in the case of continuous motions of continua.

Consider a volume V and divide it mentally into two parts V_1 and V_2 by an arbitrary cross-section S (fig. 23). Apply the momentum equation (2.2) separately to V_1 and V_2, and then to the volume as a whole. Noting that the interaction between the separate parts may be replaced by distributed mass forces and surface forces, distributed over the cross-section S moving together with the individual points of the medium one may write

$$\int_{V_1} \frac{d\boldsymbol{v}}{dt}\rho\, d\tau = \int_{V_1} \boldsymbol{F}'\rho\, d\tau + \int_{\Sigma_1} \boldsymbol{p}_n d\sigma + \int_S \boldsymbol{p}_n d\sigma\,,$$

$$\int_{V_2} \frac{d\boldsymbol{v}}{dt}\rho\, d\tau = \int_{V_2} \boldsymbol{F}''\rho\, d\tau + \int_{\Sigma_2} \boldsymbol{p}_n d\sigma + \int_S \boldsymbol{p}_{-n} d\sigma\,,$$

$$\int_{V} \frac{d\boldsymbol{v}}{dt}\rho\, d\tau = \int_{V} \boldsymbol{F}\rho\, d\tau + \int_{\Sigma} \boldsymbol{p}_n d\sigma\,,$$

where $\boldsymbol{F}'$ and $\boldsymbol{F}''$ denote the densities of the distributed mass forces, acting on V_1 and V_2, respectively. Adding the first two equations and subtracting from them the third, assuming that the law of action and reaction is fulfilled always for internal mass forces,[1)] i.e., that

[1)] One can obtain the equality (2.3) even without such an assumption, applying the equations of momentum to the infinitesimal volumes V_1, V_2 and V. Then, the law of action and reactions follows from the equation of momentum.

$$\int_{V_1} F' \rho \, d\tau + \int_{V_2} F'' \rho \, d\tau = \int_V F \rho \, d\tau \,,$$

one finds

$$\int_S (p_n + p_{-n}) \, d\sigma = 0 \,.$$

Since the volumes V, V_1 and V_2 and the cross-section S are arbitrary, it follows that

$$p_n = -p_{-n} \,. \tag{2.3}$$

Note that the assumption of the continuity of the motion is essential. For instance, as will be shown below (Chapter 7), Equation (2.3) is not satisfied, if one has inside V a surface of discontinuity S of v through which pass particles of the medium. In this case the volumes V_1 and V_2 are not individual and one must apply other relations.

The momentum equation (2.2) is applicable to any material volume V, however small. Consider the limitations which are to be imposed on the integrands in the momentum equation (2.2), in order to ensure that this equation will be applicable *to arbitrarily small individual volume elements* of a continuum. Assuming for this purpose that the characteristics of the motion are continuous and finite, construct the expression

$$\Omega = \int_V F \rho \, d\tau + \int_\Sigma p_n \, d\sigma - \int_V \frac{dv}{dt} \rho \, d\tau \,,$$

which vanishes exactly for any individual volume V.

Obviously, as V shrinks to a point, the limit

$$\lim_{V \to 0} \Omega = 0$$

holds independently of the form of the integrands. It is satisfied for any finite functions, appearing in the integrands of the expression for Ω.

Now take an arbitrary point M of a continuum at a given instant and draw from it directions parallel to the axes of a Cartesian coordinate system (fig. 25). Mark off the arbitrary infinitesimal segments $dx = MA$, $dy = MC$, $dz = MB$ and consider a volume dV in the form of the tetrahedron $MABC$. Its three faces MBC, MAB and MAC are perpendicular

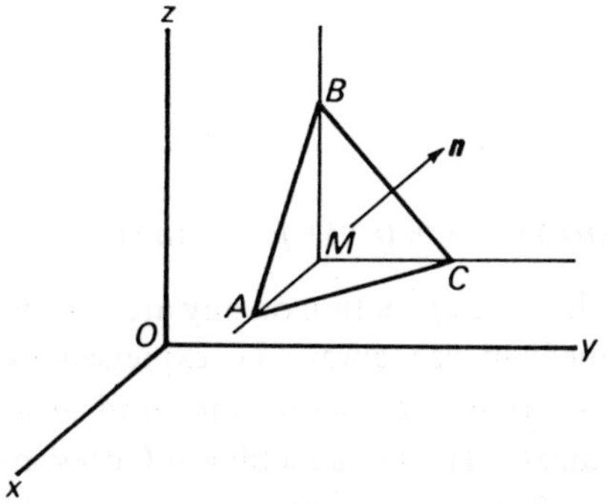

Fig. 25. On properties of internal stresses.

to the coordinate axes, while the face ABC is arbitrarily orientated and has the unit normal

$$\boldsymbol{n} = \cos(\boldsymbol{n}x)\boldsymbol{i} + \cos(\boldsymbol{n}y)\boldsymbol{j} + \cos(\boldsymbol{n}z)\boldsymbol{k} = n_i \boldsymbol{\mathfrak{z}}^i .$$

Denote the stresses acting on the areas with normals $\boldsymbol{i}, \boldsymbol{j}, \boldsymbol{k}, \boldsymbol{n}$ by $\boldsymbol{p}^1$, $\boldsymbol{p}^2$, $\boldsymbol{p}^3$ and $\boldsymbol{p}_n$, respectively, and the area of the face ABC by S. The areas of the faces MBC, MAB and MAC are then, obviously, $S\cos(\boldsymbol{n}x)$, $S\cos(\boldsymbol{n}y)$ and $S\cos(\boldsymbol{n}z)$, and the volume of the tetrahedron is $V = \frac{1}{3}hS$, where h is the height of the vertex M over the face ABC. If the tetrahedron shrinks to a point, remaining similar to itself, then h and S become infinitesimal quantities of first and second order, respectively. Calculate $\boldsymbol{\Omega}$ for the volume of a continuous medium, occupying this tetrahedron at a given instant; using Property (2.3) of the internal stresses, one obtains, obviously,

$$\boldsymbol{\Omega} = -\left(\frac{d\boldsymbol{v}}{dt}\rho\right)_M \cdot \tfrac{1}{3}Sh + (\boldsymbol{F}\rho)_M \cdot \tfrac{1}{3}Sh + \boldsymbol{p}_n S - \boldsymbol{p}^1 S\cos(nx)$$
$$- \boldsymbol{p}^2 S\cos(ny) - \boldsymbol{p}^3 S\cos(nz) + O(h^{2+\lambda}),$$

where $\lambda > 0$. Now let the tetrahedron shrink to a point, while it remains similar to itself. Then, since $\boldsymbol{\Omega} = 0$ by (2.2), one finds in the limit

$$\lim_{V\to 0} \frac{\boldsymbol{\Omega}}{h} = 0, \quad \lim_{V\to 0} \frac{\boldsymbol{\Omega}}{h^2} = 0, \quad \lim_{V\to 0} \frac{\boldsymbol{\Omega}}{h^3} = 0.$$

Clearly, the first limit will always be zero for motions with continuous and finite characteristics, i.e., no restrictions need be imposed on the integrands of $\boldsymbol{\Omega}$. It follows from the condition

$$\lim_{V \to 0} \frac{\Omega}{S} = 0$$

that one must always have

$$\boldsymbol{p}_n = \boldsymbol{p}^1 \cos(\boldsymbol{n}x) + \boldsymbol{p}^2 \cos(\boldsymbol{n}y) + \boldsymbol{p}^3 \cos(\boldsymbol{n}z), \tag{2.4}$$

which means that the stress $\boldsymbol{p}_n$ acting on any area element $\mathrm{d}\sigma$ at a point M of a continuous medium can always be expressed by (2.4) as a linear combination of the stresses $\boldsymbol{p}^i$ acting on fixed areas at M, which are parallel to the coordinate planes of a rectilinear Cartesian coordinate system.

Relation (2.4) also shows that the sum of the external surface forces

$$\int_\Sigma \boldsymbol{p}_n \mathrm{d}\sigma ,$$

acting on a volume V of a continuum bounded by the surface S may be expressed, with the aid of the Gauss–Ostrogradskii formula, in the form of the volume integral

$$\int_\Sigma \boldsymbol{p}_n \mathrm{d}\sigma = \int_V \left(\frac{\partial \boldsymbol{p}^1}{\partial x} + \frac{\partial \boldsymbol{p}^2}{\partial y} + \frac{\partial \boldsymbol{p}^3}{\partial z} \right) \mathrm{d}\tau . \tag{2.5}$$

The equation of motion of a continuous medium in Cartesian coordinates. Finally, consider the condition

$$\lim_{V \to 0} \frac{\Omega}{V} = 0 .$$

Using (2.5), represent Ω in the form

$$\Omega = \int_V \boldsymbol{F} \rho \mathrm{d}\tau + \int_V \left(\frac{\partial \boldsymbol{p}^1}{\partial x} + \frac{\partial \boldsymbol{p}^2}{\partial y} + \frac{\partial \boldsymbol{p}^3}{\partial z} \right) \mathrm{d}\tau - \int_V \frac{\mathrm{d}\boldsymbol{v}}{\mathrm{d}t} \rho \mathrm{d}\tau$$

to obtain from

$$\lim_{V \to 0} \frac{\Omega}{V} = 0$$

the equation

$$\rho \frac{\mathrm{d}\boldsymbol{v}}{\mathrm{d}t} = \rho \boldsymbol{F} + \frac{\partial \boldsymbol{p}^1}{\partial x} + \frac{\partial \boldsymbol{p}^2}{\partial y} + \frac{\partial \boldsymbol{p}^3}{\partial z} . \tag{2.6}$$

This vector equation is the fundamental differential equation of motion of a continuous medium. It is satisfied for any continuous motions of any media and, in the case of continuous motions, it is completely equivalent to the momentum equation (2.2), since it implies that $\Omega = 0$ for any volume V. It should be emphasized that Equations (2.5) and (2.6) have been derived under the assumption of continuity and differentiability of the vectors $\boldsymbol{p}^i$. Equation (2.2) may be postulated for more general cases.

Decompose the vectors $\boldsymbol{p}^i$ with respect to the base vectors $\boldsymbol{\mathfrak{z}}_1 = \boldsymbol{i}$, $\boldsymbol{\mathfrak{z}}_2 = \boldsymbol{j}$, $\boldsymbol{\mathfrak{z}}_3 = \boldsymbol{k}$ of a Cartesian coordinate system:

$$\left.\begin{aligned} \boldsymbol{p}^1 &= p^{k1}\boldsymbol{\mathfrak{z}}_k, \\ \boldsymbol{p}^2 &= p^{k2}\boldsymbol{\mathfrak{z}}_k, \\ \boldsymbol{p}^3 &= p^{k3}\boldsymbol{\mathfrak{z}}_k, \end{aligned}\right\} \quad \text{or} \quad \boldsymbol{p}^i = p^{ki}\boldsymbol{\mathfrak{z}}_k, \tag{2.7}$$

and introduce the matrix

$$\begin{Vmatrix} p^{11} & p^{12} & p^{13} \\ p^{21} & p^{22} & p^{23} \\ p^{31} & p^{32} & p^{33} \end{Vmatrix} = \| p^{ik} \| = P,$$

with nine elements. By Property (2.4) of the stress components p_n^i, one finds that

$$\boldsymbol{p}_n = p_n^1\boldsymbol{\mathfrak{z}}_1 + p_n^2\boldsymbol{\mathfrak{z}}_2 + p_n^3\boldsymbol{\mathfrak{z}}_3 = p_n^i\boldsymbol{\mathfrak{z}}_i$$

for an arbitrarily orientated plane at a given point of a medium yields the formulae

$$\left.\begin{aligned} p_n^1 &= p^{11}\cos(nx) + p^{12}\cos(ny) + p^{13}\cos(nz) = p^{1i}n_i, \\ p_n^2 &= p^{21}\cos(nx) + p^{22}\cos(ny) + p^{23}\cos(nz) = p^{2i}n_i, \\ p_n^3 &= p^{31}\cos(nx) + p^{32}\cos(ny) + p^{33}\cos(nz) = p^{3i}n_i. \end{aligned}\right\} \tag{2.8}$$

Thus, the matrix P defines the transformation from the components n_i of the vector $\boldsymbol{n} = n_i\boldsymbol{\mathfrak{z}}^i$ to the components p_n^i of the vector $\boldsymbol{p}_n$.

Nine functions p^{ik} appear in the vector equation of motion of a continuous medium (2.6), which after projection on to the axes of a Cartesian coordinate system may be rewritten in the form

$$
\begin{aligned}
\rho \frac{du}{dt} &= \rho F_x + \frac{\partial p^{11}}{\partial x} + \frac{\partial p^{12}}{\partial y} + \frac{\partial p^{13}}{\partial z}, \\
\rho \frac{dv}{dt} &= \rho F_y + \frac{\partial p^{21}}{\partial x} + \frac{\partial p^{22}}{\partial y} + \frac{\partial p^{23}}{\partial z}, \\
\rho \frac{dw}{dt} &= \rho F_z + \frac{\partial p^{31}}{\partial x} + \frac{\partial p^{32}}{\partial y} + \frac{\partial p^{33}}{\partial z},
\end{aligned}
\tag{2.9}
$$

where F_x, F_y and F_z denote the projections on the coordinate axes of the density of the mass force $\boldsymbol{F}$.

Combining these equations of motion with the equation of continuity (1.3), one obtains a system of four equations which, for given external mass forces, contains, in general, thirteen unknown functions: the density ρ, the velocity components u, v and w, and the nine components of the internal surface stresses p^{ik}.

The stress tensor. The dependence (2.4) of the stress vector $\boldsymbol{p}_n$, acting on an arbitrarily orientated elementary area, on the stress vectors $\boldsymbol{p}^1, \boldsymbol{p}^2, \boldsymbol{p}^3$, acting on the coordinate planes at the same point, can be rewritten by (2.8) in the form

$$\boldsymbol{p}_n = \boldsymbol{p}^i n_i = p^{ki}\boldsymbol{\mathfrak{z}}_k n_i = \boldsymbol{p}^i(\boldsymbol{\mathfrak{z}}_i \cdot \boldsymbol{n}) = p^{ki}\boldsymbol{\mathfrak{z}}_k(\boldsymbol{\mathfrak{z}}_i \cdot \boldsymbol{n}). \tag{2.10}$$

This equation yields a linear transformation (with coefficients p^{ki}) from the components of the vector $\boldsymbol{n}$ into the components of the vector $\boldsymbol{p}_n$. It was obtained with the aid of an orthogonal Cartesian coordinate system and, consequently, the p^{ki} were defined in an arbitrary orthogonal Cartesian coordinate system. Equation (2.10) provides a relation between the vectors $\boldsymbol{p}_n$ and $\boldsymbol{n}$, and therefore it may be written down in terms of any curvilinear coordinate system. Hence it follows that not only in orthogonal Cartesian coordinates, but also in arbitrary curvilinear coordinates one may introduce with the aid of (2.10) quantities p^{ki} which must be regarded as the contravariant components of the tensor

$$P = p^{ki}\boldsymbol{\mathfrak{z}}_k\boldsymbol{\mathfrak{z}}_i .$$

This tensor is called the internal stress tensor. The following equation will be satisfied in any coordinate system:

$$\boldsymbol{p}_n = P \cdot \boldsymbol{n} = \boldsymbol{p}^i n_i ,$$

where $\boldsymbol{p}_n$ is the stress acting on an arbitrary area with normal $\boldsymbol{n}$, and n_i are the covariant components of $\boldsymbol{n}$.

Physical components of a stress vector. Note that usually the equation $\boldsymbol{p}^i = \boldsymbol{p}_n$ is valid on corresponding coordinate planes only for an orthogonal Cartesian coordinates; it is easily verified that in an arbitrary curvilinear system $\boldsymbol{p}_i \neq \boldsymbol{p}_n$ on the corresponding coordinate planes.

In fact, consider for a given curvilinear coordinate system the plane determined by the base vectors $\mathbf{3}_{i+1}$ and $\mathbf{3}_{i+2}$ (the indices defined modulo 3); the positive direction of the normal to this plane is defined as the direction of the contravariant base vector

$$\mathbf{3}^i = \frac{\mathbf{3}_{i+1} \times \mathbf{3}_{i+2}}{\sqrt{g}},$$

the unit vector in this direction obviously being given by

$$\boldsymbol{n}^i = \frac{\mathbf{3}_i}{\sqrt{g^{ii}}},$$

where $g^{ii} > 0$ and the square root carries a positive sign. By (2.10), the stress vector $\boldsymbol{p}_n$ on such a plane, which will be denoted by $\boldsymbol{p}_i^*$, may be represented in the form

$$\boldsymbol{p}_i^* = \frac{p^{\alpha k}\mathbf{3}_\alpha(\mathbf{3}_k \cdot \mathbf{3}^i)}{\sqrt{g^{ii}}} = \frac{p^{\alpha i}\mathbf{3}_\alpha}{\sqrt{g^{ii}}},$$

and, consequently, it is not generally equal to the vector $\boldsymbol{p}^i = p^{\alpha i}\mathbf{3}_\alpha$. The stress vector $\boldsymbol{p}_i^*$ may be expanded in terms of the unit base vectors $\mathbf{3}_\alpha/\sqrt{g_{\alpha\alpha}}$ at the point under consideration:

$$\boldsymbol{p}_i^* = X^{\alpha i}\frac{\mathbf{3}_\alpha}{\sqrt{g_{\alpha\alpha}}}.$$

The quantities $X^{\alpha i}$ are referred to as the physical components of the stress vector $\boldsymbol{p}_i^*$. On the basis of the last two equations, one may write,

$$p^{\alpha i} = X^{\alpha i}\sqrt{\left(\frac{g^{ii}}{g_{\alpha\alpha}}\right)}$$

(no summation with respect to α). Hence it is clear that the physical components $X^{\alpha i}$ are not the components of any tensor.

In orthogonal Cartesian coordinates,

$$p^{\alpha i} = X^{\alpha i}$$

Equations of motion of a continuous medium in an arbitrary coordinate system. From the vector momentum equation (2.2), *viz.*,

$$\int_V \rho \frac{\mathrm{d}\boldsymbol{v}}{\mathrm{d}t}\,\mathrm{d}\tau = \int_V \boldsymbol{F}\rho\,\mathrm{d}\tau + \int_\Sigma \boldsymbol{p}_n \mathrm{d}\sigma,$$

and the Gauss–Ostrogradskii theorem

$$\int_\Sigma \boldsymbol{p}_n \mathrm{d}\sigma = \int_\Sigma \boldsymbol{p}^i n_i \mathrm{d}\sigma = \int_V \nabla_i \boldsymbol{p}^i \mathrm{d}\tau,$$

one obtains for continuous motion the equation

$$\rho \boldsymbol{a} = \rho \boldsymbol{F} + \nabla_i \boldsymbol{p}^i \quad \text{or} \quad \rho a^k = \rho F^k + \nabla_i p^{ki}, \tag{2.11}$$

which is true in any curvilinear coordinate system.

In the equations of motion (2.11),

$$a^k = \frac{\partial v^k}{\partial t} + v^i \nabla_i v^k = \frac{\partial v^k}{\partial t} + v^i \left(\frac{\partial v^k}{\partial \eta^i} + v^s \Gamma^k_{si} \right)$$

and

$$\nabla_i \boldsymbol{p}^i = \frac{\partial \boldsymbol{p}^i}{\partial \eta^i} + \boldsymbol{p}^s \Gamma^i_{si} = \nabla_i p^{ki} \boldsymbol{\mathfrak{z}}_k .$$

The vector equation (2.11) is not only valid in a moving, but also in a stationary coordinate system; in particular, it is true in a concomitant system, as well as in a frame of reference. However, one must bear in mind that the vector $\boldsymbol{a}$ is the acceleration of an individual point of the medium with respect to any inertial frame of reference and that $\boldsymbol{F}$ is the density of given mass forces. If one treats the motion and acceleration with respect to a non-inertial frame of reference, one must include inertial forces in the expression for $\boldsymbol{F}$.

Consider an infinitesimal particle of mass $\rho\,\mathrm{d}\tau$ of a continuous medium. It is acted upon by mass forces $\rho \boldsymbol{F}\,\mathrm{d}\tau$, by the force $\rho \boldsymbol{a}\,\mathrm{d}\tau$, appearing as an inertial force in the concomitant coordinate system, and by the force $\nabla_i \boldsymbol{p}^i\,\mathrm{d}\tau = (\nabla_i p^{ki})\,\boldsymbol{\mathfrak{z}}_k\,\mathrm{d}\tau$, which can be regarded as a mass force, arising from the action of the surface forces on the boundary of the particle. Equation (2.11) may be treated as an equilibrium condition with respect to the concomitant coordinate system. By (2.11), the sum of all forces acting on the particle vanishes.

Note that, if the tensor $P = P^{ki}\mathfrak{z}_k\mathfrak{z}_i$ is constant over all points of the medium, then $\nabla_i \boldsymbol{p}^i = 0$. In Cartesian coordinates, the components of the stress tensor p^{ki} appear in the equation of motion only when they depend on the coordinates x, y and z. However, the equations

$$\nabla_i \boldsymbol{p}^i = 0 \quad \text{or} \quad \nabla_i p^{ki} = 0$$

and $P = p^{ki}\,\mathfrak{z}_k\,\mathfrak{z}_i = \text{const.}$ or $\nabla_j p^{ki} = 0, i, j, k = 1, 2, 3$ are not equivalent. For example, if a medium, which is not subject to mass forces, is at rest, then

$$V_i p^{ki} = 0 \;, \qquad k = 1, 2, 3 \;.$$

This is a fundamental equation of the theory of elasticity, where one often studies problems of the equilibrium of different objects which are only stressed by external surface forces.

§3. The equation of angular momentum

As it has been pointed out previously, the system of universal equations of motion of a continuous medium, derived so far, is still incomplete. One may yet obtain further general equations which do not depend on particular properties of a moving medium. For this purpose, consider another general equation of mechanics—the equation of angular momentum.

The equation of angular momentum for a material point and a system of points. Multiplying the equation

$$m\frac{d\boldsymbol{v}}{dt} = \mathbf{F}$$

vectorially from the left by the radius vector $\boldsymbol{r}$ to the point mass m under consideration from a point O, the origin of some inertial frame of reference, one obtains the equation of angular momentum for a point;

$$\frac{d\boldsymbol{K}}{dt} = \mathfrak{M}\,, \tag{3.1}$$

where

$$\boldsymbol{K} = \boldsymbol{r} \times m\boldsymbol{v} \quad \text{and} \quad \mathfrak{M} = \boldsymbol{r} \times \mathbf{F}$$

are the angular momentum of the point mass and the torque of **F** about the point *O*, respectively. Thus, the equation of angular momentum for any point mass is a trivial consequence of Newton's second law.

Given a system of *n* point masses m_i, moving with velocities $\boldsymbol{v}_i$, one may write down the equation of angular momentum (3.1) for each point:

$$\frac{\mathrm{d}}{\mathrm{d}t}(\boldsymbol{r}_i \times m_i \boldsymbol{v}_i) = \boldsymbol{r}_i \times \mathbf{F}_i ,$$

where $\mathbf{F}_i$ is the resultant vector of all forces acting on the point m_i, including those forces which are internal to the system as a whole. Summing these equations over all *n* points of the system and defining the angular momentum of the motion of the system by

$$\boldsymbol{K} = \sum_{i=1}^{n} (\boldsymbol{r}_i \times m_i \boldsymbol{v}_i) ,$$

one obtains the obvious equation of angular momentum for a system of points

$$\frac{\mathrm{d}\boldsymbol{K}}{\mathrm{d}t} = \sum_{i=1}^{n} \boldsymbol{r}_i \times \mathbf{F}_i^{(e)} ;$$

where, by virtue of Newton's third law (fig. 26), the right-hand side is the sum of the angular momenta of only those forces which are external to the system as a whole. The derivative with respect to time of the angular momentum of a system of points with respect to some point *O* is equal to the sum of moments about *O* of all external forces acting on the system.

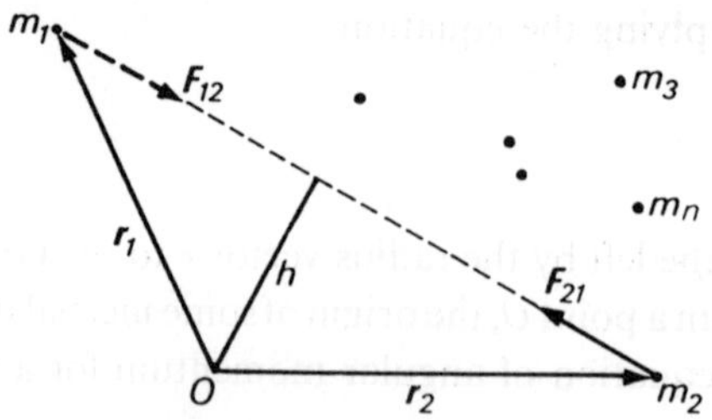

Fig. 26. The sum of the moments of all internal forces about *O* vanishes.

Note that the angular momentum of a system of point masses may be rewritten in the form

$$\boldsymbol{K} = \boldsymbol{r}^* \times m\boldsymbol{v}^* + \sum_{i=1}^{n} (\boldsymbol{r}_{i\,\mathrm{rel}} \times m_i \boldsymbol{v}_{i\,\mathrm{rel}}),$$

where $m = \Sigma_{i=1}^{n} m_i$, $\boldsymbol{r}^*$ is the radius vector of the centre of mass of the system, $\boldsymbol{v}^*$ is the velocity of the centre of mass, $\boldsymbol{r}_{i\,\mathrm{rel}}$ is the radius vector to the i-th point from the centre of mass and $\boldsymbol{v}_{i\,\mathrm{rel}}$ is the velocity of the i-th point with respect to a coordinate system, translating with the centre of mass.

The angular momentum of a finite volume of a continuous medium; intrinsic angular momenta. The angular momentum of a volume V of a continuous medium is normally defined by

$$\boldsymbol{K} = \int_V (\boldsymbol{r} \times \boldsymbol{v}) \rho \, d\tau, \tag{3.2}$$

where $\boldsymbol{r}$ is the radius vector to a point of the continuous medium from some stationary point O and $\boldsymbol{v}$ is its velocity. However, this question will be studied in greater detail.

Consider some volume τ of a continuous medium with mass m. Clearly, the velocity $\boldsymbol{v}$ of an arbitrary point M of this volume may be expressed (fig. 27) in the form of the sum

$$\boldsymbol{v} = \boldsymbol{v}^* + \boldsymbol{v}_{\mathrm{rel}},$$

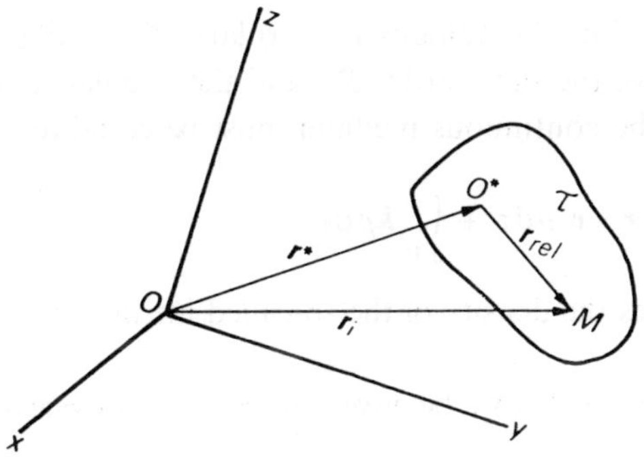

Fig. 27. On the equation of angular momentum for a finite volume of a continuous medium.

where $\boldsymbol{v}^*$ is the velocity of the centre of mass O^* of τ and $\boldsymbol{v}_{\text{rel}}$ is the velocity of the point under consideration relative to the centre of mass. Then, obviously, the angular momentum of the volume τ about some point O will be equal to the sum of the angular momentum of a point mass m, located at the centre of mass of τ with respect to the point O, and the angular momenta of all points M of τ with respect to the centre of mass O^*, i.e.,

$$\boldsymbol{K} = \boldsymbol{r}^* \times \boldsymbol{Q} + \int_\tau \boldsymbol{r}_{\text{rel}} \times \boldsymbol{v}_{\text{rel}}\, \rho \, d\tau ,$$

where $\boldsymbol{Q} = m\boldsymbol{v}^*$ is the angular momentum of a point mass m located at the centre of mass, or

$$\boldsymbol{K} = \boldsymbol{r}^* \times \boldsymbol{Q} + \boldsymbol{K}^*; \quad \boldsymbol{K}^* = \int_\tau \boldsymbol{r}_{\text{rel}} \times \boldsymbol{v}_{\text{rel}}\, \rho \, d\tau .$$

Consider now an infinitesimal volume $d\tau$. In many cases, the angular momentum $\boldsymbol{K}^*$ of an infinitesimal volume is negligible compared with $\boldsymbol{r}^* \times \boldsymbol{Q}$. Suppose, for example, that $d\tau$ is an infinitesimal homogeneous sphere of radius R rotating with angular velocity $\boldsymbol{\omega}$ about an axis passing through its centre O^*; then

$$\boldsymbol{K}^* = I\boldsymbol{\omega} = ml^2 \boldsymbol{\omega} ,$$

where I is the moment of inertia and l is the radius of inertia of this sphere with respect to its axis of rotation. Obviously, ml^2 is of order R^5, while $\boldsymbol{r}^* \times \boldsymbol{Q}$ is of order R^3 and $\boldsymbol{K}^*$, provided $\boldsymbol{\omega}$ is finite, is small compared with $\boldsymbol{r}^* \times \boldsymbol{Q}$; the angular momentum $\boldsymbol{K}$ of the volume V of the continuum medium in the limit is

$$\int_V \boldsymbol{r} \times \boldsymbol{v}\, \rho \, d\tau .$$

However, if the angular velocity $\boldsymbol{\omega}$ is so large that ωl^2 is finite, then $\boldsymbol{K}^*$ and $\boldsymbol{r}^* \times \boldsymbol{Q}$ have the same order R^3, and the angular momentum of the volume V of the continuous medium must be equal to

$$\boldsymbol{K} = \int_V \boldsymbol{r} \times \boldsymbol{v}\, \rho \, d\tau + \int_V \boldsymbol{k} \rho \, d\tau , \tag{3.3}$$

where $\boldsymbol{k}$ denotes the density of the so-called intrinsic or internal angular momentum.

This problem will now be investigated on a physically microscopic level.

Consider a system, consisting of a nucleus and, revolving about it, an electron, i.e., an atom. The electron revolves in its orbit with a velocity of the order of the speed of light. Therefore, regardless of the small size of the atom, the system nucleus–electron possesses a significant intrinsic angular momentum. The angular momentum, arising from the revolution of the electron in an orbit, is known as orbital angular momentum.

Moreover, the electron, and likewise the nucleus, have an intrinsic angular momentum, namely a spin, the origin of which cannot be explained by the introduction of corresponding mechanical motion.

In general, all atoms have, generally speaking, intrinsic angular momentum $\boldsymbol{k}$. However, in many cases, the random motion of the atoms causes the sum of these angular momenta over all atoms to vanish. On the other hand, however, the motion of the elementary particles may be ordered, for instance, by applying a magnetic field. Then the sum of the internal momenta of all atoms will differ from zero. In this case, the sum

$$\boldsymbol{K}' = \int_V \boldsymbol{k}\rho \, \mathrm{d}\tau$$

of the intrinsic angular momenta must appear in the expression for the angular momentum of a macroscopic particle of a continuous medium.

For example, if one wishes to describe in the mechanics of continua the motion of a real medium in an electromagnetic field, one must introduce into the consideration the intrinsic momenta $\boldsymbol{k}$ and define the angular momentum of a volume V of the continuum by (3.3).

Intrinsic angular momenta have only recently come to be studied in continuum mechanics, when the scope of problems in continuum mechanics was greatly expanded in connection with inquiries into modern techniques. Internal momenta $\boldsymbol{k}$ were not taken into consideration in classical problems of continuum mechanics, and the angular momentum of a volume V of a continuum was defined as

$$\int_V (\boldsymbol{r} \times \boldsymbol{v}) \rho \, \mathrm{d}\tau \, .$$

Distributed mass and surface couples. Introducing into the consideration the internal momenta $\boldsymbol{k}$, one must admit also the existence of distributed mass and surface couples.

Distributed mass and surface couples act on every particle of a continuous medium. However, it may happen that the effect of external material objects on a particle of a continuous medium cannot be replaced by a single resultant vector, but that it requires also the introduction of mass and surface couples.

Denote by $\boldsymbol{h}$ and $\boldsymbol{Q}_n$ the momenta of the mass and surface couples, respectively, referred to unit mass and unit surface. An example of distributed mass couples might be couples acting on every element of the needle of a compass, placed in the magnetic field of Earth.

The equation of angular momentum of a finite volume of a continuous medium. As a generalization of the equation of angular momentum for a single material point or for a system of points, the equation of angular momentum for a given finite volume V of a continuous medium, bounded by a surface Σ, will be formulated:

$$\frac{\mathrm{d}}{\mathrm{d}t}\left[\int_V \boldsymbol{r}\times\boldsymbol{v}\;\rho\,\mathrm{d}\tau + \int_V \boldsymbol{k}\rho\,\mathrm{d}\tau\right]$$

$$= \int_V \boldsymbol{r}\times\boldsymbol{F}\;\rho\,\mathrm{d}\tau + \int_\Sigma \boldsymbol{r}\times\boldsymbol{p}_n\,\mathrm{d}\sigma + \int_V \boldsymbol{h}\rho\,\mathrm{d}\tau + \int_\Sigma \boldsymbol{Q}_n\mathrm{d}\sigma\,. \qquad (3.4)$$

The derivative with respect to time of the angular momentum of an individual arbitrary volume V of a continuous medium (taking into account intrinsic moments) is equal to the sum of the moments of the external mass and surface forces, acting on this volume, and the sum of the moments of the distributed mass and surface couples, acting on the volume and caused by material objects outside the volume.

The equation of angular momentum, like the momentum equation, is postulated for an individual volume V of a continuous medium in the same manner as Newton's law $\boldsymbol{F} = m\boldsymbol{a}$ for a point mass. Note that the equation of angular momentum for an individual volume V of a continuous medium is not a consequence of the equation of angular momentum of the mechanics of systems of material points. It is an independent equation. All the reasoning, preceding its formulation, must be regarded as purely heuristic.

Along with the momentum equation, the Equation (3.4) for any finite, mentally separated volume is regarded as the basic vector equation of continuum mechanics. It can be applied to any continuous media and for any motions, whether continuous or in the presence of discontinuous characteristics in time or in space.

Note that presently it has been necessary to introduce into the consideration moments of higher order and to formulate new fundamental relations of continuum mechanics which are analogous to the equation of angular momentum (3.4) of first order.

The classical equation of angular momentum. In the classical case, in the absence of internal angular momentum and distributed mass and surface couples, the equation of angular momentum has the form

$$\frac{\mathrm{d}}{\mathrm{d}t}\int_V \boldsymbol{r}\times\boldsymbol{v}\,\rho\,\mathrm{d}\tau = \int_V \boldsymbol{r}\times\boldsymbol{F}\,\rho\,\mathrm{d}\tau + \int_\Sigma \boldsymbol{r}\times\boldsymbol{p}_n\,\mathrm{d}\sigma\,. \tag{3.5}$$

The derivative with respect to time of the angular momentum of an individual volume V of a continuous medium with respect to some point O (linked to an inertial system of coordinates is equal to the sum of the moments about the same point O of the external mass and surface forces, acting on the volume.

Obviously, if no external mass forces act on the body,

$$\frac{\mathrm{d}\boldsymbol{K}}{\mathrm{d}t} = 0\,,$$

and the angular momentum $\boldsymbol{K}$ is constant.

Hydromagnetic effect and the equation of angular momentum. Consider an experiment which demonstrates the fact that, in general, internal angular momentum and distributed mass couples must be taken into consideration. If an iron rod is placed in a magnetic field, it becomes magnetized, and one may show that the sum of the internal momenta $\boldsymbol{k}$ acting on it differs from zero. In fact, let this rod be freely suspended in the presence of a magnetic field in a vacuum and let it be at rest. Remove the magnetic field. Then, because of random thermal motion, the distribution of internal momenta $\boldsymbol{k}$ in the rod becomes disordered in a short time, and therefore the sum of the internal angular momenta vanishes.

Hence, since the rod is not acted upon by any external objects, the total angular momentum must be conserved. Therefore, angular momentum must appear as a consequence of the rotation of the rod as a whole, and the bar must begin to rotate.

Experiments show that, after the magnetic field is removed, such a rod actually begins to rotate.

This is the so-called hydromagnetic effect. It cannot be explained without taking into consideration internal angular momenta and distributed mass couples.

The differential form of the equation of angular momentum. In the case of continuous motion of a "continuum", one may use equation (2.4) and the Gauss–Ostrogradskii theorem and obtain for the sum of the moments of the external surface forces an expression in the form of an integral, taken over the volume V:

$$\int_\Sigma (\boldsymbol{r}\times \boldsymbol{p}_n)\mathrm{d}\sigma = \int_\Sigma (\boldsymbol{r}\times \boldsymbol{p}^i)n_i\mathrm{d}\sigma = \int_V \nabla_i(\boldsymbol{r}\times \boldsymbol{p}^i)\mathrm{d}\tau\,.$$

It can be shown that, analogously to the internal stresses $\boldsymbol{p}_n$, the moments of distributed surface couples $\boldsymbol{Q}_n$ may be represented in the form

$$\boldsymbol{Q}_n = \boldsymbol{Q}^i n_i\,.$$

Then, with the aid of the Gauss–Ostrogradskii theorem, one obtains

$$\int_\Sigma \boldsymbol{Q}_n\mathrm{d}\sigma = \int_\Sigma \boldsymbol{Q}^i n_i\mathrm{d}\sigma = \int_V \nabla_i\boldsymbol{Q}^i\mathrm{d}\tau\,.$$

Further transformations are still possible:

$$\int_V \nabla_i(\boldsymbol{r}\times \boldsymbol{p}^i)\mathrm{d}\tau = \int_V \boldsymbol{r}\times\nabla_i\boldsymbol{p}^i\,\mathrm{d}\tau + \int_V \nabla_i\boldsymbol{r}\times\boldsymbol{p}^i\,\mathrm{d}\tau$$

$$= \int_V \boldsymbol{r}\times\nabla_i\boldsymbol{p}^i\,\mathrm{d}\tau + \int_V (\boldsymbol{\mathfrak{z}}_i\times\boldsymbol{\mathfrak{z}}_k)p^{ki}\mathrm{d}\tau\,,$$

since

$$\nabla_i\boldsymbol{r} = \frac{\partial\boldsymbol{r}}{\partial x^i} = \boldsymbol{\mathfrak{z}}_i\,.$$

Now, under the condition that $\mathrm{d}m=\rho\,\mathrm{d}\tau$ is constant, the theorem of angular momentum (3.4) may be rewritten in the form

$$\int_V \left[\boldsymbol{r}\times\left(\frac{\mathrm{d}\boldsymbol{v}}{\mathrm{d}t} - \boldsymbol{F} - \frac{1}{\rho}\nabla_i\boldsymbol{p}^i\right)\right]\rho\,\mathrm{d}\tau + \int_V \frac{\mathrm{d}\boldsymbol{k}}{\mathrm{d}t}\rho\,\mathrm{d}\tau$$

$$= \int_V \boldsymbol{h}\rho\,\mathrm{d}\tau + \int_V \nabla_i\boldsymbol{Q}^i\mathrm{d}\tau + \int_V (\boldsymbol{\mathfrak{z}}_i\times\boldsymbol{\mathfrak{z}}_k)p^{ki}\mathrm{d}\tau$$

or, invoking the momentum equation (2.11), in the form

$$\int_V \frac{\mathrm{d}\boldsymbol{k}}{\mathrm{d}t}\rho\,\mathrm{d}\tau = \int_V \boldsymbol{h}\rho\,\mathrm{d}\tau + \int_V \nabla_i\boldsymbol{Q}^i\mathrm{d}\tau + \int_V (\boldsymbol{\mathfrak{z}}_i\times\boldsymbol{\mathfrak{z}}_k)p^{ki}\mathrm{d}\tau\,.$$

Since the volume V of the continuum is arbitrary, one obtains the equation of angular momentum of a continuum for the case of continuous motion in the differential form

$$\rho \frac{\mathrm{d}\boldsymbol{k}}{\mathrm{d}t} = \rho \boldsymbol{h} + \nabla_i \boldsymbol{Q}^i + (\mathbf{3}_i \times \mathbf{3}_k) p^{ki} . \tag{3.6}$$

In the classical case, in the absence of internal momenta and distributed mass and surface couples, the equation of angular momentum (3.6) acquires the form

$$(\mathbf{3}_i \times \mathbf{3}_k) p^{ki} = 0 . \tag{3.7}$$

Symmetry of the stress tensor in the classical case. Obviously, the equation of angular momentum (3.7) may be yet rewritten in the form

$$\underset{k<i}{(\mathbf{3}_i \times \mathbf{3}_k)\, p^{ki}} + (\mathbf{3}_i \times \mathbf{3}_k)\, p^{ki} = 0 .$$

Replacing in the last sum the summation indices k and i by i and k respectively, one finds

$$\underset{k<i}{(\mathbf{3}_i \times \mathbf{3}_k)\, p^{ki}} + \underset{k<i}{(\mathbf{3}_k \times \mathbf{3}_i)\, p^{ik}} = 0$$

or, from a property of the vector product,

$$\underset{k<i}{(\mathbf{3}_i \times \mathbf{3}_k)\, (p^{ki} - p^{ik})} = 0 .$$

It follows from the preceding result, that $p^{ki} = p^{ik}$ for $k \neq i$, i.e.,

$$p^{13} = p^{31} , \quad p^{12} = p^{21} , \quad p^{23} = p^{32} .$$

Thus, the moment equation in the classical case leads to the conclusion that the stress tensor is symmetric. Obviously, the equation of angular momentum (3.7) is satisfied identically, if the stress tensor is symmetric. Note the fact that the symmetry of the stress tensor follows from the equation of angular momentum, generally speaking, only in the absence of internal angular momenta and internal mass and surface couples of interaction.

Recall that earlier on four universal equations were derived which describe the motion of a continuous medium. Now, these equation may be augmented by the three equations of angular momentum. In the classical case, these three additional equations do not contain any new unknowns; they simply diminish the number of independent components of the stress tensor to six.

However, this system of equations of motion is still not complete. Below, it will be shown that in a number of cases it is possible to write down for the components of the stress tensor p^{ik} additional formulae which are connected with the physical characteristics of concrete models of continuous media. After this, one is well on the way to obtaining a closed system of equations.

Finally, some observations will be made on the subject of the concept of the vectors of the momentum $\boldsymbol{Q}$ and angular momentum $\boldsymbol{K}$. In Newtonian mechanics, the vectors $\boldsymbol{Q}$ and $\boldsymbol{K}$ may be considered to be invariant, since these quantities and the corresponding equations are conserved for a passage from one coordinate system to another Cartesian or curvilinear system which does not move with respect to the original system. However, these "invariant" objects are linked in an essential manner to the choice of the observer's frame of reference. For a passage from one reference frame to another, which is moving with respect to the first, these vectors change, even if this transition leads from one inertial system to another inertial system.

In the general case of transition to an arbitrary moving (non-inertial) coordinate system, the corresponding equations (the equations of momentum and angular momentum) change as regards the appearance of additional external inertial forces on the right-hand sides.

Note that, although the vectors $\boldsymbol{Q}$ and $\boldsymbol{K}$ are invariant in the sense stated above, they may, however, not be determined independently of the choice of a reference system in the class of systems which move with respect to each other.[1)] If one considers all processes for media in a concomitant coordinate system, then the momentum in this system vanishes.

[1)] In the general theory of relativity it is difficult to get the single-valued definition of the vectors Q and K and their points of applications as characteristics of a matter and a space in the finite volume of the Riemannian space. However, it can be done locally at any point in the form of the numerical functional relations of the local component of the vectors $\mathrm{d}\boldsymbol{Q}$ and $\mathrm{d}\boldsymbol{K}$ for the infinitesimal volumes. Integrating with respect to the volume, one can obtain the numerical integral equations, depending on the choice of the system of coordinates, representing the consequences of the local equation of momentum and of angular momentum for the finite volumes.

§4. Principal axes and principal components of a symmetric stress tensor

The tensor surface of the stress tensor. The tensor surface of the stress tensor will now be constructed. Choose any point O of a continuous medium and consider all possible area elements $d\sigma$ through it, which are characterized by the directions of their normal $\boldsymbol{n}$. A surface force with density $\boldsymbol{p}_n$ acts on each of these elements, which at times is referred to as stress. Projecting the stress $\boldsymbol{p}_n$ on to its corresponding normal $\boldsymbol{n}$, one obtains

$$p_{nn} = \boldsymbol{p}_n \cdot \boldsymbol{n} = (\boldsymbol{p}^i \cdot \boldsymbol{n}) n_i = p^{ki} n_k n_i \, .$$

For the sake of simplicity, a Cartesian coordinate system will be used, for which, as is known, the order of the indices is immaterial. Introduce the vector $\boldsymbol{r} = x_i \boldsymbol{э}^i$ from the point O and directed along $\boldsymbol{n}$. Then, obviously, $n_i = \cos(\boldsymbol{n} x_i) = x_i / r$. Choose the length of the vectors $\boldsymbol{r}$ so that

$$p_{nn} r^2 = p^{ki} x_k x_i = 2\Phi(x, y, z) = \text{const.} \, ,$$

where $2\Phi(x, y, z)$ is the quadratic form, corresponding to the symmetric stress tensor P. The geometrical locus of the points for which

$$p^{ki} x_k x_i = 2\Phi(x, y, z) = \text{const.}$$

forms a second order surface is the tensor surface of the stress tensor. The fundamental property of the internal stresses (2.4) may be rewritten in the form

$$\boldsymbol{p}_n = \boldsymbol{p}^i n_i = \boldsymbol{p}^i \frac{x_i}{r}$$

or, after projecting on to the x_k-axes of a Cartesian coordinate system, in the form

$$r p_n^k = p^{ki} x_i \, .$$

One can verify directly that

$$p^{ki} x_i \doteq \frac{\partial \Phi}{\partial x_k}$$

and, consequently

$$rp_n^k = \frac{\partial \Phi}{\partial x_k},$$

i.e.

$$r\boldsymbol{p}_n = \text{grad}\, \Phi .$$

Therefore, knowing the tensor surface $\Phi = \text{const.}$, one can use the following geometrical argument to find the direction of the stress $\boldsymbol{p}_n$, acting on the element $d\sigma$ with normal $\boldsymbol{n}$. Draw the vector $\boldsymbol{r}$, perpendicular to a given

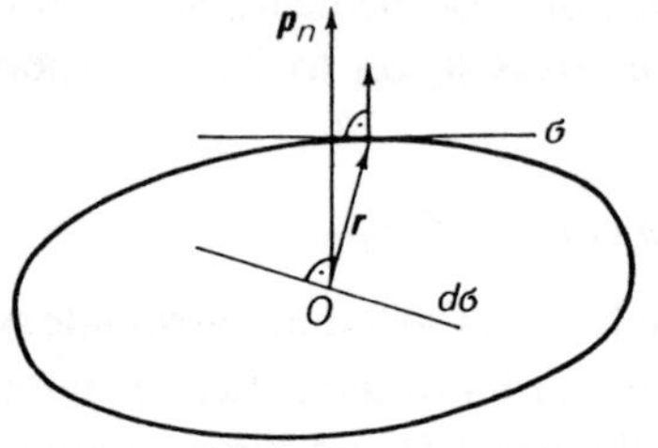

Fig. 28. The tensor surface of the stress tensor.

element at the point O (fig. 28). At the point of intersection of $\boldsymbol{r}$ with the surface $\Phi = \text{const.}$ construct the tangent plane σ to the tensor surface. Obviously, the vector $\boldsymbol{p}_n$ is perpendicular to the tangent plane σ.

The principal axes of the symmetric stress tensor. It is known that second order surfaces have at least three directions of $\boldsymbol{r}$ for which the tangent planes σ are perpendicular to $\boldsymbol{r}$. Such directions are called principal directions; obviously, for these directions, $\boldsymbol{p}_n$ is orthogonal to $d\sigma$. In the general case, there are only three such directions. They form an orthogonal trihedron and are called the principal axes of the stress tensor. If the tensor surface $\Phi = \text{const.}$ is a surface of revolution, say, a sphere, then there are infinitely many such directions.

For areas, orthogonal to the principal directions, $\boldsymbol{p}_n$ is collinear with $\boldsymbol{n}$ and therefore it must satisfy the equation

$$\boldsymbol{p}_n = p^{ki} n_i \boldsymbol{3}_k = \lambda \boldsymbol{n} = \lambda n_i \boldsymbol{3}^i \tag{4.1}$$

or

$$p_k^i n_i \boldsymbol{3}^k = \lambda \delta_k^i n_i \boldsymbol{3}^k ,$$

whence

$$(p_k^i - \lambda \delta_k^i) n_i \boldsymbol{3}^k = 0 ,$$

or

$$(p^i_k - \lambda\delta^i_k)\,n_i = 0\,. \tag{4.2}$$

One has now arrived at a homogeneous system of three algebraic equations for the determination of the direction cosines of the three principal directions n_i. This system will have a non-trivial solution if an only if

$$\Delta = \text{Det}\,\|p^i_k - \lambda\delta^i_k\| = |p^i_k - \lambda\delta^i_k| = 0\,,$$

i.e.

$$\begin{vmatrix} p^1_1-\lambda & p^2_1 & p^3_1 \\ p^1_2 & p^2_2-\lambda & p^3_2 \\ p^1_3 & p^2_3 & p^3_3-\lambda \end{vmatrix} = 0 \tag{4.3}$$

or, in expanded form,

$$-\lambda^3 + I_1\lambda^2 - I_2\lambda + I_3 = 0\,, \tag{4.4}$$

where

$$p^\alpha_\alpha = I_1\,,$$

$$\begin{vmatrix} p^2_2 & p^3_2 \\ p^2_3 & p^3_3 \end{vmatrix} + \begin{vmatrix} p^3_3 & p^1_3 \\ p^3_1 & p^1_1 \end{vmatrix} + \begin{vmatrix} p^1_1 & p^2_1 \\ p^1_2 & p^2_2 \end{vmatrix} = I_2\,,$$

$$\text{Det}\,\|p^i_k\| = I_3\,.$$

Thus a secular equation has been derived. As is known, if the tensor p^{ij} is symmetric, this equation has three real roots. The roots $\lambda_1, \lambda_2, \lambda_3$ of this equation, by (4.1), determine the stresses on the planes, orthogonal to the principal directions, (the principal planes):

$$\lambda_1 = p_{n1} = p_1\,;\quad \lambda_2 = p_{n2} = p_2\,;\quad \lambda_3 = p_{n3} = p_3\,.$$

These are referred to as the principal components of the stress tensor.

Once the p_1, p_2, p_3 are known, one can find from the system of equations (4.2) the components n_i of the vectors $\boldsymbol{n}$, determining the principal directions (where one must also use the condition $(\boldsymbol{n}\cdot\boldsymbol{n}) = 1$). Obviously, Formulae (4.1)–(4.3) are true in any curvilinear coordinate system.

Referred to principal axes x, y and z, the equation of the tensor surface $2\Phi = \text{const.}$ reduces to the canonical form

$$2\Phi = p_1x^2 + p_2y^2 + p_3z^2 = \text{const.}$$

For the components of the stress tensor, referred to principal axes, one has

$$p^{ii} = p_i^i = p_{ii} = \lambda_i = p_i$$

and

$$p^{ki} = p_k^i = p_{ki} = 0 \quad \text{for} \quad k \neq i\,.$$

On planes, perpendicular to the principal stress axes, only the normal stress components are non-zero and the tangential stress components vanish.

If $p_1 = p_2 = p_3$, the tensor surface of the stress tensor is a sphere.

The principal axes of the strain tensor, of the stress tensor and the strain rate tensor have been introduced. In general, all these axes are different. As will be seen later, the conditions under which they coincide are connected with strong physical assumptions relating to properties of a medium under consideration.

If $p_1 \neq 0$, while $p_2 = p_3 = 0$, then at a given point of a continuum one has pure stretching along the x_1-axis, if $p_1 > 0$, and pure compression, if $p_1 < 0$. Thus, any state of stress at a given point of a continuous medium can be regarded as a combination of three pure extensions or contractions along the principal axes of the stress tensor.

The coefficients of the secular equation (4.4) are invariants of the stress tensor. Obviously, they can be expressed in terms of the roots of the secular equation by the formulae

$$I_1 = p_1 + p_2 + p_3\,,$$
$$I_2 = p_2 p_3 + p_1 p_2 + p_1 p_3\,,$$
$$I_3 = p_1 p_2 p_3\,.$$

To an arbitrary symmetric tensor of the second order

$$T = T_{ij}\, \mathbf{э}^i \mathbf{э}^j = T^{i}_{\cdot k}\, \mathbf{э}_i\, \mathbf{э}^k$$

one can link the tensor surface

$$\Phi = T_{ij}\, dx^i\, dx^j = \text{const.}$$

To determine the direction of the principal axes and the principal components of tensor T one can apply the above consideration.

Chapter IV

The closed systems of mechanical equations for the simplest models of continuous media. Some results from tensor analysis

§1. Ideal fluid and gas

Any continuum in any state of continuous motion satisfies the differential equations of continuity, motion and moment of momentum. However, different real media behave differently under the same external conditions.

Consequently, even if the appropriate boundary conditions are added, these equations do not completely describe the motion of a real continuum. This fact is demonstrated by the circumstance that the numbers of equations is less than the number of unknowns appearing in them. In other words, the system is not closed.

The construction of a closed system of equations, describing the motion of a real continuum, is connected with the search for additional relations between the parameters of a given medium. The establishment of a closed system of equations is equivalent to the construction of a theoretical (mathematical) model of the medium under consideration.

The construction of new models of continuous media is an important branch of mechanics, which is referred to as rheology. This work involves the experimental study of the properties of materials. In this context, one must always also use the known general principles of mechanics and physics, for instance, thermodynamic relations. Variational principles also turn out to be useful.

In this chapter, certain of the simplest classical models of continuous media will be studied. In this context, consideration will be restricted to those cases in which the properties of a medium and the class of processes under scrutiny are such that one need not for the description of the mechanical motion determine the thermodynamic properties of the medium, i.e., in which the system of mechanical equations turns out to be closed without the inclusion of the thermodynamic equations.

In the general case of a study of all possible processes, one must likewise introduce the relations of thermodynamics.

A beginning will be made with a study of ideal fluids and gases.

Definition of an ideal fluid and an ideal gas. A medium is called an ideal fluid or an ideal gas, if the stress vector $\boldsymbol{p}_n$ of any element of area with normal $\boldsymbol{n}$ is orthogonal to that element, i.e., $\boldsymbol{p}_n \parallel \boldsymbol{n}$.

Experimental data and general physical considerations show that any fluid for very large ranges of temperature and pressures effectively has this property.

The stress tensor in an ideal fluid is spherical. In this case, the tensor surface is obviously a sphere. Consequently, $p_1 = p_2 = p_3$, i.e., the principal components of the stress tensor are identical. Let $-p$ denote these components; then p is called the pressure. The choice of sign is dictated by the desire to introduce the pressure as a positive quantity, since experiments show that, in typical cases, media for which an ideal fluid serves as a model are in a state of compression for $p > 0$.

For such a medium, any three mutually orthogonal directions are principal directions. Hence, in any Cartesian coordinate system, the matrix of the components of the stress tensor has the form

$$\begin{Vmatrix} -p & 0 & 0 \\ 0 & -p & 0 \\ 0 & 0 & -p \end{Vmatrix}.$$

In particular, one may write for the mixed components p^i_k

$$p^i_k = -p\delta^i_k \,. \tag{1.1}$$

Obviously, the components δ^i_k of the tensor $\delta^i_k \mathbf{з}^k \mathbf{з}_i$ do not change under coordinate transformations ($\delta'^i_k = \delta^i_k$), and therefore formula (1.1) for the mixed components of the stress tensor in an ideal fluid is not only true in a Cartesian, but in any curvilinear coordinate system.

The contravariant components of this tensor have the form

$$p^{ki} = g^{ks} p^i_s = -pg^{ks} \delta^i_s = -pg^{ki}, \tag{1.2}$$

and the covariant components the form

$$p_{ki} = g_{ks} p^s_i = -pg_{ks} \delta^s_i = -pg_{ki}.$$

Consequently, in an ideal fluid, the stress tensor yields a single number p, and not nine or six numbers p^{ki}, as in the general case.

For an ideal fluid,

$$P = -pG,$$

where G is the metric tensor.

Note that any tensor T, the tensor surface of which is a sphere, is called spherical. All spherical tensors have the form

$$T = kG,$$

where k is a scalar.

Equations of motion of an ideal fluid. The equations of motion of a continuum [(2.11) Chapter III] in any curvilinear coordinate system are

$$\rho a^k = \rho F^k + \nabla_i p^{ki};$$

and by (1.2), they may be rewritten for an ideal fluid in the form

$$\rho a^k = \rho F^k - g^{ki} \nabla_i p. \tag{1.3}$$

In writing down (1.3), it has been taken into consideration that the components of the tensor g^{ki} behave like constants for covariant differentiation.

These equations will now be written in vector form. Obviously, the quantities $\nabla_i p$ are covariant components of the gradient vector of p, while $g^{ki} \nabla_i p$ are its contravariant components. Therefore, equations (1.3) have the vector form

$$\rho \boldsymbol{a} = \rho \boldsymbol{F} - \operatorname{grad} p \,. \tag{1.4}$$

Their projections on Cartesian coordinate axes may be written in the form

$$\begin{aligned}
\frac{du}{dt} &= F_x - \frac{1}{\rho}\frac{\partial p}{\partial x}, \\
\frac{dv}{dt} &= F_y - \frac{1}{\rho}\frac{\partial p}{\partial y}, \\
\frac{dw}{dt} &= F_z - \frac{1}{\rho}\frac{\partial p}{\partial z}
\end{aligned} \tag{1.5}$$

or

$$\begin{aligned}
\frac{\partial u}{\partial t} + u\frac{\partial u}{\partial x} + v\frac{\partial u}{\partial y} + w\frac{\partial u}{\partial z} &= F_x - \frac{1}{\rho}\frac{\partial p}{\partial x}, \\
\frac{\partial v}{\partial t} + u\frac{\partial v}{\partial x} + v\frac{\partial v}{\partial y} + w\frac{\partial v}{\partial z} &= F_y - \frac{1}{\rho}\frac{\partial p}{\partial y}, \\
\frac{\partial w}{\partial t} + u\frac{\partial w}{\partial x} + v\frac{\partial w}{\partial y} + w\frac{\partial w}{\partial z} &= F_z - \frac{1}{\rho}\frac{\partial p}{\partial z}
\end{aligned} \tag{1.6}$$

in which they are referred to as Euler Equations.

The equations of motion of an ideal fluid in the form of Lamb–Gromeka. These equations will now be rewritten in a somewhat different form.

It is readily seen that an acceleration may always be expressed in the form

$$\frac{d\boldsymbol{v}}{dt} = \frac{\partial \boldsymbol{v}}{\partial t} + \operatorname{grad}\frac{\boldsymbol{v}^2}{2} + 2\boldsymbol{\omega} \times \boldsymbol{v}\,, \tag{1.7}$$

where ω is the vorticity vector.

In fact, in terms of Cartesian coordinates, one has for the projection of the acceleration on to the x-axis

$$\frac{du}{dt} = \frac{\partial u}{\partial t} + \frac{\partial u}{\partial x}u + \frac{\partial u}{\partial y}v + \frac{\partial u}{\partial z}w$$

$$= \frac{\partial u}{\partial t} + \frac{1}{2}\frac{\partial}{\partial x}(u^2+v^2+w^2) - \left(\frac{\partial v}{\partial x} - \frac{\partial u}{\partial y}\right)v + \left(\frac{\partial u}{\partial z} - \frac{\partial w}{\partial x}\right)w$$

$$= \frac{\partial u}{\partial t} + \frac{1}{2}\frac{\partial \boldsymbol{v}^2}{\partial x} + 2(\omega_y w - \omega_z v) = \frac{\partial u}{\partial t} + \frac{1}{2}\frac{\partial \boldsymbol{v}^2}{\partial x} + 2(\boldsymbol{\omega}\times\boldsymbol{v})_x .$$

Analogous formulae are obtained for the projections of the acceleration on to the y- and z-axes. Therefore the acceleration $d\boldsymbol{v}/dt$ may be written in the vector form (1.7) and the equation of motion of an ideal fluid becomes

$$\frac{\partial \boldsymbol{v}}{\partial t} + \tfrac{1}{2}\,\text{grad}\,\boldsymbol{v}^2 + 2\boldsymbol{\omega}\times\boldsymbol{v} = \boldsymbol{F} - \frac{1}{\rho}\,\text{grad}\,p . \tag{1.8}$$

This equation is referred to as the Euler equation of motion in the form of Lamb–Gromeka. Such a transformation of the acceleration may be applied to any continuous medium; in particular, it turns out to be very useful in the study of many problems of hydromechanics.

These three component equations of motion of an ideal fluid must be augmented by the equation of continuity

$$\frac{\partial \rho}{\partial t} + \text{div}\,\rho\boldsymbol{v} = 0 .$$

One has thus arrived at a system of four equations which for known body forces F_x, F_y, F_z contain five unknowns: u, v, w, p, ρ. *Such a system is still not closed.*

Complete system of equations of motion of an ideal incompressible (in general, inhomogeneous) fluid. In certain cases, one may assume, in addition, that the ideal fluid under consideration is incompressible, i.e., that the density of each of its particles is constant. Then this system of four equations is enlarged by the condition

$$\frac{d\rho}{dt} = \frac{\partial \rho}{\partial t} + \boldsymbol{v}\cdot\text{grad}\,\rho = 0$$

or, in Cartesian coordinates, by the equation

$$\frac{\partial \rho}{\partial t} + u\frac{\partial \rho}{\partial x} + v\frac{\partial \rho}{\partial y} + w\frac{\partial \rho}{\partial z} = 0\,.$$

This condition closes the system of equations, describing the motion of an ideal, incompressible fluid, which will now be written down in full completeness:

$$\begin{aligned} &\frac{d\boldsymbol{v}}{dt} = \boldsymbol{F} - \frac{1}{\rho}\,\mathrm{grad}\,p\,, \\ &\mathrm{div}\,\boldsymbol{v} = 0\,, \\ &\frac{d\rho}{dt} = 0\,. \end{aligned} \tag{1.9}$$

When the mass of the particle is changing, the last equation may differ.

Note that in the case of a homogeneous, incompressible fluid with the constant mass in every particle the density ρ is constant in a particle and the same for all particles; hence, in essence, it becomes a known function. The complete system of mechanical equations, in this case, consists of the Euler equations and the equation of continuity:

$$\begin{aligned} &\frac{\partial v^i}{\partial t} + v^j \nabla_j v^i = F^i - \frac{1}{\rho} g^{ij} \nabla_j p\,, \\ &\nabla_\alpha v^\alpha = 0\,. \end{aligned} \tag{1.10}$$

The closed system of equations of motion of an ideal compressible fluid (gas) in the case of barotropic processes. For a motion of a compressible fluid (gas), it is often possible to assume that

$$p = f(\rho)\,,$$

i.e., the pressure depends only on the density. *Processes in which*

$$p = f(\rho)$$

are called barotropic. An example of a barotropic process is the isothermal motion of a gas, obeying Clapeyron's equation

$$p = R\rho T\,,$$

where R is the gas constant (For isothermal motion, the temperature T is a constant parameter which is the same for all particles).

Obviously, the condition of barotropy (if $f(\rho)$ is known) permits to close the system of equations, describing the motion of an ideal compressible fluid.

The complete system of equations in this case, in Cartesian rectilinear coordinates, has the form

$$
\begin{aligned}
&\frac{\partial \rho}{\partial t}+\frac{\partial \rho u}{\partial x}+\frac{\partial \rho v}{\partial y}+\frac{\partial \rho w}{\partial z}=0, \\
&\frac{\partial u}{\partial t}+u\frac{\partial u}{\partial x}+v\frac{\partial u}{\partial y}+w\frac{\partial u}{\partial z}=F_x-\frac{1}{\rho}\frac{\partial p}{\partial x}, \\
&\frac{\partial v}{\partial t}+u\frac{\partial v}{\partial x}+v\frac{\partial v}{\partial y}+w\frac{\partial v}{\partial z}=F_y-\frac{1}{\rho}\frac{\partial p}{\partial y}, \\
&\frac{\partial w}{\partial t}+u\frac{\partial w}{\partial x}+v\frac{\partial w}{\partial y}+w\frac{\partial w}{\partial z}=F_z-\frac{1}{\rho}\frac{\partial p}{\partial z}, \\
&p=f(\rho).
\end{aligned}
\tag{1.11}
$$

Of course, in the general case of motion of a fluid or gas, the condition of barotropy may not be fulfilled. In order to describe such motion, one must introduce additional equations of a thermodynamic character.

§ 2. Linear elastic body and linear viscous fluid

Consider other particular models of continuous media: the model of a linear elastic body and the model of a linearly viscous fluid. The construction of these models can occur side by side, because, as will be seen, their methods of derivation are formally analogous. In essence, these two models describe two quite different types of mechanical behaviour of real media.

The elastic body. An elastic body is a medium in which the components of the stress tensor p^{ki} in every particle are functions of the components of the strain tensor ε_{ij}, the components of the metric tensor

g_{ij}, the temperature T and, possibly, other parameters of a physico-chemical nature (for example, the concentration of the phase):

$$p^{ij} = f^{ij}(\varepsilon_{\alpha\beta}, g^{\alpha\beta}, T, \chi_1, \ldots, \chi_n). \tag{2.1}$$

The viscous fluid. A viscous fluid is a medium in which the components of the stress tensor can be expressed in the form

$$p^{ij} = -pg^{ij} + \tau^{ij}, \tag{2.2}$$

with

$$\begin{aligned} p &= p(\rho, T, \chi_1, \ldots, \chi_n), \\ \tau^{ij} &= \varphi^{ij}(e_{\alpha\beta}, g^{\alpha\beta}, T, \chi_1, \ldots, \chi_n), \end{aligned} \tag{2.3}$$

where the $e_{\alpha\beta}$ are the components of the strain rate tensor.

In this section, special attention will be given to the dependence of the f^{ij} on $\varepsilon_{\alpha\beta}$ and $g^{\alpha\beta}$, and the dependence of the φ^{ij} on $e_{\alpha\beta}$ and $g^{\alpha\beta}$; therefore, in what follows, the parameters T and χ_i will not be explained.

The laws of Hooke and Navier–Stokes. The actual forms of the functions $f^{ij}(\varepsilon_{\alpha\beta}, g^{\alpha\beta}, T, \chi_i)$ and $\varphi^{ij}(e_{\alpha\beta}, g^{\alpha\beta}, T, \chi_i)$ may be different for different models of elastic and viscous media. Experiments show that in many solid bodies, for example, in metals under normal conditions (not very high temperatures and stresses) the stresses and strains are interrelated by Hooke's law. In many fluid media, for example, water and air, the viscous stresses and strain rates are interrelated by Navier–Stokes' law. These laws may be derived with the aid of the following argument which will be carried out for Hooke's law.

Assume that the functions f^{ij} may be expanded in Taylor series with respect to $\varepsilon_{\alpha\beta}$ and, in the absence of stresses (i.e., for $p^{ij}=0$), strains are also absent (i.e., $\varepsilon_{\alpha\beta}=0$), and vice versa.[1)]

Under these assumptions, one obtains

$$p^{ij} = f^{ij}(\varepsilon_{\alpha\beta}, g^{\alpha\beta}) = A^{ij\alpha\beta}\varepsilon_{\alpha\beta} + \ldots,$$

where the coefficients $A^{ij\alpha\beta}$ may depend on T and χ_s. If the deformations are small, one may keep only the linear terms of the series expansions of p^{ij} and write simply

[1)] Note that strains may also arise when $p^{ij} = 0$ (for example, thermal expansion). At this stage, for the sake of simplicity, consideration will be given to the case when p^{ij} are functions of ε_{ij} for $T = \text{const.}$ and $\chi = \text{const.}$

$$p^{ij} = A^{ij\alpha\beta} \varepsilon_{\alpha\beta} \,. \tag{2.4}$$

Analogous assumptions concerning the functions φ^{ij} lead to the equations

$$\tau^{ij} = B^{ij\alpha\beta} e_{\alpha\beta} \,. \tag{2.5}$$

Relations (2.4) are known as Hooke's law, and Relations (2.5) as Navier–Stokes' law (or Newton's law of viscosity).

Equations (2.4) and (2.5) have been derived under the assumptions that the $\varepsilon_{\alpha\beta}$ (for Hooke's law) and $e_{\alpha\beta}$ (for Navier–Stokes' law) are small. However, note that, in particular, for water, air, and certain other fluids, Navier–Stokes' law is applicable even in cases where the components of the strain rate tensor are not small. From general thermodynamic relations, one finds that Hooke's law is physically admissible only as an approximate law for small deformations.

The branch of continuum mechanics, in which the behaviour of continua which obey Hooke's law or the more general law (2.1), is studied, is called the theory of elasticity; the branch in which the motion of continua, which obey Navier–Stokes' law or the more general laws (2.2) and (2.3), is studied, is called the theory of viscous fluids.

It follows immediately from (2.4) and (2.5), which are invariant with respect to the choice of coordinate systems, that the $A^{ij\alpha\beta}$ and $B^{ij\alpha\beta}$ are components of fourth order tensors. They are physical characteristics of a given continuum and depend, generally speaking, on the temperature and other physico-chemical parameters, characterizing the state of the medium under consideration.

A fourth order tensor has $3^4 = 81$ components. However, owing to the symmetry of the stress tensor (in the classical case) and the symmetry of the strain and strain rate tensors, there are only 36 independent components $A^{ij\alpha\beta}$ and $B^{ij\alpha\beta}$, because the tensors A and B must be symmetric with respect to pairs of indices i and j and their symmetry with respect to the pairs of indices α and β may be assumed. If a medium, the behaviour of which is described by Hooke's or Navier–Stokes' law, has some kind of geometric properties of symmetry, then the number of independent components $A^{ij\alpha\beta}$ and $B^{ij\alpha\beta}$ will be reduced further. In particular, if a medium is isotropic, then all $A^{ij\alpha\beta}$ and $B^{ij\alpha\beta}$ are determined by two parameters.

Material properties of anisotropy, isotropy and gyrotropy. The properties of isotropic media are the same in all directions. If the properties of a medium are different in different directions, the medium is said to be anisotropic. Anisotropic media may have different types of symmetry.

A more precise mathematical definition of the property of symmetry and, in particular, of isotropy, will now be given. Mechanical and physical properties of a medium can usually be described with the aid of certain tensors and tensor equations (for example, if Hooke's law is satisfied, the elastic properties are given by means of the tensor $A^{ij\alpha\beta}$). A medium is said to possess symmetry, if there exists a group of coordinate transformations, not only consisting of the identity transformation, such that the components of the tensors, describing the properties of the medium, are invariant under the transformations in that group.

In particular, a medium is called isotropic, if the components of the tensors, defining its properties, do not change for any orthogonal transformations. Note that orthogonal transformations may be defined as transformations under which the components of the metric tensor are preserved (i.e., the scalar products of the base vectors remain unchanged):

$$g'_{ij} = g_{\alpha\beta} \frac{\partial x^\alpha}{\partial y^i} \frac{\partial x^\beta}{\partial y^j} = g_{ij} .$$

The complete orthogonal group contains the transformation of rotation (with determinant of the transformation equal to $+1$) and rotation combined with reflection (with determinant equal to -1).

If the properties of a medium are invariant only under the group of rotations, but not invariant under reflections, then the medium is said to be gyrotropic.

Consider in somewhat greater detail the meaning of the property of isotropy (and gyrotropy) for an elastic body obeying Hooke's law. Select at some point of such a continuous medium, at a given instant of time, two Cartesian coordinate systems: one, x^1, x^2, x^3, and another, y^1, y^2, y^3, which is rotated with respect to the first. Denote the components of the tensors under consideration in the system x^1, x^2, x^3 by plain symbols and in the system y^1, y^2, y^3 by primed symbols. Obviously,

$$A'^{ij\alpha\beta} = \frac{\partial y^i}{\partial x^p} \frac{\partial y^j}{\partial x^q} \frac{\partial y^\alpha}{\partial x^\lambda} \frac{\partial y^\beta}{\partial x^\mu} A^{pq\lambda\mu} . \tag{2.6}$$

Writing down Hooke's law in the system x^1, x^2, x^3, one must use the coefficients $A^{ij\alpha\beta}$, in the system y^1, y^2, y^3, the coefficients $A'^{ij\alpha\beta}$. Consider two deformed states of a medium which have the same form in the different (rotated with respect to each other) systems x^i and y^i, i.e., for which

$$\varepsilon_{ij} = \varepsilon'_{ij} .$$

Clearly, for an isotropic medium, the state of stress in this case must likewise have the same form in the systems x^i and y^i. If $A'^{ij\alpha\beta} = A^{ij\alpha\beta}$, i.e., the coefficients in Hooke's law in both coordinate systems are the same, then $p^{ij} = p'^{ij}$. In this case, the continuous medium is isotropic or gyrotropic. On the other hand, if $A'^{ij\alpha\beta} \neq A^{ij\alpha\beta}$, i.e., the coefficients in Hooke's law are different in the coordinate systems x^1, x^2, x^3 and y^1, y^2, y^3, then $p^{ij} \neq p'^{ij}$ and the medium is anisotropic. Experiments show that, for example, crystalline media with regularly ordered distributions of molecules or atoms, and also fibrous materials, are anisotropic; their properties are different in different directions.

Examples of isotropic media, for which one coordinate system has no advantage over another rotated with respect to the first, are water and other media with so-called amorphous structure, and also media consisting of small crystals, as long as these elementary crystals are distributed in a random fashion. Such materials are usually the metals used in engineering.

The laws of Hooke and Navier–Stokes for gyrotropic media. It will now be shown that for isotropic and gyrotropic elastic media there are only two independent components of the tensor $A^{ij\alpha\beta}$.[1) Direct the coordinate axes along the principal directions of the strain tensor ε_{ij}. Obviously, in this case, only coefficients of the form $A^{ij\alpha\alpha}$ appear in Hooke's law. It will be shown that the $A^{ij\alpha\alpha} = 0$ for $i \neq j$. In fact, after rotating the coordinate system about the i-th axis by 180°, one obtains a new coordinate system in which the i-th axis remains as before and the other two axes are reversed. By the transformation law (2.6) for the components of the tensor A, one has for $i \neq j$ and any α

$$A'^{ij\alpha\alpha} = -A^{ij\alpha\alpha} .$$

However, if the medium is gyrotropic or isotropic, one must have $A'^{ij\alpha\alpha} = A^{ij\alpha\alpha}$ and, consequently, $A^{ij\alpha\alpha} = 0$ for $i \neq j$. Hence, since in this coordinate system $p^{ij} = 0$ for $i \neq j$, it follows that the principal axes of the strain and stress tensors coincide in a gyrotropic, and hence in an isotropic medium, obeying Hooke's law.

1) For a fourth order tensor, the concepts of isotropy and gyrotropy coincide.

In Hooke's law, referred to principal axes, only 9 coefficients $A^{ii\alpha\alpha}$ of the 81 coefficients $A^{ij\alpha\beta}$ are essential.

The order in which the axes are enumerated is inessential, as a consequence of the property of gyrotropy of the medium.[1)] Thus,

$$A^{1111} = A^{2222} = A^{3333} = 2\mu + \lambda\,,$$
$$A^{1122} = A^{1133} = A^{2233} = \lambda\,,$$
$$A^{ii\alpha\alpha} = A^{\alpha\alpha ii}\,,$$

where $2\mu+\lambda$ and λ are the new notation for the above-mentioned two distinct and, generally speaking, non-zero components of the tensor A. [Note that the components A^{ijij}, $i \neq j$, are also non-zero. It will be seen below that these components are equal to μ; *cf.*, (2.12).]

All the foregoing reasoning may be extended to gyrotropic (and hence to isotropic) media, satisfying Navier–Stokes' law. For gyrotropic media obeying the Navier–Stokes law, the principal axes of the strain rate tensor coincide with the principal axes of the stress tensor. All the coefficients $B^{ij\alpha\beta}$ can be expressed in terms of two coefficients λ_1 and μ_1.

For isotropic media, Hooke's law

$$p^{ij} = A^{ij\alpha\beta}\varepsilon_{\alpha\beta}\,, \tag{2.7}$$

referred to the principal axes of the strain and stress tensors, has the form

$$\left.\begin{aligned} p_1 &= \lambda(\varepsilon_1+\varepsilon_2+\varepsilon_3)+2\mu\varepsilon_1\,,\\ p_2 &= \lambda(\varepsilon_1+\varepsilon_2+\varepsilon_3)+2\mu\varepsilon_2\,,\\ p_3 &= \lambda(\varepsilon_1+\varepsilon_2+\varepsilon_3)+2\mu\varepsilon_3\,; \end{aligned}\right\} \tag{2.8}$$

where λ and μ are referred to as Lamé constants.

In an analogous manner, Navier–Stokes' law for isotropic media, referred to the principal axes of the strain rate and stress tensors, may be rewritten in the form

1) For example, the x^2-axis may, obviously, be placed in the position x^1 by a rotation of 90° about the x^3-axis.

$$\begin{aligned}\tau_1 &= \lambda_1(e_1+e_2+e_3)+2\mu_1 e_1\,,\\ \tau_2 &= \lambda_1(e_1+e_2+e_3)+2\mu_1 e_2\,,\\ \tau_3 &= \lambda_1(e_1+e_2+e_3)+2\mu_1 e_3\,.\end{aligned} \tag{2.9}$$

The expression (2.8) may be written in the following invariant form

$$p_{ij} = \lambda I_1(\varepsilon)g_{ij}+2\mu\varepsilon_{ij} \tag{2.10}$$

or

$$p_{ij} = \lambda I_1(\varepsilon)g^{ij}+2\mu g^{i\alpha}g^{j\beta}\varepsilon_{\alpha\beta}\,. \tag{2.11}$$

Either of the formulae (2.10) or (2.11) represent Hooke's law for isotropic media in an arbitrary curvilinear coordinate system.

From (2.11), one readily obtains an expression for the coefficients $A^{ij\alpha\beta}$ in arbitrary curvilinear coordinates:

$$A^{ij\alpha\beta} = \lambda g^{ij}g^{\alpha\beta}+\mu(g^{i\alpha}g^{j\beta}+g^{i\beta}g^{j\alpha})\,. \tag{2.12}$$

Using an analogous argument with respect to Navier–Stokes' law, one finds that this law for isotropic media in arbitrary curvilinear coordinates has the form

$$\tau_{ij} = \lambda_1 I_1(e)g_{ij}+2\mu_1 e_{ij} \tag{2.13}$$

or

$$\tau^{ij} = \lambda_1 I_1(e)g^{ij}+2\mu_1 g^{i\alpha}g^{j\beta}e_{\alpha\beta}\,. \tag{2.14}$$

From (2.2), one obtains the following relations between the components of the stress and strain rate tensors for an isotropic viscous fluid in an arbitrary curvilinear coordinate system:

$$p^{ij} = -pg^{ij}+\lambda_1 g^{ij}\operatorname{div}\boldsymbol{v}+2\mu_1 g^{i\alpha}g^{j\beta}e_{\alpha\beta}\,. \tag{2.15}$$

In Cartesian (non-principal) coordinates, Hooke's law for isotropic media has the form

$$p_{ii} = \lambda I_1(\varepsilon)+2\mu\varepsilon_{ii} \tag{2.16}$$

and, for $i\neq j$

$$p_{ij} = 2\mu\varepsilon_{ij}\,,$$

while Navier–Stokes' law has the form

$$p_{ii} = -p+\lambda_1\operatorname{div}\boldsymbol{v}+2\mu_1\frac{\partial v_i}{\partial x^i}$$

and, for $i \neq j$

$$p_{ij} = 2\mu_1 e_{ij} = \mu_1 \left(\frac{\partial v_i}{\partial x^j} + \frac{\partial v_j}{\partial x^i} \right). \tag{2.17}$$

Young's modulus, Poisson's coefficient, coefficient of viscosity. Instead of the Lamé coefficients λ and μ, one introduces in the theory of elasticity the following characteristics of a material:

Young's modulus

$$E = \mu \frac{(3\lambda + 2\mu)}{\lambda + \mu}$$

and Poisson's coefficient

$$\sigma = \frac{\lambda}{2(\lambda + \mu)}.$$

In the theory of viscous flow, one introduces the dynamic coefficient of viscosity $\mu = \mu_1$, the kinematic viscosity $\nu = \mu/\rho$, and also a second coefficient of viscosity

$$\zeta = \lambda_1 + \tfrac{2}{3}\mu.$$

In the sequel, Lamé's coefficient λ_1 in the case of motion of a viscous fluid will be denoted simply by λ.

Numerical values of E, σ, μ and ν for certain media are given below.

In conclusion, note that for $T = \text{const.}$ and $\chi_i = \text{const.}$, the laws of Hooke and Navier–Stokes yield closed systems of equations of motion for isotropic elastic media and incompressible viscous fluids.

Navier–Stokes equations. In order to describe the complete system of equations of motion of a continuous medium in the case of viscous incompressible fluids, introduce first of all the equation of motion of a viscous, in general, compressible liquid, satisfying Navier–Stokes' law (2.15) or

$$p^{ij} = -pg^{ij} + \lambda g^{ij} \operatorname{div} \boldsymbol{v} + 2\mu e^{ij}, \tag{2.18}$$

the so-called Navier-Stokes equations. They are obtained by substituting Navier–Stokes' law (2.18) into the impulse equation.

Note beforehand that in Euclidean space

Material (at normal temperature)	$E \cdot 10^{-4}$ [kg/mm²]	σ
Steel	2.0 – 2.2	0.24 – 0.28
Iron	1.6 – 2.0	0.28
Copper	1.1	0.31 – 0.34
Aluminium	0.69	0.32 – 0.36
Bronze	1.1	0.35
Glass	0.56	0.25
Rubber	0.00008	0.47

Temperature, °C	20	100	300	500
Non-alloy steel, E [kg/mm²]	21350	21000	19800	17900

Media	Temperature °C	$\mu \cdot 10^2$ [g/sec cm]	$\nu \cdot 10^2$ [cm²/sek]	ρ, [g/cm³]
Water	5	1.514	1.514	1.00
	10	1.304	1.304	1.00
	15	1.137	1.138	0.999
	20	1.002	1.004	0.998
	50	0.548	0.554	0.998
Benzene	15	0.7	0.8	0.88
Alcohol	15	1.34	1.7	0.8
Mercury	15	1.58	0.116	13.6
Glycerine	15	23	18	1.26
Lubricating oil /average viscosity)	20	275 – 350	300 – 380	0.9 – 0.95
Air (under pressure of 1 atm)	0	$1.71 \cdot 10^{-2}$	13.2	$1.293 \cdot 10^{-3}$
	10	$1.76 \cdot 10^{-2}$	14.1	$1.247 \cdot 10^{-3}$
	15	$1.78 \cdot 10^{-2}$	14.5	$1.225 \cdot 10^{-3}$
	20	$1.81 \cdot 10^{-2}$	15.0	$1.205 \cdot 10^{-3}$
	60	$2 \cdot 10^{-2}$	18.8	$1.060 \cdot 10^{-3}$

$$\nabla_i \nabla_j v^\alpha = \nabla_j \nabla_i v^\alpha .$$

In fact, the quantities

$$T^\alpha_{ij} \equiv \nabla_i \nabla_j v^\alpha - \nabla_j \nabla_i v^\alpha$$

are the components of a third order tensor; in Cartesian system of coordinates,

$$T^\alpha_{ij} = \frac{\partial^2 v^\alpha}{\partial x^i \partial x^j} - \frac{\partial^2 v^\alpha}{\partial x^j \partial x^i} = 0$$

and, consequently,

$$T^\alpha_{ij} = 0$$

in any curvilinear coordinate system. Thus, the result of successive covariant differentiations does not depend on the order of differentiation, if one may introduce for the entire space one Cartesian system of coordinates, i.e., if the space is Euclidean.[1)]

Compute now $\nabla_j p^{ij}$, when p^{ij} are determined by (2.18) and λ and μ are constants:

$$\begin{aligned}\nabla_j p^{ij} &= -g^{ij}\nabla_j p + \lambda g^{ij}\nabla_j \operatorname{div} \boldsymbol{v} + 2\mu \nabla_j e^{ij} \\ &= -g^{ij}\nabla_j p + \lambda g^{ij}\nabla_j \nabla_\alpha v^\alpha + \mu \nabla_j g^{i\alpha} g^{i\beta}(\nabla_\alpha v_\beta + \nabla_\beta v_\alpha) \\ &= -g^{ij}\nabla_j p + \lambda g^{ij}\nabla_j \nabla_\alpha v^\alpha + \mu g^{i\alpha}\nabla_j \nabla_\alpha v^j + \mu g^{j\beta}\nabla_j \nabla_\beta v^i \\ &= -g^{ij}\nabla_j p + (\lambda+\mu) g^{ij}\nabla_j \nabla_\alpha v^\alpha + \mu \nabla^\beta \nabla_\beta v^i \\ &= -\nabla^i p + (\lambda+\mu)\nabla^i \operatorname{div} \boldsymbol{v} + \mu \Delta v^i ,\end{aligned}$$

where $\nabla^i = g^{ij}\nabla_j$ and $\Delta = \nabla^\beta \nabla_\beta = \nabla^2$ is the Laplace operator. In Cartesian coordinates,

$$\Delta v^i = \frac{\partial^2 v^i}{\partial x^2} + \frac{\partial^2 v^i}{\partial y^2} + \frac{\partial^2 v^i}{\partial z^2} .$$

In vector form, one has

$$\nabla_j p^j = -\operatorname{grad} \boldsymbol{p} + (\lambda+\mu) \operatorname{grad} \operatorname{div} \boldsymbol{v} + \mu \Delta \boldsymbol{v} . \tag{2.19}$$

[1)] In Riemannian space, the tensor T^α_{ij} is different from zero due to the curvature of the space, since in curved space Γ^k_{ij} and $\partial \Gamma^k_{ij}/\partial x^s$ cannot be made to vanish simultaneously at a given point (*cf.*, Sec. 5, Chapter II).

Thus, on the basis of [(2.11), Chapter III], the Navier–Stokes equations in an arbitrary curvilinear coordinate system have the form

$$a^i = F^i - \frac{1}{\rho} g^{ij} \frac{\partial p}{\partial x^j} + \frac{\lambda+\mu}{\rho} g^{ij} \frac{\partial \operatorname{div} \boldsymbol{v}}{\partial x^j} + \frac{\mu}{\rho} \Delta v^i .$$

In vector form, the Navier–Stokes equations can be rewritten in the form

$$\frac{\mathrm{d}\boldsymbol{v}}{\mathrm{d}t} = \boldsymbol{F} - \frac{1}{\rho} \operatorname{grad} p + \frac{\lambda+\mu}{\rho} \operatorname{grad} \operatorname{div} \boldsymbol{v} + \nu \, \Delta \boldsymbol{v} . \tag{2.20}$$

Complete system of equations of motion of incompressible viscous liquids. For an incompressible viscous liquid, the Navier–Stokes equations simplify to

$$\frac{\mathrm{d}\boldsymbol{v}}{\mathrm{d}t} = \boldsymbol{F} - \frac{1}{\rho} \operatorname{grad} p + \frac{\mu}{\rho} \Delta \boldsymbol{v} . \tag{2.21}$$

These equations and the equation of continuity

$$\operatorname{div} \boldsymbol{v} = 0$$

form a complete system of equations of motion for homogeneous incompressible viscous liquids, obeying Navier–Stokes' law, in the case when the coefficient of viscosity μ is constant.

In orthogonal Cartesian coordinates, the complete system of equations of motion of, in general, non-homogeneous incompressible viscous liquids has the form

$$\begin{gathered}
\frac{\partial \rho}{\partial t} + u \frac{\partial \rho}{\partial x} + v \frac{\partial \rho}{\partial y} + w \frac{\partial \rho}{\partial z} = 0 , \\
\frac{\partial u}{\partial x} + \frac{\partial v}{\partial y} + \frac{\partial w}{\partial z} = 0 , \\
\frac{\partial u}{\partial t} + u \frac{\partial u}{\partial x} + v \frac{\partial u}{\partial y} + w \frac{\partial u}{\partial z} = F_x - \frac{1}{\rho} \frac{\partial p}{\partial x} + \nu \left(\frac{\partial^2 u}{\partial x^2} + \frac{\partial^2 u}{\partial y^2} + \frac{\partial^2 u}{\partial z^2} \right) , \\
\frac{\partial v}{\partial t} + u \frac{\partial v}{\partial x} + v \frac{\partial v}{\partial y} + w \frac{\partial v}{\partial z} = F_y - \frac{1}{\rho} \frac{\partial p}{\partial y} + \nu \left(\frac{\partial^2 v}{\partial x^2} + \frac{\partial^2 v}{\partial y^2} + \frac{\partial^2 v}{\partial z^2} \right) , \\
\frac{\partial w}{\partial t} + u \frac{\partial w}{\partial x} + v \frac{\partial w}{\partial y} + w \frac{\partial w}{\partial z} = F_z - \frac{1}{\rho} \frac{\partial p}{\partial z} + \nu \left(\frac{\partial^2 w}{\partial x^2} + \frac{\partial^2 w}{\partial y^2} + \frac{\partial^2 w}{\partial z^2} \right) .
\end{gathered} \tag{2.22}$$

Equations of motion in terms of displacements for an elastic body; case of closed systems of equations. The equations of motion in terms of displacements for an elastic body satisfying Hooke's law, *when the displacements are small.*

$$\begin{aligned} p^{ij} &= \lambda I_1(\varepsilon) g^{ij} + 2\mu\varepsilon^{ij}, \\ \varepsilon^{ij} &= \tfrac{1}{2}(\nabla^j w^i + \nabla^i w^j), \end{aligned} \tag{2.23}$$

where w^i are the components of the displacement vector and $I_1(\varepsilon)$ is the first invariant of the strain tensor ($I_1(\varepsilon) = \nabla_i w^i$), are referred to as Lamé equations. Further, it is assumed that the Lamé constants λ and μ are constant. For the derivation of the Lamé equations, one must substitute into the impulse equations [(2.11), Chapter III] Hooke's law for the case of infinitesimal strains (2.23). This derivation is completely analogous to that of the Navier–Stokes' equations. Lamé's equations have the form

$$(\lambda + \mu) \text{ grad div } \boldsymbol{w} + \mu\Delta\boldsymbol{w} + \rho\boldsymbol{F} = \rho\boldsymbol{a} . \tag{2.24}$$

In the Cartesian system of coordinates, these equations may be rewritten in the form

$$\begin{aligned} \rho a_x &= (\lambda+\mu)\frac{\partial}{\partial x}\left(\frac{\partial u}{\partial x} + \frac{\partial v}{\partial y} + \frac{\partial w}{\partial z}\right) \\ &\quad + \mu\left(\frac{\partial^2 u}{\partial x^2} + \frac{\partial^2 u}{\partial y^2} + \frac{\partial^2 u}{\partial z^2}\right) + \rho F_x , \\ \rho a_y &= (\lambda+\mu)\frac{\partial}{\partial y}\left(\frac{\partial u}{\partial x} + \frac{\partial v}{\partial y} + \frac{\partial w}{\partial z}\right) \\ &\quad + \mu\left(\frac{\partial^2 v}{\partial x^2} + \frac{\partial^2 v}{\partial y^2} + \frac{\partial^2 v}{\partial z^2}\right) + \rho F_y , \\ \rho a_z &= (\lambda+\mu)\frac{\partial}{\partial z}\left(\frac{\partial u}{\partial x} + \frac{\partial v}{\partial y} + \frac{\partial w}{\partial z}\right) \\ &\quad + \mu\left(\frac{\partial^2 w}{\partial x^2} + \frac{\partial^2 w}{\partial y^2} + \frac{\partial^2 w}{\partial z^2}\right) + \rho F_z , \end{aligned} \tag{2.25}$$

where u, v, w denote the components of the displacement vector $\boldsymbol{w}$.

These equations have been derived on the assumption that the deformations are small and, in particular, that the changes in density are small ($\rho = \rho_0 + \rho'$, $\rho' \ll \rho_0$). Therefore one must write in these equations, accurate to first order small quantities, ρ_0 instead of ρ.

The system of Lamé equations for dynamical problems remains closed, if one adds to it the definition of the acceleration:

$$\boldsymbol{a} = \frac{\mathrm{d}\boldsymbol{v}}{\mathrm{d}t} = \left(\frac{\partial \boldsymbol{v}}{\partial t}\right)_{x^i} + v^\alpha \frac{\partial \boldsymbol{v}}{\partial x^\alpha},$$

$$\boldsymbol{v} = \frac{\mathrm{d}\boldsymbol{w}}{\mathrm{d}t}.$$

For infinitesimal deformations, in the theory of elasticity, the equation of continuity need not be considered. It serves for the determination of ρ', which does not enter into the basic Lamé equations.

Note that equations (2.24) have been established for small deformations, when the displacements, velocities and accelerations may be finite. However, one often considers the case when *not only the deformations, but also the displacements, velocities and accelerations themselves are small.* Then, after neglecting non-linear terms, one obtains

$$\boldsymbol{a} = \frac{\partial^2 \boldsymbol{w}}{\partial t^2},$$

and Lamé's equations reduce to

$$\left(\frac{\partial^2 \boldsymbol{w}}{\partial t^2}\right)_{x^i} = \boldsymbol{F} + \frac{\lambda + \mu}{\rho_0} \operatorname{grad} \operatorname{div} \boldsymbol{w} + \frac{\mu}{\rho_0} \Delta \boldsymbol{w}. \tag{2.26}$$

In problems of the theory of elasticity, as a rule, one must find the displacements of individual particles of a medium, for example, changes of shape of the external boundaries of a "solid" body. Therefore, in the theory of elasticity, one employs usually a Lagrangian point of view and Lagrangian coordinates.

It has been explained earlier that one may employ two Lagrangian coordinate systems, an actual and an initial one (Chapter II). All equations are formed for a state of a medium at a definite actual instant of time. Therefore it is obvious that both the impulse equations for a continuum and the Lamé equations derived on their basis in terms of components, corresponding to a Lagrangian actual system of coordinates, have the same form as in a reference system.

For transition from an actual Lagrangian coordinate system to an initial Lagrangian coordinate system, the equations change their form. This arises out of the fact that the formulae of transition for the components of vectors and tensors in the initial Lagrangian coordinate system to the corresponding components in the actual Lagrangian coordinate system do not coincide with the usual transformation formulae for the components of tensors for transitions from one coordinate system to another in the one and same space. The spaces of the initial and actual states with identical coordinates ξ^1, ξ^2, ξ^3 must be considered to be different with different metrics, $ds^2 \neq ds_0^2$, a result of deformation.[1)]

However, if the deformations and displacements are small, then the initial and the actual Lagrangian coordinate systems differ very little and therefore, accurate to first order small quantities, one may assume that the equations in terms of components, corresponding to actual and initial Lagrangian coordinate systems, coincide. The employment of an initial Lagrangian coordinate system may prove to be more convenient than that of an actual system, since with an actual Lagrangian system one must also determine for a complete solution of a problem the position of this system with respect to a reference system.

On the necessity of the construction of other models. Large sections of hydrodynamics and the theory of elasticity have been developed within the framework of these simple models. Nevertheless, the motion of real media is by far not always sufficiently exactly described with the help of these very simple models.

For example, for a study of the motion of ionized or natural gases in the absence of barotropy, one must use more complicated models.

In many cases, Hooke's law for "solids" is not applicable; for example, when residual stresses remain in "solids" after removal of external stresses (asphalt, putty, and also metals under high loads). Therefore, one must construct models which take into consideration plasticity, creep and other properties. For complex models of continuous media, the

[1)] This question has been discussed in detail in L. I. Sedov, *Foundations of the non-linear mechanics of continua,* Pergamon Press, 1966, p. 88.

above discussion of concepts and characteristics, arising from the motion and states of particles of a medium, is insufficient.

It is necessary to consider also other characteristics, such as the temperature T, internal energy U, entropy S, residual deformations, characteristics of electromagnetic fields and many others. In these cases, the system of mechanical equations must be augmented and additional relations of physics and, in particular, of thermodynamics employed.

§3. Examples of equations in curvilinear coordinates and additional results from tensor analysis

For applications, it is useful to have readily available the equation of continuity and the equations of motion in different concrete curvilinear coordinate systems.

The Christoffel symbols in orthogonal and the equation of continuity in arbitrary coordinate systems. Write down the formulae expressing the Christoffel symbols $\Gamma^{\gamma}_{\alpha\beta}$ in terms of the components of the metric tensor g_{ij} in arbitrary orthogonal coordinates. In Euclidean and Riemannian spaces, the Christoffel symbols are defined by

$$\Gamma^{j}_{\alpha\beta} = \tfrac{1}{2} g^{js} \left(\frac{\partial g_{\alpha s}}{\partial x^{\beta}} + \frac{\partial g_{\beta s}}{\partial x^{\alpha}} - \frac{\partial g_{\alpha\beta}}{\partial x^{s}} \right), \tag{3.1}$$

whence, in the orthogonal system of coordinates ($g_{ij} = 0$ for $i \neq j$), one readily obtains

$$\Gamma^{\alpha}_{\alpha\beta} = \tfrac{1}{2} g^{\alpha\alpha} \frac{\partial g_{\alpha\alpha}}{\partial x^{\beta}}, \tag{3.2}$$

$$\Gamma^{\alpha}_{\beta\beta} = -\tfrac{1}{2} g^{\alpha\alpha} \frac{\partial g_{\beta\beta}}{\partial x^{\alpha}}, \qquad \alpha \neq \beta \tag{3.3}$$

$$\Gamma^{\alpha}_{\alpha\alpha} = \tfrac{1}{2} g^{\alpha\alpha} \frac{\partial g_{\alpha\alpha}}{\partial x^{\alpha}}, \tag{3.4}$$

(no summation with respect to α in (3.2)–(3.4))

$$\Gamma^{\alpha}_{\beta\gamma} = 0 \quad \text{for} \quad \alpha \neq \beta, \quad \beta \neq \gamma, \quad \alpha \neq \gamma. \tag{3.5}$$

With the help of (3.2), one obtains for $\Sigma_\alpha \Gamma^\alpha_{\alpha\beta}$ the following formula in terms of orthogonal coordinates:

$$\Gamma^\alpha_{\alpha\beta} = \frac{1}{2g}\frac{\partial g}{\partial x^\beta} = \frac{1}{\sqrt{g}}\frac{\partial\sqrt{g}}{\partial x^\beta}, \tag{3.6}$$

where g is the determinant of the matrix $\|g_{ij}\|$.

Consider the proof of (3.6) in *an arbitrary coordinate system.* One has

$$g = |g_{ij}| = |(\mathfrak{z}_i, \mathfrak{z}_j)| \quad \text{and} \quad \frac{\partial \mathfrak{z}_k}{\partial x^\beta} = \Gamma^\omega_{k\beta}\mathfrak{z}_\omega .$$

For the evaluation of the derivative $\partial g/\partial x^\beta$, one must differentiate in every element of the determinant g scalar products of two vectors $\mathfrak{z}_i$ and $\mathfrak{z}_j$.

Differentiating the first factor $\mathfrak{z}_i$, one arrives at three determinants in each of which the terms of the row i are replaced by terms of the form $\Gamma^\omega_{i\beta} g_{\omega j}$. It is easily seen that each of these determinants is equal to $\Gamma^i_{i\beta}\, g$, where i is the fixed index equal to the number of the corresponding row (determinants for fixed $\omega \neq i$ are equal to zero). The sum of the three determinants is equal to $\Sigma^3_{i=1} \Gamma^i_{i\beta}\, g$. From the symmetry of g_{ij} it is obvious that differentiation of the second factors $\mathfrak{z}_j$ yields exactly the same sum.

Thus it follows that

$$\frac{\partial g}{\partial x^\beta} = 2g\Gamma^i_{i\beta} ,$$

where the double index i in this formula implies a summation.

The expression for the divergence of any vector in an arbitrary coordinate system may now be written down in the form

$$\operatorname{div} \mathbf{v} = \nabla_\alpha v^\alpha = \frac{\partial v^\alpha}{\partial x^\alpha} + v^\beta \Gamma^\alpha_{\alpha\beta} = \frac{\partial v^\alpha}{\partial x^\alpha} + \frac{v^\beta}{\sqrt{g}}\frac{\partial\sqrt{g}}{\partial x^\beta} = \frac{1}{\sqrt{g}}\frac{\partial v^\alpha\sqrt{g}}{\partial x^\alpha} . \tag{3.7}$$

The equation of continuity in arbitrary curvilinear coordinates assumes the form

$$\sqrt{g}\frac{\partial\rho}{\partial t} + \frac{\partial\rho v^1\sqrt{g}}{\partial x^1} + \frac{\partial\rho v^2\sqrt{g}}{\partial x^2} + \frac{\partial\rho v^3\sqrt{g}}{\partial x^3} = 0 . \tag{3.8}$$

Recall that v^α are the components of the vector $\boldsymbol{v}$ in its decomposition along the covariant base vectors $\mathbf{з}_\alpha$, which, generally speaking, are not unit vectors.

Physical components of vectors and tensors. For the velocity vector, $\boldsymbol{v}$ one may also write

$$\boldsymbol{v} = v^i \mathbf{з}_i = u^1 \frac{\mathbf{з}_1}{\sqrt{(g_{11})}} + u^2 \frac{\mathbf{з}_2}{\sqrt{(g_{22})}} + u^3 \frac{\mathbf{з}_3}{\sqrt{(g_{33})}}, \tag{3.9}$$

where $\boldsymbol{e}_i = \mathbf{з}_i/\sqrt{(g_{ii})}$ are unit vectors. If the coordinate system is orthogonal, then the components

$$u^i = v^i \sqrt{(g_{ii})}$$

(no summation with respect to i) are equal to the projections of the velocity vector $\boldsymbol{v}$ on to the tangents to the coordinate lines; they are called the physical components of the velocity vector. Obviously, for orthogonal coordinate systems, the quantities $u_i = v_i\sqrt{(g^{ii})}$ (no summation with respect to i) coincide with the physical components u^i above. Analogously, one may introduce the physical components of any vector, for example, of the acceleration $\boldsymbol{a}$ or of grad p, and, in general, of a tensor of any order.[1)]

With the use of physical components, the equation of continuity (3.8) in an arbitrary orthogonal coordinate system becomes

$$\sqrt{g}\,\frac{\partial \rho}{\partial t} + \frac{\partial \rho\sqrt{(g_{22}g_{33})}u^1}{\partial x^1} + \frac{\partial \rho\sqrt{(g_{11}g_{33})}u^2}{\partial x^2} + \frac{\partial \rho\sqrt{(g_{11}g_{22})}u^3}{\partial x^3} = 0. \tag{3.10}$$

Equation of continuity in cylindrical and spherical coordinates. In the case of cylindrical coordinates:

$$x^1 = r\,, \quad x^2 = \varphi\,, \quad x^3 = z\,,$$

$$ds^2 = dr^2 + r^2 d\varphi^2 + dz^2\,,$$

$$|\mathbf{з}_1| = 1\,, \quad |\mathbf{з}_2| = r\,, \quad |\mathbf{з}_3| = 1\,,$$

1) For example, for the tensor $T = T^i{}_j\, \mathbf{з}_i \mathbf{з}^j$, the physical components may be defined by

$$T = T^i{}_j \sqrt{(g_{ii})}\sqrt{(g^{jj})}\,\frac{\mathbf{з}_i}{\sqrt{(g_{ii})}}\,\frac{\mathbf{з}^j}{\sqrt{(g^{jj})}} = (T^i{}_j)_{\text{phys.}}\, \boldsymbol{e}_i \boldsymbol{e}^j\,.$$

i.e.,

$$g_{11}=1\,,\quad g_{22}=r^2\,,\quad g_{33}=1\,,\quad g_{ij}=0 \text{ for } i\neq j\,.$$

The equation of continuity in cylindrical coordinates assumes the form

$$r\frac{\partial\rho}{\partial t}+\frac{\partial\rho u_r r}{\partial r}+\frac{\partial\rho u_\varphi}{\partial\varphi}+r\frac{\partial\rho u_z}{\partial z}=0\,.$$

In the case of spherical coordinates:

$$x^1=r\,,\quad x^2=\theta \text{ (latitude)}\,,\quad x^3=\lambda \text{ (longitude)}\,,$$

$$ds^2 = dr^2 + r^2 d\theta^2 + r^2\sin^2\theta\, d\lambda^2\,,$$

$$g_{11}=1\,,\quad g_{22}=r^2\,,\quad g_{33}=r^2\sin^2\theta\,.$$

The equation of continuity assumes the form

$$r^2\sin\theta\frac{\partial\rho}{\partial t}+\sin\theta\frac{\partial\rho u_r r^2}{\partial r}+r\frac{\partial\rho u_\theta\sin\theta}{\partial\theta}+r\frac{\partial\rho u_\lambda}{\partial\lambda}=0\,.$$

Components of acceleration in orthogonal coordinates. In order to write down Euler's equations of motion in terms of orthogonal, curvilinear coordinates, the components of the acceleration dv^i/dt will now be expressed in terms of g^{ij} and v^i. One has for the components of the acceleration dv/dt

$$a^j=\frac{\partial v^j}{\partial t}+v^i\nabla_i v^j=\frac{\partial v^j}{\partial t}+v^i\frac{\partial v^j}{\partial x^i}+v^i v^\beta\Gamma^j_{i\beta}\,.$$

The last term in this formula can be rewritten

$$v^i v^\beta\Gamma^j_{i\beta}=\underset{\substack{\beta\neq j\\ \text{no summation over } j}}{2v^j v^\beta\Gamma^j_{\beta j}+(v^\beta)^2\Gamma^j_{\beta\beta}}=\underset{\substack{\beta\neq j\\ \text{no summation over } j}}{2v^j v^\beta\Gamma^j_{\beta j}}+(v^j)^2\Gamma^j_{jj}+\underset{j\neq\beta}{(v^\beta)^2\Gamma^j_{\beta\beta}}\,.$$

In orthogonal coordinates $g^{ii}=1/g_{ii}$ and one has with the aid of Formulae (3.2) — (3.4) for the Christoffel symbols $\Gamma^j_{\alpha\beta}$ in terms of g_{ij}

$$a^j=\frac{\partial v^j}{\partial t}+\frac{\partial v^j}{\partial x^i}v^i+\underset{\beta\neq j}{\frac{v^j v^\beta}{g_{jj}}\frac{\partial g_{jj}}{\partial x^\beta}}+\frac{1}{2}\frac{(v^j)^2}{g_{jj}}\frac{\partial g_{jj}}{\partial x^j}-\frac{(v^\beta)^2}{2g_{jj}}\frac{\partial g_{\beta\beta}}{\partial x^j}\,. \qquad (3.11)$$

no summation over j

The physical components of the acceleration in cylindrical coordinates can now be written down, if in place of v^i in (3.9) one introduces the physical velocity components u^i:

$$a_r = \frac{\partial u_r}{\partial t} + u_r \frac{\partial u_r}{\partial r} + \frac{u_\varphi}{r}\frac{\partial u_r}{\partial \varphi} + u_z \frac{\partial u_r}{\partial z} - \frac{u_\varphi^2}{r},$$

$$a_\varphi = \frac{\partial u_\varphi}{\partial t} + u_r \frac{\partial u_\varphi}{\partial r} + \frac{u_\varphi}{r}\frac{\partial u_\varphi}{\partial \varphi} + u_z \frac{\partial u_\varphi}{\partial z} + \frac{u_r u_\varphi}{r}, \qquad (3.12)$$

$$a_z = \frac{\partial u_z}{\partial t} + u_r \frac{\partial u_z}{\partial r} + \frac{u_\varphi}{r}\frac{\partial u_z}{\partial \varphi} + u_z \frac{\partial u_z}{\partial z}.$$

The physical components of the acceleration in spherical coordinates can be written in the form

$$a_r = \frac{\partial u_r}{\partial t} + u_r \frac{\partial u_r}{\partial r} + \frac{u_\theta}{r}\frac{\partial u_r}{\partial \theta} + \frac{u_\lambda}{r\sin\theta}\frac{\partial u_r}{\partial \lambda} - \frac{u_\theta^2 + u_\lambda^2}{r},$$

$$a_\theta = \frac{\partial u_\theta}{\partial t} + u_r \frac{\partial u_\theta}{\partial r} + \frac{u_\theta}{r}\frac{\partial u_\theta}{\partial \theta} + \frac{u_\lambda}{r\sin\theta}\frac{\partial u_\theta}{\partial \lambda} + \frac{u_r u_\theta - u_\lambda^2 \operatorname{ctg}\theta}{r},$$

$$a_\lambda = \frac{\partial u_\lambda}{\partial t} + u_r \frac{\partial u_\lambda}{\partial r} + \frac{u_\theta}{r}\frac{\partial u_\lambda}{\partial \theta} + \frac{u_\lambda}{r\sin\theta}\frac{\partial u_\lambda}{\partial \lambda} + \frac{u_r u_\lambda + \operatorname{ctg}\theta\, u_\theta u_\lambda}{r}. \qquad (3.13)$$

Components of the gradient vector of a scalar function in orthogonal coordinates. The projections of the vector grad p on to the axes of orthogonal coordinate systems will now be derived. One has

$$\text{grad } p = \frac{\partial p}{\partial x^i}\, \mathbf{3}^i = (\nabla_i p)\, \mathbf{3}^i = A_i \frac{\mathbf{3}^i}{\sqrt{(g^{ii})}},$$

whence the physical components A_i of the vector grad p in an orthogonal coordinate system will be

$$A_i = \frac{\partial p}{\partial x^i}\sqrt{(g^{ii})} = \frac{1}{\sqrt{(g_{ii})}}\frac{\partial p}{\partial x^i}.$$

no summation over i

The physical components of the vector grad p in cylindrical coordinates are now

$$\begin{aligned}
\text{grad } p|_r &= \frac{\partial p}{\partial r}, \\
\text{grad } p|_\varphi &= \frac{1}{r}\frac{\partial p}{\partial \varphi}, \\
\text{grad } p|_z &= \frac{\partial p}{\partial z},
\end{aligned} \tag{3.14}$$

and in spherical coordinates

$$\begin{aligned}
\text{grad } p|_r &= \frac{\partial p}{\partial r}, \\
\text{grad } p|_\theta &= \frac{1}{r}\frac{\partial p}{\partial \theta}, \\
\text{grad } p|_\lambda &= \frac{1}{r\sin\theta}\frac{\partial p}{\partial \lambda}.
\end{aligned} \tag{3.15}$$

Euler's equations in cylindrical and spherical coordinates. Euler's equations in cylindrical and spherical coordinates are readily written down using (3.12), (3.14) and (3.13), (3.15), respectively. In cylindrical coordinates, they assume the form

$$\begin{aligned}
&\frac{\partial u_r}{\partial r}u_r + \frac{u_\varphi}{r}\frac{\partial u_r}{\partial \varphi} + \frac{\partial u_r}{\partial z}u_z - \frac{u_\varphi^2}{r} + \frac{\partial u_r}{\partial t} = F_r - \frac{1}{\rho}\frac{\partial p}{\partial r}, \\
&\frac{u_r u_\varphi}{r} + \frac{\partial u_\varphi}{\partial r}u_r + \frac{1}{r}\frac{\partial u_\varphi}{\partial \varphi}u_\varphi + \frac{\partial u_\varphi}{\partial z}u_z + \frac{\partial u_\varphi}{\partial t} = F_\varphi - \frac{1}{r\rho}\frac{\partial p}{\partial \varphi}, \\
&\frac{\partial u_z}{\partial r}u_r + \frac{1}{r}\frac{\partial u_z}{\partial \varphi}u_\varphi + \frac{\partial u_z}{\partial z}u_z + \frac{\partial u_z}{\partial t} = F_z - \frac{1}{\rho}\frac{\partial p}{\partial z},
\end{aligned} \tag{3.16}$$

in spherical coordinates, the form

$$\frac{\partial u_r}{\partial t} + u_r\frac{\partial u_r}{\partial r} + \frac{u_\theta}{r}\frac{\partial u_r}{\partial \theta} + \frac{u_\lambda}{r\sin\theta}\frac{\partial u_r}{\partial \lambda} - \frac{u_\theta^2 + u_\lambda^2}{r}$$

$$= F_r - \frac{1}{\rho}\frac{\partial p}{\partial r},$$

$$\frac{\partial u_\theta}{\partial t} + u_r \frac{\partial u_\theta}{\partial r} + \frac{u_\theta}{r}\frac{\partial u_\theta}{\partial \theta} + \frac{u_\lambda}{r \sin\theta}\frac{\partial u_\theta}{\partial \lambda} + \frac{u_r u_\theta}{r} - \frac{\operatorname{ctg}\theta}{r} u_\lambda^2$$

$$= F_\theta - \frac{1}{r\rho}\frac{\partial p}{\partial \theta},$$

$$\frac{\partial u_\lambda}{\partial t} + u_r \frac{\partial u_\lambda}{\partial r} + \frac{u_\theta}{r}\frac{\partial u_\lambda}{\partial \theta} + \frac{u_\lambda}{r \sin\theta}\frac{\partial u_\lambda}{\partial \lambda} + \frac{u_r u_\lambda}{r} + \frac{\operatorname{ctg}\theta u_\theta u_\lambda}{r}$$

$$= F_\lambda - \frac{r}{\rho r \sin\theta}\frac{\partial p}{\partial \lambda}. \tag{3.17}$$

Laplace operator for a scalar function in orthogonal coordinates. Setting

$$\boldsymbol{v} = \operatorname{grad} \Phi,$$

one obtains readily from (3.7) the expression for the Laplace operator of a scalar function Φ in any orthogonal coordinate system:

$$\operatorname{div}\operatorname{grad}\Phi = \nabla^2\Phi = \Delta\Phi$$

$$= \frac{1}{\sqrt{g}}\left[\frac{\partial \sqrt{\left(\frac{g_{22}g_{33}}{g_{11}}\right)}\frac{\partial\Phi}{\partial x^1}}{\partial x^1} + \frac{\partial \sqrt{\left(\frac{g_{33}g_{11}}{g_{22}}\right)}\frac{\partial\Phi}{\partial x^2}}{\partial x^2} + \frac{\partial \sqrt{\left(\frac{g_{22}g_{11}}{g_{33}}\right)}\frac{\partial\Phi}{\partial x^3}}{\partial x^3}\right] \tag{3.18}$$

Hence it follows that in cylindrical coordinates

$$\Delta\Phi = \frac{1}{r}\frac{\partial}{\partial r}\left(r\frac{\partial\Phi}{\partial r}\right) + \frac{1}{r^2}\frac{\partial^2\Phi}{\partial\varphi^2} + \frac{\partial^2\Phi}{\partial z^2}, \tag{3.19}$$

and in spherical coordinates

$$\Delta\Phi = \frac{1}{r^2}\left[\frac{\partial}{\partial r}\left(r^2\frac{\partial\Phi}{\partial r}\right) + \frac{1}{\sin\theta}\frac{\partial}{\partial\theta}\left(\sin\theta\frac{\partial\Phi}{\partial\theta}\right) + \frac{1}{\sin^2\theta}\frac{\partial^2\Phi}{\partial\lambda^2}\right]. \tag{3.20}$$

These examples demonstrate the application of the general formulae to important particular cases.

Components of the strain rate tensor and the Navier-Stokes equation in cylindrical and spherical coordinates. For completeness, the expressions for the physical components of the strain rate tensor will now be written down for cylindrical

$$e_{rr} = \frac{\partial u_r}{\partial r}\,, \qquad 2e_{r\varphi} = \frac{1}{r}\frac{\partial u_r}{\partial \varphi} + r\frac{\partial}{\partial r}\left(\frac{u_\varphi}{r}\right),$$

$$e_{\varphi\varphi} = \frac{1}{r}\frac{\partial u_\varphi}{\partial \varphi} + \frac{u_r}{r}\,, \qquad 2e_{\varphi z} = \frac{\partial u_\varphi}{\partial z} + \frac{1}{r}\frac{\partial u_z}{\partial \varphi}\,,$$

$$e_{zz} = \frac{\partial u_z}{\partial z}\,, \qquad 2e_{rz} = \frac{\partial u_z}{\partial r} + \frac{\partial u_r}{\partial z}\,,$$

and spherical coordinates

$$e_{rr} = \frac{\partial u_r}{\partial r}\,, \qquad e_{\theta\theta} = \frac{1}{r}\frac{\partial u_\theta}{\partial \theta} + \frac{u_r}{r}\,,$$

$$e_{\lambda\lambda} = \frac{1}{r\sin\theta}\frac{\partial u_\lambda}{\partial \lambda} + \frac{u_r}{r} + \frac{u_\theta \operatorname{ctg}\theta}{r}\,,$$

$$2e_{r\theta} = \frac{1}{r}\frac{\partial u_r}{\partial \theta} + \frac{\partial u_\theta}{\partial r} - \frac{u_\theta}{r}\,, \qquad 2e_{\lambda r} = \frac{\partial u_\lambda}{\partial r} + \frac{1}{r\sin\theta}\frac{\partial u_r}{\partial \theta} - \frac{u_\lambda}{r}\,,$$

$$2e_{\lambda\theta} = \frac{1}{r\sin\theta}\frac{\partial u_\theta}{\partial \lambda} + \frac{1}{r}\frac{\partial u_\lambda}{\partial \theta} - \frac{u_\lambda \operatorname{ctg}\theta}{r}\,,$$

and also the Navier-Stokes equations for incompressible fluid in cylindrical

$$a_r = F_r - \frac{1}{\rho}\frac{\partial p}{\partial r} + \nu\left(\Delta u_r - \frac{u_r}{r^2} - \frac{2}{r^2}\frac{\partial u_\varphi}{\partial \varphi}\right),$$

$$a_\varphi = F_\varphi - \frac{1}{\rho r}\frac{\partial p}{\partial \varphi} + \nu\left(\Delta u_\varphi - \frac{u_\varphi}{r^2} + \frac{2}{r^2}\frac{\partial u_r}{\partial \varphi}\right),$$

$$a_z = F_z - \frac{1}{\rho}\frac{\partial p}{\partial z} + \nu \Delta u_z \quad \left(\Delta = \frac{\partial^2}{\partial r^2} + \frac{1}{r}\frac{\partial}{\partial r} + \frac{1}{r^2}\frac{\partial^2}{\partial \varphi^2} + \frac{\partial^2}{\partial z^2} \right),$$

and spherical coordinates

$$a_r = F_r - \frac{1}{\rho}\frac{\partial p}{\partial r} + \nu \left(\Delta u_r - \frac{2u_r}{r^2} - \frac{2u_\theta}{r^2} \operatorname{ctg}\theta - \frac{2}{r^2 \sin\theta}\frac{\partial u_\lambda}{\partial \lambda} - \frac{2}{r^2}\frac{\partial u_\theta}{\partial \theta} \right),$$

$$a_\theta = F_\theta - \frac{1}{\rho r}\frac{\partial p}{\partial \theta} + \nu \left(\Delta u_\theta - \frac{u_\theta}{r^2 \sin^2\theta} - \frac{2\cos\theta}{r^2 \sin^2\theta}\frac{\partial u_\lambda}{\partial \lambda} + \frac{2}{r^2}\frac{\partial u_r}{\partial \theta} \right),$$

$$a_\lambda = F_\lambda - \frac{1}{r_\theta \sin\theta}\frac{\partial p}{\partial \lambda} + \nu \left(\Delta u_\lambda - \frac{u_\lambda}{r^2 \sin^2\theta} + \frac{2\cos\theta}{r^2 \sin^2\theta}\frac{\partial u_\theta}{\partial \lambda} + \frac{2}{r^2 \sin\theta}\frac{\partial u_r}{\partial \lambda} \right)$$

$$\left(\Delta = \frac{\partial^2}{\partial r^2} + \frac{2}{r}\frac{\partial}{\partial r} + \frac{1}{r^2}\frac{\partial^2}{\partial \theta^2} + \frac{\operatorname{ctg}\theta}{r^2}\frac{\partial}{\partial \theta} + \frac{1}{r^2 \sin^2\theta}\frac{\partial^2}{\partial \lambda^2} \right),$$

where the physical components of the acceleration are given by (3.12) and (3.13).

Note, that the expressions in the two last systems of the Navier-Stokes equation, which are multiplied by ν, are the physical components of the vector $\nabla^2 v$, obtained as the result of application of the Laplace operator to the velocity vector v. Comparing those expressions with (3.19) and (3.20), respectively, it is evident, that application of the Laplace operator to the components of the vector and to the scalar functions, in general, gives different formulas.

Equivalence in three-dimensional space of an anti-symmetric second order tensor to an axial vector. In three-dimensional space, there corresponds to every anti-symmetric second order tensor $A_{ij}\mathbf{э}^i\mathbf{э}^j$ an axial vector $\boldsymbol{B}=B^i\mathbf{э}_i$, the contravariant components of which are defined by the formulae

$$B^\gamma = \frac{1}{\sqrt{g}} A_{\alpha\beta} , \tag{3.21}$$

where the three indices α, β, γ are obtained from 1, 2, 3 by cyclic permutation, and $g = \mathrm{Det}\,\|g_{\alpha\beta}\|$. In order to prove this statement, examine the transformation formulae for the quantities B^γ for transition from the coordinate system x^i to the system y^i. For this purpose, use the transformation formulae for the components of the tensor $g_{ij}\mathbf{э}^i\mathbf{э}^j$ and the components of the tensor $A_{ij}\mathbf{э}^i\mathbf{э}^j$, and also the property of anti-symmetry of the A_{ij}.

The determinant g is given by [1)]

$$g = |g_{ij}| = |g'_{kl}| \left|\frac{\partial y^k}{\partial x^i}\right| \left|\frac{\partial y^l}{\partial x^j}\right| = g' \Delta^2 ,$$

where Δ is the determinant of the matrix $\|\partial y^k/\partial x^i\|$. Therefore

$$\frac{1}{\sqrt{g}} A_{ij} = \frac{1}{\sqrt{g'}} \frac{1}{|\Delta|} A'_{pq} \frac{\partial y^p}{\partial x^i} \frac{\partial y^q}{\partial x^j}$$

$$= \frac{1}{\sqrt{g'}} A'_{\alpha\beta} \frac{\dfrac{\partial y^\alpha}{\partial x^i}\dfrac{\partial y^\beta}{\partial x^j} - \dfrac{\partial y^\beta}{\partial x^i}\dfrac{\partial y^\alpha}{\partial x^j}}{|\Delta|} \tag{3.22}$$

(summation only over α, $\beta = 1, 2, 3$ such that $\alpha > \beta$). It is easily seen that the quantities

$$\frac{\partial y^\alpha}{\partial x^i}\frac{\partial y^\beta}{\partial x^j} - \frac{\partial y^\beta}{\partial x^i}\frac{\partial y^\alpha}{\partial x^j}$$

1) In what follows, the transformation formula of g is important; the link between g and the metric is, in general, unimportant.

are the minors corresponding to the element $\partial x^k/\partial y^\gamma$ of the matrix, inverse to $\|\partial y^i/\partial x^j\|$, and, consequently,

$$\frac{1}{|\Delta|}\left[\frac{\partial y^\alpha}{\partial x^i}\frac{\partial y^\beta}{\partial x^j} - \frac{\partial y^\beta}{\partial x^i}\frac{\partial y^\alpha}{\partial x^j}\right] = \frac{\partial x^k}{\partial y^\gamma}\frac{\Delta}{|\Delta|},$$

where the indices α, β, γ and i, j, k must be cyclic permutations of 1, 2, 3.

Therefore (3.22) can be rewritten in the form

$$B^k = B'^\gamma \frac{\partial x^k}{\partial y^\gamma}\frac{\Delta}{|\Delta|}. \tag{3.23}$$

These formulae coincide with the ordinary transformation formulae for the contravariant components of a vector only when $\Delta > 0$. Therefore the vector $\boldsymbol{B} = B^i \boldsymbol{3}_i$, defined by (3.21), is referred to as axial vector or pseudo-vector in contrast to ordinary polar vectors.

Under the transformation of inversion, for example, for the transformation

$$y^i = -x^i,$$

the components of the polar vector $\boldsymbol{b}$ change sign:

$$b'^k = b^i \frac{\partial y^k}{\partial x^i} = -b^k,$$

and the components of the axial vector $\boldsymbol{B}$, as is evident from (3.23), do not change sign:

$$B'^k = -B^i \frac{\partial y^k}{\partial x^i} = B^k.$$

Determination of the vector curl $\boldsymbol{A}$ *in curvilinear coordinates.* The vector $\boldsymbol{c} = \text{curl}\,\boldsymbol{A}$ is derived, by definition, from the formula

$$\boldsymbol{c} = c^\gamma \boldsymbol{3}_\gamma,$$

$$c^\gamma = \frac{1}{\sqrt{g}}(\nabla_\alpha A_\beta - \nabla_\beta A_\alpha) = \frac{1}{\sqrt{g}}\left(\frac{\partial A_\beta}{\partial x^\alpha} - \frac{\partial A_\alpha}{\partial x^\beta}\right)$$

($\alpha, \beta, \gamma = 1, 2, 3$ in cyclic permutation). Obviously, if $\boldsymbol{A}$ is a polar vector, then curl $\boldsymbol{A}$ is an axial vector.

In Cartesian coordinates, the components of the vector $\boldsymbol{c} = \operatorname{curl} \boldsymbol{A}$ are determined by

$$c^\gamma = \frac{\partial A_\beta}{\partial x^\alpha} - \frac{\partial A_\alpha}{\partial x^\beta}$$

($\alpha, \beta, \gamma = 1, 2, 3$ in cyclic permutation).

The Levi–Civita tensor. Obviously, the rotation vector $\boldsymbol{A} = A_\alpha \boldsymbol{3}^\alpha$ may be considered to be the result of a contraction of the tensor $\nabla_\alpha A_\beta \boldsymbol{3}^\alpha \boldsymbol{3}^\beta$ with the, antisymmetric in all indices, third order pseudo-tensor $\varepsilon = \varepsilon^{\alpha\beta\gamma} \boldsymbol{3}_\alpha \boldsymbol{3}_\beta \boldsymbol{3}_\gamma$, the component of which are given by

$$\varepsilon^{\alpha\beta\gamma} = \begin{cases} \dfrac{1}{\sqrt{g}}, & \text{if } \alpha, \beta, \gamma \text{ form an even permutation of } 1, 2, 3\,, \\[2ex] -\dfrac{1}{\sqrt{g}}, & \text{if } \alpha, \beta, \gamma \text{ form an odd permutation of } 1, 2, 3\,, \\[2ex] 0\,, & \text{if two of the indices } \alpha, \beta, \gamma \text{ are equal.} \end{cases} \qquad (3.24)$$

It is readily verified directly that (3.24) actually defines the components of a pseudo-tensor, i.e., that the transformation formulae of the $\varepsilon^{\alpha\beta\gamma}$ for the transition to another coordinate system y^i may be written in the form

$$\varepsilon'^{ijk} = \frac{\Delta}{|\Delta|} \varepsilon^{\alpha\beta\gamma} \frac{\partial y^i}{\partial x^\alpha} \frac{\partial y^j}{\partial x^\beta} \frac{\partial y^k}{\partial x^\gamma},$$

where ε'^{ijk} are represented in terms of $g' = \operatorname{Det} \| g'_{ij} \|$ by formulae analogous to (3.24). The terminology of Levi–Civita tensor for the pseudo-tensor is connected with the presence in the transformation formula of the sign, determined by the multiplier $\Delta/|\Delta|$. For proper transformations ($\Delta > 0$), pseudo-tensors do not differ from ordinary tensors. Thus, the components of the vector curl A may be written in the form

$$c^\gamma = \varepsilon^{\gamma\alpha\beta} \nabla_\alpha A_\beta\,.$$

Components of the vector product in curvilinear coordinates. Consider yet the operation of the vector product of the two vectors $\boldsymbol{A}$ and $\boldsymbol{B}$ in curvilinear coordinates. By definition, let the components c^γ of the vector product $\boldsymbol{A} \times \boldsymbol{B}$ be represented by the formulae

$$c^{\gamma} = \varepsilon^{\gamma\alpha\beta} A_{\alpha} B_{\beta} ,$$

i.e.,

$$c^{\gamma} = \frac{1}{\sqrt{g}} (A_i B_j - A_j B_i)$$

($i, j, \gamma = 1, 2, 3$ in cyclic permutation). Hence it is seen, in particular, that the vector product of two polar vectors is an axial vector (pseudo-vector).

Examples of axial vectors. Physical examples of axial vectors, which are, in essence, anti-symmetric second order tensors, are the vorticity vector $\boldsymbol{\omega} = \frac{1}{2}$ curl $\boldsymbol{v}$, the magnetic field $\boldsymbol{H}$, the magnetic induction $\boldsymbol{B}$, *etc.*

The equivalence of an anti-symmetric tensor with two indices to an axial vector in these indices is valid only in three-dimensional space.

In n-dimensional space, for $n > 3$, a similar equivalence does not exist.

In contemporary physics, side by side with three-dimensional space with space variables x^i, immediate physical significance attaches to four-dimensional space with the coordinates of a space point x^1, x^2, x^3 and time $t = x^4$.

For the formulation of the fundamental physical equations, one must consider vectors and tensors in four-dimensional space with coordinates x^1, x^2, x^3, x^4 and assume that the coordinates of points of four-dimensional space, including time, are interlinked and in some sense equivalent.

On the correspondence in four-dimensional space of an anti-symmetric second order tensor to an axial and a polar vector. In such a four-dimensional space, one encounters in the physical equations anti-symmetric second order tensors which play a fundamental role.

Let F_{ik} denote an anti-symmetric tensor in four-dimensional space. By definition,

$$F_{ik} = -F_{ki}$$

and, consequently,

$$F_{ii} = 0 .$$

Under the coordinate transformation

$$x^i = f^i(y^k) , \qquad i, k = 1, 2, 3, 4 , \tag{3.25}$$

the components of the tensor F_{ik} transform in accordance with the ordinary formulae

$$F_{ik}^{(y)} = F_{pq}^{(x)} \frac{\partial x^p}{\partial y^i} \frac{\partial x^q}{\partial y^k} . \tag{3.25'}$$

The transformation formulae (3.25′) and the results derived below do not depend on the method of introduction of the metric in four-dimensional space. However, it is convenient to consider together with four-dimensional space a three-dimensional ordinary space x^1, x^2, x^3 with its metric defined by the ordinary method, *viz.*,

$$ds^2 = g_{\alpha\beta}\, dx^\alpha dx^\beta , \qquad \alpha, \beta = 1, 2, 3 .$$

For the four-dimensional matrix of the tensor F_{ik}, one may write

$$\|F_{ik}\| = \begin{Vmatrix} 0 & F_{12} & F_{13} & F_{14} \\ F_{21} & 0 & F_{23} & F_{24} \\ F_{31} & F_{32} & 0 & F_{34} \\ F_{41} & F_{42} & F_{43} & 0 \end{Vmatrix} = \begin{Vmatrix} 0 & \sqrt{(g)}\, H^3 & -\sqrt{(g)}\, H^2 & E_1 \\ -\sqrt{(g)}\, H^3 & 0 & H^1\sqrt{(g)} & E_2 \\ \sqrt{(g)}\, H^2 & -\sqrt{(g)}\, H^1 & 0 & E_3 \\ -E_1 & -E_2 & -E_3 & 0 \end{Vmatrix} \tag{3.26}$$

where

$$g = \operatorname{Det} \|g_{\alpha\beta}\| .$$

The letters H^α and E_α $(\alpha = 1, 2, 3)$ denote the corresponding elements of the matrix $\|F_{ik}\|$. The transformation formulae (3.25′) for the components F_{ik} can be viewed also as transformation formulae for the quantities H^α and E_α.

If, together with the general coordinate transformation (3.25), one considers the particular coordinate transformation

$$\begin{aligned} x^\alpha &= f^\alpha(y^1, y^2, y^3) \qquad (\alpha = 1, 2, 3) , \\ x^4 &= y^4 , \end{aligned} \tag{3.27}$$

in which only the space coordinates transform and the time coordinate remains unchanged, then the general transformation formulae (3.25′) give special transformation formulae for the quantities, denoted in the second matrix (3.26) by H^α and E_α. It follows from these formulae that for the special transformation (3.27) the quantities H^α and E_α may be considered as contravariant components of the three-dimensional axial vector

$$H = H^{\gamma}\mathbf{3}_{\gamma} \qquad (\gamma = 1, 2, 3)\,,$$

$$H^{\gamma} = \frac{1}{\sqrt{g}}\, F_{\alpha\beta}$$

(α, β, $\gamma = 1, 2, 3$ in cyclic permutation) and as covariant components of the polar vector

$$E = E_{\gamma}\mathbf{3}^{\gamma} \qquad (\gamma = 1, 2, 3)\,,$$

$$E_{\gamma} = F_{\gamma 4}\,,$$

respectively. Obviously, the transformation formulae for H^{γ} and E_{γ} in the general case of the four-dimensional transformation (3.25) are not vectorial. From the point of view of four-dimensional space, the three-dimensional vectors $\boldsymbol{H}$ and $\boldsymbol{E}$ are not invariant objects.

It will be seen below that the electric and magnetic field vectors may be considered as vectors $\boldsymbol{E}$ and $\boldsymbol{H}$ for a corresponding four-dimensional tensor

$$F = F_{ik}\mathbf{3}^{i}\mathbf{3}^{k}\,.$$

Note that this link between the F_{ik}, E_{γ} and H^{γ} is independent of the metric of the four-dimensional space. In (3.26), instead of the three-dimensional metric tensor $g_{\alpha\beta}\mathbf{3}^{\alpha}\mathbf{3}^{\beta}$, one could have used any other three-dimensional second order tensor for which the determinant composed of its components does not vanish.

Chapter V

Basic thermodynamic concepts and equations

§1. Kinetic energy theorem and work of internal surface forces

One of the most important general consequences of the dynamic equations of motion of continuous media is the kinetic energy theorem.

Let V denote an arbitrary finite volume which moves together with the particles of the medium, and Σ the surface enclosing this volume. Suppose that inside the volume V the components of the stress tensor $P = p^{ij}\mathbf{3}_i\mathbf{3}_j$ and of the velocity vector $\boldsymbol{v} = v^i\mathbf{3}_i = v_i\mathbf{3}^i$ are continuous differentiable functions of the space coordinates and time.

Select the vector $\mathrm{d}\boldsymbol{r} = \boldsymbol{v}\mathrm{d}t$, the displacement vector of an infinitesimal volume of the continuum $\mathrm{d}\tau$ during time $\mathrm{d}t$; multiplying the impulse equation scalarly by $\mathrm{d}\boldsymbol{r}$ and integrating over V, one obtains

$$\int_V \rho \boldsymbol{a} \cdot \boldsymbol{v} \mathrm{d}t \mathrm{d}\tau = \int_V \rho \boldsymbol{F} \cdot \mathrm{d}\boldsymbol{r} \mathrm{d}\tau + \int_V (\nabla_j p^{ij}) v_i \mathrm{d}t \mathrm{d}\tau . \tag{1.1}$$

Each of the integrals in this relation will now be transformed.

Kinetic energy of the volume V of a continuous medium. The scalar (invariant) quantity $\boldsymbol{v} \cdot \boldsymbol{a}$ may be calculated using any system of coordinates. For example, in Cartesian coordinates, one easily obtains

$$\boldsymbol{v} \cdot \boldsymbol{a} = \frac{\mathrm{d}x}{\mathrm{d}t}\frac{\mathrm{d}}{\mathrm{d}t}\left(\frac{\mathrm{d}x}{\mathrm{d}t}\right) + \frac{\mathrm{d}y}{\mathrm{d}t}\frac{\mathrm{d}}{\mathrm{d}t}\left(\frac{\mathrm{d}y}{\mathrm{d}t}\right) + \frac{\mathrm{d}z}{\mathrm{d}t}\frac{\mathrm{d}}{\mathrm{d}t}\left(\frac{\mathrm{d}z}{\mathrm{d}t}\right) = \frac{1}{2}\frac{\mathrm{d}}{\mathrm{d}t}\boldsymbol{v}^2 .$$

Since the mass $dm = \rho d\tau$ is constant, one has, obviously,

$$\int_V \boldsymbol{a}\cdot\boldsymbol{v}\rho\, d\tau\, dt = \int_M d\left(\frac{v^2}{2}\right) dm = d\int_M \frac{v^2}{2}\, dm = dE\,,$$

where, by definition,

$$E = \int_V \frac{\rho \boldsymbol{v}^2}{2}\, d\tau \tag{1.2}$$

is the kinetic energy of the volume V of the medium.

Work of internal and external mass forces. Mass forces $\boldsymbol{F}$ can be divided into two groups: Internal forces $\boldsymbol{F}^{(i)}$ and external forces $\boldsymbol{F}^{(e)}$, which are external with respect to the entire volume V. Then

$$\int_V \rho\boldsymbol{F}\cdot d\boldsymbol{r}\, d\tau = \int_V \rho\boldsymbol{F}^{(e)}\cdot d\boldsymbol{r}\, d\tau + \int_V \rho\boldsymbol{F}^{(i)}\cdot d\boldsymbol{r}\, d\tau = dA_m^{(e)} + dA_m^{(i)}\,, \tag{1.3}$$

where $dA_m^{(e)}$ and $dA_m^{(i)}$ represent the elementary work done by the mass forces, which are external and internal with respect to V, on the volume V over an infinitesimal displacement.

Note that the sum of all internal mass forces, acting on the entire volume V, is always equal to zero, and that the work of these forces may be different from zero.

The last integral in (1.1) can be rewritten in the form of the two integrals

$$\int_V (\nabla_j p^{ij}) v_i\, dt\, d\tau = \int_V \nabla_j (p^{ij} v_i)\, dt\, d\tau - \int_V p^{ij}\nabla_j v_i\, dt\, d\tau\,. \tag{1.4}$$

Transforming the first and second integrals on the right-hand side of (1.4) by means of the Gauss–Ostrogradskii Theorem and the obvious identity

$$\nabla_j v_i = \tfrac{1}{2}(\nabla_j v_i + \nabla_i v_j) + \tfrac{1}{2}(\nabla_j v_i - \nabla_i v_j) = e_{ij} + \omega_{ij}\,,$$

respectively, one arrives at

$$\int_V (\nabla_j p^{ij}) v_i\, dt\, d\tau = \int_\Sigma p^{ij} v_i n_j\, d\sigma\, dt - \int_V p^{ij} e_{ij}\, dt\, d\tau - \int_V p^{ij}\omega_{ij}\, dt\, d\tau\,, \tag{1.5}$$

where Σ is the surface, enclosing the volume V, and the n_j are the covariant components of the unit vector of the outward normal to V and Σ.

Note that, by virtue of the antisymmetry of the tensor $\boldsymbol{\omega}_{ij}\mathbf{3}^i\mathbf{3}^j$, one has

$$p^{ij}\omega_{ij} = \underset{[i<j]}{p^{ij}\omega_{ij} + p^{ji}\omega_{ji}} = \underset{[i<j]}{(p^{ij} - p^{ji})\omega_{ij}} \tag{1.5'}$$

Therefore, in the classical case when the stress tensor is symmetric ($p^{ij} = p^{ji}$), the last integral in (1.5) vanishes.

Work done by external surface forces. Since $p^{ij}n_j\mathbf{3}_i = \boldsymbol{p}_n$, one may write

$$\int_\Sigma p^{ij}v_i n_j \mathrm{d}\sigma \mathrm{d}t = \int_\Sigma \boldsymbol{p}_n \cdot \mathrm{d}\boldsymbol{r} \mathrm{d}\sigma = \mathrm{d}A_\Sigma^{(e)}, \tag{1.6}$$

where $\mathrm{d}A_\Sigma^{(e)}$ denotes the work done by the external surface forces acting over the surface Σ bounding the volume V, during the infinitesimal displacements $\mathrm{d}\boldsymbol{r} = \boldsymbol{v}\mathrm{d}t$ of points of Σ.

Definition of work done by internal surface forces. The last integral in (1.4), an invariant quantity, by definition, is referred to as the work of the internal surface stress forces[1)] on the volume V:

$$-\int_V p^{ij}\nabla_j v_i \mathrm{d}t \mathrm{d}\tau = \mathrm{d}A_\Sigma^{(i)}. \tag{1.7}$$

The kinetic energy theorem for a finite volume of a continuum. Thus, Equation (1.1) may be rewritten in the form

$$\mathrm{d}E = \mathrm{d}A_m^{(e)} + \mathrm{d}A_m^{(i)} + \mathrm{d}A_\Sigma^{(e)} + \mathrm{d}A_\Sigma^{(i)}, \tag{1.8}$$

i.e., for real motion, the differential of the kinetic energy of a finite individual volume of a continuum is equal to the sum of the elements of work done by the external mass, the internal mass and the external and internal surface forces, acting on this volume. This statement is known as the theorem on kinetic energy in applications to a finite volume of a continuous deformable medium.

[1)] Note, that even in the case of continuous displacement, in spite of the validity of the law of action and reaction for internal stresses, the work done by the forces of interaction is in general, different from zero due to the presence of deformations.

Note that in the formulation of the kinetic energy theorem (1.8) the quantity $\mathrm{d}E$ is the total differential of the function E, the kinetic energy of a volume of the continuous medium, and the remaining terms $\mathrm{d}A_m^{(e)}$, $\mathrm{d}A_m^{(i)}$, $\mathrm{d}A_\Sigma^{(e)}$ and $\mathrm{d}A_\Sigma^{(i)}$, in the general case, are simply infinitesimal quantities, the elements of work of the corresponding forces over the system of continuous infinitesimal displacements

$$\mathrm{d}\boldsymbol{r} = \boldsymbol{v}\,\mathrm{d}t\,,$$

defined at every point of the medium.

The work of internal surface forces when the stress tensor is symmetric. It follows from (1.4), (1.5) and (1.5′) that the expression for the work done by the internal surface forces can be rewritten in the form

$$\mathrm{d}A_\Sigma^{(i)} = -\int_V p^{ij} e_{ij}\,\mathrm{d}t\,\mathrm{d}\tau - \int_V p^{ij}\omega_{ij}\,\mathrm{d}t\,\mathrm{d}\tau\,, \tag{1.9}$$

or, when the stress tensor is symmetric,

$$\mathrm{d}A_\Sigma^{(i)} = -\int_V p^{ij} e_{ij}\,\mathrm{d}t\,\mathrm{d}\tau\,. \tag{1.10}$$

As is well-known, one may always establish a correspondence between an antisymmetric tensor ω_{ij} in three-dimensional space and an axial vector $\boldsymbol{\omega}$, the vorticity vector of the velocity (*cf.*, Chapter IV, §3). It follows from the foregoing reasoning that the presence of vorticity in a moving continuous medium in the case of a symmetric stress tensor (in particular, in the absence of internal angular momentum and internal mass and surface couples) does not influence directly the magnitude of the elementary work of the internal surface forces, and consequently also not the change of kinetic energy.

Kinetic energy theorem for an infinitesimal volume of a continuum. The kinetic energy theorem for an infinitesimal volume of a continuum will now be written down.

For this purpose, one can choose the small volume ΔV with a considered point of continuum inside.

Shrinking the volume to this point, the equation (1.1) written for ΔV and divided by the mass Δm of a particle inside the volume, by (1.2) and (1.4) gives:

$$\frac{dv^2}{2} = \boldsymbol{F}\cdot d\boldsymbol{r} + \frac{1}{\rho}\nabla_j(p^{ij}v_i)dt - \frac{1}{\rho}p^{ij}\nabla_j v_i dt\,.$$

The quantity $v^2/2$ may be called the kinetic energy density and the quantities $\boldsymbol{F}\cdot d\boldsymbol{r}$, $\rho^{-1}\nabla(p^{ij}v_i)dt$, $-\rho^{-1}p^{ij}\nabla_j v_i dt$ the densities of the work done by the mass, external and internal surface forces.

In the kinetic energy theorem for an infinitesimal volume of a continuum, the elementary work done by the internal mass forces does not appear, since it tends to zero as V is shrinking to the point and $M \to 0$. It follows directly from the assumption of the existence of the density of the mass forces, as a limit of the ratio of the internal mass forces acting on the volume to the mass occupying this volume.

For example, let the internal mass forces be due to Newtonian gravitational attraction between the particles of the volume V. Then the work of the internal mass forces in the volume V with mass M can, obviously, be written in the form

$$\int_M \int_M \frac{dm_1\,dm_2}{(\boldsymbol{r}_1-\boldsymbol{r}_2)^2}\,\frac{\boldsymbol{r}_1-\boldsymbol{r}_2}{|\boldsymbol{r}_1-\boldsymbol{r}_2|}\cdot d\boldsymbol{r}\,.$$

The limit of this expression, divided by M, tends to zero, as V is shrinking to the point.

Thus, the kinetic energy theorem, which applies to every infinitesimal particle, can be formulated as follows: At every point of a continuum, the differential of the kinetic energy density is equal to the sum of the densities of the elements of work done by the external mass and the external and internal surface forces, acting on this infinitesimal particle.

It is seen that the kinetic energy theorem is an immediate consequence of the impulse equation and represents the balance of mechanical energy. The kinetic energy theorem relates to energy, but, in the general case, it is not equivalent to the principal of conservation of energy. Only when the mechanical energy of the system under consideration does not transform into heat or some other form of energy, the kinetic energy theorem coincides with the principle of conservation of energy (within the framework of mechanical formulations of problems). Note that the general principle of conservation of energy in this case falls into two parts—the separate principles of conservation of mechanical and non-mechanical energies.

Next, expressions will be derived for the work densities of the internal surface forces $(dm)^{-1}\,dA_\Sigma^{(i)}$ in some particular cases. One has

$$\frac{1}{dm}\,dA_{\Sigma}^{(i)} = -\frac{1}{\rho}\,p^{ij}e_{ij}dt - \frac{1}{\rho}\,p^{i}\omega_{ij}dt$$

$$= -\frac{1}{\rho}\,p^{ij}e_{ij}dt - \frac{1}{\rho}\underset{[i<j]}{(p^{ij}-p^{ji})\omega_{ij}}dt\,.$$

If the medium moves as a rigid body, then all $e_{ij}=0$ and

$$\frac{1}{dm}\,dA_{\Sigma}^{(i)} = -\frac{1}{\rho}\underset{[i<j]}{(p^{ij}-p^{ji})\omega_{ij}}\,.$$

If the stress tensor is not symmetric ($p^{ij}\neq p^{ji}$), then, in the case of rigid body motion, the work of the internal surface forces may be non-zero, since the angular velocity ω and, consequently, the ω_{ij} may differ from zero (for rotation).

If the stress tensor is symmetric, then

$$\frac{1}{dm}\,dA_{\Sigma}^{(i)} = -\frac{1}{\rho}\,p^{ij}e_{ij}dt\,, \tag{1.11}$$

i.e., in this case, the work of the internal surface forces is, in general, due to deformations. If a medium with a symmetric stress tensor moves like an absolutely rigid body, then the work of the internal surface forces in it is always equal to zero.

Work density of internal forces in the case of ideal fluids. For an ideal medium $p^{ij}=-pg^{ij}$, and therefore

$$\frac{1}{dm}\,dA_{\Sigma}^{(i)} = \frac{p}{\rho}\,g^{ij}e_{ij}dt = \frac{p}{\rho}\,e^{i}_{\cdot i}dt = \frac{p}{\rho}\,\mathrm{div}\,\boldsymbol{v}dt\,.$$

Replacing div $\boldsymbol{v}$ with the help of the equation of continuity, one obtains

$$\frac{1}{dm}\,dA_{\Sigma}^{(i)} = -\frac{p}{\rho^2}\frac{d\rho}{dt}\,dt = p\,d\,\frac{1}{\rho} = p\,dV\,, \tag{1.12}$$

where V is the specific volume.

For infinitesimal elements of an ideal medium, the kinetic energy theorem in regions of continuous motion of the medium assumes the form

$$d\,\frac{v^2}{2} = \frac{1}{dm}\,dA_{m}^{(e)} + \frac{1}{dm}\,dA_{\Sigma}^{(e)} + p\,d\,\frac{1}{\rho}$$

$$= \boldsymbol{F}^{(e)} \cdot \boldsymbol{v}\mathrm{d}t - \frac{1}{\rho} \nabla_k (pv^k)\mathrm{d}t + p\mathrm{d}\frac{1}{\rho}. \qquad (1.13)$$

It is worthwhile to note that for motion of a continuum relative to a moving or fixed coordinate system, the quantity of the density of the work done by the internal forces is not, in general, equal with opposite sign to the work density of all external surface and mass forces.

§ 2. The first law of thermodynamics (principal of conservation of energy) and the equation of heat flow

Variables of state. A start will be made with an explanation of the concepts underlying thermodynamics and hence all continuum mechanics, namely of the concepts of "state" of a system and of "variables of state". The state of a system (for example, of some volume of a continuous medium) will be said to be given, if the values of certain parameters $\mu^1, \mu^2, \ldots, \mu^n$ are known, which completely determine all relevant characteristics of the system (medium). The determining parameters μ^i, which may, in general, assume arbitrary values within certain intervals, are referred to as variables of state.

The choice of the variables of state and their number differ for different models of continua.

What is actually meant by saying that one knows the state of a medium? This question may be answered in the following manner. A body as a whole consists of atoms and molecules and, if at every instant of time the positions and motions of all elementary particles constituting the body are known, then the state of the entire body is also known. However, this answer may not be satisfactory. In fact, if one wishes to know, for example, the state of one cubic centimetre of motionless air, then one must specify $3.27 \cdot 10^{19}$ functions of time, the coordinates of the molecules (conceived as material points) contained in this volume, since the molecules of even a gas at rest are in motion. At the same time, it is well known that from a macroscopic point of view, in many cases, the state of air (and other gases) at rest is determined by altogether only two parameters, namely the pressure p and the density ρ.

In the macroscopic approach, only processes, effects and properties are studied which are essential for finite bodies and which are observed or employed under various conditions in nature and engineering.

The by no means trivial problem of the transition from a large number of parameters, defining a state of a medium viewed as a discrete system, to a smaller number of variables similar to ρ and p for a gas and determining the macroscopic state of a medium, constitutes one of the most important objectives of the physics of liquids, gases and solids. The solution of this problem is always connected with supplementary hypotheses, such as the laws of statistics and other natural phenomena which must be verified and studied by experiments and observations.

Macroscopic parameters may be constructed as statistical averages, computed under certain assumptions with respect to an aggregate of a large number of molecules, which are in motion and are, in general, distributed arbitrarily. For example, in gases, the macroscopic velocity $\boldsymbol{v}$ may be introduced as the velocity of the centre of gravity of the set of molecules in a physically small volume, the temperature T as the mean energy of the random motion of atoms and molecules relative to macroscopic motion, taking place with one degree of freedom, the stress $\boldsymbol{p}_n$, acting on some plane, as the mean characteristic impulse, exerted by the molecules on this face during their random motion, etc.

In the general case, determining parameters are introduced for definite classes of problems under consideration with the help of hypotheses which are guided by experimental data and theoretical investigations. In many complicated cases, the problem of the introduction of determining parameters is still open and an object for study: for example, for models of visco-plastic solids, for non-equilibrium phenomena in complicated physical, chemical and biological systems, for different types of phenomena of radiation and in many other situations.

On the number of variables of state for continuous media. The internal state of a small particle of a material continuum may, generally speaking, be characterized by a finite and much smaller number of determining parameters than the discrete system of elementary particles. For example, the internal state of a particle of a deformable rigid solid in the classical theory of elasticity is characterized by only seven variables, namely the six components of the strain tensor ε_{ij} and the temperature T, as well as by physical parameters which are constant for a particular medium, namely Young's modulus E, Poisson's coefficient σ and the thermal heat capacity c. At the same time, the possibility is not excluded that the number of parameters determining the state even of an infinitesimal particle of a continuum, in any of the models of a continuum (continuous medium), will be infinite.

Examples of such models are those of bodies with heredity. For such models, one assumes that the stresses depend not only on the deformation and temperature at a given instant, but on the entire pre-history of deformation of the body, i.e., on the functions $\varepsilon_{ij}(t)$ and $T(t)$. This statement is equivalent to saying that the p^{ij} depend on the ε_{ij}, T and all their time derivatives, i.e., that the number of variables of state for such media is infinite. Other more complicated examples may be taken from continua encountered in the kinetic theories developed in statistical physics, for example, a gas which is described by Boltzmann's equation. However, such models are very complicated and the experience of theory and practice shows that in the majority of practically important cases one may overcome the problem of specification of the state of small particles by use of a finite and, in general, not large number of parameters. In complicated kinetic theories, one employs often for the construction of solutions approximate methods which, from a physical point of view, are equivalent to a transition to models with finite numbers of degrees of freedom for infinitesimal particles.

Note that for the determination of the state of a finite volume of a continuous medium it is, in general, always necessary to introduce functions (and not a finite set of numbers), namely distributions of deformations, temperatures, etc.

The specification of these functions is equivalent to giving infinite numbers of parameters (for example, the Fourier coefficients for these functions). Therefore the number of determining parameters for a finite volume in the general case of any model of a continuous medium is always infinite.

However, in the small, all functions, determining the state of a body, may, in first approximation, be assumed to be either linear or quadratic or a low degree polynomial. Therefore, the coefficients of these polynomials form a finite number of parameters characterizing the state of infinitesimal elements of a continuum.

In the development of continuum mechanics, infinitesimal particles are considered as thermodynamic systems for which the mechanical concepts of position and characteristics of motion as well as the physical concept of internal state are defined.

It will be shown below that for infinitesimal particles there exists a finite system of characteristics, determining parameters, which are specified numerically in the coordinate system and the system of units employed.

Certain of these parameters may be geometrical or mechanical as, for example, are space coordinates, velocity, density, characteristics of deformation, etc; others may be physical or chemical, as, for example, temperature, concentration of different components, structural parameters, phase characteristics of substances, coefficients of thermal conductivity, viscosity, elastic moduli, etc.

It will be assumed that the symbols μ^i denote parameters which in a given reference system may vary, and the symbols k^i physical constants. Certain of the parameters μ^i and k^i may be components of different vectors and tensors.

Complete systems of determining parameters. By definition, for a fixed small particle, the quantities $\mu^1, \mu^2, \ldots, \mu^n, k^1, k^2, \ldots, k^m$ form a base (a complete system of determining parameters), i.e., they may be specified independently and, within known ranges, arbitrarily; their set has the property that all other characteristics of state and motion, considered in a given class of problems, may be expressed in terms of them in a universal form, which does not depend on a particular concrete problem.

For example, it will be shown below that density and temperature for gas particles, within known limits, may be assigned arbitrarily, whereas other thermodynamic functions, for example, entropy and pressure are determined in terms of these quantities.

One must make a distinction between a system of determining parameters for a given problem and a system of parameters, determining a state of a medium. In the first case, one is dealing with a set of parameters, characterizing the conditions of the problem, which singles out a unique global phenomenon for finite bodies on the basis of a system of equations and supplementary boundary and other conditions (the choice of which is connected with the formulation of particular problems); in the second case, one has the characteristics of a state for which one must construct equations to be fulfilled for all possible particular problems and processes.

The fixing of a system of parameters, determining the physical state of the elements of a medium, is important and, in a logical sense, the initial stage in the determination of a model of a continuous medium, which is intended for the description of the motion of certain real media under certain definite types of external conditions.

From the mathematical point of view, the parameters of state $\mu^1, \mu^2, \ldots, \mu^n, k^1, k^2, \ldots, k^m$ represent the arguments of functions, entering into the closed system of equations describing the behaviour of a medium.[1)] Clearly, these functions may depend on the choice of the independent variables and, in accordance with this system of determining parameters, fixing a given model of a continuous medium, may consist of different quantities. For example, in the case of a gas, these quantities may be p and ρ, or p and T, or ρ and T, etc.

From the physical point of view, however, these different systems of determining parameters for a given model of a continuum may not be equivalent. As will be seen below, in specifications of the internal energy as functions of ρ and the entropy S or of p and ρ, different types of information are involved. For the introduction of systems of determining parameters, one must understand and bear in mind the system of quantities and characteristics which are to be considered as dependent variables. Obviously, a system of determining variables, as regards their number and structure, may, in general, differ when the quantities determined are different.

On holonomic and non-holonomic thermodynamical systems. One may introduce the analogy between the concepts of numbers of degrees of freedom in theoretical mechanics and numbers of variables of state in continuum mechanics.

Indeed, the number of degrees of freedom usually is determined as the number of independent parameters determining the position of a mechanical system. For example, an absolutely rigid body has six degrees of freedom. Note that, if an absolutely rigid body is considered as a physical system, then one must give for its specification values to ten constant parameters, viz., mass, position of centre of mass in the body and the components of the inertia tensor referred to the centre of mass.

In theoretical mechanics, non-holonomic systems are studied. In continuum mechanics, one may also encounter a situation when the

1) In its own sense, the concept of the determining parameters implies that the factual functional links establish corresponding laws, hypotheses and direct definitions.

determining parameters μ^i change arbitrarily within definite limits and their increments $\delta\mu^i$, under the conditions of the class of problems under consideration, are interlinked, for example, by m non-integrable relations of the form

$$A_i \delta\mu^i = 0 \,,$$

where A_i are certain functions of the determining parameters. Then the number $n-m$ of independent increments $\delta\mu^i$ will be smaller than the number of independent variables of determining parameters μ^i, and the thermodynamic system is said to be non-holonomic.

For non-holonomic systems, the number of degrees of freedom is defined as the number of independent increments $\delta\mu^i$.

Consider, as an example, the non-holonomic relations which are only a consequence of the definition of the parameters μ^i. Let $\mu^1(\xi^1, \xi^2, \xi^3, t)$ be some scalar parameter (for example, the density of a particle ρ), t the time, ξ^1, ξ^2, ξ^3 the Lagrangian coordinates of the centre of a separate particle, μ^2 and μ^3 the first and second derivatives of μ^1 with respect to t, viz.,

$$\mu^2 = \frac{\partial \mu^1}{\partial t}\,, \qquad \mu^3 = \frac{\partial \mu^2}{\partial t} = \frac{\partial^2 \mu^1}{\partial t^2}\,. \tag{2.1}$$

As a natural physical condition, assume that the time t does not enter explicitly into the system μ^i.

Clearly, under different external conditions, there may occur states with different, arbitrary within certain limits, variables μ^1, μ^2 and μ^3, while the increments $d\mu^1$ and $d\mu^2$ for time changes dt and constant ξ^1, ξ^2, ξ^3 will always be interrelated by the same non-holonomic relation

$$\mu^2 d\mu^2 - \mu^3 d\mu^1 = 0\,, \tag{2.2}$$

which in the presence of (2.1) is an identity. It follows from (2.1) that for arbitrary $\delta\mu^1$ and $\delta\mu^2$ the relation (2.2) is not fulfilled. Such types of non-holonomic relations are encountered in the utilization of models of continuous media for which one has among the determining parameters successive derivatives with respect to time. Equations (2.1) and (2.2) are not connected with observations or experiments, but one must consider as experimental result the fact that successive derivatives may and expediently enter into the set of determining parameters for the construction of models of continua and that they must be viewed as arguments of certain functions which in the end are obtained from data based on experiments.

Space of states. Introduce now into the consideration the space of states, i.e., the space the coordinates of which are the variables of state μ^i (viz., phase space). Obviously, different states of thermodynamic systems will correspond to different points of the space of states.

Processes and cycles. The set of states of a medium, corresponding to some sequence of values of the variables of state, is said to be a process. Special significance have physically real processes, i.e., processes in which within the framework of applicable models under consideration a sequence of states may be realized in the course of time. Depending on external and internal interactions, one may consider different real processes.

Processes may be continuous, when the set of states for given particles $\mu^1, \mu^2, \ldots, \mu^n$ form in the space a continuous curve. In the theory, one also encounters processes with discontinuous values of the variables of state μ^i and, in particular, discontinuous processes, consisting of segments of continuous curves in the space of states. In continuum mechanics, one studies continuous processes and discontinuous processes with distinct points of discontinuity on continuous curves in the space of states.

Between two sets of data relating to identical states one may, in general, construct many different processes which are either continuous or discontinuous. The family of curves, corresponding to real processes which may be caused by various external conditions, possesses, generally speaking, great arbitrariness; however, in certain cases, for example for non-holonomic systems, the corresponding curves are characterized by certain, readily displayed, special properties. In the example considered above, it follows from (2.1) that continuous processes with $\mu^1 = \text{const.}$ for $\mu^2 \neq 0$ are impossible. However, in this case, the presence of Equality (2.1) in the space of states does not exclude a real continuous process between any two points with arbitrary given coordinates $\mu^{1\prime}, \mu^{2\prime}, \mu^{3\prime}$ and $\mu^{1\prime\prime}, \mu^{2\prime\prime}, \mu^{3\prime\prime}$.

The number of variable parameters and their character for different types of processes may differ. For example, processes may be purely mechanical, when all parameters of a non-mechanical nature remain constant.

Processes as a result of which a system returns in the space of states to its original position are said to be cyclic.

In the case of continuous processes, a closed curve corresponds to a cycle in the space of states.

In this chapter, consideration will be given to continuous processes; processes with points of discontinuity will be studied in Chapter VII.

One may fix some state A and consider all possible continuous cycles which pass through the state A and some arbitrary state B. Different external conditions correspond to different processes or cycles. This circumstance arises from the fact that the equations determining $\mu^1, \mu^2, \ldots, \mu^n$ contain certain functions which may be different; one may dispose of them in a different manner and thus influence the processes under consideration. Known examples of models of continuous media show that for a fixed state A the state B may, generally speaking, coincide with all possible states in the phase volume which is determined by the physically admissible values of the determining parameters.

On the interaction of systems with external objects. During any process, in the general case, a system interacts with external bodies and fields. The basic problem in the construction of a model of a continuous medium consists of the establishment of the laws and mechanisms of interaction of separate particles of the continuum with bodies, which are external with respect to it, and with fields, in particular with adjacent particles of the same medium.

Also, in continuum mechanics, one requires for applications macroscopic relations with small numbers of determining parameters. Frequently, such relations depend on the ideas at a microscopic level on the molecules, atoms and other particles, on their distributions, motions and on forces of interaction between them in the body. However, all details of such representations are never known completely. It is also important to emphasize that even all known details cannot, and, in principle, need not be taken into account. Therefore, in the construction of models of continua, one must always formulate in one form or another and utilize phenomenological hypotheses which after verification of the usefulness for the description of observations during experiments are referred to as laws of nature.

In physics and, in particular, in continuum mechanics, great significance attaches to considerations of the interchange of energy between given particles (thermodynamic systems) and adjoining particles, external bodies and external fields. The concept of energy is closely linked to representations of different forms of energy. These forms may be the kinetic energy of particles, the potential energy which is conditioned by the relative distribution of particles, heat energy, electromagnetic energy, the energy of chemical bonds and certain other forms of energy. For a more detailed study at a microscopic level, the concepts of different forms (and of the number of forms) of energy change. However, experience shows that on a macroscopic scale one may make a distinction between certain phenomenological symptoms listed above and other forms of energy and speak of conversions of energy from one form to another.

A start will be made at a basic physical position regarding the existence of symptoms, which permit to distinguish, at a macroscopic level, the forms of energy of a system and the forms of influx into the system of energy as the result of its interaction with external bodies and fields and to require the taking into account of conversions of energy from one form to another.

Consider a system which is characterized by a finite number of determining parameters, for example, an infinitesimal particle of a continuum or a finite volume V under the condition that all particles of this volume undergo identical processes (the variables of state in this case are constants with respect to volume).

It will be implied that, from a point of view of data on the characteristics of the internal state of a particle $\mu^1, \mu^2, \ldots, \mu^n$ and their infinitesimal changes $d\mu^1, d\mu^2, \ldots, d\mu^n$, one may try out different overall macroscopic inflows of energy to a particle from outside. Data on these inflows, as functions of the elementary process with consideration of increments $d\mu^1, d\mu^2, \ldots, d\mu^n$, may and, generally speaking, must be considered as a description of the properties of a model which forms an important part of the constructive development of the model. Naturally, one may select instead of this information on the properties of different inflows of energy to a particle from outside as data, which enter into the definition of a model, other details (and this is actually done in practice) from which one may deduce the interchange of energy with the aid of certain sets of relations which are universal or specially suited for a given model.

In mechanics, until recently, main significance has attached to flows to a particle of energy of a mechanical nature, i.e., the work done by external macroscopic volume or mass and external surface forces on particles, and inflows of heat energy which a particle may obtain on account of heat conduction, radiation, chemical conversions, flux of electric current and other mechanisms. (The energy corresponding to these flows, which may be returned or received by a particle, can be converted from one to another inside or outside a particle.)

At the present time, in many cases, one must take into consideration electromagnetic interactions; thus there arises the necessity to consider energy interchange of particles with external media taking into account more complex mechanisms of interaction such as, for example, the work done by distributed surface couples, energy transfer with consideration of chemical, structural and phase changes, etc.

Note that in modern times a beginning has been made with the study of new macroscopic mechanisms of energy transfer between individual particles and surrounding media and of laws of energy transfer between elementary particles. At a microscopic level, and in many cases at a macroscopic level (properties of metals, interactions inside a body at low temperatures, interaction of laser beams with ordinary bódies, etc.), the essence of the mechanism of interaction may only be understood within the framework of quantum mechanics, when the necessary phenomenological formulation of these interactions may be given in complicated models of continuous media within the framework of Newtonian mechanics.

The total influx of energy for an elementary process $d\mu^1, d\mu^2, \ldots, d\mu^n$ to a small particle may be represented in the form

$$dA^{(e)} + dQ^{(e)} + dQ^{**}, \tag{2.3}$$

where $dA^{(e)}$ is the work done by external macroscopic mass and surface forces, $dQ^{(e)}$ is the flux of heat and dQ^{**} is the external flux of energy to the particle which arises by taking into account different mechanisms of interactions which differ from the work done by macroscopic forces and heat transfer; for example, this situation occurs if one takes into consideration interactions with an electromagnetic field including energy which can be spent on magnetization and electric polarization of a medium, and other phenomena.

For the elementary work done by external forces in correspondence with the basic meaning of the system of determining parameters and in connection with a number of assumptions which enter into the definitions of models of continuous media, one may write down for an infinitesimal process corresponding to changes in the parameters $d\mu^1, d\mu^2, \ldots, d\mu^n$ a formula of the form

$$dA^{(e)} = P_i(\mu^1, \mu^2, \ldots, \mu^n, k^1, k^2, \ldots, k^m)d\mu^i \tag{2.4}$$

which expresses the work $dA^{(e)}$ done by the external forces in terms of the internal parameters of the particle under consideration and of their increments. The form of the functions P_i, in essence, is connected with the formulation of the basic postulates which are required in the definition of a model.

Formula (2.4) for a small particle of a continuum may be considered as a generalization of the formula

$$dA^{(e)} = m\mathbf{v}d\mathbf{v} \tag{2.5}$$

for a material point mass m, moving with velocity $\mathbf{v}$, or of the formula for an absolutely rigid body of arbitrary finite dimensions

$$dA^{(e)} = m\mathbf{v}^* d\mathbf{v}^* + Apdp + Bqdq + Crdr\,, \tag{2.6}$$

where m is the mass of the body, $\mathbf{v}^*$ is the velocity of its centre of mass, A, B and C are its moments of inertia with respect to its central axes, p, q and r are the projections of its instantaneous angular velocity on to these axes.

For an ideal fluid, in which the pressure is given as a function of variables of state, on the basis of the kinetic energy theorem (1.13), one may write

$$\frac{1}{dm}dA^{(e)} = vdv - pd\frac{1}{\rho}\,. \tag{2.7}$$

Each of the relations (2.4)–(2.7) may be conceived as a definition of $dA^{(e)}$ in terms of the internal parameters of a medium.

Each of these relations may be reduced to an equation which determines the values of the parameters in a concrete system, provided one knows from supplementary studies the laws for $dA^{(e)}$ yielding the energy transfer between a given particle and external bodies as functions of the external conditions. Thus, for a material point, one may utilize an expression for $dA^{(e)}$ different from (2.5), namely

$$\mathrm{d}A^{(e)} = \boldsymbol{F} \cdot \mathrm{d}\boldsymbol{r},$$

where $\boldsymbol{F}$ is a force. If the force $\boldsymbol{F}$ is known (viz., if one knows the mechanism of interaction of the point with external objects), then (2.5) yields an equation which determines the motion of the point. It must be emphasized that the laws of interaction, characterizing a given model, may be established on the basis of observations during experiments in which the right-hand side in (2.4) is changed after a corresponding treatment and generalization of the results of these observations.

In an analogous manner to (2.4), on the basis of physical phenomena and, in general, on the basis of special physical assumptions entering into the definition of the model of a medium under consideration, one may also write

$$\mathrm{d}Q^* = \mathrm{d}Q^{(e)} + \mathrm{d}Q^{**} = Q_i(\mu^1, \mu^2, \ldots, \mu^n, k^1, \ldots, k^m)\mathrm{d}\mu^i. \tag{2.8}$$

For example, for an undeformable rigid body or for an ideal incompressible fluid, it may be assumed that

$$\mathrm{d}Q^* = \mathrm{d}Q^{(e)} = c(T)\mathrm{d}T, \tag{2.9}$$

where $c(T)$ is the coefficient of thermal heat capacity and T the temperature. Relation (2.9) may serve for the experimental determination of $\mathrm{d}Q^{(e)}$. If the law of heat transfer has been established for $\mathrm{d}Q^{(e)}$, then (2.8) with $\mathrm{d}Q^{**} = 0$ becomes the equation of heat diffusion.

In more general cases, Formulae (2.4) and (2.9) become more complicated. For example, for a viscous incompressible fluid, one has on the right-hand side of (2.4) a certain positive term and the same term with a negative sign enters into (2.8). This term corresponds to the dissipative work done by the viscous stress forces; this work by internal forces becomes heat (*cf.*, for more detail, see §7, this chapter). Such a situation is typical for all cases in which there occurs inside a particle conversion from one form of energy to another. It follows from the later work that the greatest significance attaches to quantities represented by the sums $P_i + Q_i$.

The fluxes of energy $\mathrm{d}Q^*$, $\mathrm{d}Q^{(e)}$ and $\mathrm{d}Q^{**}$, as well as the elementary work done by the external macroscopic forces $\mathrm{d}A^{(e)}$, are not, in the general case, differentials of certain functions and represent themselves infinitesimal quantities.

The principle of conservation of energy—The first law of thermodynamics. Let there be given a process proceeding in the space of states from the point A with the values of the parameters of state μ_0^i along the

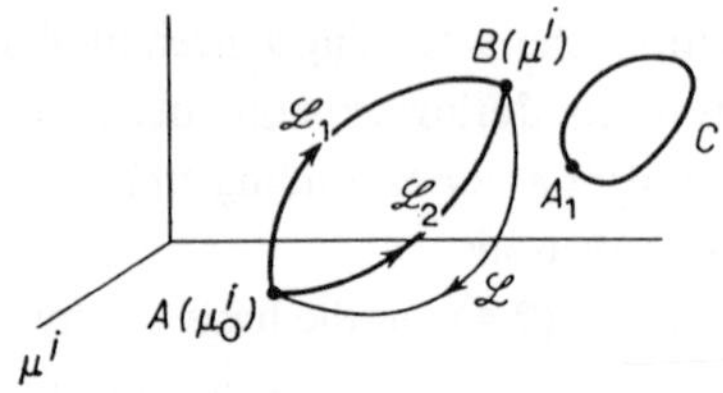

Fig. 29. On the principle of conservation of energy.

curve $\mathscr{L}_1$ to the point B with the values of the parameters μ^i (Fig. 29). Introduce the concept of the total influx of energy which the system receives during this process from outside. Obviously, this influx is given by

$$A^{(e)} + Q^* = \int_{AB(\mathscr{L}_1)} P_i \mathrm{d}\mu^i + \int_{AB(\mathscr{L}_1)} Q_i \mathrm{d}\mu^i , \tag{2.10}$$

and, at a first glance, it must depend on the process, i.e., on the path of integration $\mathscr{L}_1$ in the space of states.

The first principle of thermodynamics, or the law of conservation of energy, may be formulated as the impossibility of a realization of perpetual motion of the first kind, i.e., of a cyclically operating machine which could provide a source of useful energy without utilisation of any source of energy, external to this machine.

This statement must be conceived as a law which is confirmed by all known experimental data.

Now, let a system execute a cycle, for example the cycle C. Then the first principle of thermodynamics, or the law of conservation of energy, reduces to the statement that the total influx of energy from outside into a system performing any realizable cycle is equal to zero, i.e.,

$$\oint_C (P_i + Q_i) \mathrm{d}\mu^i = 0 . \tag{2.11}$$

Hence it follows directly that the total influx of energy (2.10) into the system from outside does not depend on the process $\mathscr{L}_1$; it depends only on the initial and final states of the system. In fact, introduce between the states A and B, in addition to the arbitrary process under consideration $\mathscr{L}_1$, another process $\mathscr{L}_2$ and a process $\mathscr{L}$ which proceeds from the state B to the state A. The processes $\mathscr{L}_1\mathscr{L}$ and $\mathscr{L}_2\mathscr{L}$ form closed cycles, and by the law of conservation of energy it follows directly that

$$A^{(e)}+Q^* = \int_{\mathscr{L}_1} (P_i+Q_i)\mathrm{d}\mu^i = \int_{\mathscr{L}_2} (P_i+Q_i)\mathrm{d}\mu^i$$

$$= -\int_{\mathscr{L}} (P_i+Q_i)\mathrm{d}\mu^i\,. \tag{2.12}$$

Total energy of a system. Therefore, if the initial state A of a system is fixed, then for all realizable processes the total influx of energy into the system from outside depends only on the final state of the system, i.e.,

$$A^{(e)}+Q^* = \mathscr{E}(\mu^1, \ldots, \mu^n)-\mathscr{E}(\mu_0^1, \ldots, \mu_0^n)\,,$$

where $\mathscr{E}$ is a single-valued function of the parameters of state of the system, known as its total energy. Thus, it follows from the first law of thermodynamics that there exists a function of state $\mathscr{E}(\mu^1, \mu^2, \ldots, \mu^n)$ the total differential of which for realizable processes is equal to the sum of the elements of work $\mathrm{d}A^{(e)}$ done by the external mass and surface macroscopic forces and the elementary influx into the system from outside of other forms of energy

$$\mathrm{d}\mathscr{E} = \mathrm{d}A^{(e)}+\mathrm{d}Q^* = (P_i+Q_i)\mathrm{d}\mu^i\,. \tag{2.13}$$

It is easily seen that the total energy of the system $\mathscr{E}(\mu^1, \ldots, \mu^n)$ is defined exactly apart from an additive constant, namely the value of $\mathscr{E}$ in the initial state of the system.

If the external influx of energy into a system is known, then the first law of thermodynamics (2.11) may serve as a basis for the determination of the total energy of the system $\mathscr{E}$. Conversely, if the energy is known from somewhere or other, then the law of conservation of energy may be utilized to elucidate the mechanism of interaction of the particles under consideration with external bodies, i.e., for the determination of $\mathrm{d}A^{(e)}+\mathrm{d}Q^*$.

For the determination of the total energy of the system $\mathscr{E}(\mu^1, \ldots, \mu^n)$, one must, generally speaking, know the functions P_i and Q_i. By (2.13),

P_i and Q_i may not be arbitrary functions of the parameters of state.

In fact, Formula (2.13) may be rewritten in the form

$$\left(P_i+Q_i-\frac{\partial \mathscr{E}}{\partial \mu^i}\right)d\mu^i=0. \tag{2.14}$$

If a system is holonomic, i.e., if all $d\mu^i$ are independent (in particular, if the parameters μ^i do not include successive derivatives with respect to time), then it follows from (2.14) that

$$P_i+Q_i=\frac{\partial \mathscr{E}}{\partial \mu^i},$$

and, consequently, that P_i+Q_i must satisfy the conditions of integrability

$$\frac{\partial(P_i+Q_i)}{\partial \mu^k}=\frac{\partial(P_k+Q_k)}{\partial \mu^i}. \tag{2.15}$$

For non-holonomic systems, one may write

$$P_i+Q_i=\frac{\partial \mathscr{E}}{\partial \mu^i}+R_i.$$

For all processes for which the first law of thermodynamics is fulfilled, the quantities R_i must satisfy

$$R_i d\mu^i=0, \tag{2.16}$$

i.e., for all realizable processes, the quantities R_i do not contribute to the energy balance. However, the R_i may differ from zero. For certain important cases, the general form of the functions R_i which satisfy (2.16) may be stated.[1)]

For non-holonomic systems, there must be fulfilled for realizable processes instead of (2.15) the conditions

$$\left.\begin{aligned}&\frac{\partial(P_i+Q_i-R_i)}{\partial \mu^k}=\frac{\partial(P_k+Q_k-R_k)}{\partial \mu^i},\\ &R_i d\mu^i=0.\end{aligned}\right\} \tag{2.17}$$

[1)] *cf.*, L. I. Sedov, *Some problems of the construction of new models of continuous media.* Proceedings of XI. Congress of Applied and Theoretical Mechanics. Munich 1964.

Note that, since (2.15) or (2.17) are necessary and sufficient conditions for the existence of a function $\mathscr{E}(\mu^1, \mu^2, \ldots, \mu^n)$ which satisfies (2.13) for realizable processes, they may also be conceived as one of the possible formulations of the law of conservation of energy.

The conditions (2.15) may be utilized either in the verification of experimental results for the definition of P_i+Q_i or for a reduction of the numbers of experimental measurements, when certain of the quantities P_i+Q_i are determined experimentally and others are computed from (2.15).

Internal energy of a system. Previously, the function $E=v^2/2$ has been introduced, the density of the kinetic energy of a medium; generally speaking, the expression $E\rho\,\mathrm{d}\tau$ does not coincide with the function $\mathscr{E}$, the total energy of a particle. Let

$$\mathscr{E} = (E+U)\rho\,\mathrm{d}\tau \, ,$$

where U is a scalar function of the parameters of state, the so-called internal energy density. The internal energy density or the internal energy per unit mass or the specific internal energy U, just as the total energy of the system $\mathscr{E}$, is determined exactly apart from an additive constant and exists for every thermodynamic system.

The specific internal energy U does not depend explicitly on the space coordinates and time, if the space and time may be assumed to be homogeneous. The property of homogeneity signifies that at all points of space and at all instants of time under identical external conditions in a given thermodynamic system processes proceed in an identical manner.

The total energy and the internal energy may be introduced for a body as a whole as well as for its separate parts. The internal energy of a finite part of a body or of a body as a whole does not possess, generally speaking, the property of addition, i.e., the internal energy of an entire body is not equal to the sum of the internal energies of its constituent parts.

For example; under identical conditions (identical temperature, etc.), the internal energy of two small drops of water will not be equal to the internal energy of one large drop the mass of which is equal to the sum of the masses of the two small drops, if one takes into account the energy

arising from surface tension.[1)]
Obviously, the internal energy, arising from mutual attractions between parts of a body in accordance with the law of universal gravitation, is also not additive.

However, in many cases, the internal energy may be assumed to be additive; in particular, this will be true for water in those cases when the surface tension need not be taken into account, or for an elastic body which obeys Hooke's Law. If the internal energy is additive, then the total energy of an arbitrary finite volume V is given by

$$\mathscr{E} = \int_V \rho\left(\frac{v^2}{2} + U\right) d\tau .$$

The further reasoning concerns basically infinitesimal small particles and therefore is valid for cases when the internal energy is additive as well as for cases when it does not have this property.

Note that the concept of internal energy, like all other thermodynamic relationships and concepts, is necessary in the general case of a study of motion of continuous media, but there exist certain particular examples of continua, in particular those listed in Chapter IV, where the concept of internal energy is not necessary for a completion of the system of equations, describing continuous motion. The concept of internal energy in explicit form is not required in the study of mechanical motion of an ideal incompressible fluid; similarly, in the theory of elastic bodies, one may proceed without this concept provided thermal effects are not considered.

Equation of the law of conservation of energy. Thus, the universal relation expressing the law of conservation of energy may be presented in the form

$$dE + dU_m = dA^{(e)} + dQ^{(e)} + dQ^{**} , \tag{2.18}$$

[1)] It is useful in this context to warn against vagueness which may arise in connection with the fact that, if one considers spontaneous processes of slow merging of two isolated drops into a single one, by the law of conservation of energy, the internal energy of one large drop at rest will, of course, be equal to the sum of the energies of two small drops, which at first are also at rest. However, the temperature of the single large drop will under these conditions be larger than the common temperature of the two small drops until merging occurs.

where dU_m is the change in internal energy of the body under consideration, dE is the change in its kinetic energy, $dA^{(e)}$ is the element of work done by external macroscopic forces, $dQ^{(e)}$ is the elementary flux of heat into the body from outside and dQ^{**} is the elementary flux from outside of other nonthermal forms of energy which differ from the work of macroscopic mechanical forces.

Equation of heat flow. Subtracting from this relation (1.8), expressing the kinetic energy theorem for continuous media, one obtains

$$dU_m = -dA^{(i)} + dQ^{(e)} + dQ^{**}$$

or (2.19)

$$dU_m = -dA^{(i)} + dQ^{*} ,$$

which is known as the equation of heat flow and may take the place of the law of conservation of energy.

If a process is very smooth, so that accelerations can be neglected, then $dE=0$, and therefore for such processes one may assume that the work done by the external forces is equal to the work done by the internal forces, taken with the opposite sign:

$$dA^{(e)} = -dA^{(i)} .$$

Thus, for such processes, for example for quasistatistic processes, the equation of heat flow may be rewritten in the form

$$dU_m = dA^{(e)} + dQ^{*} .$$

The differential equation of heat flow. The equation of heat flow (2.19) may be written down for any volume of a continuum, imagined to have been isolated. Construct it for an infinitesimal particle of a continuous medium. For the density U of the internal energy, one has

$$U = \lim_{\Delta m \to 0} \frac{U_{\Delta m}}{\Delta m}$$

(assuming the limit to exist). For a small particle $dU_{\Delta m} = \Delta m\, dU$. Analogously, one can introduce elementary fluxes of external energy per unit mass of a medium

$$dq = \lim_{\Delta m \to 0} \frac{dQ^{(e)}}{\Delta m},$$

$$dq^{**} = \lim_{\Delta m \to 0} \frac{dQ^{**}}{\Delta m}.$$

Dividing (2.19) by Δm, letting $\Delta m \to 0$ and recalling that the density of work done by the internal surface forces is equal to

$$-\frac{1}{\rho} p^{ij} \nabla_j v_i dt,$$

the differential equation of heat flux may be written in the form

$$dU = \frac{1}{\rho} p^{ij} \nabla_j v_i dt + dq + dq^{**}. \tag{2.20}$$

It is known from the theory of deformation[1] that the components of the strain rate tensor may be presented in the form

$$\hat{e}_{ij} = \tfrac{1}{2}(\hat{\nabla}_j \hat{v}_i + \hat{\nabla}_i \hat{v}_j) = \tfrac{1}{2} \frac{d\hat{g}_{ij}}{dt};$$

if the metric of the initial state $\mathring{g}_{ij}$ does not depend on the time, then

$$\hat{e}_{ij} dt = d\hat{\varepsilon}_{ij},$$

where the differential components of the strain tensor are determined in the concomitant coordinate system. Therefore the equation of heat flux for $p^{ij} = p^{ji}$ may be rewritten in the form

$$dU = \frac{1}{\rho} \hat{p}^{ij} d\hat{\varepsilon}_{ij} + dq + dq^{**}. \tag{2.21}$$

The differential components of the strain tensor $d\hat{\varepsilon}_{ij}$, evaluated in the concomitant coordinate system, as also the components of the strain tensor $\hat{\varepsilon}_{ij}$ may be studied in any arbitrary coordinate system. Denote the components of the tensor $d\hat{\varepsilon}_{ij}$ in an arbitrary coordinate by $d\varepsilon_{ij} = e_{ij} dt$. In an arbitrary (not concomitant) coordinate system, the components $d\varepsilon_{ij}$ introduced in this manner will not be differential components of the strain tensor ε_{ij} in the same coordinate system. Taking this fact into

1) *cf.*, Chapter II, §6.

account, Equation (2.21) may be written in an arbitrary coordinate system in the form

$$dU = \frac{1}{\rho} p^{ij} d\varepsilon_{ij} + dq + dq^{**} , \tag{2.22}$$

where

$$d\varepsilon_{ij} = e_{ij} dt .$$

In every state, for different motions of a continuous medium, the quantities $d\varepsilon_{ij}$, as well as the increments of the determining parameters on which depend dU, dq and dq^{**}, may assume, to a known degree, arbitrary values. The circumstance is linked to the fact that (2.22) does not contain external mass forces and does not depend explicitly on boundary and other external conditions which may differ and which exert an essential influence on the increments of the determining parameters, entering into (2.22). At the same time, Equation (2.22) is a universal equation which holds for all possible processes.

The arbitrariness, connected with the linear independence of the corresponding set of differentials of determining parameters for all possible processes may be utilized for obtaining from the single equation (2.22) several equations of the type of equations of state.

§3. Thermodynamic equilibrium; reversible and irreversible processes

Thermodynamic equilibrium. As is well known, one may study the mechanical equilibrium of an absolutely rigid body. A body is said to be in a position of mechanical equilibrium with respect to a given reference system, if it may remain in this position when all the external conditions are conserved for an indefinite period. A thermodynamic equilibrium state of a system is said to be a state in which all characteristics of the internal state of the system (including mechanical) may conserve their values for an indefinite period while the external conditions are preserved. In the space of states, a thermodynamic equilibrium state is represented by a point.

A thermodynamic equilibrium state for a small volume of a system, is essentially linked to the presence of its characteristic temperature *cf.*, pp. 237–239.

It must be added, that there exists a number of parameters connected with the description of the non-equilibrium states (a time of relaxation, the different ⟨⟨ temperatures ⟩⟩ of the components of mixtures etc.) which are not between the essential parameters in an equilibrium state. If one has among the determining parameters substantial derivatives with respect to the time, then these parameters vanish in an equilibrium state. In the determination of the conditions for an equilibrium state, only the parameters which characterize such a state can appear.

It may happen, that the body as a whole is not in equilibrium but its parts are in equilibrium.

Equilibrium and non-equilibrium processes. Thermodynamic processes may proceed quickly as well as slowly. One may examine the limiting case of a process flowing so slowly that the rates of change of all its parameters are infinitely small. In the space of states, such a process is represented by a curve each point of which is an equilibrium point. Infinitely slow processes in which every intermediate state is an equilibrium state are referred to as equilibrium processes; in the relations which describe equilibrium processes, the magnitudes of the rates of change of parameters are not essential, but the directions of change of the determining parameters in an equilibrium process may be essential.

Processes which proceed at finite rates (provided the rates influence the physical bonds) are said to be non-equilibrium.

When it is said that a system undergoes a certain process, one has in mind a definite substantial material object the variables of state of which change, i.e., follow a Lagrangian point of view. Obviously, the definitions of equilibrium and stationary processes in the presence of motion of a medium do not coincide in the general case. A process may be stationary, i.e., all parameters of state of the system may not change with time at a given point of geometrical space ($\partial\mu^i/\partial t = 0$), and at the same time it may be non-equilibrium, i.e., it may involve finite rates of change of parameters ($d\mu^k/dt \neq 0$) which exert an essential influence on the process.

Reversible and irreversible processes. A process, proceeding from some state A to a state B, is said to be reversible, if for each intermediate state all equations *for infinitesimal increments of the parameters* are likewise satisfied when the signs of these increments are reversed. Thus, if some sequence of states forms a reversible process in the space of states, then this sequence may be performed in both directions, where corresponding to each element of the path the external fluxes of energy $dA^{(e)}$, $dQ^{(e)}$ and dQ^{**} in the direct and the reverse processes only differ in sign. If a process does not have such a property, it is said to be irreversible.

Note that there appear among the defining parameters for irreversible processes in an essential manner quantities which characterize the directions of change of certain of the determining parameters; for reversible processes, the directions of change of the thermodynamic parameters are not essential. As a rule, one studies reversible processes which simultaneously are also equilibrium processes.[1)] However, one may consider also reversible processes which are not composed of thermodynamic equilibrium states.

The concepts of reversibility and irreversibility of phenomena are fundamental. In the modern sciences, they are used as characteristic and important features, defining the nature of the reological and other physical laws in the variety of the microscopic and macroscopic phenomena.

It can be observed, that in the analytical mechanics of a material system, every motion is reversible if all acting forces are invariant, when the direction of a velocity is reversed. In particular, it is valid when all acting forces are the functions of coordinates of the substantial points of a system. If a motion is reversible, then all points of the system can move on some of its trajectories with the identical, with respect to magnitude, velocities in both directions, direct and reverse.

For example, the motion of celestial bodies — when only the Newtonian gravity forces are taken into account — represents the reversible phenomena. If in the Cauchy problem for the system of differential equations, one preserves the initial position and the magnitude for all celestial bodies and

[1)] The concepts of equilibrium and reversible processes differ in the general case. However, infinitely slow equilibrium processes, for which in the finite relations between the determining parameters not only rates, but also, in general, directions of change of these parameters are inessential (are not essential arguments), may be considered to be reversible.

On the other hand, as an example of an irreversible process for a system as a whole, one may cite the phenomenon of stationary heat transfer in a heat conducting medium at rest; in this case, the states of all small particles of the medium may be considered to be equilibrium states.

reverses the direction of a velocity only, it is obtained that in the solution a future and past replace their positions.

If the acting forces and the interactions in the considered macroscopic system depend essentially on the direction of a rate of a change of the state parameters (for example the friction force), then the corresponding phenomena are irreversible.

It must be emphasized, that the reversible and irreversible phenomena can consist of the consecutive series of the reversible and irreversible states of the parts of a matter. This situation will be illustrated below, on the phenomena appearing in the real models of continuum media.

Note, that the irreversible phenomena as a whole, for the bodies of finite dimensions can appear simultaneously with the reversible processes in all physically infinitesimal particles, forming the body under consideration. Strictly speaking, all real processes with macroscopic scales proceed at finite rates the directions of which are essential; therefore, in reality, they are irreversible, but in practice, many processes may be assumed to be thermodynamically reversible. From practical and theoretical points of view, one may, in a number of applications, build models of real phenomena using the concept of reversible processes.

Example of a process which may be practically considered to be reversible. For example, studies have shown that sometimes in practice one may assume that even the very fast moving process of outflow of gas particles from the nozzle of a jet engine is reversible. Here gas particles change over, during a time interval of the order of one thousandth of a second, from an effective state of rest at pressures of the order of 70 atmospheres, in the combustion chamber of a jet engine, to a state of motion with speeds of the order of 3000 m/sec and almost zero pressure in free space. Such is the motion of gases during high altitude flights of vehicles with rocket engines. In this process, the exchange of heat energy between the gas particles at different temperatures does not find time to take place; the variable thermodynamic characteristics in the particles are, in practice, interrelated as in equilibrium.

On equilibrium and most probable states. Consider some set of macroscopic parameters of state, introduced as statistical mean of corresponding characteristics of microscopic motion of molecules, for example, the temperature T and density ρ. Obviously, to all concrete values of T and ρ there may correspond many distributions of characteristics of microscopic motion. It turns out that there correspond to equilibrium values of macroscopic parameters the greatest number of possible different microstates. Therefore, if a thermodynamic system is left to itself, then the most probable of all states, in which it may find itself, is the equilibrium state. In this context, all isolated systems either are already in an equilibrium state or are tending to such a state.

Macroscopic characteristics, mean probabilities and irreversibility. Note still that all known microscopic laws, describing motions and interactions of elementary particles, for example, the Newtonian law of gravitational attraction and the laws of electromagnetic interaction, are reversible. However, irreversibility appears only as a consequence of statistical laws, true for large ensembles of particles, and in its way is the price to be paid for the possibility of introducing instead of a complex system of huge numbers of particles (with known, generally speaking, the approximate laws of interaction and initial conditions) a simple dummy system which may be described by means of a not large number of macroscopic characteristics connected with the most probable states.

Thanks to the large numbers of particles in practically small volumes, the probability distributions have very sharp peaks; this means that there are chances for the realization of only completely defined values of mean quantities.

For irreversible processes, the probability distributions depend, in general, on the time.

On the concept of temperature. One of the basic characteristics of a state of physical bodies is temperature. In daily life, the primary idea of the temperature of a body is directly connected with the senses. A body A is said to have higher temperature than a body B ($T_A > T_B$), if on contact of the body A with the body B there arises of necessity a transfer of heat energy from A to B. Two bodies in thermodynamic equilibrium, when brought into contact, have the same temperature, if there develops no flow of thermal energy between them. It is known from experiment that, if any two bodies A and B are brought into contact and then this system is left to itself, then there takes place in it a process as a result of which the temperatures of A and B become aligned. This fact makes it possible to construct thermometers, instruments with the help of which one may quantitatively measure the temperature in some scale, for example in Celsius, the characteristic points of which are the boiling and freezing points of water at atmospheric pressure.

The concept of temperature makes no sense in analytical mechanics for systems with not a large number of degrees of freedom. In practice, temperature may be ascribed to all possible bodies consisting of large numbers of particles.

In contrast to what is done in the mechanics of material points and absolutely rigid bodies, in continuum mechanics it is impossible, generally speaking, to manage without the concept of temperature. The expression for the external heat flux $dq^{(e)}$ enters into the equation of heat flow and into the law of conservation of energy; therefore one must study the mechanism of heat transfer and, consequently, also introduce the concept of temperature. A detailed and thorough study of the concept of temperature is closely linked to the study of molecular-kinetic theory. However, in this connection, it must be noted that a very perfect concept of temperature and of methods of its measurement was introduced long ago into science independently of a profound understanding of temperature within the framework of statistical physics.

Section 5 presents an account of an outstanding thermodynamic macroscopic theory in which, on the basis of the second law of thermodynamics, there is given a strict definition of absolute temperature for thermodynamic equilibrium states of a body.

It is well known from molecular-kinetic theory that the temperature T may be regarded as a quantity which is proportional to the mean energy of the random thermal motion of molecules per one degree of freedom. If different types of elementary particles have on the average different energies, or if particles of one kind have different mean energies, arising from different degrees of freedom, then, for sufficiently slowly progressing processes, the interaction of microscopic particles leads to equalization of mean energies. For sharply marked non-equilibrium processes, when inside macroscopically small particles there is not sufficient time for the occurrence of statistical equalization of energy between different types of particles, the concept of temperature of macroscopic particles as a whole loses its basic meaning.

In non-equilibrium cases, one may sometimes ascribe to a medium several temperatures, for example, temperatures of vibrational, rotational, translational degrees of freedom of molecules or temperatures of ions and electrons in a plasma, if the ions and electrons are separately in equilibrium states, etc.

In the presence of thermodynamic equilibrium in small volumes of a body, the temperature for small particles is uniquely defined. However, even in this case, the concept of temperature may lose its meaning for bodies of finite dimensions, if there does not exist thermal equilibrium between different parts of the body.

For example, what is meant by the temperature of Earth? One may speak of the temperatures of the tropics, temperate zones, at its poles, at its centre. However, it is difficult to define the temperature of Earth as a whole nor is it always expedient to do so.

As a rule, one considers the temperature of sufficiently small parts of a body and studies the heat flows in the body. Experiments have shown that in many practical questions one may often assume that thermodynamic equilibrium exists in small volumes of a system. In applications, absence of equilibrium and irreversibility frequently occur only when taking into account an absence of equilibrium in large volumes of a body for non-uniform distributions of temperature and other thermodynamic characteristics (such as concentrations of chemical components of mixtures, etc.) over particles.

§4. Two-parameter media. Perfect gas. Carnot cycle

A medium all the thermodynamic functions of which depend only on two thermodynamic parameters of state is referred to as a two-parameter medium. If these two parameters are the pressure p and the density ρ, then the specific internal energy of such a medium must be expressible in terms of them, viz., $U = U(p, \rho)$.

Equation of heat flow for an ideal gas. If a medium is an ideal compressible fluid (gas), then the work done by the internal surface forces, referred to unit mass, has the form

$$\frac{1}{\mathrm{d}m}\,\mathrm{d}A_{\Sigma}^{(i)} = p\mathrm{d}\,\frac{1}{\rho}\,, \tag{1.12}$$

and the equation of heat flow, on the assumption that $\mathrm{d}q^{**} = 0$, may be written in the form

$$\mathrm{d}U + p\mathrm{d}\,\frac{1}{\rho} = \mathrm{d}q^{(e)}\,. \tag{4.1}$$

Equation of state of a perfect gas. In a perfect gas, the pressure, density and temperature are interrelated by Clapeyron's equation

$$p = \rho RT\,, \tag{4.2}$$

where R is some constant, which is referred to as the gas constant and differs for different gases. An equation of the type (4.2), interrelating pressure, temperature, density and, possibly, other physical characteristics of a medium is called an equation of state.

For air,

$$R = 287.042\,\frac{\mathrm{m}^2}{\sec^2 \mathrm{C}^\circ}\,.$$

One may introduce a universal (constant for all gases) gas constant R_0 and Boltzmann's constant k by the relation

$$R = \frac{R_0}{\mathcal{M}} = \frac{k}{m}\,.$$

Here m is the mean mass of molecules in grams and $\mathcal{M}$ is the mean mass of one gram-mole of gas, determined by

$$n\mathcal{M} = n_1 M_1 + n_2 M_2 + \ldots n_N M_N = \sum_{i=1}^{N} n_i M_i$$

where n is the total number of molecules in a given volume of a mixture, n_i is the number of molecules and M_i the corresponding mass of a gram-mole of the separate kind of gas,

$$R_0 = 8.3144 \cdot 10^7\,\frac{\mathrm{erg}}{\mathrm{mole}\,\mathrm{C}^\circ}\quad k = 1.38 \cdot 10^{-16}\,\frac{\mathrm{erg}}{\mathrm{C}^\circ}$$

Internal energy of a perfect gas. A perfect gas can be defined as a gas in which the molecules interact only in collisions. Therefore it may be assumed that the internal energy of a mono-atomic perfect gas represents the total kinetic energy of the random motion of the atoms.

The expression for the internal energy per unit mass may be written in the form

$$U = \frac{1}{M} \sum_{i=1}^{N} \frac{m_i v_i^2}{2} + \text{const.},$$

where M is the total mass of the atoms, m_i and v_i are the masses and velocities of individual atoms relative to their common centre of mass and N is the number of atoms in the small volume under consideration. If it is assumed that all atoms of a gas are identical, then $M = Nm$ and

$$U = \frac{v_{\text{mean}}^2}{2} + \text{const.},$$

where v_{mean}^2 is the mean value of the square of the velocities of the atoms in random motion. For a perfect gas, in accordance with the definition of temperature as a characteristic of the mean kinetic energy, contained in a single degree of freedom in the random thermal motion of the atoms, the specific internal energy U may be presented in the form

$$U = c_V T + \text{const.}, \tag{4.3}$$

where c_V denotes the dimensional coefficient of proportionality between $v_{\text{mean}}^2/2$ and T.

Specification of the internal energy U in the form (4.3), together with Clapeyron's equation, fixes a definite model of a continuous medium, known as a perfect gas. Comparison with experimental data shows that the motions of real gases under normal conditions are sufficiently well described by such a model.

Specific heat capacities at constant volume and pressure. Mayer's formula. On the basis of equation of heat flow (4.1) for a perfect ideal gas in the case of a process taking place at constant specific volume ($d(1/\rho) = 0$), one readily finds

$$(dq^{(e)})_{V=\text{const}} = dU = c_V dT,$$

or

$$\left(\frac{dq^{(e)}}{dT}\right)_{V=\text{const.}} = c_V .$$

Consequently, the coefficient represents the amount of heat which must be supplied to unit mass of a medium in order to raise at constant volume its temperature by 1°C; therefore c_V is referred to as specific heat capacity at constant volume.[1)] For a process at constant pressure, one finds from the equation of heat flow for an ideal perfect gas

$$(dq^{(e)})_{p=\text{const.}} = dU + p d\frac{1}{\rho} = c_V dT + d\frac{p}{\rho} = (c_V + R) dT. \quad (4.4)$$

The amount of heat which must be supplied to unit mass of a medium in order to raise at constant pressure its temperature by 1°C is referred to as specific heat capacity at constant pressure and denoted by c_p:

$$c_p = \left(\frac{dq^{(e)}}{dT}\right)_{p=\text{const.}}$$

Therefore (4.4) yields the following formula which links for a perfect gas the specific heat capacities at constant pressure and volume and the gas constant R:

$$c_p - c_V = R , \quad (4.5)$$

which is known as Mayer's formula. The equation of heat flow in the general case involves the external heat flow $dq^{(e)}$. In certain cases, the equation of heat flow may be used for the determination of the required or attained heat flow, if the motion and sequence of states of the continuous medium are given or known. In problems in which it is desired to determine motions and states of a medium, the laws determining the external heat flow must be given.

Physical mechanisms of supply of heat to a medium. The flow of heat energy into or out of a medium may be caused by different physical phenomena. In applications, the following physical mechanisms of heat supply are most important.

1) If the number of particles changes during the collisions, then (4.3) must be replaced by $U = \int c_V(T)\, dT$.

1. *Thermal conductivity*, a phenomenon of heat transfer between parts of a medium, in direct contact with each other, which takes place at the expense of mechanical interactions and collisions during the thermal motion of molecules, atoms, electrons and other particles constituting a medium. Emission of heat, caused by thermal conductivity, is linked in an essential manner to the macroscopic non-uniform distribution of temperature through the volume of a body.

2. *Radiation and absorption of heat*, a phenomenon caused by changes of possible states of elementary particles (molecules, atoms, electrons, etc.) constituting a medium.

3. *Production of heat* caused by electrical dissipative processes and, in particular, Joule heat, which is generated inside a body in the presence of an electric current.

4. Sometimes, with the aid of a supplementary condition, one may attribute to external heat flux $dq^{(e)}$ a certain portion of increase in the internal energy dU and work done by internal forces $dA^{(i)}$ by transfer of these terms to the right-hand side of the heat flow equation. For example, one may replace a change in internal energy due to *chemical conversions* or *phase transitions,* linked to emission or absorption of heat, by an external heat flow and consider only the change in internal energy due to changes in temperature, mechanical parameters and, possibly, other changing properties of a medium.

The solution of concrete problems by use of an equation of heat flow which makes allowance for heat flow, as a rule, are mathematically very difficult. In applications, one frequently employs additional assumptions and, in particular, widely utilizes the following processes.

Adiabatic processes. I. Ideal processes which do not involve influx of external heat or heat exchange between neighbouring particles, i.e., for which $dq^{(e)} = 0$, are said to be adiabatic processes. The idea of adiabatic processes is connected with a consideration of thermally isolated bodies or of rapidly proceeding (but sometimes reversible) processes, when there is not enough time for heat exchange to develop in an essential manner.

Isothermal processes. II. As another example may serve an ideal process in which heat exchange, due to heat conductivity or radiation, is so intensive and the change of state takes place so slowly that one may assume the temperature of all parts of the system to be constant. Such processes are said to be isothermal.

The equation of an isothermal process has the form

$$\frac{dT}{dt} = 0 .$$

This equation, together with the equation of state of a medium, replaces the heat flow equation, a fact which, generally speaking, greatly simplifies the theoretical solution of the problem of finding the motion of a medium.

One may compute from the heat flow equation the amount of heat $dQ^{(e)}$ which must flow to every particle of a medium for the realization of an isothermal process.

Note that the condition $dT/dt=0$ only means that the temperature is constant in time inside every individual particle of a medium and that the temperatures of different individual particles may be different under this condition. However, frequently when speaking of isothermal processes, it is assumed that the temperature is constant in space and time, i.e.,

$$T = \text{const.} .$$

Finally, at times, processes are said to be isothermal in which the temperature of particles may vary in time, but is the same for all particles. In this case, instead of the equation $dT/dt=0$, one requires that

$$\text{grad } T = 0 \quad \text{or} \quad T = f(t) .$$

Obviously, a clear understanding of the formulation of a problem excludes the possibility of any confusion arising from the existence of various definitions of isothermal processes.

III. For two-parameter continuous media, one may take as the relation fixing a process, instead of the heat flow equation, directly some link between the density and the pressure. If this relationship is the same for all particles, then such a process is called *barotropic.*

Polytropic processes. In particular, a process is said to be polytropic, if

$$p = C\rho^n ,$$

where n is a constant, the index of polytropy, and C an arbitrary constant.

With the aid of the heat flow equation, for a given function $p=f(\rho)$, one can readily determine the amount of external heat flow corresponding to this relationship.

If a gas is perfect and a process polytropic, then one finds from the equation of heat flow for $n>1$

$$dq = dU + C\rho^n d\frac{1}{\rho} = c_V dT - \frac{dRT}{n-1},$$

whence, by Mayer's formula

$$R = c_p - c_V$$

for constant R, one obtains for the heat flow the simple formula

$$dq = c_V \frac{n - \dfrac{c_p}{c_V}}{n-1} dT = c^* dT.$$

If $n > c_p/c_V > 1$, then with an increase in temperature i.e., $dT > 0$, heat is supplied. If $1 < n < c_p/c_V$, then $dq < 0$ for $dT > 0$ and, consequently, during an increase in temperature heat is released.

If $n = c_p/c_V$, then $dq = 0$, i.e., such a polytropic process is adiabatic. These properties characterize the physical meaning of the index of polytropy.

Isotherms of a perfect gas. Consider the space of states of a two-parameter medium specified by p and $V = 1/\rho$, for example, a perfect gas. All thermodynamic functions of such a medium and, in particular, the temperature which will now be denoted by θ must be functions of p and $1/\rho$:

$$\theta = \theta\left(p, \frac{1}{\rho}\right).$$

Consider an isothermal (θ=const.) equilibrium process taking place in such a medium. Draw in the space of state (p, $1/\rho$) the curves θ=const. (i.e., the isotherms, Fig. 30a).

Obviously, in the case of a perfect gas, the isotherms in the plane (p, $1/\rho$) will be the hyperbolae

$$\frac{p}{\rho} = \text{const}. \tag{4.6}$$

From the equation of heat flow

$$dU + p\,d\frac{1}{\rho} = dq\,,$$

one can always compute the flow of heat dq which must be supplied to the system, in order that the process will be isothermal. For an ideal perfect gas, this heat flow is equal to

$$(dq)_{\text{isoth}} = p\,d\frac{1}{\rho} = R\theta\rho\,d\frac{1}{\rho}\,.$$

For a perfect gas $dq^{(e)} > 0$ for isothermal expansion and $dq^{(e)} < 0$ for isothermal compression. For an arbitrary gas, the form of the isotherms in the plane $(p, 1/\rho)$ depends on the form of the equation of state.

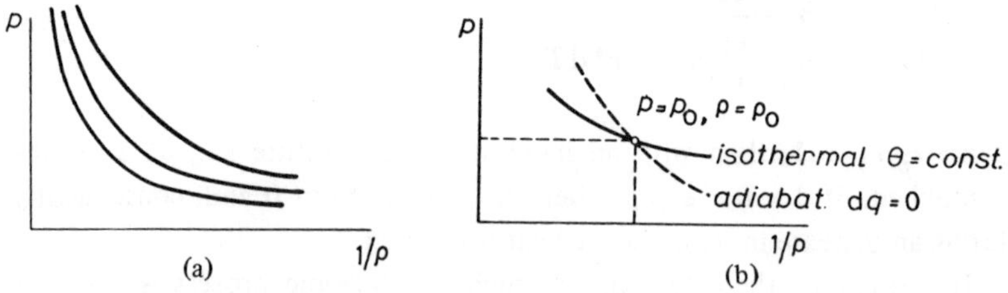

Fig. 30. (a) Isotherms of a perfect gas. (b) Mutual disposition of Poisson's adiabats and isotherms for a perfect gas.

Note that on one and the same isotherm $\theta = \text{const.}$ one may find points which correspond, for example, to the boiling and freezing points of water, since the boiling and freezing temperatures of water depend on the pressure.

Poisson's adiabat. For adiabatic processes ($dq = 0$), the equation of heat flow is easily seen to take the form

$$dU + p\,d\frac{1}{\rho} = 0\,, \tag{4.7}$$

whence, if the internal energy $U(p, 1/\rho)$ is known, one may find the dependence of p on ρ in the case of continuous adiabatic processes.

For a perfect gas, Equation (4.7) assumes the form

$$\frac{c_V}{R}\,d\frac{p}{\rho} + p\,d\frac{1}{\rho} = 0$$

or, with $\gamma = c_p/c_V$,

$$\frac{1}{\gamma-1}\left(\frac{\mathrm{d}p}{\rho}+p\mathrm{d}\frac{1}{\rho}\right)+p\mathrm{d}\frac{1}{\rho}=0\,,$$

whence

$$\frac{1}{\rho}\,\mathrm{d}p+\gamma p\,\mathrm{d}\frac{1}{\rho}=0\;;$$

integration now yields

$$\frac{p}{\rho^{\gamma}}=\text{const}\,. \tag{4.8}$$

This curve in the plane $(p, 1/\rho)$ is referred to as Poisson's adiabat, and $\gamma = c_p/c_V$ is called the index of the adiabat. Obviously, an isotherm (4.6) and an adiabat (4.8) may be drawn through every point $p_0, 1/\rho_0$ of the plane of states $(p, 1/\rho)$.

Mutual disposition of isotherms and adiabats for a perfect gas. Consider now the disposition at every point in the plane $(p, 1/\rho)$ of the isotherms and Poisson's adiabats for a perfect gas. For points along an isotherm, passing through the point p_0, $1/\rho_0$, one will have

$$\frac{p}{\rho}=\frac{p_0}{\rho_0}\,,\quad \text{i.e.}\quad \frac{p_{\text{isoth}}}{p_0}=\frac{\rho}{\rho_0}\,,$$

and along the adiabat passing through the same point,

$$\frac{p}{\rho^{\gamma}}=\frac{p_0}{\rho_0^{\gamma}}\,,\quad \text{i.e,}\quad \frac{p_{\text{ad}}}{p_0}=\left(\frac{\rho}{\rho_0}\right)^{\gamma}.$$

Since the adiabatic index $\gamma = c_p/c_V > 1$,

$$p_{\text{isoth}} > p_{\text{ad}} \quad \text{for} \quad \rho/\rho_0 < 1\,, \qquad p_{\text{isoth}} < p_{\text{ad}} \quad \text{for} \quad \rho/\rho_0 > 1\,,$$

i.e., the isotherms in the plane $(p, 1/\rho)$, on the right-hand and left-hand sides of the point p_0, $1/\rho_0$, pass above and below the adiabats, respectively (Fig. 30b).

Note that this property of the isotherms and adiabats has been established for a perfect gas. It is preserved also for many other media, but it is not true, for example, for water in the temperature +4° C.[1)]

Work performed by a system. It must be emphasized once more that the work done by the internal forces $\int p\,\mathrm{d}\,1/\rho$ may always be computed, if the function $p(\rho)$ is given, i.e., a curve in the plane $(p, 1/\rho)$. This means that the work done by the internal forces $\int p\,\mathrm{d}\,1/\rho$ can be computed for any process $\mathscr{L}_1$ between points A and B in the plane of states. However, the work done by the internal forces during an infinitely slow process is set equal, with an opposite sign, to the work done by the external forces over the system or, with the same sign, to the work which the system itself performs over the external bodies. Thus,[2)] the integral

$$\int_{AB(\mathscr{L}_1)} p\,\mathrm{d}\frac{1}{\rho} = \frac{1}{m}A\,, \tag{4.9}$$

evaluated along the path $\mathscr{L}_1$ in the plane $(p, 1/\rho)$, if $A > 0$, represent the total work which the thermodynamic system performs over the external bodies during the equilibrium process $\mathscr{L}_1$ or, if $A < 0$, the total work done by the external forces, which must be applied over the system to realize the process $\mathscr{L}_1$.

The total heat flow which is supplied to a system from outside. Analogously, for any process $\mathscr{L}_1$ $(p = p(1/\rho))$, if the internal energy of the medium $(U = U(p, 1/\rho))$ is given, then one can compute the total heat flow

$$Q^{(e)} = \int_{AB(\mathscr{L}_1)} \mathrm{d}Q^{(e)}\,. \tag{4.10}$$

1) For the water, the derivative $(\partial(1/\rho)/\partial T)_p$ is positive when the temperature is above +4°C, negative when the temperature is below +4° C and vanishes when the temperature is equal to +4° C. The isotherm, corresponding to +4° C coincides with the adiabat, and the adiabat, connecting the isotherms of the temperature above and below +4° C, does not exist.
(*cf.*, for example, Rumer Ju. B., Ryvkin M, Sz, Thermodynamics. Statistical physics and kinetics. Part I, Chapter I, § 17).

2) This is the consequence of the law of the balance of forces of action and reaction, when the displacement is continuous through the boundary between the considered system and the surrounding medium. In the presence of discontinuity of these displacements, the total energy flow will also be continuous and the flow of a work will be, in general, discontinuous on the boundary (*cf.*, p. 458 for a case of jumps and the example on page 269.

which must be supplied to a system from an external medium (if $Q^{(e)} > 0$) or drawn from the system into the external medium (if $Q^{(e)} < 0$) to realize the process $\mathscr{L}_1$.

On the basis of the first law of thermodynamics,

$$Q^{(e)} = \int_{AB(\mathscr{L}_1)} \left(\mathrm{d}U + p\mathrm{d}\,\frac{1}{\rho} \right) \mathrm{d}m = \int_A \mathrm{d}U_m + A = U_{mB} - U_{mA} + A\,. \qquad (4.11)$$

Carnot cycle. Consider the following important, reversible, closed, equilibrium process which is referred to as reversible Carnot cycle. Let the work performing body, i.e., the medium which undergoes this cycle, be a perfect gas or any other two-parameter medium with determining parameters p and $1/\rho$.[1] From an arbitrary point $M(p_0, 1/\rho_0)$ of the space of states, the gas infinitely slowly expands along the isotherm

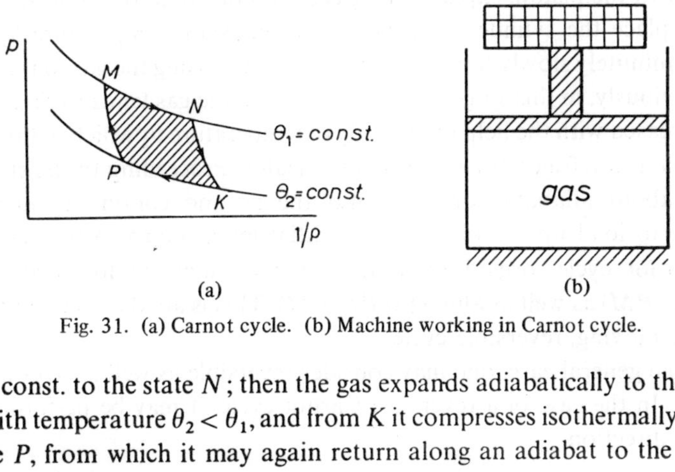

(a) (b)

Fig. 31. (a) Carnot cycle. (b) Machine working in Carnot cycle.

$\theta_1 = $const. to the state N; then the gas expands adiabatically to the state K with temperature $\theta_2 < \theta_1$, and from K it compresses isothermally to the state P, from which it may again return along an adiabat to the initial state M (Fig. 31a).

Example of an engine which does work over a reversible Carnot cycle. A system, executing a Carnot cycle, is called an engine. For example, one can imagine this engine to be realized in the following manner. Take a volume of gas with temperature θ_1 and contain it in a cylinder,

[1] Instead of $1/\rho$, one may use the equivalent quantity $V = m/\rho$, since all arguments apply to the substantial volume V with constant mass m, and the density is assumed (in the example) uniform in all points of the volume.

one end of which is closed by a fixed wall and the other end fitted with a movable piston, initially in equilibrium (Fig. 31b). To start with, one must force the gas in the cylinder to expand from M to N for θ_1 = const. For this purpose, assume that the lateral walls of the cylinder and piston are thermally insulated and that the bottom conducts heat well and stands on a heater, a body of high heat capacity held at constant temperature θ_1. The expansion of the gas will be carried out by gradually removing with the piston an infinitesimal load so that the piston infinitely slowing rises, the temperature θ of the gas has time to become equal to the temperature θ_1 of the heater and during all the time of the raising of the piston corresponds to θ_1. Under these conditions, the pressure p decreases and the volume of the gas increases. Reaching by such a path the state N, one takes the cylinder with the heater, shuts the bottom by an additional lid which does not conduct heat and again, taking continuously an infinitesimal load with the piston, expands the gas adiabatically to the state K. Then again place the cylinder on a body with constant temperature θ_2 and begin infinitely slowly to load the piston, compressing the gas to the state P. Obviously, in this stage, the temperature of the gas tends to rise, but it is decreased with the help of the body of temperature θ_2 which in this case will act as a refrigerator rather than a heater. Proceeding to the state P, bring about adiabatic compression of the gas and, continuing infinitely slowly to load up the piston to its initial load, return to the state M. A Carnot cycle, organized in this manner, may be followed along ($MNKPM$) as well as along ($MPKNM$). This is an idealized, infinitely slowly moving, reversible cycle.

In the general case, one may consider reversible as well as irreversible cycles. In the case of a reversible Carnot cycle, it may be performed in either direction.

The system executing a Carnot cycle, as a heat engine or a refrigerator. Integrate the equation of heat flow (4.1) (for $dq^{**} = 0$) along the complete Carnot cycle. Since the internal energy is a single-valued function of state, one has $\oint dU = 0$ and

$$\oint p\,d\frac{1}{\rho} = \oint dq$$

or

$$A = Q^{(e)}, \tag{4.12}$$

where A is the total work "done" by the system as a result of the Carnot cycle and $Q^{(e)}$ is the total "inflow" of heat from outside into the system.

Since the density of mechanical work A/m of a two-parameter medium, working over any closed cycle, is equal to $\oint p\,\mathrm{d}\,1/\rho$, then, obviously, this quantity is numerically equal to the area bounded by the curves, describing the cyclic process in the plane $(p, 1/\rho)$ and consequently, generally speaking, differs from zero. In the case of the Carnot cycle under consideration, the work density A/m is equal to the area of $MNKPM$ and $A>0$; if the cycle is performed in the opposite direction, one has $A<0$. If $A>0$, then the system performs in the cycle mechanical work and, by (4.12), one must supply the heat $Q^{(e)}$ to this system in order to obtain this work. One has a heat engine which works over a Carnot cycle; it takes heat from outside and delivers mechanical work. On the other hand, if $A<0$, then the external forces perform work on the system and, by (4.12), one obtains heat from the system. Along the segments NK and PM of the adiabats, one has $\mathrm{d}q = 0$. The system exchanged heat with the external medium only during the segments MN and KP of the isotherms of the process.

It has been established above that, in order to achieve isothermal expansion or compression of a gas, one must, respectively, supply or extract heat from the system. Therefore, along the segment MN of the isotherm (expansion), one must introduce heat, denoted by $Q_1>0$, and along the segment KP (compression), a quantity of heat $Q_2>0$ must be drawn, which is equivalent to a supply of heat $-Q_2<0$. Since the sections NK and PM are adiabats, one finds for the total heat $Q^{(e)}$ supplied during a complete Carnot cycle in a clockwise direction

$$Q^{(e)} = Q_1 - Q_2 .$$

By (4.12), one may write

$$A = Q^{(e)} = Q_1 - Q_2 . \tag{4.13}$$

In this case, the system performing a Carnot cycle is a heat engine which receives the heat Q_1 from a hotter body, delivers a part Q_2 of this heat to a colder body and performs mechanical work at the expense of the quantity of heat $Q_1 - Q_2$. If the Carnot cycle is operated in the opposite direction, then along PK the heat $Q_2 > 0$ is supplied, along NM the negative quantity of heat $-Q_1 < 0$. The general quantity of heat supplied in the reverse cycle is $Q^{(e)} = A_1 < 0$ (negative) and determined by

$$A_1 = Q^{(e)} = Q_2 - Q_1 < 0 .$$

In this case, the engine operating in a Carnot cycle works like a refrigerator, i.e., it takes the heat Q_2 from a less heated reservoir and, at the expense of mechanical work obtained from outside, it passes the heat

$$Q_1 = Q_2 - A_1$$

on to a hotter reservoir.

§ 5. The second law of thermodynamics and the concept of entropy

Consider now the second law of thermodynamics which, like the first law, represents a universal statement confirming all known experimental data and all theoretical notions of the mechanisms of physical phenomena. The second law of thermodynamics asserts that a device is impossible which would transfer heat from a body at lower temperature to a body at higher temperature without any changes in other bodies. It is assumed, that the bodies M and N can exchange with the external devices, only heat[1)] (the first formulation of the second law of thermodynamics).

The second law of thermodynamics can also be formulated as follows: It is impossible to construct the so-called perpetual motion of the second kind, i.e., an engine which, operating in accordance with the first law of thermodynamics in some cycle, would periodically perform work only at the expense of cooling some other heat source with a fixed temperature (withdrawal of heat from a reservoir at constant temperature) (second formulation).

1) If the bodies M and N can exchange, with the external devices, not only heat, then heat can be transferred from a less to a more heated body, without any changes in other bodies. For example, this process can be performed with the help of a cooling engine, operating on a Carnot cycle and drawing the necessary energy from the bodies M and N also.

It will be shown below that these two formulations of the second law of thermodynamics are equivalent.

It will first be shown that a study of the Carnot cycle will yield important consequences and a quantitative formulation of the second law of thermodynamics.

Efficiency of Carnot cycle. The concept of the coefficient of useful work or of the efficiency η of a heat engine, operating on a Carnot cycle, will now be introduced. By definition, the efficiency η of a Carnot cycle is the ratio of the mechanical work $A > 0$, obtained in the realization of a cycle, to the quantity of heat $Q_1 > 0$, supplied to the system during the cycle. By (4.13), one has for the efficiency of a Carnot cycle the formula

$$\eta = \frac{A}{Q_1} = 1 - \frac{Q_2}{Q_1} < 1 . \tag{5.1}$$

The above property of η, namely $\eta < 1$, for a Carnot cycle is a consequence of the first law of thermodynamics.

Carnot's theorem. A remarkable consequence of the second law of thermodynamics is the following theorem due to Carnot on the properties of the efficiency of a Carnot cycle.

For every reversible Carnot cycle, the quantity η depends only on the temperatures θ_1 and θ_2, given on the isotherms MN and KP (Fig. 32), and neither on the properties of the working body, partaking in the Carnot cycle (in the above detailed example, one could take a perfect gas) nor on the method of organization of the cycle, determined by, for example, the dimensions of the working body and the degree of expansion along the isotherm.

It will now be shown that η depends only on θ_1 and θ_2 and is an absolute characteristic of a reversible Carnot cycle, i.e., a universal function $\eta(\theta_1, \theta_2)$. Simultaneously, it will be demonstrated that, if the temperatures θ_1 and θ_2 are fixed, then the efficiency η' of an engine working in an irreversible Carnot cycle (i.e., any heat engine working on a cycle of the type shown in Fig. 32 which draws heat only from reservoirs with constant temperatures θ_1 and θ_2) may not be larger than the efficiency η of an engine, working over the corresponding reversible Carnot cycle, i.e.,

$$\eta' \leqslant \eta . \tag{5.2}$$

Thus, the efficiency of a Carnot cycle has a maximum for a reversible process and along no paths can it attain the value unity, since, in order to obtain mechanical work A, it is necessary not only to take the heat Q_1 from a surrounding medium to achieve isothermal expansion, but one must also supply to the surrounding medium a part Q_2 of the extracted heat for the purpose of isothermal compression.

Consider (5.2) first. Assume that there exist two Carnot cycles: One which is irreversible with efficiency η' and another which is reversible with efficiency η, where the heater and refrigerators in these cycles have, respectively, identical temperatures θ_1 and θ_2, with $\theta_1 > \theta_2$. Assume that $\eta' > \eta$ and show that this assumption leads to a contradiction to the second law of thermodynamics. In fact, let the heat engine with η' operate in the proper direction and perform the mechanical work A'. Construct a second (reversible) heat engine which works in the opposite direction, i.e., use it as a refrigerator. Then, for the engine with efficiency η' one has $Q'_1 > 0$, $Q'_2 > 0$ and $A' = Q'_1 - Q'_2 > 0$, for the engine with efficiency η, one has $Q_1 > 0, Q_2 > 0$ and $A = Q_2 - Q_1 < 0$ ($-A > 0$ being the work performed by the refrigerator).

Select the reversible Carnot cycle in such a manner that $-A = A'$, i.e., $Q'_1 - Q'_2 = Q_1 - Q_2$, and join these two engines together.[1] Thus one obtains an engine for which

$$A_0 = A' + A = Q'_1 + Q_2 - Q_1 - Q'_2 = 0\,.$$

The sole effect produced by this engine will be a redistribution of heat between bodies which act as heater and refrigerator.

The engine takes the heat $(Q_1 - Q'_1) = (Q_2 - Q'_2)$ from one of them and delivers it to the other. Select reversible and irreversible cycles such that $|A| = A'$, whence $\eta Q_1 = \eta' Q'_1$. Hence it follows that, as a consequence of the assumption $\eta' > \eta$, one has the inequality

$$Q'_1 < Q_1$$

or

$$(Q_1 - Q'_1) = (Q_2 - Q'_2) > 0\,. \tag{5.3}$$

[1] Such a choice of reversible engines with the same efficiency η is always possible by a simple choice of the dimensions of the working body, since the quantities of work and heat in the case under consideration are proportional to the mass of the body.

The positive quantity $(Q_2 - Q_2')$ is equal to the total amount of heat, taken from the reservoir at temperature θ_2, and is equal to the positive quantity $(Q_1 - Q_1')$, the heat supplied to the reservoir at the temperature $\theta_1 > \theta_2$. Thus, the engine, without expenditure of external energy, will supply heat from a cold reservoir to a hot reservoir, which is impossible by the first formulation of the second law of thermodynamics.

Consequently, the earlier assumption $\eta' > \eta$ leads to a contradiction with the second law of thermodynamics and must be rejected. One can only have

$$\eta' < \eta \quad \text{or} \quad \eta' = \eta \,. \tag{5.4}$$

If an engine with efficiency η' is also reversible, then, interchanging the places of η' and η in the preceding reasoning, one obtains

$$\eta < \eta' \quad \text{or} \quad \eta' = \eta \,. \tag{5.5}$$

Relations (5.4) and (5.5) are compatible only when

$$\eta' = \eta \,.$$

This result proves the equality of the efficiencies of any two reversible Carnot cycles for identical θ_1 and θ_2. If an engine with efficiency η' is irreversible, then it is impossible to force it to work with the same results in the opposite direction, and therefore one cannot prove (5.5).

Consequently, if a cycle with efficiency η' is irreversible, then, generally speaking, one has the inequality

$$\eta' \leqslant \eta \,.$$

The coefficient η characterises the degree of utilization of the heat energy Q_1, imparted by a hot body to a working engine; only that part of this energy, which is determined by η, is converted by the engine into mechanical work. A reversible engine is most efficient, since, generally speaking, one has always for irreversible engines $\eta' < \eta$. In this sense, it is said that irreversibility leads to an additional loss of part of usable energy.

In order to prove the equality of the efficiencies of all reversible Carnot cycles, one should not use properties of the working body nor any particular properties of the cycle; consequently, the efficiency of a reversible Carnot cycle does not depend on the properties of the working substance nor on the degrees of expansion, it is only a universal function of θ_1 and θ_2: $\eta = \eta(\theta_1, \theta_2)$.

Next, this universal function $\eta(\theta_1, \theta_2)$ will be determined. By the definition of the efficiency of a Carnot cycle, one has

$$\eta(\theta_1, \theta_2) = \frac{A}{Q_1} = 1 - \frac{Q_2}{Q_1}.$$

Introduce instead of $\eta(\theta_1, \theta_2)$ the function

$$f(\theta_1, \theta_2) = 1 - \eta(\theta_1, \theta_2),$$

i.e.,

$$f(\theta_1, \theta_2) = \frac{Q_2}{Q_1},$$

and obtain a functional equation for $f(\theta_1, \theta_2)$. For this purpose, consider three bodies with large thermal heat capacities at temperatures θ_1, θ_2 and θ_3 and three reversible Carnot cycles in which these bodies serve as heaters or refrigerators. Obviously,

$$f(\theta_1, \theta_2) = \frac{Q_2}{Q_1} = \frac{Q_2}{Q_3}\frac{Q_3}{Q_1} = f(\theta_3, \theta_2) \cdot f(\theta_1, \theta_3), \tag{5.6}$$

where, for example, $f(\theta_3, \theta_2) = 1 - \eta(\theta_3, \theta_2)$ for the Carnot cycle in which the body at temperature θ_3 acts as a heater and the body at temperature θ_2 as a refrigerator, etc. Note that the order of stating the arguments of the functions is essential; in the first place, one has always the temperature of the heater and in the second that of the refrigerator of the Carnot cycle under consideration.

For $\theta_1 = \theta_2$, Equation (5.6) reduces to

$$1 = f(\theta_3, \theta_1) \cdot f(\theta_1, \theta_3),$$

i.e., an interchange of the arguments transforms the function f into $1/f$. Using this property of f, one finds from (5.6)

$$\frac{Q_2}{Q_1} = f(\theta_1, \theta_2) = \frac{f(\theta_3, \theta_2)}{f(\theta_3, \theta_1)}. \tag{5.7}$$

The ratio Q_2/Q_1 does not depend on θ_3, it depends only on the values of θ_1 and θ_2.

Since θ_3 may be assumed to be constant for all possible θ_2 and θ_1, the solution of the functional equation (5.7) has the form

$$f(\theta_1, \theta_2) = \frac{\omega(\theta_2)}{\omega(\theta_1)}.$$

Consequently

$$\frac{Q_2}{Q_1} = \frac{\omega(\theta_2)}{\omega(\theta_1)}.$$

The value of the function $\omega(\theta)$ is called the absolute temperature[1] T, so that

$$\frac{Q_2}{Q_1} = \frac{T_2}{T_1}, \tag{5.8}$$

i.e., the ratio of the heat Q_2, returned by a thermodynamic system during a Carnot reversible cycle to the refrigerator, to the heat Q_1, obtained by the system from the heater, is equal to the ratio of the absolute temperatures of the refrigerator and heater. This result established the link between the concepts of temperatures, as characteristics of isotherms, and the energies received and given up by the corresponding Carnot cycle.

Quantitative formulation of the second law of thermodynamics in conformity with a reversible Carnot cycle. Equation (5.8) for a reversible process may be rewritten in the form

$$\frac{Q_1}{T_1} - \frac{Q_2}{T_2} = 0.$$

Further, in correspondence with the general definitions, it will be agreed to assume that the quantity of heat $Q_1 = Q_1^{(e)}$, obtained by a system, is positive and the quantity of heat $-Q_2 = Q_2^{(e)}$, given up by the system, is negative; then the preceding equation assumes the form

$$\frac{Q_1^{(e)}}{T_1} + \frac{Q_2^{(e)}}{T_2} = 0. \tag{5.9}$$

This universal statement follows from the second law of thermodynamics and can serve as its quantitative formulation for any reversible Carnot cycle in which the working body may be an arbitrary two-parameter medium.

[1] It is easily verified by direct calculation that, if one uses as working body in a Carnot cycle a perfect gas with the equation of state $p = R\rho T$ (where T is in $°K$ (Kelvin)), then $Q_2/Q_1 = T_2/T_1$. Consequently, the absolute temperature introduced here is proportional to the Kelvin temperature.

Quantitative formulation of the second law of thermodynamics in conformity with an arbitrary reversible cycle. Consider some reversible cycle $\mathscr{L}$, represented in the space of states $(p, 1/\rho)$ by a stepped curve which coincides with the outer boundaries of the sum of reversible Carnot cycles (Fig. 32a). Since (5.9) is valid for every, individual selected Carnot cycle, one may add these equalities for all Carnot cycles to obtain

$$\sum_i \frac{Q_i}{T_i} = 0 ;$$

obviously, the terms Q_i/T_i, corresponding to paths which are internal with respect to $\mathscr{L}$, drop out on summation, since each of these paths, for example AB, will be traversed twice in opposite directions, once when the body with temperature T_i will act as refrigerator and a second time when it acts as heater. Therefore, finally, one obtains

$$\sum \frac{Q_i}{T_i} = 0 , \tag{5.10}$$

where the summation is only carried out over heat flows along the broken curves bounding the total cycle $\mathscr{L}$.

Now, let $\mathscr{L}$ be an arbitrary reversible cycle, performed by a two-parameter thermodynamic system. For the realization of such a cycle, one requires a large number of heat reservoirs with temperatures which

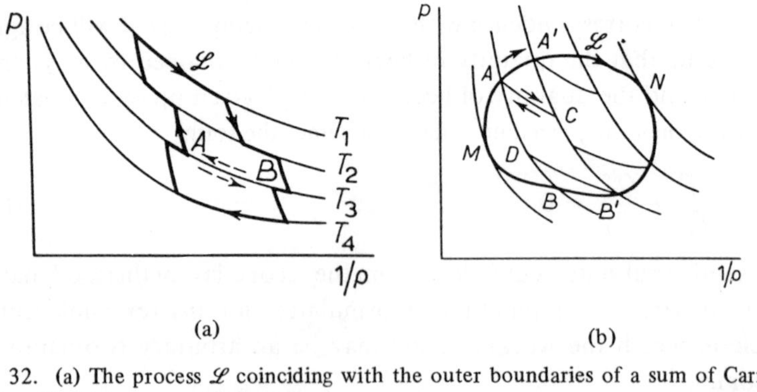

(a) (b)

Fig. 32. (a) The process $\mathscr{L}$ coinciding with the outer boundaries of a sum of Carnot cycles. (b) Arbitrary reversible cycle.

differ infinitely little from each other. Successively, the system comes into contact with that reservoir the temperature of which coincides with that of the system during a given element of the cycle; at the same time, the system undergoes infinitely slow compression or expansion. On the infinitesimal segment AA' of the closed curve $\mathscr{L}$ (Fig. 32b), let the system receive an elementary amount of heat $\Delta Q^{(e)}$. Draw through the point A an isotherm AC and through the point A' an adiabat $A'C$ and denote by $\Delta Q_{\text{isoth.}}$ the heat which the system would have received if it had undergone the infinitesimal isothermal process AC. A relationship between $\int_{AA'} dQ^{(e)} = \Delta Q^{(e)}$ and $\Delta Q_{\text{isoth.}}$ may be obtained by study of the small cycle $AA'CA$. In fact, apply to this cycle the law of conservation of energy $\Delta Q^{(e)} - \Delta Q_{\text{isoth.}} = \Delta A$ (the segment of the isotherm AC in the cycle $AA'CA$ is traversed in the direction CA whence one has the term $(-\Delta Q_{\text{isoth.}})$ in the law of conservation of energy for the cycle $AA'CA$). The quantities of heat $\Delta Q^{(e)}$ and $\Delta Q_{\text{isoth.}}$ for infinitesimal elements of a cycle are infinitesimal to the first order, whereas the work ΔA, performed during a small cycle, is represented by the area $AA'CA$ and therefore is an infinitesimal quantity of second order, i.e., it is infinitesimal compared with $\Delta Q^{(e)}$ and $\Delta Q_{\text{isoth.}}$. Extend the adiabat $A'C$ to its intersection with $\mathscr{L}$ at a second point B' and draw an adiabat through the point A. Then, with the same degree of approximation, the heat $\Delta Q^{(e)}$, obtained by the system along the section BB' of the process $\mathscr{L}$, is equal to the heat which corresponds to the segment of the isotherm $B'\Delta$. Hence the two elements of heat $\Delta Q^{(e)}$ along the parts AA' and BB' of the process are, to small second order quantities, equal to the amount of heat $\Delta Q_{\text{isoth.}}$ which the system would have received from a heater and refrigerator, if it had been the working body of a Carnot engine performing the reversible Carnot cycle $ACB'DA$.

If one subdivides the entire region, lying inside the curve $\mathscr{L}$, with the aid of a system of adiabats (Fig. 32,b) and constructs the corresponding isotherms, then one arrives at a process $\mathscr{L}'$ which will be represented in the space of states by a broken line consisting of segments of adiabats and isotherms. One may apply to this process Equation (5.10):

$$\sum_i \frac{\Delta Q_{i\,\text{isoth.}}}{T_i} = 0\,, \tag{5.11}$$

where the summation is carried out over the heat flows occurring along the boundary $\mathscr{L}'$.

If the number of the adiabats drawn tends to infinity and the segments of the cycle $\mathscr{L}$, through the ends of which pass the adiabats, tend to zero, then $\mathscr{L}' \to \mathscr{L}$, and

$$\Delta Q_{\text{isoth.}} \to \Delta Q^{(e)};$$

since the difference $\Delta Q^{(e)} - \Delta Q_{\text{isoth.}}$ is a small quantity of second order (the area of an infinitesimal curvilinear triangle), one obtains from (5.11) in the limit

$$\oint_{\mathscr{L}} \frac{dQ^{(e)}}{T} = 0, \tag{5.12}$$

which is exactly satisfied for any reversible cycle performed by a two-parameter medium.

From

$$\int_C \frac{dQ^{(e)}}{T} = 0,$$

it follows for any reversible cycle C that the integral

$$\int_{AB(\mathscr{L})} dQ^{(e)}/T$$

for any reversible process $\mathscr{L}$ between states A and B does not depend on the path of integration $\mathscr{L}$.

Introduction of entropy with the help of reversible processes for two-parameter media. Fixing the point of the initial state of a system A for any state B of a two-parameter medium in which one may go from A along reversible paths, one may introduce a function of the parameters of state, i.e., of the coordinates of the point B,

$$S(B) = S\left(p, \frac{1}{\rho}\right) = \int_A^B \frac{dQ^{(e)}}{T} + S(A), \tag{5.13}$$

which is known as the entropy. By (5.13), the entropy is determined exactly apart from an additive constant $S(A)$. By (5.13), during an increase of entropy for any changes of the coordinates of the point B, one has

$$dS = \frac{dQ^{(e)}}{T}.$$

Thus, although the elementary heat flow, expressible in terms of the parameters of state and their differentials, is not, in general, a perfect differential, one has for it the integrating factor $1/T(p, \rho^{-1})$, a quantity which is the inverse of the absolute temperature.

Using the equation of heat flow, one obtains for the differential of the entropy the expression

$$\mathrm{d}S = \frac{\mathrm{d}Q^{(e)}}{T} = \frac{\mathrm{d}U_m + \mathrm{d}A^{(i)}}{T} \tag{5.14'}$$

or, referred to unit mass,

$$\mathrm{d}s = \frac{\mathrm{d}q^{(e)}}{T} = \frac{\mathrm{d}U + p\,\mathrm{d}\dfrac{1}{\rho}}{T}, \tag{5.14}$$

which may be used for the evaluation of the entropy of a two-parameter medium, when the internal energy U of the medium is known as a function of the parameters of state.

The entropy for a perfect gas. For example, for a perfect gas with constant heat capacity ($p = \rho R T$, $U = c_V T$), one will have

$$\mathrm{d}s = \frac{c_V \mathrm{d}T}{T} + \frac{R\,\mathrm{d}\dfrac{1}{\rho}}{\dfrac{1}{\rho}}$$

or

$$s = c_V \log \frac{T}{\rho^{\gamma-1}} + \text{const.} = c_p \log - \frac{T}{p^{(\gamma-1)/\gamma}} + \text{const.}_1 =$$

$$= c_V \log \frac{p}{\rho^\gamma} + \text{const.}_2 = c_V \log \frac{p}{\rho^\gamma} - c_V \log \frac{p_0}{\rho_0^\gamma} + s_0\,, \tag{5.15}$$

where p_0, ρ_0 and s_0 are corresponding constants.

Conditions imposed by the existence of entropy on the form of the equation of state. Equation (5.14) imposes a limitation of the functions $U(p, \rho)$ and $T(p, \rho)$, i.e., on the basic thermodynamic functions of state of a medium. Since ds must be a perfect differential, the condition of integrability (5.14) assumes the form

$$\frac{\partial}{\partial \rho}\left(\frac{1}{T}\frac{\partial U}{\partial p}\right) = \frac{\partial}{\partial p}\left(\frac{1}{T}\frac{\partial U}{\partial \rho} - \frac{p}{\rho^2 T}\right)$$

or

$$\frac{\partial T}{\partial \rho}\frac{\partial U}{\partial p}=\frac{\partial T}{\partial p}\left(\frac{\partial U}{\partial \rho}-\frac{p}{\rho^2}\right)+\frac{T}{\rho^2}. \tag{5.16}$$

For a given function $U(p, \rho)$, the functions $T(p, \rho)$ must be solutions of (5.16); consequently, such functions are not arbitrary, although there exist many different solutions of the partial differential equation (5.16).

Consider next the formulation of the second law for an irreversible Carnot cycle.

Quantitative formulation of the second law of thermodynamics applicable to an irreversible Carnot cycle. Let two reservoirs with temperatures T_1 and $T_2 (T_1 > T_2)$ act as heater and refrigerator in two Carnot cycles, one of which is reversible (with efficiency η) and the other irreversible (with efficiency η').

Then, since $\eta' \leqslant \eta = 1 - T_2/T_1$, one has

$$1-\frac{Q_2'}{Q_1'}\leqslant 1-\frac{T_2}{T_1}, \quad \text{or} \quad \frac{Q_2'}{Q_1'}\geqslant\frac{T_2}{T_1},$$

whence

$$\frac{Q_1'}{T_1}-\frac{Q_2'}{T_2}\leqslant 0.$$

Assuming the heat Q_1', flowing into the system, to be positive, and the heat Q_2', given up by the system, negative, one obtains

$$\sum_{i=1}^{2}\frac{Q_i'}{T_i}\leqslant 0. \tag{5.17}$$

This is a quantitative formulation of the second law of thermodynamics for an irreversible Carnot cycle.

Example illustrating the character of variations of the entropy in irreversible processes. Consider a particular example which demonstrates how one may introduce the entropy for a system as a whole and how the entropy varies in irreversible processes.

Let there be given two infinitesimal volumes I and II of an incompressible fluid at identical pressures and different temperatures. Let the temperatures of Volumes I and II be T_1 and T_2, respectively $(T_2 > T_1)$.

If these volumes are brought into contact, then exchange of heat will take place between them, when at every instant of time one may ascribe to each of volumes I and II a definite value of the temperature (since volumes I and II are small).

The process of heat transfer from II to I is irreversible, since, by the second law of thermodynamics, heat may be transferred, without expenditure of energy from outside, from a particle at temperature T_2 to a particle at temperature T_1 only if $T_2 > T_1$.

Denote by $\mathrm{d}Q$ the positive quantity of heat which passes from particle II to particle I during the time $\mathrm{d}t$ and consider, for the sake of simplicity, the case when the system, consisting of the set of particles I and II, does not exchange heat with an external medium. If the irreversibility is connected only with the process of heat transfer from one particle to another, for which the states and the processes in every individual particle may be assumed to be reversible, then one may write for the separate particles

$$\mathrm{d}S_{\mathrm{I}} = \frac{\mathrm{d}Q}{T_1}, \qquad \mathrm{d}S_{\mathrm{II}} = \frac{-\mathrm{d}Q}{T_2}.$$

The change in the entropy of the whole system I+II can be computed, assuming that the total entropy S is an additive function, i.e.,

$$S = S_{\mathrm{I}} + S_{\mathrm{II}};$$

consequently,

$$\mathrm{d}S = \mathrm{d}S_{\mathrm{I}} + \mathrm{d}S_{\mathrm{II}} = \mathrm{d}Q\,\frac{T_2 - T_1}{T_1 T_2} > 0.$$

Thus, although in the above example there is no heat flow from an external body into the system I+II, the entropy of this system increases as a result of the irreversible internal process.

Quantitative formulation of the second law of thermodynamics applicable to a multi-parameter medium. Previously, starting from the Carnot cycle, the entropy has been introduced as a function of state only for a two-parameter medium. Consider how one may introduce the entropy for media the states of which are determined by n variable defining parameters $\mu^1, \mu^2, \ldots, \mu^n$; it will be shown that for an arbitrary irreversible cycle C, for which at all intermediate states the temperature

T may be defined, one has the inequality

$$\int_C \frac{dQ^{(e)}}{T} \leqslant 0 .$$

For this purpose, select an arbitrary (possibly, irreversible) cycle C, performed by an arbitrary system, over infinitesimal elements dl_i on each of which one may assume the temperature of the system to be constant (T_i). Denote by dQ_i the heat which the system obtains during the process dl_i.

For the duration of each elementary process dl_i, a multi-parameter medium may be employed as heat reservoir in an elementary reversible Carnot cycle, realized by an auxiliary two-parameter medium; these Carnot cycles may be selected in such a manner that in each of them the auxiliary two-parameter system returns to the multi-parameter system under consideration an amount of heat which is equal to dQ_i. Therefore one may imagine that the entire heat, which the multi-parameter medium receives from outside during the process C, is obtained by means of contact with auxiliary two-parameter media. Select as second heat reservoir for all auxiliary Carnot cycles some body with constant temperature T_0.

For each elementary Carnot cycle, one has

$$-\frac{dQ_i^{(e)}}{T_i} + \frac{dQ_0}{T_0} = 0 , \tag{5.18}$$

where $(-dQ_i^{(e)})$ and dQ_0 are the amounts of heat obtained by the two-parameter media along the isotherms T_i and T_0, respectively. Integrating (5.18) along the entire cycle C, one obtains

$$Q_0 = \oint_C dQ_0 = T_0 \oint_C \frac{dQ^{(e)}}{T} .$$

Consider now a thermodynamic system consisting of a set of two-parameter and multi-parameter media. This system receives heat only from a reservoir at temperature T_0. The total heat flow from outside in the cycle C is Q_0.

By the law of conservation of energy, Q_0 is equal to the work A which this system performs on external bodies:

$$Q_0 = T_0 \oint \frac{dQ^{(e)}}{T} = A .$$

The work A may not be positive, since in that case one would have a cyclically working engine which onlv on account of the heat flow from a single reservoir at fixed temperature T_0 would perform mechanical work on external bodies. The impossibility of the existence of such an engine is one of the formulations of the second law of thermodynamics.

Equivalence of the formulations of the second law of thermodynamics. This formulation is referred to as the second formulation of the second law of thermodynamics (*cf*., p. 252). One can show, that this formulation is equivalent to the first one. For this purpose, one can assume, that the first formulation is valid, while the second is not, i.e., it is possible to construct a cyclically working engine, which exchanges heat with one body of the fixed temperature and performs the positive work.

Thus, the work obtained A, could be utilized in a refrigerating engine, working over a Carnot cycle, to transfer heat from some reservoir at temperature $T^* < T_0$ to a reservoir at temperature T_0.

By the equation

$$A = Q_1 - Q_2 = Q_0 > 0\,,$$

this engine could take heat Q_2 from a reservoir at temperature T^* and transfer heat $Q_1 > Q_0$ to the reservoir at temperature T_0. As a result, in a thermodynamic system consisting of this refrigerating engine and the engine considered above performing the work A, since $Q_1 - Q_0 > 0$ and $Q_2 > 0$, heat would flow without expenditure of any energy from outside from a body at lower temperature T^* to a body at the higher temperature T_0 and the bodies with temperature T^* and T_0 are exchanging heat with the engine only, which is impossible according to the first formulation of the second law.

In order to complete the reasoning on the equivalence of the two formulations of the second law of thermodynamics, note that, if, in contrast, one assumes that the first formulation is not valid, then there will follow from this assumption the possibility of a construction of perpetual motion of the second kind.

In fact, assume that heat passes, without expenditure of work from outside, from a heat reservoir with temperature T_2 to a heat reservoir with temperature $T_1 (T_1 > T_2)$. Take some amount of heat Q_1 passing from the body at temperature T_2 to the body at temperature T_1, and utilize it in a heat engine, working, for example, in a Carnot cycle which takes this heat Q_1 from the body with temperature T_1, passes a part $Q_2 (Q_2 < Q_1)$ of it back to the body at temperature T_2 and produces some mechanical work $A > 0$.

If one considers both these processes (the direct one and that in the heat engine) as a single one, then clearly, as a result of this process, mechanical work will be generated periodically only at the expense of the heat flow $Q_1 - Q_2$ from the single reservoir at temperature T_2, which contradicts the second formulation of the second law of thermodynamics.

Quantitative formulation of the second law of thermodynamics with respect to irreversible processes in any medium. Thus, it has been shown that the assumption

$$\oint_C \frac{dQ^{(e)}}{T} > 0$$

contradicts the second law of thermodynamics and, consequently, that one must have

$$\oint_C \frac{dQ^{(e)}}{T} \leqslant 0$$

for any cycle C performed by all multi-parameter (including two-parameter) media.

Quantitative formulation of the second law of thermodynamics with respect to reversible processes in any medium. If the basic cycle C is reversible, then, repeating the reasoning with respect to the cycle C, proceeding in the opposite direction, one arrives at the conclusion that the assumption $A < 0$ likewise contradicts the second law of thermodynamics. For a reversible cycle C, in any medium, one has only the possibility

$$\oint_C \frac{dQ^{(e)}}{T} = 0 .$$

Introduction of entropy for multi-parameter media with the aid of reversible processes. It follows now that in the case of reversible processes the integral $\int_A^B dQ^{(e)}/T$ does not depend on the path of integration and that for a fixed initial state A: it is only a function of the final state B of the medium. Consequently, one may introduce, with the aid of reversible processes for multi-parameter media, just as with the aid of reversible processes for two-parameter media, a single-valued function of state

$$S(B) = \int_A^B \frac{dQ^{(e)}}{T} + \text{const.},$$

which is referred to as entropy.

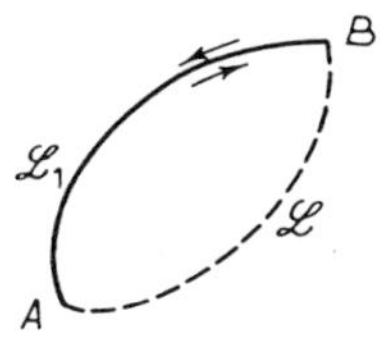

Fig. 33. Introduction of entropy with the aid of reversible processes.

However, if the process $\mathscr{L}$ transforming the medium from the state A to the state B is irreversible, then the entropy in the state B may be computed along a corresponding arbitrary[1] reversible process $\mathscr{L}_1$ between A and B, if such a reversible process exists. Then

$$S(B) - S(A) \geqslant \int_{AB(\mathscr{L})} \frac{dQ^{(e)}}{T}.$$

Actually, consider two states A and B of some medium. Let there exist between these two states two processes one of which ($\mathscr{L}_1$) is reversible, the other ($\mathscr{L}$) irreversible.

With the aid of the reversible process $\mathscr{L}_1$, one may compute the entropy

$$S_B - S_A = \int_{AB(\mathscr{L}_1)} \frac{dQ^{(e)}}{T}$$

However, if one considers the cycle $C = A\mathscr{L}B\mathscr{L}_1A$, then, obviously, it will be irreversible and

$$\oint_C \frac{dQ^{(e)}}{T} \leqslant 0,$$

[1] Obviously, the value of the difference $S(B) - S(A)$ is the same for each of the different reversible paths between A and B.

whence

$$S_B - S_A \geqslant \int_{AB(\mathscr{L})} \frac{dQ^{(e)}}{T}.$$

Attention must be drawn to the fact that the last integral is taken along the irreversible path $\mathscr{L}$.

Uncompensated heat. For an irreversible process, linking two infinitely close states A and B, one has

$$dS \geqslant \frac{dQ^{(e)}}{T}.$$

Thus, in the case of irreversible processes,

$$TdS \geqslant dQ^{(e)}$$

or

$$TdS = dQ^{(e)} + dQ',$$

where dQ', the so-called uncompensated heat, is always larger or equal to zero. In the case of reversible processes,

$$dQ' = 0.$$

Note, however, that for $dQ' = 0$ processes may, in general, be irreversible.

The example of equilibrium, irreversible process. It has been shown previously, that from the assumption of existence (together with irreversible processes) of the reversible processes between two arbitrary states A and B, and from the second law of thermodynamics follows the existence of entropy, representing itself the function of parameters, determining the states A and B. Just as these considerations and the established properties of entropy have the fundamental significance in all macroscopic theories, it is important to introduce, similarly to the concept of energy, the concept of entropy as a thermodynamical function for arbitrary mechanical, physical and general macroscopic model in natural sciences. From this point of view, the above developed theory needs to be generalized and supplemented for the case when, generally speaking, two arbitrary states A and B in some models, according to their definitions, cannot be linked by reversible processes.

For a long time, many authors believed that such models do not exist in physics, but these conceptions, existing till now, are not true. In particular,

Thus, in kinematics, a continuous medium may be conceived as an abstract geometrical object, and not merely as a material body. The motion of a medium can be governed by various laws. When considering the motion of a material body, these may be known physical laws or, when speaking, for example, of prices, they may be mathematical laws of economics which are at that time applicable.

Continuity of the functions specifying laws of motion. In a study of the mechanics of deformable media, one will wish to rely on the apparatus of differential and integral calculus. Therefore assume that the functions entering into the laws of motion of a continuum are continuous and possess continuous partial derivatives with respect to all variables. This assumption is general enough, but it limits greatly the class of phenomena which may be studied.

In fact, water, for example, may splash. Then particles, which lie infinitely close to each other at one instant, will not be close to each other at the following instant of time. It is impossible to describe such phenomena and retain the assumption of continuity of the laws of motion. At a later stage, it will be seen that in many cases the assumption of continuity of a motion must be dropped and that one has to consider also motions which are such that they themselves or their derivatives suffer discontinuities on some distinct surfaces. Motions of this sort, for example, shock waves, will be considered in the sequel. However, note that the investigation of discontinuous motions will still be based on the theory of continuous motions.

Single-valuedness of the laws of motion. On the grounds of physical considerations, assume that at every instant of time $t=\text{const.}$ the functions $x^i = x^i(\xi^1, \xi^2, \xi^3, t)$ are single-valued. As is known, in that case the Jacobian

$$\Delta = \begin{vmatrix} \dfrac{\partial x^1}{\partial \xi^1} & \dfrac{\partial x^1}{\partial \xi^2} & \dfrac{\partial x^1}{\partial \xi^3} \\ \dfrac{\partial x^2}{\partial \xi^1} & \dfrac{\partial x^2}{\partial \xi^2} & \dfrac{\partial x^2}{\partial \xi^3} \\ \dfrac{\partial x^3}{\partial \xi^1} & \dfrac{\partial x^3}{\partial \xi^2} & \dfrac{\partial x^3}{\partial \xi^3} \end{vmatrix}$$

the load, in general, is not equal to the mechanical work, produced by the load on the external bodies. It is quite obvious and well known, that from the law of action and reaction, one cannot prove the validity of the analogous law of exhange of pure mechanical energies in non-conservative system or even in the case of the interaction of a given mechanical system with an electromagnetic field.

As a continuation of the above stated theory, the more detailed interpretation of the concept of an entropy is given in the article: Kameniarzh Ia. A. and Sedov L. I., "Macroscopic introduction of entropy with relaxed assumptions about realizable processes". PMM Vol. 43, No. 1, 1979, pp. 2–6.

On the work done by a heat engine for $\mathrm{d}Q^{**} \neq 0$. Now obtain the basic energy equation for a heat engine working with $\mathrm{d}Q^{**} \neq 0$. An example of such a machine is a device in which use is made of the so-called magneto-thermal effect. It is known that adiabatic magnetization and demagnetization of paramagnetic substances is analogous to adiabatic compression and expansion of two-parameter systems, similar to a perfect gas. For adiabatic magnetization, one must consume external energy $\mathrm{d}Q^{**}$; the internal energy of the medium increases during this process and its temperature rises. For adiabatic demagnetization, the system gives up energy $\mathrm{d}Q^{**}$; its internal energy decreases and its temperature falls.

The magneto-thermal effect could be used to produce very low temperatures (in fact, temperatures of $0.0044° K$ have been obtained in this manner).

For $\mathrm{d}Q^{**} \neq 0$, the heat flow equation has the form

$$\mathrm{d}U_m = -\mathrm{d}A^{(i)} + \mathrm{d}Q^{(e)} + \mathrm{d}Q^{**} .$$

For a closed cycle C, one finds

$$\oint_C (\mathrm{d}A^{(i)} - \mathrm{d}Q^{**}) = \oint_C \mathrm{d}Q^{(e)} = Q^{(e)} , \tag{5.19}$$

where $\mathrm{d}Q^{(e)}$ denotes the total heat flux into the working body of the engine.

From the kinetic energy theorem $\mathrm{d}E = \mathrm{d}A^{(i)} + \mathrm{d}A^{(e)}$, one has for a closed cycle ($\oint \mathrm{d}E = 0$)

$$\oint_C dA^{(i)} = -\oint_C dA^{(e)} .$$

Hence Equation (5.19) may be rewritten

$$-\oint_C (dA^{(e)} + dQ^{**}) = Q^{(e)} .$$

This relation is also the energy equation describing the work done by the heat engine in the case when an engine, performing a closed cycle and receiving from outside the medium the heat $Q^{(e)}$, may spend it not only on the production of mechanical work on external bodies, but also on transmission to them of energy which is not of a mechanical (work done by macroscopic forces) and thermal nature.

It is seen that the derivations, relating to the work done by heat engines, are also applicable in the case $dQ^{**} \neq 0$ provided dA is interpreted as the sum $dA^{(e)} + dQ^{**}$, i.e., if it is assumed that

$$A = \oint_C (dA^{(e)} + dQ^{**}) .$$

On the states with the same internal energy and different entropy. Finally, note that there may exist processes as a result of which the energy of a system does not change, while its entropy varies. This circumstance is connected with the fact that the internal energy and entropy may depend on different thermodynamic parameters and that processes may be closed with respect to the energy parameters and open with respect to the parameters of entropy.

For example, in a perfect gas,

$$U = c_V T, \quad s = c_p \log \frac{T}{p^{(\gamma - 1)/\gamma}} + \text{const.}$$

and any process, taking place between two states at identical temperature and different pressures, is of this type.

Thus, if a gas, contained in a balloon under high pressure p_0, is released and if the temperature of the gas does not escape into an atmosphere, then its internal energy does not change and s increases (on account of the decrease in p, Fig. 35). In this process, the gas releases energy in the form of mechanical work and receives exactly the same amount of energy in the form of heat.

medium some coordinate lines ξ_1, ξ_2, ξ_3 consisting of points of the continuous medium, then at subsequent instants of time they become together with points of the continuum coordinate lines of the concomitant system. However, if they were chosen to be straight lines at the initial instant, then, generally speaking, they will be curved at subsequent instants (fig. 3).

Hence, a coordinate system, linked to the particles of a continuous medium, varies with time. The choice of such a coordinate system at any given instant of time is arbitrary; however, at subsequent instants, it is already determined, since it is "frozen into" the medium and deformed together with it. Such a coordinate system, which is fixed in a medium, has been defined earlier as a concomitant system. All points of a continuum are always at rest with respect to the moving concomitant system, since, by definition, their coordinates ξ^1, ξ^2, ξ^3 do not vary in the concomitant system.

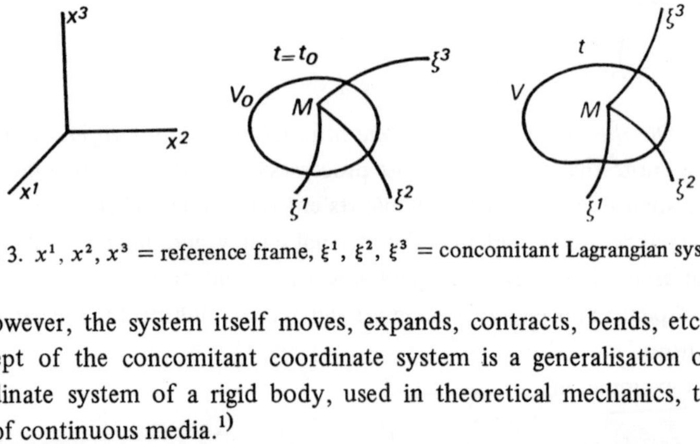

Fig. 3. x^1, x^2, x^3 = reference frame, ξ^1, ξ^2, ξ^3 = concomitant Lagrangian system.

However, the system itself moves, expands, contracts, bends, etc. The concept of the concomitant coordinate system is a generalisation of the coordinate system of a rigid body, used in theoretical mechanics, to the case of continuous media.[1)]

Thus, always in considering the motion of a continuous medium, one must individualize points and, consequently, use Lagrangian coordinates. Therefore, in investigations of motions of continuous media, one must

[1)] Obviously, for each coordinate system and, in particular, for the reference system of an observer, one may always introduce an imagined idealized medium for which the coordinate system under consideration is concomitant.

For the same medium one can introduce different concomitant coordinate systems. Then, sometimes, one can separate the concept of the reference frame of an observer as of the same invariant object, from the concept of the generally used coordinate system, which may differ for the observer in the given reference system.

Euclidean space. When can one introduce a single Cartesian coordinate system for an entire space? For the sake of simplicity, consider two-dimensional space. Obviously, one can always introduce for the entire plane a single Cartesian coordinate system. On the surface of a sphere, the curvature of which is not zero, this cannot be done, i.e., it is impossible to introduce on the sphere a coordinate system such that the distance between two arbitrary points in it is defined by a formula of the type (2.1). However, in the neighbourhood of each point on the sphere, one may approximate with the aid of a Cartesian coordinate system located in the local tangent plane. For three-dimensional spaces, it is also not always possible to introduce a simple, Cartesian coordinate system suitable for an entire space.

In the sequel, only such spaces will be considered for which it is possible to introduce a single Cartesian coordinate system. Such spaces are referred to as Euclidean spaces and the corresponding mechanics is called Newtonian mechanics. Experience shows that within not very large scales one may assume with a high degree of accuracy that real physical space is Euclidean.

Absolute time. The concept of time is linked to experience and is indispensable in mechanics. Any mechanical phenomenon is always described from a point of view of an observer. Time, generally speaking, may depend on the system of measurement of the observer.

Assume that time flows independently for all observers—in a train, in an aircraft, in an auditorium. Hence, use will be made of absolute time—an idealization, which is convenient for the description of reality only if the effects of the theory of relativity can be negelected.

Hence, it will be assumed that one is dealing with the motion of continuous media—of continua in Euclidean space—and absolute time will be used. Thus, three fundamental hypotheses have been introduced above with the aid of which the theory of motion of deformable bodies will be developed. Derivations from this theory, based on these hypotheses, very often, but not always, agree with experiment. If necessary, this model of space and time can be refined and generalized. However, all further generalizations will rest heavily on Newtonian mechanics, based on the fundamental hypotheses above. The essence of these hypotheses will become clearer as the theory is developed further.

§6. Thermodynamic potentials of two-parameter media

Consider yet an important problem from the thermodynamics of two-parameter media, with, in general, reversible processes (dq^{**} being assumed to vanish).

It has been stated above that for two-parameter media the three functions of state U, s and T cannot be arbitrary. For example, if the internal energy U is given as a function of p and ρ, then $T(p, \rho)$ must satisfy (5.16) which, for given $U(p, \rho)$, must be considered as a linear, first order partial differential equation for the determination of $T(p, \rho)$.

This differential equation is known to have many solutions, i.e., many expressions $T(p, \rho)$ for a given $U(p, \rho)$ and, consequently, also the thermodynamic properties of a medium are not determined uniquely. In order to remove this lack of single-valuedness, one must select one of the particular solutions of (5.16), i.e., decide on the concrete form of the equation of state $T = T(p, \rho)$. Subsequently, the entropy (apart from an additive constant) can be determined from (5.14).

As determining thermodynamic variable of a two-parameter medium, one may often and conveniently take different pairs of variables, for example, ρ and s, p and s, ρ and T, etc. There arises the question: Is it possible to ascribe to the internal energy U a function of such variables that as a result of this specification the other thermodynamic functions are determined completely and uniquely? It will be shown that this step is possible.

Internal energy and entropy as thermodynamic potentials. Let U be given as a function of ρ and s. Then one finds, by differentiation and use of the heat flow equation, taking into account the second law of thermodynamics for reversible processes (5.14),

$$dU = \left(\frac{\partial U}{\partial s}\right)_\rho ds + \left(\frac{\partial U}{\partial \rho}\right)_s d\rho = T ds - p d\frac{1}{\rho}, \tag{6.1}$$

whence, in view of the arbitrariness of ds and $d\rho$,

$$T = \left(\frac{\partial U}{\partial s}\right)_\rho, \qquad p = \rho^2\left(\frac{\partial U}{\partial \rho}\right)_s, \tag{6.2}$$

i.e., T and p are determined uniquely as functions of ρ and s. In this case, the internal energy is referred to as thermodynamic potential. It is also seen from (6.1) that, if one gives the entropy s as a function of U and ρ, then

$$\frac{1}{T}=\left(\frac{\partial s}{\partial U}\right)_\rho, \quad \frac{p}{T}=\left(\frac{\partial s}{\partial 1/\rho}\right)_U,$$

i.e., then the entropy is for the variables U and ρ a thermodynamic potential.

Free energy. If the determining thermodynamic variables are ρ and T, then (6.1) is written more conveniently in the form

$$\mathrm{d}(U-Ts)=-s\mathrm{d}T+\frac{p}{\rho^2}\mathrm{d}\rho$$

or

$$\mathrm{d}F=-s\mathrm{d}T+\frac{p}{\rho^2}\mathrm{d}\rho\,, \tag{6.3}$$

where F denotes the function of state

$$F\equiv U-Ts\,, \tag{6.4}$$

called the free energy. If F is known as a function of ρ and T, then p and s are determined uniquely by (6.3). In fact, it follows from (6.3) that

$$s=-\left(\frac{\partial F}{\partial T}\right)_\rho, \quad p=\rho^2\left(\frac{\partial F}{\partial \rho}\right)_T. \tag{6.5}$$

In the case of utilization of the variables ρ and T, the free energy $F(\rho, T)$ is a thermodynamic potential.

Heat content or enthalpy. Analogously, if the determining parameters are the pressure p and the entropy s, then (6.1) is conveniently rewritten as

$$\mathrm{d}\left(U+\frac{p}{\rho}\right)=T\mathrm{d}s+\frac{\mathrm{d}p}{\rho}$$

or

$$\mathrm{d}i=T\mathrm{d}s+\frac{\mathrm{d}p}{\rho}, \tag{6.6}$$

where the function of state

Laws of motion of a continuum. For any point of a continuum, specified by the coordinates *a, b, c*, one may write down the law of motion which contains not only functions of a single variable, as in the case of the motion of a point, but of four variables: the initial coordinates *a, b, c* and the time *t*.

$$\left.\begin{aligned} x^1 &= x^1(a, b, c, t)\,, \\ x^2 &= x^2(a, b, c, t)\,, \\ x^3 &= x^3(a, b, c, t) \end{aligned}\right\} \quad \text{or} \quad x^i = x^i(a, b, c, t)\,, \tag{1.2}$$

If in (1.2) a, b, c are fixed and t varies, then (1.2) describes the law of motion of one selected point of the continuum. If a, b, c vary, and t is fixed, then (1.2) gives the distribution of the points of the medium in space at a given instant of time. If a, b, c and t vary, then one may interpret (1.2) as a formula which determines the motion of the continuous medium and, by definition, the functions (1.2) yield the law of motion of the continuum.

Lagrangian coordinates. The coordinates a, b, c or ξ^1, ξ^2, ξ^3 (or sometimes definite functions of these variables), which individualize the points of a medium, and the time t are referred to as Lagrangian coordinates.

The fundamental problem of continuum mechanics consists of the determination of the functions (1.2).

In the sequel, the concept of the law of motion will always be used explicitly or implicitly.

For the sake of generality, note that a medium represents an accumulation of points, but it need not be a material body. For instance, it may sometimes be agreed to represent by points in a plane the prices of some products and to study by the methods of the kinematics of continuous media the motion of prices in economics.

Also, and this is done frequently, one may study the laws of displacement in space of different states of motion of material particles, and not of the particles themselves. For example, waves are observed on a rye field in windy weather, and one may speak of the displacements in space of the highest points or of the depressions of the rye surface, and not of the ears themselves.

velocity vector: $\boldsymbol{a} = (\partial \boldsymbol{v}/\partial t)_{\xi^i}$, and in the evaluation of its components it must be kept in mind that the point of the medium moves in space with time and that the base vector $\mathbf{э}_i$ of a curvilinear system vary in space from point to point.

In Cartesian coordinate systems, one has also

$$a^1 = \frac{\partial^2 x}{\partial t^2}, \quad a^2 = \frac{\partial^2 y}{\partial t^2}, \quad a^3 = \frac{\partial^2 z}{\partial t^2}.$$

In many studies of motions of continua, the basic problem of finding the laws of motion may be replaced by the problem of the determination of the functional dependence of the components of the velocity v^i or acceleration a^i on ξ^1, ξ^2, ξ^3 and t.

Note especially that the Lagrangian approach to the study of motion of continua is fundamental for physical laws, since they are concerned with motions of individual material points.

§ 2. Eulerian approach to the study of motion of continua

The essence of the Eulerian approach. Assume now that not the history of the motion of individual points of a continuum is of interest, but what happens at different times at a given geometrical point of space which is linked to the reference system of an observer. Concentrate attention on a given point of space at which arrive at different times different particles of the medium. In essence, this is the Eulerian approach to the study of the motion of continua. For instance, the motion of water in a river can be studied either by following the motion of each particle from the upper reaches to the mouth of the river (the Lagrangian approach), or by observation of the changes of the flow of water at definite locations in the river without tracing the motion of individual water particles along the entire river (the Eulerian approach).

Eulerian variables. An Eulerian approach is often employed in applications. The geometric space coordinates x^1, x^2, x^3 and the time t are referred to as Eulerian variables. In the Eulerian approach, a motion is considered to be known, if its velocity, acceleration, temperature and

The heat flow equation and (6.13) yield yet the equation

$$c_p - c_V = -\frac{T}{\rho^2}\left(\frac{\partial p}{\partial T}\right)_\rho \left(\frac{\partial \rho}{\partial T}\right)_p , \tag{6.14}$$

which, on the basis of the relation

$$\left(\frac{\partial p}{\partial T}\right)_\rho = -\left(\frac{\partial p}{\partial \rho}\right)_T \left(\frac{\partial \rho}{\partial T}\right)_p ,$$

following from the formula

$$\mathrm{d}p = \left(\frac{\partial p}{\partial T}\right)_\rho \mathrm{d}T + \left(\frac{\partial p}{\partial \rho}\right)_T \mathrm{d}\rho ,$$

reduces to the form

$$c_p - c_V = \frac{T}{\rho^2}\left(\frac{\partial p}{\partial \rho}\right)_T \left(\frac{\partial \rho}{\partial T}\right)_p^2 . \tag{6.15}$$

Equations (6.13)–(6.15) are true for arbitrary two-parameter media. Using data from experimental measurements of c_p and c_V and measured values of the coefficients of thermal density changes for constant pressure $(\partial\rho/\partial T)_p = k_\rho$ and of pressure change for constant volume $(\partial p/\partial T)_\rho = k_p$, one may determine the derivatives of the internal energy and heat content from the formulae

$$\left.\begin{aligned} \left(\frac{\partial U}{\partial \rho}\right)_T &= \frac{c_p - c_V}{k_\rho} + \frac{p}{\rho^2} = -\frac{1}{\rho^2}\left[T\left(\frac{\partial p}{\partial T}\right)_\rho - p\right], \\ \left(\frac{\partial U}{\partial T}\right)_\rho &= c_V , \end{aligned}\right\} \tag{6.16}$$

$$\left.\begin{aligned} \left(\frac{\partial i}{\partial T}\right)_p &= c_p , \\ \left(\frac{\partial i}{\partial p}\right)_T &= -\frac{c_p - c_V}{k_p} + \frac{1}{\rho} = \frac{1}{\rho^2}\left[T\left(\frac{\partial \rho}{\partial T}\right)_p + \rho\right]. \end{aligned}\right\} \tag{6.17}$$

Obviously, as a consequence of the first and second laws of thermodynamics, one must satisfy in addition the integrability conditions

$$-\frac{T}{\rho^2}\left(\frac{\partial^2 p}{\partial T^2}\right)_\rho = \left(\frac{\partial c_V}{\partial \rho}\right)_T , \tag{6.18}$$

$$\left(\frac{\partial c_p}{\partial p}\right)_T = -T\left(\frac{\partial^2 \frac{1}{\rho}}{\partial T^2}\right)_p, \tag{6.19}$$

which may be used to reduce the number of experiments or to verify experimental results.

It has been established above that for every thermodynamic system always two functions of state may be introduced: the internal energy U and the entropy s, and for equilibrium processes yet one more function of state: the absolute temperature T; a new universal equation, the heat flow equation, has been obtained:

$$dU = \frac{p^{ij}}{\rho}\nabla_j v_i dt + dq + dq^{**} \tag{6.20}$$

or for $p^{ij} = p^{ji}$

$$dU = \frac{p^{ij}}{\rho} e_{ij}\, dt + dq + dq^{**}$$

and the second law of thermodynamics

$$T dS = dQ^{(e)} + dQ', \qquad dQ' \geqslant 0,$$

or (per unit mass)

$$T ds = dq + dq', \qquad dq' = \frac{dQ'}{dm} \geqslant 0, \tag{6.21}$$

has been considered which, in general, is required for the construction of concrete models of continuous media.

These results will now be applied to the development of models of continua.

§7. Examples of ideal and viscous fluids and their thermodynamic properties. Heat conduction

In order to study particular problems of motion of continua with the aid of the earlier obtained system of universal equations (equation of continuity, proceed from the Eulerian to Lagrangian variables in all formulae for $\boldsymbol{a}$, T, etc. Consequently, the transition from Eulerian to Lagrangian variables for a given velocity field is, in general, connected with an integration of ordinary differential equations.

Clearly, specifications of a motion of a continuum from Lagrangian and Eulerian points of view are in a mechanical sense equivalent.

§3. Scalar and vector fields and their characteristics

Definition of scalar and vector fields. In the study of the motion of a continuum, one must introduce into the consideration scalar and vector quantities: the temperature T, the velocity $\boldsymbol{v}$, etc. In general, these can be considered in different coordinate systems: In an observer's coordinate system and in a system concomitant with the medium. They may be functions of x^1, x^2, x^3 or functions of ξ^1, ξ^2, ξ^3. In each of these coordinate systems, one can select a finite or infinite region and associate with every point of these regions a number, for instance, a temperature T, or a vector, for instance, a velocity $\boldsymbol{v}$, or, as will be seen later, some other more complicated characteristics.

The set of values of this or that quantity, which is given at every point of a region under consideration, is called the field of this quantity. If the quantity in question is a scalar, i.e., a number the value of which at a given point is independent of the choice of the coordinate system, the field is said to be scalar. Examples of scalar fields are temperature, density, etc. If a quantity in question is a vector, for instance, velocity, acceleration, etc., the field is called a vector field. The velocity in every coordinate system x^i has three components v^i and, consequently, at a given point and in a given coordinate system it is determined by three numbers. Hence, the velocity field and any vector field is equivalent to the three fields of the projections of the vector under consideration. However, although a vector itself does not depend on the coordinate system, its projections depend on the system of coordinates.

Next, certain general characteristics of scalar and vector fields will be studied using the temperature field T and the velocity field $\boldsymbol{v}$ as examples.

always postulate the existence of a reference system x^1, x^2, x^3 with respect to which a motion is studied, and that of a concomitant coordinate system.

In the Lagrangian approach to the study of motions of continuous media, one uses ξ^1, ξ^2, ξ^3 and t as independent variables. In other words, it operates on the complete history of every individual point of the medium separately. In practice, such a description is often too detailed and too complicated; however, it is always implied in the formulation of physical laws. Besides the concept of laws of motion, one must still introduce for the description of the motion of continuous media certain other concepts, in particular, those of the velocity and acceleration of particles of a continuous medium.

Velocity. Let some point of a medium at the instant t occupy the point M, and at the instant $t+\Delta t$ the point M', and let $\overline{MM'}=\Delta \boldsymbol{r}$.

By the quantity $\Delta \boldsymbol{r}$ is understood the infinitesimal oriented displacement of the individual point of the medium during the time interval Δt. When one may introduce the radius vector $\boldsymbol{r}$ in the space (and in Euclidean space this is always possible), the quantity $\Delta \boldsymbol{r}$ is, obviously, the differential of the radius vector of the material point under consideration.

The limit of the ratios of the two infinitesimal quantities $\Delta \boldsymbol{r}$ and Δt for $\Delta t \to 0$ in the case of non-Euclidean space or the partial derivative of the radius vector of the point of the medium with respect to time $\partial \boldsymbol{r}/\partial t$ in the case of Euclidean space is referred to as the velocity of the point of the medium. The symbol $\boldsymbol{v}$ will be used to denote the velocity.

The radius vector $\boldsymbol{r}$ depends in the general case on the three parameters ξ^1, ξ^2, ξ^3, individualizing the point of the medium, and on the time t. The velocity is evaluated for individual points of the medium, i.e., for fixed ξ_1, ξ_2, ξ_3, therefore it is equal to the partial derivative of $\boldsymbol{r}$ with respect to t

$$\boldsymbol{v} = \frac{\partial \boldsymbol{r}}{\partial t}\,.$$

Velocity is evaluated with respect to a frame of reference. Obviously, the medium is at rest with respect to the concomitant coordinate system, and hence the velocity with respect to the concomitant system always vanishes.

Thus, from the point of view of continuum mechanics, an ideal incompressible fluid is specified if one knows the density and the heat capacity $c(T)$ only.

In addition, for the solution of concrete problems, one must state the external mass force F, the inflow of heat dq and the supplementary boundary, initial or other conditions which are required for a unique assignment of a solution of the system of partial differential equations.

Independence of mechanical problem from the heat problem and the link between the heat problem and the mechanical problem in the motion of an ideal incompressible fluid. In conclusion, note that the solution of the mechanical problem of the determination of the motion of an ideal incompressible fluid under the action of given forces neither depends on the solution of the problem of the temperature distribution in the fluid volume nor that is it required to know the internal energy.

Conversely, Equation (7.3) for the determination of the temperature distribution T in the case of a given heat flow becomes definite only after the velocity distribution $\mathbf{v}(x^i, t)$ has been found by solution of the mechanical problem.

Consequently, in the presence of motion of a medium, the solution of the thermal problem depends on that of the mechanical problem.

(b) Model of an ideal gas, i.e., of an ideal compressible fluid

Model of an ideal gas. Define an ideal gas, firstly, as a medium in which the stress tensor is spherical:

$$p^{ij} = -pg^{ij} ;$$

secondly, as a two-parameter medium in which the internal energy depends only on two parameters, for example, on ρ and s:

$$U = U(\rho, s) ;$$

thirdly, as a medium in which in the case of continuous motion all mechanical processes are reversible[1] and, consequently,

$$dq' = 0 .$$

These three assumptions, under the condition that $U(\rho, s)$ is given, completely fix the model of an ideal gas or of an ideal compressible fluid in a thermodynamic as well as in a mechanical sense.
In fact, if the mass forces F and the external heat flow dq are given, then the two equations (6.2), the so-called equations of state, the second law of thermodynamics

$$T\, ds = dq$$

or the heat flow equation

$$dU = -p\, d\frac{1}{\rho} + dq ,$$

the equation of continuity

$$\frac{d\rho}{dt} + \rho \operatorname{div} \boldsymbol{v} = 0 ,$$

and the three Euler equations

$$a_i = F_i - \frac{1}{\rho}\frac{\partial p}{\partial x^i}$$

represent a closed system for the determination of the seven unknown functions: ρ, v_i, p, s and T.

The equations of state (6.2) are valid for any process. Instead of $U(\rho, s)$, any of the potentials $F(\rho, T), i(p, s)$ or $\Psi(p, T)$ may be specified.

Complete system of equations of motion of an ideal compressible fluid or gas in the case of adiabatic processes. If in each element of a body a process is reversible and adiabatic, then $dq' = 0$ and $dq = 0$; therefore

[1] Note that at times one considers irreversible physico-chemical processes in an ideal gas, for example chemical reactions; then dq', as a consequence of the irreversibility of such processes, may differ from zero. Moreover, let us note, that as it has been shown in Chapter IX (*cf.*, Vol. II, pp. 1000, 1001) from the assumption of the form of the internal (or free) energy and of the reversibility of a process, follows that the stress tensor is spherical.

The concept of a vector. Certain vectors have already been considered, for instance, the velocity vector $\boldsymbol{v}$, the radius vector $\boldsymbol{r}$, the displacement vector $d\boldsymbol{r}$. What is a vector? It is not a scalar; however, at the same time, like a scalar, it is invariant, being independent of the choice of coordinate system. In defining vectors, it is often said that they represent triplets of numbers, the so-called components of the vector, which transform in a definite manner during the transition from one coordinate system to another. However, this definition is insufficient, since a vector is always related to a definite base and, in stating its components, one must always specify the base to which these components refer.

In a Cartesian coordinate system, the components of a vector are related to $\boldsymbol{i}, \boldsymbol{j}, \boldsymbol{k}$, in an arbitrary curvilinear coordinate system, to the base vectors $\mathfrak{z}_i$ which vary in space from point to point. Hence, in contrast to the components of a vector in a Cartesian system, the components of a vector in a curvilinear coordinate system are essentially linked to the point at which they are considered.

For instance, in the case of the velocity vector $\boldsymbol{v}$ at every point of space, one must consider the numbers v^1, v^2, v^3 together with the base vectors $\mathfrak{z}_1$, $\mathfrak{z}_2$, $\mathfrak{z}_3$ and define the vector $\boldsymbol{v}$ by (1.6) through the base vectors one may represent by analogous methods every vector in a given coordinate system.

Acceleration. Beside the velocity $\boldsymbol{v}$, one must still consider the acceleration vector $\boldsymbol{a}$ of a point of the continuum

$$\boldsymbol{a} = \left(\frac{\partial \boldsymbol{v}}{\partial t}\right)_{\xi^i} = a^i \mathfrak{z}_i ,$$

with the components $a^i = a^i(\xi^1, \xi^2, \xi^3, t)$. The acceleration $\boldsymbol{a}$, as the velocity $\boldsymbol{v}$, is evaluated for individual points of the continuum. This definition of the acceleration depends on the observer's choice of the coordinate system x^i which is used in the description of the law of motion (1.2) and which may move.

Note that the relations

$$a^1 = \frac{\partial v^1}{\partial t}, \quad a^2 = \frac{\partial v^2}{\partial t}, \quad a^3 = \frac{\partial v^3}{\partial t}$$

are valid only in Cartesian and not in curvilinear coordinate systems. In fact, the acceleration vector is defined as the time derivative of the

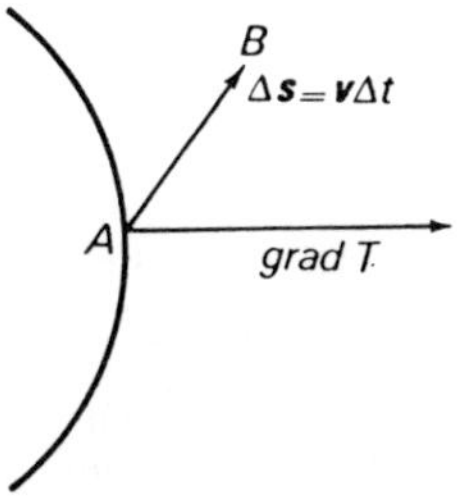

Fig. 5. On the concept of the convective derivative.

from point to point (such a field is said to be homogeneous) or for motion along a level surface $T(x, y, z, t) = \text{const.}$ at $t = \text{const.}$

Formulae for the determination of acceleration components in a Cartesian system. The formulae for the determination of the acceleration components in the case when the velocity is given in Eulerian variables are closely related to the concepts of individual local and convective derivatives. Let the velocity components v^1, v^2, v^3 be given in a Cartesian coordinate system as functions of Eulerian variables.

How are the acceleration components determined? The acceleration is defined for particles of a continuum; hence the components of the acceleration will be determined as the individual derivatives with respect to time of the corresponding velocity components, i.e.,

$$a_x = \left(\frac{\partial u}{\partial t}\right)_{\xi^i} = \frac{du}{dt} = \left(\frac{\partial u}{\partial t}\right)_{x^i} + u\frac{\partial u}{\partial x} + v\frac{\partial u}{\partial y} + w\frac{\partial u}{\partial z},$$

$$a_y = \left(\frac{\partial v}{\partial t}\right)_{\xi^i} = \frac{dv}{dt} = \left(\frac{\partial v}{\partial t}\right)_{x^i} + u\frac{\partial v}{\partial x} + v\frac{\partial v}{\partial y} + w\frac{\partial v}{\partial z},$$

$$a_z = \left(\frac{\partial w}{\partial t}\right)_{\xi^i} = \frac{dw}{dt} = \left(\frac{\partial w}{\partial t}\right)_{x^i} + u\frac{\partial w}{\partial x} + v\frac{\partial w}{\partial y} + w\frac{\partial w}{\partial z}.$$

Note the fact that these formulae are valid only in Cartesian coordinate systems (where all $\mathbf{э}_i$ are independent of coordinates and time).

Steady and unsteady motions. Different processes and motions are said to be steady or stationary, if all their characteristics in terms of Eulerian variables depend only on x^1, x^2, x^3 and not on the time t. Thus, for steady processes and motions, the local derivatives of all characteristics vanish, i.e.,

$$\Psi_m = m\left(i_0 - T s_0 + \int_{T_0}^{T} c_p \, dT - T\int_{T_0}^{T} \frac{c_p}{T} dT + RT\ln\frac{\rho}{\rho_0}\right).$$

In the formula (7.8) the heat capacities $c_v(T)$ and $c_v(T)$ are given per unit of mass. The constants U^0, S^0, i^0 are the values of U, S and i at temperature T_0, pressure p_0 and density ρ_0. For different gases, the quantities U^0, S^0, i^0, c_p, c_v, and R may in many cases be taken as inversely proportional to the molecular mass M.

The equations (7.8) become particularly simple in the practically important case when the heat capacity c_v is constant, that is when c_v does not depend upon the temperature. In this case we infer from the formula of Mayer

$$R = c_p - c_v$$

that the heat capacity c_v is also constant.

For the perfect gas when the heat capacity c_v and c_p are constant, then from (7.8) it follows that

$$U_m = mc_V T + \text{const.}, \quad i_m = mc_p T + \text{const.},$$

$$S = m\left(c_V \ln\frac{T\rho_0^{\gamma-1}}{T_0\rho^{\gamma-1}} + s_0\right) = m\left(c_V \ln\frac{p\rho_0^{\gamma-1}}{p_0\rho^{\gamma-1}} + s_0\right). \tag{7.9}$$

consequently

$$p = p_0\left(\frac{\rho}{\rho_0}\right)^{\gamma} e^{\frac{s-s_0}{c_V}},$$

$$\frac{U_m}{m} = U = c_V T_0\left(\frac{\rho}{\rho_0}\right)^{\gamma-1} e^{\frac{s-s_0}{c_V}} + \text{const.}, \tag{7.10}$$

where the non-dimensional constant c_p/c_v is called *Poisson's coefficient.*

The relations (7.10) between S, T and ρ and those between S, ρ, and p can be regarded for any process as equations of state, analogous to the equation of Clapeyron $p = \rho RT$. It is clear, that the model of the perfect gas with constant heat capacity can be fully described by assuming only one function $U(\rho, s)$

$$U = c_V\, T_0 \left(\frac{\rho}{\rho_0}\right)^{\gamma - 1} e^{\frac{s - s_0}{c_V}} + \text{const.} \tag{7.11}$$

The equation of state (7.5) does not correspond to real conditions in highly compressed gases, if their density is very high. The equation also applies neither to states which are near to the points of condensation of gases into liquids nor to liquids themselves.

Furthermore, we may note that at very low temperatures, the equation of state (7.5) and the formulae (7.8) cease to be in accord with the general laws of thermodynamics (the second law, the theorem of Nernst and its consequences) governing the behaviour of matter near the absolute zero.

Van der Waals' gas. Let us consider an ideal gas which is governed by the *van der Waals equation* of state

$$p = \frac{\rho \boldsymbol{R} \mathrm{T}}{1 - b\rho} - a\rho^2 , \tag{7.12}$$

where $\boldsymbol{R}$, b and a are positive physical constants; this is a refinement of Clapeyron's equation. It describes processes near the condensation point of gases and describes the real relations for some intervals of the liquid phase. The denominator $1 - b\rho = 1 - \rho/\rho^*$ brings about a steep rise of the pressure at high density for ρ nearing ρ^*; the additional term $-a\rho^2$ is also only appreciable at high density. This term is connected with the repulsion forces between molecules which appear only in the case of their very close approach; this happens only for very high density.

From equation (7.12) and the formula (6.16) we have

$$\left(\frac{\partial U}{\partial \rho}\right)_T = -a \quad \text{and} \quad \left(\frac{\partial U}{\partial \mathrm{T}}\right)_\rho = c_v(\mathrm{T}) . \tag{7.13}$$

From the first equation of (7.13) it follows that the heat capacity c_v for a van der Waals medium depends only on the temperature.

The function $c_V(\mathrm{T})$ is defined by the properties of the medium. The form of this function yields information additional to the equation of state (7.12) for the medium.

From Equation (7.13) it follows that for van der Waals gases the internal energy is given by

$$\frac{dx^i}{d\lambda} = v^i(x^1, x^2, x^3, t), \qquad i = 1, 2, 3, \tag{3.2}$$

i.e., the set of differential equations for the stream lines. They differ from the differential equations, defining the law of motion or the trajectories of the motion of the particles of a continuous medium, which, obviously, have the form

$$\frac{dx^i}{dt} = v^i(x^1, x^2, x^3, t). \tag{3.3}$$

The time enters on the right-hand as well as on the left-hand side of Equations (3.3). In Equations (3.2), the derivatives are taken with respect to λ and the right-hand side depends on t. In the integration of (3.2), the variable t must be considered to be a constant parameter, whereas in Equation (3.3) it is a variable.

Thus, generally speaking, the stream lines do not coincide with the particle trajectories. The family of stream lines $x^i = x^i(c^1, c^2, c^3, \lambda, t)$ depends on the time and varies with it. However, the parameter t appears on the right-hand side of (3.2) and (3.3) only in the case of unsteady motions. In the case of steady motion, the distinction between Equations (3.2) and (3.3) disappears; it reduces to different notation for the differentiation parameter, a fact which has no significance. Therefore stream lines and trajectories coincide for steady motions.

Examples of stream lines and trajectories. Consider some examples. A motion of a rigid body is said to be translatory, if any straight segment, selected inside the body, moves parallel to itself. All points of a rigid body in translation have at a given instant of time the same velocity in magnitude and direction. Consequently, the stream lines are in this case always straight lines. What about the trajectories? Translation of a rigid body can proceed along any trajectories and hence also along a circle (fig. 6). Therefore, in the general case of translation of a rigid body, the stream lines and trajectories do not coincide.

In the case of arbitrary motion of a rigid body, the stream lines are spirals and the trajectories may be arbitrary.

Individual and local derivatives with respect to time. A distribution of temperature may be given from a Lagrangian point of view by $T(\xi^i, t)$ and from an Eulerian point of view by $T(x^i, t)$. If the distribution T is given in Lagrangian variables, then the change of the temperature T in a particle of the continuum with time is readily computed. It is given by the derivative

$$\left(\frac{\partial T}{\partial t}\right)_{\xi^i}.$$

How does one evaluate this quantity, if the distribution of the temperature is given in terms of Eulerian variables by $T(x^1, x^2, x^3, t)$? Obviously, for this purpose, one must step over from Eulerian to Lagrangian variables and apply the chain rule. Then

$$T(x^1,x^2,x^3,t)=T[x^1(\xi^1,\xi^2,\xi^3,t),\ x^2(\xi^1,\xi^2,\xi^3,t),\ x^3(\xi^1,\xi^2,\xi^3,t),t]$$

and

$$\left(\frac{\partial T}{\partial t}\right)_{\xi^i} = \left(\frac{\partial T}{\partial t}\right)_{x^i} + \frac{\partial T}{\partial x^1}\left(\frac{\partial x^1}{\partial t}\right)_{\xi^i} + \frac{\partial T}{\partial x^2}\left(\frac{\partial x^2}{\partial t}\right)_{\xi^i} + \frac{\partial T}{\partial x^3}\left(\frac{\partial x^3}{\partial t}\right)_{\xi^i},$$

where the derivatives $(\partial x^1/\partial t)$, $(\partial x^2/\partial t)$, $(\partial x^3/\partial t)$ are evaluated for fixed ξ^1, ξ^2, ξ^3, and, consequently, they are the corresponding velocity components v^1, v^2, v^3. Hence

$$\left(\frac{\partial T}{\partial t}\right)_{\xi^i} = \left(\frac{\partial T}{\partial t}\right)_{x^i} + v^i\frac{\partial T}{\partial x^i}.$$

Note that for a given $T(x^1, x^2, x^3, t)$ for the evaluation of $(\partial T/\partial t)_{\xi^i}$ one need not know completely the law of motion, but only the velocity field $\boldsymbol{v}$.

The derivative $(\partial T/\partial t)_{\xi^i}$ characterizes the time variation of the temperature at a given point of the medium; it is referred to as the individual or total derivative of the temperature T with respect to the time t and often denoted by $\mathrm{d}T/\mathrm{d}t$. The derivative $(\partial T/\partial t)_{x^i}$ characterizes the rate of change of the temperature T at a given point of space x^i. It is referred to as the local derivative and denoted by $\partial T/\partial t$. In the general case, the total derivative $\mathrm{d}T/\mathrm{d}t$ is not equal to the local derivative $\partial T/\partial t$; it differs from it by the quantity, which depends on the motion of the particles and is called the convective derivative. Thus,

$$\frac{\mathrm{d}T}{\mathrm{d}t} = \frac{\partial T}{\partial t} + v^i\frac{\partial T}{\partial x^i}.$$

The equation of motion of a linear viscous fluid with constant coefficients of viscosity, the Navier–Stokes equation, has the form

$$\frac{d\boldsymbol{v}}{dt} = \boldsymbol{F} - \frac{1}{\rho}\,\mathrm{grad}\, p + \left(\frac{\zeta}{\rho} - \frac{\nu}{3}\right)\mathrm{grad\,div}\,\boldsymbol{v} + \nu\Delta\boldsymbol{v}\,, \tag{7.23}$$

where $\nu = \mu/\rho$ is the kinematic viscosity.

The equation of heat flow, on account of the second law of thermodynamics, may be rewritten in the form

$$dU = -\frac{dA^{(i)}}{dm} + T\,ds - dq'\,. \tag{7.24}$$

Work done by the internal forces in a viscous fluid. The element of work done by the internal stresses, referred to unit mass of a viscous fluid, will now be evaluated. Substituting p^{ij} from (7.19) into the general expression for the work done by the internal surface forces, one finds

$$\begin{aligned}\frac{dA^{(i)}}{dm} &= -\frac{p^{ij}}{\rho}e_{ij}dt = \frac{p}{\rho}g^{ij}e_{ij}dt - \frac{\tau^{ij}}{\rho}e_{ij}dt \\ &= \frac{p}{\rho}\,\mathrm{div}\,\boldsymbol{v}\,dt - \frac{\tau^{ij}e_{ij}}{\rho}dt\,.\end{aligned} \tag{7.25}$$

As a consequence of the equation of continuity

$$\mathrm{div}\,\boldsymbol{v} = -\frac{1}{\rho}\frac{d\rho}{dt}\,,$$

one finds

$$\frac{1}{dm}dA^{(i)} = p\,d\frac{1}{\rho} - \frac{\tau^{ij}e_{ij}}{\rho}dt\,. \tag{7.26}$$

Pressure and temperature in a viscous fluid. By (7.24) and (7.26)

$$dU = -p\,d\frac{1}{\rho} + \frac{\tau^{ij}e_{ij}}{\rho}dt + T\,ds - dq'\,. \tag{7.27}$$

By definition, for a viscous compressible fluid (viscous gas), as for an ideal compressible fluid (ideal gas), the pressures p and temperatures T for any processes are determined by[1)]

$$dU = -p\,d\frac{1}{\rho} + T\,ds\,, \tag{7.28}$$

i.e., one has

$$p = -\left(\frac{\partial U}{\partial\, 1/\rho}\right)_s \qquad T = \left(\frac{\partial U}{\partial s}\right)_\rho . \tag{6.2}$$

A certain basis for the introduction of a similar model may be provided by the fact that the parameters p and T in a viscous fluid at rest, i.e., for $e_{ij}=0$, must coincide with the corresponding parameters in an ideal fluid at rest; the same is true when the fluid moves as a rigid body. Viscous stresses only appear for motion with deformation.

Expression for uncompensated heat in a viscous fluid. In the introduced model of a viscous fluid, the pressure, temperature, entropy and internal energy are linked by relations which do not depend on the property of viscosity. Comparing (7.27) and (7.28), one obtains an expression for the uncompensated heat which is true in the general case of a non-linear viscous fluid:

$$dq' = \frac{\tau^{ij} e_{ij}}{\rho}\,dt\,. \tag{7.29}$$

Dissipation of mechanical energy in a viscous fluid. It will be shown that the presence of dq' causes dissipation of mechanical energy when a viscous fluid moves. In fact, the kinetic energy theorem may be rewritten for a viscous fluid in the form

$$d\frac{v^2}{2} = \frac{dA^{(e)}}{dm} + pq\frac{1}{\rho} - \frac{\tau^{ij} e_{ij}}{\rho}\,dt\,.$$

The quantity $-(1/\rho)\tau^{ij}e_{ij}dt = -dq'$, representing the work per unit mass done by the viscous stresses, is always negative (or vanishes when $e_{ij}=0$),

1) This important assumption, referred to as Gibb's formula, permits immediate establishment of a formula for the uncompensated heat dq'.

Note that $\partial T/\partial x$, $\partial T/\partial y$, $\partial T/\partial z$ can be interpreted as the projections of a vector, since

$$dT = \frac{\partial T}{\partial x}\,dx + \frac{\partial T}{\partial y}\,dy + \frac{\partial T}{\partial z}\,dz$$

is invariant and dx, dy, dz are the components of the vector d$\boldsymbol{r}$.

The convective derivative. The expression $\Sigma_{i=1}^{3} v^i\,\partial T/\partial x^i$ has been called the convective derivative of the temperature T with respect to the time. Using the concepts of the gradient vector of the temperature and of the scalar product, this expression may be rewritten

$$v^i\frac{\partial T}{\partial x^i} = \boldsymbol{v}\cdot \text{grad } T\,.$$

The word derivative is always interpreted as the limit of the ratio of the increment of a function to the increment of an argument as the latter tends to zero. However, what increment of a function is taken in the case of the definition of the convective derivative? Rewrite the convective derivative of the temperature in the form

$$\frac{\boldsymbol{v}\,\Delta t\cdot \text{grad } T}{\Delta t}\,.$$

Clearly, the quantity $\boldsymbol{v}\Delta t$ is equal to the displacement Δs (fig. 5) and

$$v^i\frac{\partial T}{\partial x^i} = \lim_{\Delta t\to 0}\frac{\Delta \boldsymbol{s}\cdot \text{grad } T}{\Delta t} = \lim_{\Delta t\to 0}\frac{(\text{grad } T)_s \Delta s}{\Delta t} =$$

$$= \frac{\partial T}{\partial s}\lim_{\Delta t\to 0}\frac{\Delta s}{\Delta t} = \lim_{\Delta t\to 0}\frac{\Delta T}{\Delta t}\,,$$

where the increment ΔT of the temperature occurs as a particle of the continuum is displaced in the direction $\boldsymbol{s}$ with velocity $\boldsymbol{v}$ at time Δt from one point of space to another (the points A and B of fig. 5).

In the general case, the convective derivative differs from zero, since the values of the temperature at the point A and B differ. It can be equal to zero, for example, in the absence of motion, or of a temperature gradient, i.e., when the temperature at a given instant of time does not vary in space and the latter by three functions $v^1(x, y, z, t)$, $v^2(x, y, z, t)$, $v^3(x, y, z, t)$.

Source and sink in space. Consider another example of potential flow which will be important later on. Let

$$\varphi = -\frac{Q}{4\pi r}, \tag{3.7}$$

where $r = \sqrt{(x^2 + y^2 + z^2)}$ and $Q = \text{const.}$ or $Q = Q(t)$. Obviously, the equipotential surfaces $\varphi = \text{const.}$ are in this case given by $r = \text{const.}$, i.e., they are concentric spheres with centre at the origin of coordinates. The velocity $\boldsymbol{v} = \text{grad}\ \varphi$ is orthogonal to these spheres, i.e., it is directed along the radii, emanating from the origin. The stream lines are rays along these radii. Let $Q > 0$; then, since grad φ points in the direction of increasing φ,

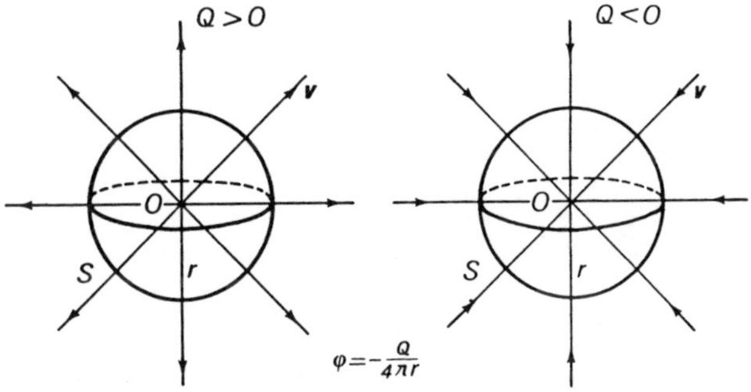

Fig. 8. Flows from point source and point sink in space.

the velocity $\boldsymbol{v}$ has the same direction as $\boldsymbol{r}$. If $Q < 0$, then $\boldsymbol{v}$ has the direction of $-\boldsymbol{r}$ (fig. 8). The magnitude of the velocity is

$$|(\text{grad}\ \varphi)_r| = \left|\frac{\partial\varphi}{\partial r}\right| = \frac{|Q|}{4\pi r^2}.$$

It tends to zero for $r \to \infty$ and becomes infinite for $r \to 0$. Thus, the origin and the point at infinity are critical points. For $Q > 0$, the liquid flows out of the origin of coordinates in all directions, and the flow is said to be due to a point source. For $Q < 0$, the liquid flows into the origin, i.e., one has a sink. In the first case, the point at infinity is a sink, in the second case it is a source.

in unit time through a unit area at right angles to this direction. The amount of heat passing through an arbitrary orientated area element $d\sigma$ in time dt will obviously be equal to $q \cdot n\, d\sigma\, dt$ (where $\boldsymbol{n}$ is normal to $d\sigma$), and the total heat flow $dQ^{(e)}$ to the volume V may be presented in the form (Fig. 36)

$$dQ^{(e)} = -\int_\Sigma \boldsymbol{q} \cdot \boldsymbol{n}\, d\sigma\, dt\,,$$

with $\boldsymbol{n}$ the outward normal to Σ. By the theorem of Gauss–Ostrogradskii,

$$dQ^{(e)} = -\int_V \operatorname{div} \boldsymbol{q}\, d\tau\, dt\,.$$

The amount of heat flowing into an infinitesimal volume $d\tau$ in time dt is equal to

$$dQ^{(e)} = -\operatorname{div} \boldsymbol{q}\, d\tau\, dt\,,$$

and into unit mass of a medium

$$dq = -\frac{1}{\rho} \operatorname{div} \boldsymbol{q}\, dt\,. \tag{7.31}$$

Fourier's law of heat conduction. The laws for the definition of $\boldsymbol{q}$ may differ.

A law, which is basic, most popular and in many practical cases well founded is Fourier's law of heat conduction which has the form

$$\boldsymbol{q} = -\varkappa \operatorname{grad} T\,.$$

The heat flow vector and temperature gradient naturally have opposite directions, whence $\varkappa > 0$. The coefficient $\varkappa$ is referred to as the conductivity. Consider the particular important cases when $\varkappa$ is constant or a function of the temperature T.

Expression for heat flow due to heat conduction obeying Fourier's law. In the practically important simple case when $\varkappa = 0$, the heat flux dq per unit mass of a medium is given by

$$\frac{dq}{dt} = \frac{\varkappa}{\rho} \operatorname{div} \operatorname{grad} T = \frac{\varkappa}{\rho} \nabla_i g^{ij} \nabla_j T = \frac{\varkappa}{\rho} \nabla_i \nabla^i T\,,$$

or

$$\frac{dq}{dt} = \frac{\varkappa}{\rho} \nabla^2 T = \frac{\varkappa}{\rho} \Delta T ,$$

where ΔT is the Laplace operator applied to the temperature.

In Cartesian coordinates, this equation assumes the form

$$\frac{dq}{dt} = \frac{\varkappa}{\rho} \left(\frac{\partial^2 T}{\partial x^2} + \frac{\partial^2 T}{\partial y^2} + \frac{\partial^2 T}{\partial z^2} \right).$$

Equation of heat flow for a viscous heat conducting fluid. Thus, the equation of heat flow for a viscous fluid, in the case when the heat flux into it is due to heat conduction in accordance with Fourier's law, will have the form

$$\frac{dU}{dt} = -p \frac{d\, 1/\rho}{dt} + \frac{1}{\rho} \tau^{ij} e_{ij} + \frac{\varkappa}{\rho} \Delta T \tag{7.32}$$

or, by (7.28),

$$T \frac{ds}{dt} = \frac{1}{\rho} \tau^{ij} e_{ij} + \frac{\varkappa}{\rho} \Delta T .$$

For a fluid obeying the Navier-Stokes' law this equation, by (7.30), may be rewritten

$$T \frac{ds}{dt} = \frac{\zeta}{\rho} (\operatorname{div} \mathbf{v})^2 + \frac{2\mu}{\rho} [e^{ij} e_{ij} - \tfrac{1}{3} (\operatorname{div} \mathbf{v})^2] + \frac{\varkappa}{\rho} \Delta T . \tag{7.33}$$

It may serve for the determination of the temperature distribution in a fluid. The internal energy U or the entropy s must be known for this purpose as functions of the temperature and density. For example, if

$$U = \int c_V \, dT + \text{const.},$$

then in Cartesian coordinates

$$\frac{dU}{dt} = c_V \frac{dT}{dt} = c_V \left(\frac{\partial T}{\partial t} + u \frac{\partial T}{\partial x} + v \frac{\partial T}{\partial y} + w \frac{\partial T}{\partial z} \right).$$

The heat flow equation (7.17) in the case of a fluid at rest coincides with the normal heat conduction equation always be implied.

Thus, near any given point, there exists a relationship between the coordinate increments $d\zeta^i$ and $d\eta^i$. The derivatives $\partial\zeta^i/\partial\eta^j$ are functions of a point; however, in (4.2), at a given point, they are constants and one has there linear relations between the coordinate increments $d\zeta^i$ and $d\eta^i$. Let

$$\frac{\partial\zeta^i}{\partial\eta^j} = a^{i\cdot}_{\cdot j} \,.$$

(In the sequel, it will be shown that the arrangement of the upper and lower indices and their order of writing is very important). The quantities $a^{i\cdot}_{\cdot j}$ form the matrix

$$\|a^{i\cdot}_{\cdot j}\| = \mathbf{A},$$

where the first index i corresponds to its rows and the second j to its columns. It follows from the mutual single-valuedness that the Jacobian of the transformation, equal to the determinant of the matrix $\|a^{i\cdot}_{\cdot j}\|$, does not vanish, i.e., that $\Delta = |a^{i\cdot}_{\cdot j}| \neq 0$. Since $\Delta \neq 0$, the linear relations (4.2) can be solved for the $d\eta^i$ and one can write down together with (4.2) the formulae

$$d\eta^i = \frac{\partial\eta^i}{d\zeta^j}\, d\zeta^j \,. \tag{4.3}$$

Introduce the matrix

$$\mathbf{B} = \|b^{i\cdot}_{\cdot j}\| \,,$$

where

$$b^{i\cdot}_{\cdot j} = \frac{\partial\eta^i}{\partial\zeta^j} \,.$$

The matrices **A** and **B**, introduced for the transformation and its inverse, are mutually inverse, i.e., their product is equal to the unit matrix. In fact,

$$\mathbf{A}\cdot\mathbf{B} = \|a^{i\cdot}_{\cdot j}\| \cdot \|b^{j\cdot}_{\cdot k}\| = \|a^{i\cdot}_{\cdot j} b^{j\cdot}_{\cdot k}\| \;;$$

however,

$$a^{i\cdot}_{\cdot j} b^{j\cdot}_{\cdot k} = \frac{\partial\zeta^i}{\partial\eta^j}\frac{\partial\eta^j}{\partial\zeta^k} = \frac{\partial\zeta^i}{\partial\zeta^k} = \begin{cases} 1 \text{ for } i=k \,, \\ 0 \text{ for } i\neq k \,, \end{cases}$$

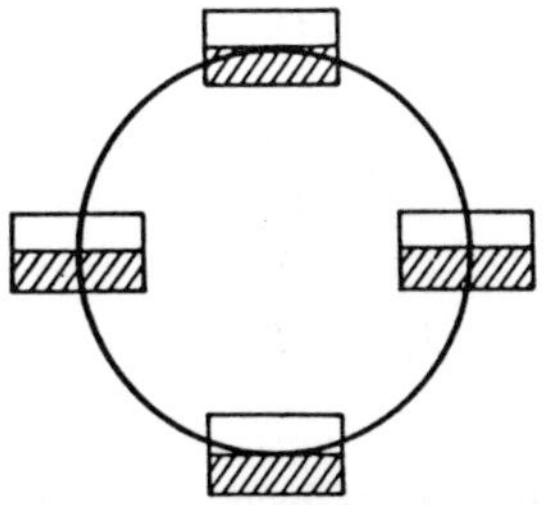

Fig. 6. Translation of a rigid body along a circle.

Do there exist unsteady motions for which the stream lines coincide with the trajectories? For instance, consider rectilinear translation of a rigid body with variable velocity. In this case, stream lines and trajectories are straight lines, and the motion itself is, of course, unsteady. Analogously, stream lines and trajectories will coincide in the case of the revolution of a rigid body with variable angular velocity about a fixed axis. In general, the stream lines and trajectories will coincide with each other during those unsteady motions in which the magnitude of the velocity varies at a given space point in time, but not its direction.

Consequently, stream lines and trajectories coincide for fields $\boldsymbol{v}(x^1, x^2, x^3)$ and $\boldsymbol{v}_1 = f(x^1, x^2, x^3, t)\boldsymbol{v}(x^1, x^2, x^3)$, where $f(x^1, x^2, x^3, t)$ is a scalar function of its arguments.

The existence of stream lines. The differential equations (3.2) of the stream lines may be rewritten in the form

$$\frac{dx^2}{dx^1} = \frac{v^2}{v^1}, \quad \frac{dx^3}{dx^1} = \frac{v^3}{v^1}, \tag{3.4}$$

and the Cauchy problem for this system posed, i.e., the problem of finding, with t being a constant parameter those solutions $x^2(x^1)$, $x^3(x^1)$ which for given $x^1 = x^1_0$ become given functions x^2_0 and x^3_0. In other words, one should draw the integral curve of the system of Equations (3.4) (a stream line) through a given point x^1_0, x^2_0, x^3_0. As is well-known from the general theory of ordinary differential equations, Cauchy problem has always a unique solution in every case when the right-hand sides in (3.4) and their derivatives, with respect to x^i, are continuous and single-valued. (This theorem is valid under some more general assumptions.) Thus, under very general limitations, through each point of a given volume, occupied by a continuous medium in motion, one may draw a unique stream line.

of thermodynamics, the scalar product of the vectors $\boldsymbol{q}$ and grad T is always negative.

By Fourier's law, $\boldsymbol{q} = -\varkappa$ grad T, $\varkappa > 0$. If this law applies, Equation (8.5) assumes the form

$$\frac{dS}{dt} = \int_V \frac{\varkappa}{T^2} |\mathrm{grad}\, T|^2 d\tau . \tag{8.6}$$

Thus, in spite of the absence of heat flow into a body as a whole from outside, one obtains under the conditions $T ds = dq$ and $dq' = 0$ that the entropy of the body as a whole grows. The above reasoning represents a generalization of the example of §5 of this chapter. It is clear from the above arguments that the condition $dq' = 0$ is not a sufficient condition for the reversibility of a process.

Criteria for irreversibility. A sufficient criterion for the reversibility of processes may be obtained as follows. Let

$$dS = d_e S + d_i S , \tag{8.7}$$

where dS is the differential of the entropy, $d_e S$ and $d_i S$ are infinitesimal terms, the first of which determines the increase of the entropy due to the influx of entropy from outside taking place as a consequence of energy exchange with external bodies. The quantity $d_i S$ is, by definition, essentially positive in the presence of irreversibility and causes growth of the entropy as a consequence of internal irreversible processes. For reversible processes, $d_e S$ may have any sign and $d_i S = 0$. In the preceding example, one has for a body as a whole

$$\frac{dS}{dt} = \frac{d_e S}{dt} + \frac{d_i S}{dt} = -\int_\Sigma \frac{q_n}{T} d\sigma - \int_V \frac{\boldsymbol{q} \cdot \mathrm{grad}\, T}{T^2} d\tau .$$

On the basis of the definitions above, one must set

$$\frac{d_e S}{dt} = -\int_\Sigma \frac{q_n}{T} d\sigma = -\int_V \mathrm{div} \frac{\boldsymbol{q}}{T} d\tau ,$$

$$\frac{d_i S}{dt} = -\int_V \frac{\boldsymbol{q} \cdot \mathrm{grad}\, T}{T^2} d\tau ,$$

whence one obtains for the density of entropy

$$\mathrm{d}_e s = -\frac{1}{\rho}\operatorname{div}\frac{\boldsymbol{q}}{T}\,\mathrm{d}t\,, \qquad \mathrm{d}_i s = -\frac{\boldsymbol{q}\cdot\operatorname{grad} T}{\rho T^2}\,\mathrm{d}t\,. \tag{8.8}$$

It follows from (8.8) that in the general case

$$T\,\mathrm{d}_i s \neq \mathrm{d}q'\,.$$

In certain cases, in the absence of a temperature gradient, the relation $T\,\mathrm{d}_i s = \mathrm{d}q'$ is valid.

Formulae for the generation of entropy. As a rule, the quantities $\mathrm{d}_e s$ and $\mathrm{d}_i s$ are determined by formulae of the form

$$\frac{\mathrm{d}_e s}{\mathrm{d}t} = -\frac{1}{\rho}\operatorname{div}\boldsymbol{S}\,, \qquad \rho\frac{\mathrm{d}_i s}{\mathrm{d}t} = \sigma = \sum_\alpha X_\alpha \chi^\alpha \geqslant 0\,, \tag{8.9}$$

where $\boldsymbol{S}$ is the flow vector of the entropy, σ determines the amount of irreversible growth of entropy due to internal processes, χ^α are referred to as generalized flows and X_α as generalized thermodynamic "forces".

In the case of heat conduction,

$$\boldsymbol{S} = \frac{\boldsymbol{q}}{T}, \qquad \chi^\alpha = q^\alpha, \quad X_\alpha = -\frac{1}{\rho T^2}\frac{\partial T}{\partial x^\alpha}\,.$$

In existing theories of irreversible processes, the basic problem is the establishment of formulae of the type (8.9). In many cases, it is assumed that there exist links between the flows χ^α and the "forces" X_α. Different principles have been established for the determination of these links. In the case of motion of a viscous heat conducting fluid, the second formula (8.9) for $\mathrm{d}_i s/\mathrm{d}t$ has the form

$$\frac{\mathrm{d}_i s}{\mathrm{d}t} = \frac{1}{\rho}\frac{\tau^{ij} e_{ij}}{T} - \frac{\boldsymbol{q}\cdot\operatorname{grad} T}{\rho T^2} = \frac{1}{T}\frac{\mathrm{d}q'}{\mathrm{d}t} - \frac{\boldsymbol{q}\cdot\operatorname{grad} T}{\rho T^2}\,.$$

In this case, the functional dependence of τ^{ij} and $\boldsymbol{q}$ on e_{ij} and grad T, respectively, determines the properties of viscosity and heat conductivity of a medium.

Navier-Stokes law (7.6) and Fourier's law present particular examples of the links between generalized flows and thermodynamic "forces".

$\varphi=$const., and it is larger, wherever the equipotential surfaces are more closely spaced. The projection of the velocity $\boldsymbol{v}$ on to an arbitrary direction $\boldsymbol{s}$ is the derivative of the potential φ in this direction: $v_s=\partial\varphi/\partial s$.

Necessary and sufficient conditions for the existence of a potential. Consider the expression

$$u\mathrm{d}x+v\mathrm{d}y+w\mathrm{d}z\,. \tag{3.6}$$

If the flow is potential, then (3.6) is the total differential of the function φ (with respect to the coordinates x, y, z). In fact,

$$u\mathrm{d}x+v\mathrm{d}y+w\mathrm{d}z=\frac{\partial\varphi}{\partial x}\mathrm{d}x+\frac{\partial\varphi}{\partial y}\mathrm{d}y+\frac{\partial\varphi}{\partial z}\mathrm{d}z=\mathrm{d}\varphi\,.$$

The converse is also true: If the expression (2.6) is a total differential, then the flow is potential. It is well-known that it is necessary and sufficient, for (3.6) to be a total differential, that

$$\frac{\partial u}{\partial y}=\frac{\partial v}{\partial x},\quad \frac{\partial u}{\partial z}=\frac{\partial w}{\partial x},\quad \frac{\partial v}{\partial z}=\frac{\partial w}{\partial y},$$

which are the necessary and sufficient conditions for a flow to be potential.

It will be shown later that potential flow plays an important role; certain examples of potential flow will now be studied.

Translatory flow. Translatory flow with constant velocity u_0 along the x-axis is an example of potential flow. In this case, $u=u_0$, $v=w=0$; since

$$\frac{\partial\varphi}{\partial x}=u_0\,,\quad \frac{\partial\varphi}{\partial y}=\frac{\partial\varphi}{\partial z}=0\,,$$

the potential is given by $\varphi=u_0x+$const. This example easily leads to two obvious conclusions. Firstly, a potential is always determined exactly, apart from an additive constant, and, secondly, any translatory flow is always potential. In fact, in the general case of translatory flow: $u=u_0$, $v=v_0$, $w=w_0$ and $\varphi=u_0x+v_0y+w_0z+C$, where u_0, v_0, w_0 and C may depend on t.

Note that it is simpler to study flows with potentials than those without potentials, since the first are determined by single functions $\varphi(x, y, z, t)$

instead of the first of the equalities (8.11) when $\chi^\alpha(\partial\sigma/\partial\chi^\alpha) \neq 0$. In addition, from (8.10) and (8.12) one obtains

$$\Omega_\alpha \chi^\alpha = 0 . \tag{8.15}$$

One may consider the quantity Ω_α as a part of the ⟨⟨force⟩⟩ X_α, which does not cause dissipation. The equality (8.15) is satisfied identically, if

$$\Omega_\alpha = k_{\alpha\beta}\chi^\beta ,$$

where $\| k_{\alpha\beta} \|$ — the arbitrary antisymmetric matrix ($k_{\alpha\beta} = - k_{\beta\alpha}$).
In the presence of links between χ^β one can introduce to the definition of model, the non-vanishing quantities Ω_α, which satisfy (8.15) only for the equations of links $\varphi_\varkappa(\chi^\beta) = 0$, $\varkappa = 1, 2, \ldots, s \geqslant 1$.

If the dissipation function is homogeneous in its arguments, then, on the basis of Euler's theorem on homogeneous functions, λ and m are constants. If the dissipation function is a quadratic form of its arguments, then

$$\lambda = m = \tfrac{1}{2} .$$

In this case, Relation (8.11) determines a linear link between the "forces" and flows. From the Formula (8.11), when σ is a quadratic form of χ^α it follows that the matrix of the constant coefficients in this linear links is symmetric.

On the theory of Onsager. A theory applicable to a number of models, based on the assumptions that the dissipation function is a quadratic form of its arguments and that the relations (8.11) are true, constitutes the substance of Onsager's theory. Fourier's and Navier-Stokes' laws are particular examples of Onsager's theory.

In this case, one obtains for the dissipation function on the basis of the Fourier and Navier-Stokes laws the formula

$$\sigma = \varkappa \frac{|\mathrm{grad}\ T|^2}{T^2} + (\zeta(e^i_i)^2 + 2\mu[e^{ij}e_{ij} - \tfrac{1}{3}(e^i_i)^2])\frac{1}{T} . \tag{8.16}$$

It is clear from the statements above that the function σ must be positive, because $\varkappa > 0$, $\zeta > 0$, $\mu > 0$.

It will be shown later on in this course that in the theory of ideal plasticity Equation (8.9) may be studied with the function σ homogeneous of first degree in χ^α; then $\lambda = 1$ and the formula $X^\alpha = (\partial\sigma/\partial\chi^\alpha)$ may be valid, but quantities X_α are linked by the relations of the type $f_\omega(X_\beta) = 0$, $\omega \geqslant 1$. The corresponding formulae for $\chi^\alpha(X_\beta)$ are changing their form by (8.11).

§9. The introduction to the theory of models of gas and fluid mixtures with chemical interaction and the diffusion of the components

On formulation of a problem of motion of mixture as a whole. In many important phenomena it is necessary to investigate the processes; in which there occurs the motion of the mixture or various gases and fluids, or gases and fluids with solid particles, or fluids only; accompanied by chemical, phase or some other transitions and the phenomena of diffusion, representing itself the internal relative motion of substances constituting the mixture.

In some cases, the corresponding processes can be regarded as reversible, but in the general case it is necessary to take into account the irreversibility and the multi-valued internal force and energetic interactions. For the description of the corresponding phenomena one can construct the various models conformable to some separate classes of mixtures and motions with the corresponding physico-chemical transitions.

Next, for the construction of particular models, we will follow the arrangement of the problem of the multicomponent continuous media[1)], given in §1, Chapter III. In connection with it, for the infinitesimal volume ΔV one takes into consideration the corresponding masses of components $m_i = \rho_i$ and the diffusion flux vectors $I_i = \rho_i(v_i - v)$ for n different components, describing the mixture and determined physically and chemically different, interacting between themselves substances, which occupy simultaneously one and the same volume ΔV.

On the characteristics of macroscopic particles of a mixture. In the general case, for description of interaction between the components of the

[1)] In some cases the separate consideration of the following theory will be added, when the components of a mixture occupy the adjacent volumes.

mixture it is necessary to define and to select the parameters of the components or of the mixture as a whole, which characterize mechanical, physical and chemical properties, essential for description and formulation of the fundamental natural laws, in the frame of the point of view, conditioned by the foundation of the considered problems. For example, as such defining parameters one can introduce the entropy s, taken with respect to the unit of mass, the density of the mixture as a whole and the density of the separate components $\rho_1, \rho_2, \ldots, \rho_N$. In the absence of the thermodynamical equilbrium in the particles of the mixture between the different components and in the presence of the equilibrium inside of each component, one can use the absolute temperatures $T_1, T_2, \ldots, T_N$ of the separate components as the characteristics of the state.[1)] Here one can add the different kinds of introduced and studied earlier geometrical and kinematical characteristics of properties of the deformation of the particle and for the diffusion of components the vectors of the flux of diffusion I_i, for example. In a number of examples, such characteristics of atoms and molecules are essential as the magnitude of their charges, in the case when ionization is taken into account, their electromagnetic characteristics, or the energy of the excitation states of the molecules and atoms etc. In the general case, for formulation of the laws of interaction, transformation of energy and the kinetic equations, one must take into account not the magnitude of the mentioned parameters only, but also their gradients, i.e., their derivatives of various order taken with respect to the coordinates and the time.

The construction of the model of the mixture is applicable also to the investigations of the phenomena of the motion of plasma, representing itself the mixture of ions, electrons and neutral particles. For plasma the macroscopic electromagnetic interaction between the components inside the particle, and for the particles of plasma *as a mixture as a whole* with the external electromagnetic fields, are very important.

The macroscopic mixture can be considered as one continuum with complicated properties, characterized by the system of internal parameters, to which one can refer the magnitude of densities $\rho_1, \rho_2, \ldots, \rho_N$ of the

[1)] The presence of the equilibrium temperature of each component inside the particle, in the absence of the equilibrium with respect to temperature between different components occupying one and the same volume inside the particle, is the arbitrary strong assumption in the considered phenomena, the validity of which is necessary to be substantiated by the complementary physical considerations.

components of the mixture. For the determination of the model of the mixture it is necessary, besides the above stated universal mechanical and thermodynamic equations, to establish the complementary equations as a formulation of laws for determination of the internal characteristics of the state and, in particular, for the determination of the density ρ_i of the components.

Equations of the balance of masses for the components of the mixtures. In the general case, besides the equations of mechanics for the velocity v and density ρ, one can write also the equation of the balance of mass, stated in §1, Chapter III, containing the densities ρ_i and the velocities v of the mixture as a whole. These equations[1] have the form:

$$\frac{1}{\Delta V}\frac{dm_i}{dt} = \frac{d\rho_i}{dt} + \rho_i \operatorname{div} v = \varkappa_i - \operatorname{div} \boldsymbol{I}_i . \tag{9.1}$$

The magnitudes $\varkappa_i\,\Delta V$ and div $I_i\,\Delta V$ describe, in the moving individual volume ΔV of the mixture as a whole, the production of the mass m_i of the component with the number i per unit of time, due to the physico-chemical reactions and diffusion. In order to obtain from (9.1) the effective equations for ρ_i and c_i, it is necessary to base on the preliminary obtained physico-chemical laws, determining $\varkappa_i$ and I_i. Similarly as for the foundation of laws determining the forces in dependence on the type and conditions of interaction, one uses the second law of thermodynamics. For determination of laws for $\varkappa_i$ and I_i, corresponding to the reality, it is necessary to base on the hypothesis, which then should be verified in the experiments, taking into account the equation (9.1).

Equation of the balance of masses for the physico-chemical reactions. For example, in such a way in chemistry, without taking motion into account, it is established and then extrapolated on the case of presence of

[1] In practice, it is often more convenient to consider the mass concentrations $c_i = \rho_i/\rho$ $(i = 1, 2, \ldots, N)$ instead of the density components ρ_i. Obviously, the following formula are valid for c_i:

$$\frac{1}{\Delta V}\frac{dm_i}{dt} = \rho\,\frac{dc_i}{dt} = \varkappa_i - \operatorname{div} \boldsymbol{I}_i \quad (i = 1, 2, \ldots, N) . \tag{9.1'}$$

motion, the fundamental equations of the balance of mass for the phase transition and chemical reaction, these equations have the form

$$\varkappa_i = \frac{1}{\Delta V}\frac{dm_i^1}{dt} = M_i \sum_{\alpha=1}^{r} \nu_{i\alpha}\,\omega_\alpha \quad (i = 1, 2, \ldots, N). \tag{9.2}$$

Here r $(\alpha = 1, 2, \ldots, r)$ is the number of the simultaneously independent[1)] reactions or phase transitions. The quantity dm_i^1/dt gives the change of the mass m_i of the i-component of the matter in the small volume ΔV, due to the physico-chemical interaction; M_i are the molar masses and $\nu_{i\alpha}$ are the stochiometric coefficient, which are negative for the components which enter into the reaction and positive for the products of the reaction when the reaction goes into the positively assigned direction i.e., when $\omega_\alpha > 0$; if a given component with number i does not participate in the reaction with number α, then $\nu_{i\alpha} = 0$; the products $M_i\,\nu_{i\alpha}$ determine themselves the mass ratios of the corresponding components, which participate in the reaction with number α; the quantity ω_α characterizes the speed of the reaction with number α in the given mixture. Obviously, the production of mass of the components of the mixture in the considered reaction per unit of time and unit of volume, is proportional to the quantity ω_α. If the reaction with number α is taking place, then $\omega_\alpha \neq 0$; otherwise $\omega_\alpha = 0$. The law of conservation of mass for the mixture as a whole and for each component separately yields the equations

$$\sum_{i=1}^{N} M_i\,\nu_{i\alpha} = 0, \quad \alpha = 1, 2, \ldots, r \quad \text{and} \quad \sum_{i=1}^{N} \varkappa_i = 0\,. \tag{9.3}$$

The formulas (9.2) give the relations for the quantities $\varkappa_i$ $(i = 1, 2, \ldots, N)$ expressed by the speed of reactions ω_α $(\alpha = 1, 2, \ldots, r)$. In (9.2) we have the quantities $\varkappa_i$ which must be determined from additional physico-chemical laws for the r functions — the velocities ω_α of the conversions. For $r < N$ the number of complementary relations for $\varkappa_i$, in accordance with (9.2), reduces to $N - r$.

1) Each of the independent reactions corresponds to the different formula of reaction, linking the various components and can take place when all other reactions are absent.
The question of the possible number of the independent reactions during the phase transition in equilibrium (the rule of Gibbs' gas) is considered in the physical chemistry.

If in the N-component mixture ($N \geqslant 2$) takes place only one phase transition between the components, enumerated by the indices 1 and 2, and the molecular weights of phases are equal, then

$$r = 1, \quad i = 1, 2, \ldots, N, \quad M_1 = M_2 = M\,,$$

$$\nu_{11} = -1, \quad \nu_{21} = 1, \quad \nu_{31} = \ldots = \nu_{N1} = 0\,,$$

$$\varkappa_1 = -M\omega_1, \quad \varkappa_2 = M\omega_1, \quad \varkappa_3 = \ldots = \varkappa_N = 0\,.$$

Let us consider the example of the mixture, made of the hydrogen H_2 with the molecular weight $M_1 = 2$, the oxygen O_2 with the molecular weight $M_2 = 32$, streams of the water H_2O with the molecular weight $M_3 = 18$ and the same inert gas. Assume, that only one chemical reaction takes place consisted in formation of streams of water in accordance with the chemical equation

$$2H_2 + O_2 = 2H_2O\,.$$

In this case, the following numerical values and relations are obvious

$$r = 1, \quad N = 4, \quad \nu_{11} = -2, \quad \nu_{21} = -1, \quad \nu_{31} = 2, \quad \nu_{41} = 0,$$

$$\varkappa_1 = -4\omega_1, \quad \varkappa_2 = -32\omega_1, \quad \varkappa_3 = 36\omega_1, \quad \varkappa_1 = 0.$$

For irreversible processes it is usual to refer to the relations expressing dependence of the velocities ω_α on the parameters defining the state and the composition of the mixture as the equations of the kinetics of the chemical reactions and of the physical processes, as long as we refer only to the redistribution of the component unrelated to any change in the basic nature of the particles forming the components. If on the base of the complementary laws or specified with maintenance of the balance of mass, quantities ρ_i and v are determined in terms of other parameters, then in the absence of diffusion, Equations (9.2) and (9.1) can be used to determine the quantities ω_α for $r < N$. It will be shown below, how physico-chemical processes take place in the reversible way.

The problem of laws, determining the diffusion flux vectors I_i are in the centre of the thermodynamics of irreversible processes; these laws are considered below, in relation to the formula for the production of entropy, stipulated by the internal dissipation of energy.

The internal energy of the mixture as a whole. The introduction of the specific models of mixture is essentially linked with the choice of the system of determining parameters (arguments) and with the definition of internal energy for the mixture as a whole, as a fixed function of these parameters (arguments). The two-parameter ideal compressible fluids or gases, for which the internal energy of the small particle is determined as some function of the specific entropy s, density ρ and is proportional to the mass m of particle, was considered in §6. Considering the mixture as a whole one can take as the fundamental assumption that the internal energy U_m for *a small particle of mixture* is represented in the form

$$U_m = U_m(s, \rho, m_1, m_2, \ldots, m_N, \chi^1, \chi^2, \ldots, \chi^l), \tag{9.4}$$

where $\chi^1, \chi^2, \ldots, \chi^l$ — the same parameters, characterizing the internal processes and the state of the mixture. For example, the different temperatures of components, the characteristics of the internal relative macroscopic motions of components and, in particular the diffusion flux vectors or the characteristics of interactions, stipulated by the arrangement and structure of components, the components of the deformation tensors of the different components of the mixture, the parameters characterizing the electromagnetic properties of components, and other quantities. Due to the changes of parameters χ^l one can consider the change of internal energy, connected with the work done by the internal macroscopic forces of interaction between components, during the relative motion of component on the account of diffusion.

The simple generalization of the two-parameter models of ideal continua, considered in §6, to the case of mixture, one can obtain by narrowing the form of function (9.4) and using the following fundamental assumption:

$$U_m = U_m(s, \rho, m_1, m_2, \ldots, m_N). \tag{9.5}$$

The absence of other parameters of type χ^s in formula (9.5) leads to the number of simplifications, but together with it some effects are excluded from considerations, the effects which can appear in real phenomena in the mixtures with complicated properties, for example in a plasma.

On the basis of the formula (9.5) one can write

$$dU_m = \frac{\partial U_m}{\partial s} ds + \frac{\partial U_m}{\partial (1/\rho)} d\frac{1}{\rho} + \sum_{k=1}^{N} \frac{\partial U_m}{\partial m_k} dm_k . \tag{9.6}$$

In formulae (9.5) and (9.6) the arguments and the increments ds, $d(1/\rho)$ and N increments dm_k can be independent and in some limits arbitrary.

The change of the internal energy of individual particle during its motion or the changes of U_m due to the transition from one particle to the other and, generally speaking, during many other real processes may correspond to the increment dU_m calculated on the base of formula (9.6). Together with formulae (9.5) and (9.6) one can consider the following functions

$$m\theta(s, \rho, m_1, m_2, \ldots, m_N) = \frac{\partial U_m}{\partial s},$$

$$mp'(s, \rho, m_1, m_2, \ldots, m_N) = -\frac{\partial U_m}{\partial (1/\rho)}, \tag{9.7}$$

$$\mu'_k(s, \rho, m_1, m_2, \ldots, m_N) = \left(\frac{\partial U_m}{\partial m_k}\right)_{s,\rho},$$

where $m = m_1 + m_2 + \ldots + m_N$.

In accordance to formula (6.2) the equalities $\theta = T$ and $p' = p$, where T denotes the absolute temperature and p pressure are valid for the reversible processes in the two-parameter medium when $dq^{**} = 0$. Obviously, the meaning of θ and p' will be the same in a given case for the reversible processes when $m_k = \text{const.}$ and $dq^{**} = 0$.

In the more general cases, with variable m_k for the irreversible processes and when dq^{**} differs from zero, the formulas (9.7) for $\theta = T$ and $p' = p$ can be considered as definitions of the temperature and pressure in these complicated cases. The determination of the function (9.5) corresponding to the experimental data can be reduced to the determination, on the base of experiments of the function $\theta(s, \rho, m_1, m_2, \ldots, m_N)$, $p'(s, \rho, m_1, m_2, \ldots, m_N)$ and $\mu'_k(s, \rho, m_1, m_2, \ldots, m_N)$, which should satisfy the integrability conditions.

Functions

$$\mu_k = \mu'_k - \theta s + p' \frac{1}{\rho}, \tag{9.8}$$

are referred to as the chemical potentials. On the basis of (9.6) and (9.7) it is easy to check the validity of the equality

$$\mu_k = \left(\frac{\partial U_m}{\partial m_k}\right)_{S_m, \Delta V}, \tag{9.8'}$$

where $S_m = ms$ and $\Delta V = m/\rho$.

The free energy and thermodynamic potential of the mixture. For a more detailed explanation of the physical meaning of the quantity μ_k, we will consider (together with the internal energy of a particle of a mixture U_e) the free energy F_m and the thermodynamic potential Ψ_m, defined by formulae

$$\begin{aligned} F_m(s, \rho, m_1, m_2, \ldots, m_N) &= U_m - m\theta s\,, \\ \Psi_m(s, \rho, m_1, m_2, \ldots, m_N) &= U_m - m\theta s + mp' \frac{1}{\rho}. \end{aligned} \tag{9.9}$$

On the basis of the equalities (9.7), one can include the arguments ρ and θ in the function Γ_m, the arguments θ and p' in the function Ψ_m, instead of variables s and ρ. On the basis of formula (9.9), it is easy to check the validity of the equalities

$$\begin{aligned} \mu_k &= \left(\frac{\partial \Psi_m}{\partial m_k}\right)_{p', \theta} = \left(\frac{\partial U_m}{\partial m_k}\right)_{s, 1/\rho} - \theta s + p' \frac{1}{\rho} \\ &= \left(\frac{\partial F_m}{\partial m_k}\right)_{\rho, \theta} + p' \frac{1}{\rho}, \end{aligned} \tag{9.10}$$

and equalities[1)]

$$\mu_k = \left(\frac{\partial \Psi_m}{\partial m_k}\right)_{p', \theta} = \left(\frac{\partial U_m}{\partial m_k}\right)_{S_m, \Delta V = \frac{m}{\rho}} = \left(\frac{\partial F_m}{\partial m_k}\right)_{\theta, \Delta V} . \tag{9.11}$$

In some case it is more convenient to replace the determination of the function $U_m(s, \rho, m_k)$ in experiments, by the determination of function $F_m(\theta, \rho, m_k)$ or $\Psi_m(\theta, p', m_k)$. The direct determination of the chemical potentials μ_k as functions of the arguments $\theta = T$, $p' = p$ and μ_k on the basis of experiments, is more convenient than the determination of the function $\mu_k'(s, \rho, m_k)$.

[1)] Note that the formulae (9.7), (9.9), (9.10) and (9.11) follow mathematically from the definition (9.5).

Homogenity of thermodynamic functions with respect to mass. If the non-additivity of the internal energy with respect to the mass and, in particular the surface energy of a particle or the energy which depends on the form of a particle, can be neglected, then it can be accepted as a basic experimental result that for the identical specific entropy s and density ρ, the internal energy will increase by a factor n where all the masses m_k, constituting the mixture, also increases by a factor n. In particular, this assumption is valid for all the bodies, in which it is possible to compute the internal energy of the all body equal to the sum of internal energies of its particles. The assumption gives the following bond on the function (9.5)

$$nU_m(s,\rho,m_1,m_2,\ldots,m_N) = U_m(s,\rho,nm_1,nm_2,\ldots,nm_N)\,, \tag{9.12}$$

i.e., the internal energy is the homogenous function of the mass m_i of first degree. The assumption (9.12) can be fulfilled also in the case when the internal energy is non-additive.

Supposing that $n^{-1} = m_1 + m_2 + \ldots + m_N = m$, we obtain by (9.12)

$$U_m(s,\rho,m_1,m_2,\ldots,m_N) = mU_m(s,\rho,c_1,c_2,\ldots,c_N)\,, \tag{9.13}$$

where $c_1 = (m_i/m) = (\rho_i/\rho)$ — the mass concentration of the components of a mixture.

From relations (9.12) and (9.9) follows, that the functions $F_m(\theta,\rho,m_k)$ and $\Psi_m(p',\theta,m_k)$ also fulfil the equalities of the form (9.12), i.e., they are also the homogenous functions of the first degree with respect to m_k. In this case on the basis of Eulers' theorem for the homogenous functions, and from equalities (9.10) it follows

$$\begin{aligned}
\Psi_m &= \sum_{k=1}^{N} m_k \left(\frac{\partial \Psi_m}{\partial m_k}\right)_{p',\theta} = \sum_{k=1}^{N} m_k \mu_k(\theta,p',c_k)\,,\\
U_m &= \sum_{k=1}^{N} m_k \left(\frac{\partial U_m}{\partial m_k}\right)_{s,\rho} = \sum_{k=1}^{N} m_k \left[\mu_k(s,\rho,c_k) + \theta s - \frac{p'}{\rho}\right],\\
F_m &= \sum_{k=1}^{N} m_k \left(\frac{\partial F_m}{\partial m_k}\right)_{\rho,\theta} = \sum_{k=1}^{N} m_k \left[\mu_k(\theta,\rho,c_k) - \frac{p'}{\rho}\right].
\end{aligned} \tag{9.14}$$

Note, that in the general case μ_k are not only the functions of s, ρ, or θ, ρ or θ and p', but they can depend on the ratios $(m_1/m) = c_1$, $(m_1/m) = c_2, \ldots, (m_N/m) = c_N$.

The mixture of perfect gases. As an example we consider the formulae for the thermodynamic functions for the mixture of perfect gases. The formulae (7.8) which give the expressions for the thermodynamic functions of perfect gas, can be considered as the formulae for the thermodynamic functions of the separate components, with masses m_k, for which the corresponding quantities ρ_k, T_k, p_k, $c_{Vk}(T_k)$, $c_{p_k}(T_k)$ and the values of all constants are determined. On the other hand, these formulae can be considered as determination of the characteristic thermodynamic functions of the mixture as a whole in an equilibrium state with the corresponding density ρ, temperature T and pressure p, if the mixture as a whole can be considered also as a perfect gas.

To obtain the thermodynamic functions and all constants for the mixture as a whole, in dependence on the mass, m_k and other characteristics of components, the following mental experiment can be conducted. If we consider two different, separated by the impenetrable partition, volumes with the masses m_1 and m_2 of one and the same perfect gas and each of these volumes is in equilibrium, then after removal of the partition, the equilibrium will not be destroyed assuming that the temperature and pressure of the gases are the same. Due to the equality of the pressure the motion will not appear and the equality of the temperature excludes the possibility of the heat exchange. On the basis of the equation of state, from the equality of pressure and temperature follows the equality of a density. From the formulae (7.9) which are valid for the first and second gas with multipliers m_1 and m_2, respectively, follow the same formulae, with the same constants c_V and c_p, for the system of two gases as a whole, not only with the multiplier $m_1 + m_2$. Obviously, the analogous situation is preserved also in the case of contact of the great number of masses m_1, $m_2, \ldots, m_N$ of one and the same gas.

If we consider the same experiment with volumes of two different gases with masses m_1 and m_2, then in spite of the equality of pressure and temperature before removal of the partition, the state of the system of two gases as a whole will not be in equilibrium after removal of the partition. Due to the diffusion after removal of the partition, the state of the mixed gases with the uniform distribution substances of each of them possesses, with respect to the total volume, greater probability than the initial state with non-uniform distribution of masses. In the absence of the external heat fluxes, the transition from the initial (just after removal of the partition) non-equilibrium state with the entropy $S_{m_1+m_2}$, equal to the sum of the entropy of the separated components, to the state with entropy $S_{m_1+m_2}$ of the final equilibrium (more probably) state, gives the increase of the entropy. Therefore, for the mixture as a whole must be

$$S_{m_1+m_2} > S_{m_1}(p,T) + S_{m_2}(p,T).$$

But this conclusion is essentially linked with the fact, that the different gases are mixed. If they are mixed with the identical gases, then the initial state coincides with final state and

$$S_{m_1+m_2} = S_{m_1}(p,T) + S_{m_2}(p,T).$$

For the mixture of N different, with respect to the same essential physico-chemical indication, ideal gases one can use the assumption, that the mixture is composed from the N different gases, occupying one and the same volume, and one can introduce the characteristics and the state equations for the equilibrium states of each gas separately and independently of the presence of other gases in a volume. In accordance with this assumption, the interaction energy of gases is neglected, therefore by (7.8) the following equality is satisfied

$$\begin{aligned} U_m &= U_{m_1}(T_1) + U_{m_2}(T_2) + \ldots + U_{m_N}(T_N) \\ &= m \sum_{k=1}^{N} \frac{m_k}{m} \left(U_{0k} + \int_{T_{0k}}^{T_k} c_{Vk}(T)\, dT \right). \end{aligned} \tag{9.15}$$

In addition, for the N ideal gases in the mixture, we introduce the initial pressures $p_1, p_2, \ldots, p_N$ and the pressure p of the mixture as a whole, determined by the state equations

$$p_k = \rho_k R_k T_k = \rho_k \frac{R_0}{M_k} T_k$$

(R_0 — the universal gas constant) and by Dalton's law

$$p = \sum_{k=1}^{N} p_k = \sum_{k=1}^{N} \rho_k R_k T_k = \rho \sum_{k=1}^{N} \frac{R_k m_k}{m} T_k. \tag{9.16}$$

If one assumes, that the probability of the state of each component of gas does not depend on the probability of the state of other components, then the entropy of a mixture represents itself as a sum of entropy of components for the corresponding states; therefore we introduce the entropy of the mixture as a whole in accordance with the equality

$$S_m(p, T, m_1, m_2, \dots, m_N) = \sum_{k=1}^{N} S_{m_k}(p_k, T_k, m_k),$$

where by (9.7)

$$S_{m_k} = m_k \left(s_{ok} + \int_{T_{ok}}^{T_k} \frac{c_{p_k}}{T}\, dT - R_k \ln \frac{p_k}{p_{ok}} \right). \tag{9.17}$$

In correspondence with the statement of the problem, connected with the basic formula (9.5) for the internal energy U_m, in which the possibility of presence of different temperatures is not taken into account, we assume that the temperatures of components in the infinitesimal particle are the same, i.e.,

$$T_1 = T_2 = \dots = T_N = T.$$

From the Dalton's law, it also follows that

$$\frac{mp}{\rho} = \sum_{k=1}^{N} \frac{m_k p_k}{\rho_k},$$

and therefore for the thermodynamic potential of the mixture as a whole the following formula is valid

$$\begin{aligned} \Psi_m &= U_m - TS_m + \frac{mp}{\rho} = \sum_{k=1}^{N} U_{m_k} - T \sum_{k=1}^{N} S_{m_k} \\ &+ \sum_{k=1}^{N} \frac{m_k p_k}{\rho_k} = \sum_{k=1}^{N} \Psi_{m_k}(p_k, T) = \sum_{k=1}^{N} m_k \mu_k . \end{aligned} \tag{9.18}$$

From formulas (9.15), (9.17) and (9.18) the following formulae are obtained for chemical potentials:

$$\begin{aligned} \mu_k(T, p_k) = \Bigg[& U_{ok} + R_k T_0 - s_{ok} T + \int_{T_0}^{T} c_{p_k}\, dT \\ & - T \int_{T_0}^{T} \frac{c_{p_k}\, dT}{T} + R_k T \ln \frac{p_k}{p_{ok}} \Bigg], \end{aligned} \tag{9.19}$$

where it was also taken into account that $c_{p_k} - c_{V_k} = R_k$.

Just as, taking (9.16) into account and when $T_1 = T_2 = \ldots = T_N$ we have

$$p_k = p\,\frac{R_k \rho_k}{R\rho} = p\,\frac{R_k m_k}{Rm} = p\,\frac{R_k m_k}{\sum_{i=1}^{N} R_i m_i},$$

then, from formula (9.17), it follows, that

$$S_m(p, T, m_1, m_2, \ldots, m_N) = \sum_{k=1}^{N} S_{m_k}(T, p) - \sum_{k=1}^{N} R_k m_k \ln \frac{R_k m_k}{\sum_{i=1}^{N} R_i m_i}. \tag{9.20}$$

Obviously, that $S_m > \sum_{k=1}^{N} S_{m_k}(p, T)$, just as

$$\Delta S_m = -\sum_{k=1}^{N} R_k m_k \ln \frac{R_k m_k}{\sum_{i=1}^{N} R_i m_i} = -\sum_{k=1}^{N} R_0 n_k \ln \frac{n_k}{\sum_{i=1}^{N} n_i} > 0, \tag{9.21}$$

where n_i is the number of moles in the ith component of a given particle with mass $m = \sum_{k=1}^{N} m_i$.

If we consider the initial separation of N gases with the masses m_k and with the same pressures p and temperatures T by the partitions, then immediately after removal of partition, the entropy will be equal to $\sum_{k=1}^{N} S'_{m_k}(T, p)$. After mixing, with constant volume and without supply of external heat, one obtains the equilibrium state of the mixture with the same values[1] of pressure and temperature and with the entropy $S_m(p, T, m_1, m_2, \ldots, m_N)$. The quantity ΔS_m (9.21) gives the increase of entropy, due to the irreversible process of diffusion. It is important to note, that the increase of entropy

[1] The consequence of the preservation of energy and of Dalton's law for the equilibrium.

(9.21) due to the diffusion does not depend on the nature of mixing gases, but depends only on the number of mixing gases.

Gibbs' paradox. The theory of determination of the thermodynamic functions for the mixtures of various ideal gases has been developed above. If we formally adopt this theory for the N parts of one and the same gas, then we will come to the contradiction. Indeed, it is directly seen, that in this case the entropy of the "mixture" should be equal

$$\sum_{k=1}^{N} S_{m_k}(T, p)\,,$$

when, the above developed theory gives for the entropy

$$S_m > \sum_{k=1}^{N} S_{m_k}(T, p)\,.$$

This contradiction is referred to as the Gibbs' paradox. Its explanation is based on the fact, that in the above developed theory of mixtures, the assumption of the physico-chemical distinction between the components is essential. This distinction can be small, but it is always physically determined as some finite distinction. As a consequence of this distinction, during unitization (immediately after removal of partitions) of the various volumes of the components of gas with the identical pressure and temperatures p and T, the common volume appears, in which the gas is in the essentially non-equilibrium state. Therefore, this state in the case of a constant volume and without the heat supply, turns into the equilibrium state with the same values p and T, with the aid of the diffusion process. From the statistical considerations it follows directly, that thermodynamic functions and, in particular, the entropies for one and the same number of fractions, and fractions indentical to each other are different.

In the gas with the identical molecules, occupying the volume ΔV, it is physically impossible to separate the different substantial components, occupying the same volume ΔV, therefore the previous method of calculation of the entropy for the mixture of the physically different components is impossible in the case of gas with identical particles.

Equations of the first and second law of thermodynamics. On the base of the second law of thermodynamics (*cf.*, §2)

$$T\,dS = dQ^{(e)} + dQ' , \tag{9.22}$$

and the heat flow equation for a mixture as a whole (*cf.*, §5)

$$dU_m = \frac{m}{\rho} p^{ij}\, \nabla_j v_i\, dt + dQ^{(e)} + dQ^{**} , \tag{9.23}$$

we have

$$dU_m = \frac{m}{\rho} p^{ij}\, \nabla_j v_i\, dt + T\,dS + dQ^{**} - dQ' . \tag{9.24}$$

In the general case, in order to fix a model, besides the function of internal energy (9.4) or (9.5), it is necessary to specify the quantities dQ' and dQ^{**}. For the reversible processes $dQ' = 0$, for the irreversible $dQ' \geqslant 0$. The quantity dQ^{**} represents itself the part of the flux from outside of other nonthermal forms of energy which differs from the work done by external forces on the displacements of the medium as a whole, and is conditioned by the external, with respect to the given particle, objects, in particular the neighbouring particles of a medium which are in contact with considered particle along its boundary Σ, and possibly, also the flux of energy due to interaction with the electromagnetic field (*cf.*, Chapter VI). Just as the interaction of the given particle with the neighbouring particles (on account of the mass forces and distribution of stresses p^{ij} on the surface Σ, and also on account of the external heat flux) do not participate in dQ^{**}, then in the absence of diffusion, external fields and generalized forces on Σ (for example the distribution of surface couples, *cf.*, Chapter III, §3) one can assume, that $d\theta^{**} = 0$. However, in the presence of diffusion, the flux of energy dQ^{**} may differ from zero, an account of the flux to the given particle the nonthermal energy of masses of different components, entering on account of diffusion through the boundary Σ of the substantial particle determined for the mixture as a whole. The quantity dQ^{**} may also differ from zero, if in formula (9.4) the complementary arguments χ^s are taken into account and, in particular the kinetic energy of the relative motion of components due to diffusion is also taken into consideration. Taking into account the diffusion, it is necessary to include in the fluxes of energy the elementary work done by external (to the given particle) macroscopic mass or surface forces on relative displacements $(v_i - v)\,dt$ of the particles of components, forming the mixture.

§ 10. The modelling of a mixture for reversible processes.

By the definition of the reversible processes, we have

$$dQ' = 0 \quad \text{and} \quad I_i = 0 \qquad (i = 1, 2, \dots, N). \tag{10.1}$$

We will consider here only such a model, in which the formulae of the form (9.5) are valid for the internal energy U_m of the mixture as a whole and, in addition, we assume also, that the following equality takes place

$$dQ^{**} = 0. \tag{10.2}$$

The state equations of a mixture for reversible processes. In this case the heat flux equation for the mixture as a whole (9.24), taking into account the second law of thermodynamics, may be rewritten in the form

$$dU_m = \frac{m}{\rho} p \operatorname{div} v \, dt + \frac{m}{\rho} (p^{ij} - pg^{ij}) \nabla_j v_i \, dt + mT \, ds ,$$

or

$$\begin{aligned} &\frac{\partial U_m}{\partial (m/\rho)} \, d\frac{m}{\rho} + \frac{\partial U_m}{\partial S_m} \, dS_m + \sum_{k=1}^{N} \left(\frac{\partial U_m}{\partial m_k} \right)_{\frac{m}{\rho}, S_m} dm_k \\ &\quad = - p d \frac{m}{\rho} + T \, dS_m + \frac{m}{\rho} \tau^{ij} \nabla_j v_i \, dt . \end{aligned} \tag{10.3}$$

Hence, if by definition

$$p = -\left(\frac{\partial U_m}{\partial (m/\rho)} \right)_{S_m, m_k} = -\left(\frac{\partial U}{\partial (1/\rho)} \right)_{s, m_k} , \tag{10.4}$$

and if one assumes that the components of the tensor $\tau^{ij} = p^{ij} - pg^{ij}$ do not depend on the velocity gradient $\nabla_j v_i$, then in view of the fact that dS_m and dm_k are independent, the following relations are obtained

$$T = \left(\frac{\partial U_m}{\partial S_m} \right)_{\frac{m}{\rho}, m_k} = \left(\frac{\partial U}{\partial s} \right)_{\frac{1}{\rho}, m_k} \quad \text{and} \quad \tau^{ij} = 0 , \tag{10.5}$$

and taking into account the formulas (9.8) for the chemical potentials and the definition (9.2) for $\varkappa_k$, the relations

$$\sum_{k=1}^{N} \left(\frac{\partial U_m}{\partial m_k}\right)_{\frac{m}{\rho},\, S_m} \mathrm{d}m_k = \sum_{k=1}^{N} \mu_k \,\mathrm{d}m_k = \sum_{k=1}^{N} \mu_k \varkappa_k \,\mathrm{d}t\, \Delta V = 0\,, \tag{10.6}$$

are obtained.

The equalities (10.4) and (10.5) represent themselves the equation of state. By specification of the functions (9.5) for the internal energy, in accordance with (10.4) and (10.5), the pressure $p(1/\rho, s, m_1/m, m_2/m, \ldots, m_N/m)$, the temperature of a mixture $T(1/\rho, s, m_1/m, m_2/m, \ldots, m_N/m)$ are defined and, in addition, it is obtained that $\tau^{ij} = 0$. In particular, from the above assumptions follows that the constructed model of the medium — is the ideal fluid or ideal gas.

The conditions of the chemical equilibrium. In the absence of the chemical and phase transitions, we have $\mathrm{d}_{m_k} = \varkappa_k \,\mathrm{d}l\, \Delta V = 0$, and therefore the equalities (10.6) are satisfied identically. In the presence of the chemical or phase transitions and if there is no diffusion, the quantities may be considered in the finite periods of time as characteristics of fixed particle of the medium and therefore they are functions of Lagrangian coordinates (identical for all components, just as $v_i = v$) and time t. In this case the equations of the balance of mass for the physico-chemical processes (9.2) for each particle can be integrated with respect to time t and yield

$$m_i - m_{0i} = M_i \sum_{\alpha=1}^{r} \nu_{i\alpha}\, \omega_\alpha$$

or, taking into account the equality

$$\left.\begin{aligned} m &= m_0 = \text{const.}\,, \\ \rho_i &= \frac{\rho_{0i}}{\rho_0}\rho + M_i \sum_{\alpha=1}^{r} \nu_{i\alpha} \frac{\omega_\alpha}{\Delta V}\,, \end{aligned}\right\} \quad (i = 1, 2, \ldots, N)\,, \tag{10.7}$$

where, taking into account the continuity equations for a medium as a whole, the following notation was introduced

$$\omega_\alpha = \int_{t_0}^{t} \omega_\alpha \Delta V \,\mathrm{d}t = m \int_{t_0}^{t} \frac{\omega_\alpha}{\rho} \mathrm{d}t\,.$$

On the basis of the equations of balance for physico-chemical processes, the relation (10.6) can be rewritten in the form

$$\sum_{\alpha=1}^{r}\left(\sum_{i=1}^{r}\mu_i M_i \nu_{i\alpha}\right) w_\alpha \, dt = 0 \,. \tag{10.8}$$

Just as r reactions are independent, then quantities $w_\alpha \, dt$ can be considered as linearly independent quantities and it yields to the r equilibrium conditions for the r reactions.

$$\sum_{i=1}^{N}\mu_i M_i \nu_{i\alpha} = 0, \quad \alpha = 1, 2, \ldots, r \,. \tag{10.9}$$

These equations represent themselves as the fundamental equations of thermodynamic equilibrium for chemical reactions or phase transitions.

The mixture as an ideal two-parameter medium. Just as the chemical potentials μ_i may be considered as functions of ρ_i, ρ and T, then, on the basis of the integrated equations of balance of mass for the physico-chemical processes (10.7), the conditions for equilibrium (10.9) may be considered as the r equations, determining the r quantities $w_\alpha/\Delta V$ as functions of ρ, T and ⟨⟨initial⟩⟩ concentrations ρ_{0i}/ρ of components, constituting the mixture. Thus, for reversible processes the concentrations of components $\rho_i/\rho = c_i$ are generally determined in terms of the density of mixture ρ, temperature T and constants c_{0i} — the ⟨⟨initial⟩⟩ characteristics of the state of a mixture.

For the reversible processes, the equation of state for the mixture as a whole

$$U_m = mU\left(\rho, s, \frac{m_1}{m}, \frac{m_2}{m}, \ldots, \frac{m_N}{m}\right) \,,$$

$$p = p\left(\frac{1}{\rho}, T, \frac{m_1}{m}, \frac{m_2}{m}, \ldots, \frac{m_N}{m}\right) \,,$$

$$s = s\left(\frac{1}{\rho}, T, \frac{m_1}{m}, \frac{m_2}{m}, \ldots, \frac{m_N}{m}\right),$$

can be rewritten in the form

$$U_m = mU(\rho, T, c_{01}, c_{02}, \ldots, c_{0N}),$$

$$p = p\left(\frac{1}{\rho}, T, c_{01}, c_{02}, \ldots, c_{0N}\right), \tag{10.10}$$

$$s = s\left(\frac{1}{\rho}, T, c_{01}, c_{02}, \ldots, c_{0N}\right).$$

The constitutive equations for the mixture have the form of the constitutive equations for a two-parameter medium, but maintain the constant parameters c_{0i}, which for different particles in the concrete examples can have different values.

In the case of irreversible phenomena or in the pressure of diffusion, the above stated theory does not hold generally. However, if chemical reactions go on very fast and in each instant of time, we can assume that the relations, corresponding to the chemical equilibrium take place, then in this case we can use the equation of state (10.10). But, due to the presence of diffusion, which may change the relative structure of components in each given particle, it is obtained, that the quantities c_{0i} may change. For determination of the variable (in this case) parameters c_{0i} which appear in the equation of state (10.10), it is necessary to use the equations representing the law of diffusion.

In the case of equilibrium for two phases, when the molecules of each phase are the same, we have $\tau = 1$, $i = 1$, $\nu_{11} = +1$, $\nu_{21} = -1$, $M_1 = M_2$, and therefore the equilibrium equations (10.9) reduces to

$$\mu_1 = \mu_2 .$$

Obviously, in the case of three-phase equilibrium two equations

$$\mu_1 = \mu_2 \quad \text{and} \quad \mu_2 = \mu_3 ,$$

must be satisfied simultaneously.

The conditions that have been obtained for phase equilibrium will now be applied to solids, liquids and gases.

The law of Guldberg and Waage. As the second example we will consider the important problem of the reversible chemical reactions, in particular, of dissociation and recombination in the moving mixture of perfect gases. With the aid of formula (9.19) for $\mu_i(T, \rho_k)$, the equations of chemical equilibrium (10.9) of spontaneous mixture of the N perfect gases, for reversible reactions can be written in the form:

$$\sum_{i=1}^{N} \left(U_{0i} + R_i T_0 - T s_{0i} + \int_{T_0}^{T} c_{p_i}(T)\, dt \right.$$
$$\left. - T \int_{T_0}^{T} \frac{c_{p_i}}{T} dT + \frac{R_0}{M_i} T \ln \frac{p_i}{p_{0i}} \right) M_i \nu_{i\alpha} = 0 .$$

This equation can be given by formula

$$\left(\frac{p_1}{p_{01}}\right)^{\nu_{1\alpha}} \left(\frac{p_2}{p_{02}}\right)^{\nu_{2\alpha}} \cdots \left(\frac{p_N}{p_{0N}}\right)^{\nu_{N\alpha}} = k_1^{\nu_{1\alpha}} k_2^{\nu_{2\alpha}} \cdots k_N^{\nu_{N\alpha}} \tag{10.11}$$

$$(\alpha = 1, 2, \ldots, r) ,$$

where

$$k_i(T) = \exp\left\{ -\frac{1}{R_i T} \left[U_{0i} + R_i T_{0i} - T s_{0i} \right.\right.$$
$$\left.\left. + \int_{T_0}^{T} c_{p_i}\, dT - T \int_{T_0}^{T} \frac{c_{p_i}}{T} dT \right]\right\}. \tag{10.12}$$

Taking into account the Clapeyron's equation, the equilibrium equations (10.11) together with equations of balance of masses (10.7), forms the system of $r + N$ equations for determination of the $r + N$ quantities $\omega_\alpha / \Delta V$ and ρ_i / ρ_{0i}, specifying the structure of the mixture and progress of chemical reaction.

Equation (10.11) expresses the law of mass action or the Guldberg-Waage law. The quantities $k_i(T)$ depend on T_0, s_{0i} and U_{0i}, which correspond to ⟨⟨ initial ⟩⟩ state of the mixture with the densities ρ_{0i} of components. If the heat capacities c_{p_i} and c_{V_i} are constant, the quantities $k_i(T)$ are given by the formulae

$$k_i(T) = \left(\frac{T}{T_0}\right)^{\frac{\gamma_i}{\gamma_i - 1}} e^{\left(-\frac{s_{0i} - c_{p_i}}{R_i} - \frac{U_{0i} - c_{V_i} T_0}{R_i T}\right)} , \tag{10.13}$$

where $\gamma_i = c_{p_i}/c_{Vi}$. In the general case, $k_i(T)$ are determined from the experimental data.

The equation (10.11) can be put also in the following form:

$$(R_0 T)^{\nu''_\alpha} \prod_j \left(\frac{\rho_j}{M_j p_{0j} k_j}\right)^{\nu''_{j\alpha}} - (R_0 T)^{\nu'_\alpha} \prod_i \left(\frac{\rho_i}{M_i p_{0i} k_i}\right)^{\nu'_{i\alpha}} = 0 , \tag{10.14}$$

where

$$\nu'_{i\alpha} = \nu_{i\alpha} > 0, \quad \nu''_{j\alpha} = -\nu_{j\alpha} > 0 , \quad \nu''_\alpha = \sum_j \nu''_{j\alpha} , \quad \nu'_\alpha = \sum_i \nu'_{i\alpha} .$$

After determination of the parameters $\omega_\alpha/\Delta V$ and the composition of the mixture as functions of ρ and T, it is easy to determine the finite values of ⟨⟨velocity⟩⟩ of the chemical processes through the individual derivatives $d\rho/dt$ and dT/dt.

Sacha's equation. This method used for the study of reversible processes in mixtures can be applied to ionized gases; electrons and ions of atoms or molecules excited in different ways can be regarded as separate components of a mixture. The degree of ionization is characterized by corresponding parameters ω_α. In this case the equations (10.7) and (10.8) or, in the approximation of perfect gas, Equation (10.11) defines the degree of ionization and is called Sacha's equation when applicable to this phenomena.

On the properties of air as the mixture of gases. On the Figures 37, 38 and 39 there are constructed the dependences of the concentration of components, molecular weight, effective heat capacity c_V, specific internal energy and entropy and density for the equilibrium state of air as the mixture of gases, on the temperature and pressure, obtained as the result of calculations carried in accordance with the above stated theory of equilibrium states. These diagrams illustrate the dependence of the type (10.10), when instead of the density, the more convenient for practice variable: pressure, was taken.

The complete system of equations of motion of a mixture for reversible processes. The complete system of equations for the reversible processes in the mixture consists of the equation of continuity of the

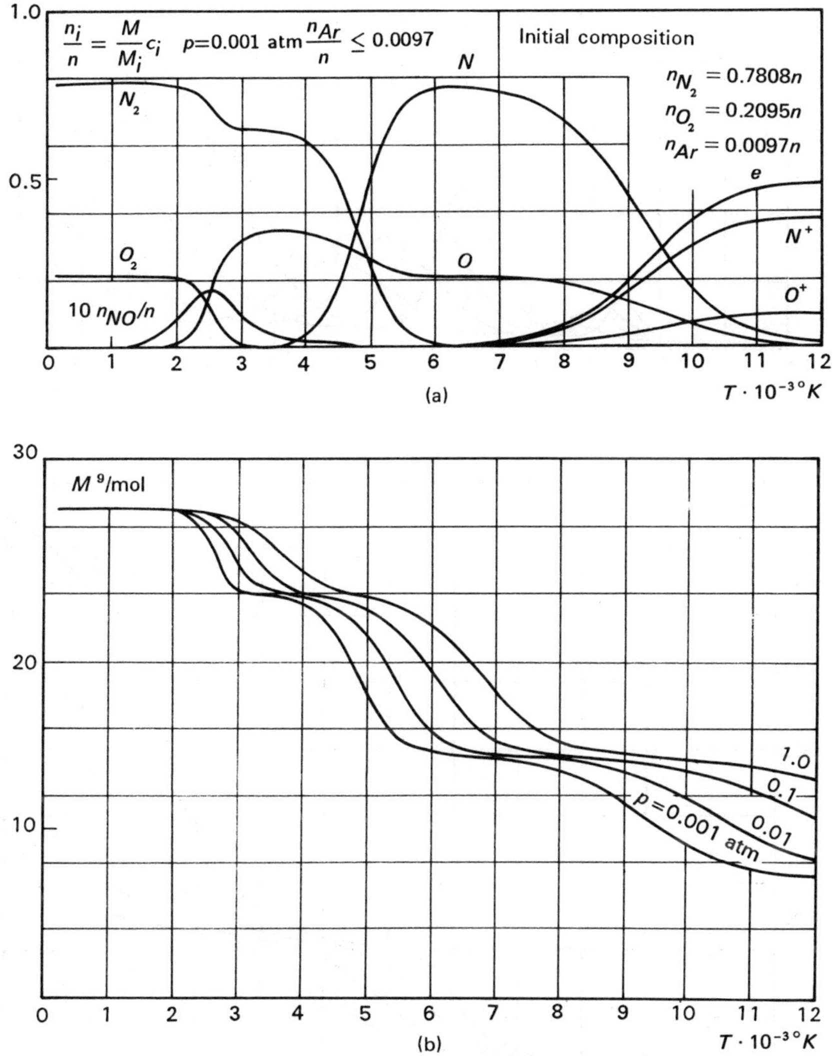

Fig. 37. (a) Molecular structure of air in dependence on the temperature for the pressure $p = 0.001$ atm. The changes of the molecular structure occur noticeably at the expense of dissociation over 2000°K, and at the expense of ionization — starting from 7000°K. N_2, O_2 and NO — molecules of nitrogen, oxygen and nitric oxide: N and O — atoms of nitrogen and oxygen; O^+ and N^+ — the single ionized atoms, e — electrons. The concentration of argon $n_{Ar/n}$ is always small. (b) The mean molecular weight of air in dependence on the temperature and pressure.

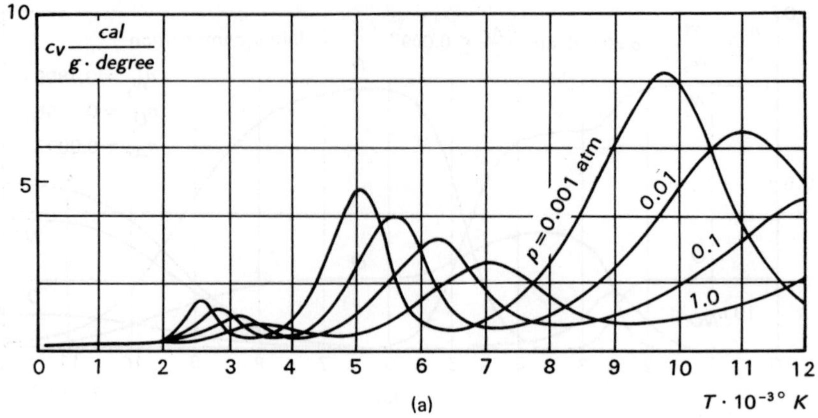

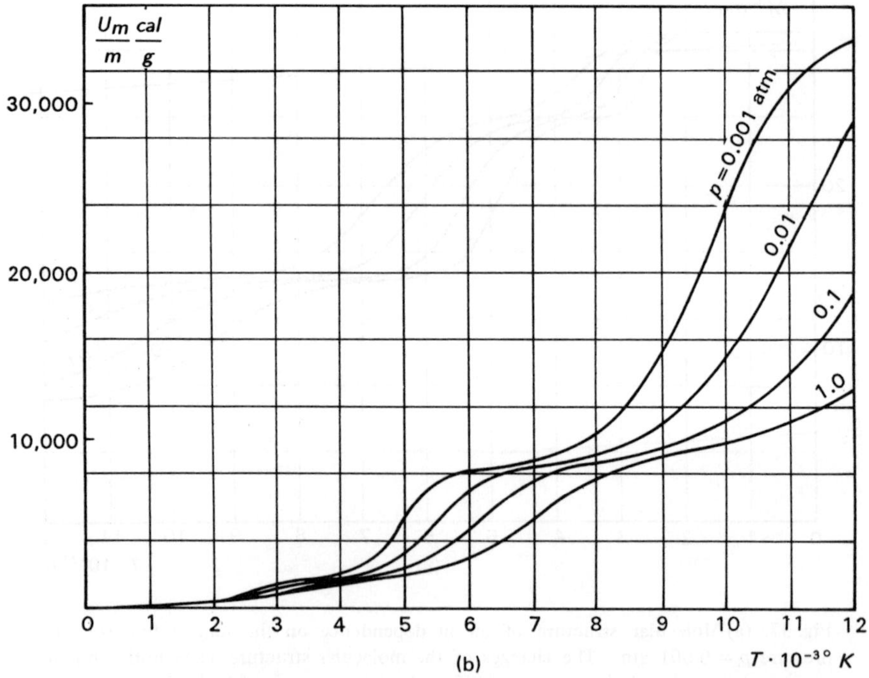

Fig. 38. (a) The heat capacity of air as the function of the temperature and pressure. The noticeable changes of the heat capacity occur for the temperatures over 2000°K. (b) The specific internal energy of air as the function of the temperature and pressure.

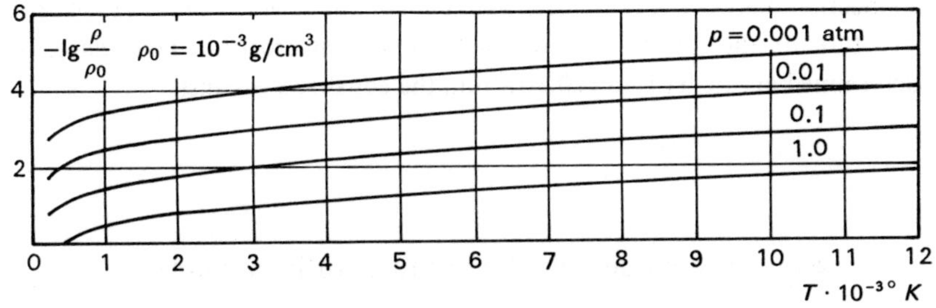

(a)

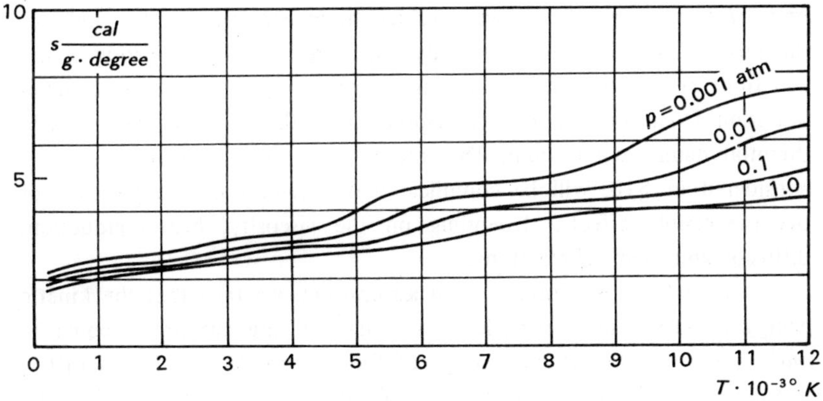

(b)

Fig. 39. (a) The dependence of the density of air on the temperature and pressure. (b) The specific entropy of air as a function of temperature and pressure.

mixture as a whole, three Euler's equations for ideal compressible medium with the equations of state (10.10) and the equations of heat flux or the second law of thermodynamics in the form

$$dU = p d \frac{1}{\rho} + dq \quad \text{or} \quad T ds = dq \, ,$$

where dq is the external input of heat which must be given as supplementary information. For the adiabatic processes $dq = 0$. In such a case the connection between the density and temperature, which depends on the structure of the mixture, in which the chemical processes take place, is determined on the basis of (10.10) from the equality $s = \text{const}$.

§11. The modelling of a mixture for irreversible processes

The general foundation of a problem of movement of a mixture for irreversible processes. Now, we consider the construction of the model of the mixture, in which the thermodynamic irreversibility occurring in the equations of state and other relations characterizing the model, gives small deviation from the relations corresponding to the reversible processes in thermodynamic equilibrium. The models for the reversible processes, considered in §10, will be complicated on account of the presence of the irreversible effects, depending on the viscosity, heat conduction, diffusion and chemical reactions.

Just as before, as a fundamental assumption, we take that the kinetic energy of infinitesimal particle of the mass of the mixture is equal to $mu^1/2$, and for the internal energy of the mixture U_m the formula (9.5) is valid or in the particular case the formula (9.15) is valid, when the additional conditions of equilibrium of the temperatures of components of the mixture hold

$$T_1 = T_2 = \ldots = T_N = T. \tag{11.1}$$

In other words, we assume, that in each particle the temperature equilibrium between components constituting mixture takes place. In some applications and, in particular for description of processes with the very fast change of the state in the mixture, the assumption (11.1) is too strong, therefore in the number of cases it is necessary to construct other models with different temperatures of components. The neglect of the kinetic energy of the components in their motion, with respect to the particle as a whole, is linked with the assumption that this energy is small in comparison with the mean energy of the chaotic thermal motion, characterizing comparatively higher temperature introduced in the internal energy U_m.

The general equation for the change of the entropy. The fundamental equations of thermodynamics for the increment of entropy (9.22) and the heat flux equation (9.23) can be obviously written in the form

$$T\,dS_m = dQ^{(e)} + dQ' = dQ^{(e)} + dQ^{**} + dQ' - dQ^{**}\,, \tag{11.2}$$

and

$$\begin{aligned} dU_m &= \frac{\partial U_m}{\partial (m/\rho)}\, d\frac{m}{\rho} + \frac{\partial U_m}{\partial S_m}\, dS_m + \sum_{k=1}^{N} \mu_k\, dm_k \\ &= -p\, d\frac{m}{\rho} + T\, dS_m + \tau^{ij} e_{ij}\, \Delta V\, dt + dQ^{**} - dQ'\,. \end{aligned} \tag{11.3}$$

Here, the symmetry of the stress tensor, i.e., $p^{ij} = p^{ji}$ and formulae $e_{ij} = \frac{1}{2}(\nabla_j u_i + \nabla_i u_j)$ have been taken into account. With the aid of (11.3), on the basis of the definition of the pressure p and the assumption that the quantity $dQ^{**} - dQ^{*}$ does not contain the term of form[1] $A\, dS$, where A is the same function of determining parameters, we arrive at the constitutive equations:

$$p = -\frac{\partial U_m}{\partial (m/\rho)} = -\frac{\partial U}{\partial (1/p)}\,, \qquad T = \frac{\partial U_m}{\partial S_m} = \frac{\partial U}{\partial s}\,, \tag{11.4}$$

and taking into account the equations of balance of mass (9.1) and (9.2), we arrive at

$$dQ' - dQ^{**} = \left\{ \tau^{ij} e_{ij} - \sum_{k=1}^{N} \left[\mu_k \left(M_k \sum_{\alpha=1}^{r} \nu_{k\alpha} w_\alpha - \operatorname{div} I_k \right) \right] \right\} \Delta V\, dt. \tag{11.5}$$

[1] This assumption is equivalent to the supposition that for the irreversible processes the formula determining the temperature $T = \partial U/\partial s$, obtained earlier for the reversible processes, is valid.

On the basis of this equality, the equation (11.2) for the increase of entropy can be rewritten in the form

$$T\,dS_m = dQ^{(e)} + dQ^{**} + \left[\tau^{ij} e_{ij} + \sum_{k=1}^{N} \mu_k \operatorname{div} I_k - \sum_{k,\alpha=1}^{N,r} \mu_k M_k \nu_{k\alpha} w_\alpha\right] \Delta V\, dt\,. \tag{11.6}$$

Note, that the rather general form of the last formula and the constitutive equation (11.4) is essentially linked with the assumption (9.5) that the internal energy U_m depends only on the density and entropy of the mixture and mass of component m_k.

Supposition on the fluxes of energy. Assume next, that the total internal influx of energy to the infinitesimal particle of the mixture from the volume ΔV, bounded by the surface Σ moving with the velocity $\boldsymbol{v}$, can be determined by the formula

$$\begin{aligned} dA^{(e)} + dQ^{(e)} + dQ^{**} &= \sum_{k=1}^{N} \rho_k \boldsymbol{F}_k \cdot \boldsymbol{v}_k\, \Delta V\, dt \\ &+ \sum_{k=1}^{N} \int_\Sigma p_k^{ij} n_j v_{ki}\, dt\, d\sigma - \int_\Sigma \boldsymbol{q}_0 \cdot \boldsymbol{n}\, dt\, d\sigma + dQ^{(e)}_{\text{mass}} \\ &= \sum_{k=1}^{N} \rho_k \boldsymbol{F}_k \cdot \boldsymbol{v}\, \Delta V\, dt + \int_\Sigma p^{ij} n_j v_i\, dt\, d\sigma + \sum_{k=1}^{N} \boldsymbol{F}_k \cdot \boldsymbol{I}_k\, \Delta V\, dt \\ &+ \nabla_j \left(\sum_{k=1}^{N} \frac{p_k^{ij} I_{ki}}{\rho_k} \right) dV\, dt - \nabla_j q_0^j\, \Delta V\, dt + dQ^{(e)}_{\text{mass}}\,, \end{aligned} \tag{11.7}$$

where $\rho_k \boldsymbol{F}_k \Delta V$ — the external mass force, acting on the kth component of the mixture, $p_k^{ij} n_j\, d\sigma$ — the components of the surface forces, acting on the area $d\sigma$ of the surface Σ on the kth component of the mixture. In addition, it is accepted that $p^{ij} n_j\, d\sigma\, \boldsymbol{\mathfrak{z}}_i = \sum_{k=1}^{N} p_k^{ij} n_j\, d\sigma\, \boldsymbol{\mathfrak{z}}_i$ is the vector of the stress force on the area $d\sigma$ of the surface Σ for a medium as a whole; $\boldsymbol{q}_0$ — the vector of external influx of energy on the expense of the heat conductivity and diffusion; $dQ^{(e)}_{\text{mass}}$ — the external mass influx of heat to the particle, in particular in $dQ^{(e)}_{\text{mass}}$ can take part of the absorbed heat flux due to the external radiation or the Joule's heat, produced at the expense of the electric currents. The quantity $\sum\limits_{k=1}^{N} \boldsymbol{F}_k \cdot \boldsymbol{I}_k \, \Delta V \, dt$ is equal to the elementary work done by the external mass force $\boldsymbol{F}_k$, in general different for the diverse components, on the displacements of internal motion due to diffusion. If the force $\boldsymbol{F}_k$, calculated per unit of mass, is the same for all components, then $\sum\limits_{k=1}^{N} \boldsymbol{F}_k \cdot \boldsymbol{I}_k \, \Delta V \, dt = 0$. For example, for the gravitation force $\boldsymbol{F}_k = \boldsymbol{g}$, $k = 1, 2, \ldots, N$ ($\boldsymbol{g}$ — the acceleration of the gravitational force), just as $\sum\limits_{k=1}^{N} \boldsymbol{I}_k = 0$,

$$\sum_{k=1}^{N} \boldsymbol{F}_k \cdot \boldsymbol{I}_k \, \Delta V \, dt = \boldsymbol{g} \cdot \left(\sum_{k=1}^{N} \boldsymbol{I}_k \right) \Delta V \, dt = 0 \, .$$

In accordance with the definition of displacement of the points of a medium as a whole and with the equality (11.7) we have

$$\begin{aligned} dA^{(e)} &= \sum_{k=1}^{N} \rho_k \boldsymbol{F}_k \cdot \boldsymbol{v} \, \Delta V \, dt + \int_{\Sigma} p^{ij} n_j v_i \, dt \, d\sigma \, , \\ dQ^{(e)} + dQ^{**} &= - \nabla_j \left[q_0^j - \sum_{k=1}^{N} \frac{p_k^{ij} I_{ki}}{\rho_k} \right] \Delta V \, dt \\ &\quad + \sum_{k=1}^{N} \boldsymbol{F}_k \cdot \boldsymbol{I}_k \, \Delta V \, dt + dQ^{(e)}_{\text{mass}} \, . \end{aligned} \tag{11.8}$$

On the basis of this formula, the equation (11.6) for the total change of the specific entropy $s = S_m/m$ of the mixture can be rewritten in the form

$$\begin{aligned} \frac{dS_m}{dt} = \rho \, \Delta V \, \frac{ds}{dt} &= \left[- \frac{1}{T} \nabla_j q^j + \frac{1}{T} \sum_{k=1}^{N} \mu_k \, \nabla_i I_k^i + \frac{1}{T} \sum_{k=1}^{N} \boldsymbol{F}_k \cdot \boldsymbol{I}_k \right. \\ &\quad \left. - \sum_{k,\alpha=1}^{N,r} \frac{\mu_k M_k \nu_{k\alpha} w_\alpha}{T} + \frac{\tau^{ij} e_{ij}}{T} \right] \Delta V + \frac{dQ^{(e)}_{\text{mass}}}{T \, dt} . \end{aligned} \tag{11.9}$$

Here, the following designation has been used

$$q = q^j \mathbf{\mathfrak{z}}_j = q_0 - \sum_{k=1}^{N} \frac{p_k^{ij} I_{kj}}{\rho_k} \mathbf{\mathfrak{z}}_i = q_0 - \sum_{k=1}^{N} p_k^{ij} (v_{kj} - v_j) \mathbf{\mathfrak{z}}_i \, .$$

The vector q characterizes the total flow of energy through the elements of the surface Σ due to the heat conductivity, diffusion and the work done by the surface forces on the relative motions of components of the mixture.

On the complete system of motion of the mixture and on the kinetic equations. The equations of the balance of mass, the impulse equation for the mixture as a whole, the equations of the state (11.4) and the equation for the entropy (11.9) form the complete system of equations of motion of a mixture, provided that they are supplemented by the information about the heat flux $dQ^{(e)}_{\text{mass}}$ and by the macroscopic kinetic equations, determining the quantities

$$q, I_k, \omega_\alpha \text{ and } \tau^{ij} \, .$$

The establishment of this kinetic equation is the fundamental task of the thermodynamics of irreversible processes.[1)]

The solution of this problem is always linked with the acceptance of various kinds of hypothesis about the properties and mechanisms of irreversible effects, accompanying the motion of a medium. These hypothesis can vary in dependence on the concrete properties of the considered media and separated class of motion, and may be connected with the assumptions of mathematical character, for example, on the basis on the small deviation of the state of the particles from equilibrium.[2)] Finally, these hypothesis must always be made on the basis of the experiment. Also the various

[1)] For determination of the motion of a mixture as a whole it is not necessary to find the above introduced quantities q_0, p_k^{ij}, v_k; there are no corresponding equations for their determination, they play the auxiliary role, they serve only for heuristic foundations of equations (11.9), which can also be obtained using, instead of (11.7), other hypothesis about the external fluxes of energy.

[2)] In the paper of Sedov, L. I., "On the general form of the kinetic equations of the chemical reaction in gases." (Dokl. Akad. Nauk. SSSR, Vol. IX, No. 1, 1948) the same general forms of the kinetic equations for the irreversible chemical reactions were proposed.

empirical formulae, obtained from the treatment of the experimental data may be useful for the foundation of the kinetic equations.

In the case of motions and processes which differ slightly from reversible processes, to obtain the kinetic relations it is possible to use the Onsager's method (*cf.*, §8 and below). The relations for the considered media and phenomena, obtained in such a manner, must also be verified or confirmed by the experiment, even if these relations may be additionally based on the theoretical considerations taken from the statistical theories.

Formula for the production of entropy. For the application of the Onsager's theory it is necessary to represent the total change of the entropy $m(\mathrm{d}s/\mathrm{d}t)$ in the form of sum

$$m\,\frac{\mathrm{d}s}{\mathrm{d}t} = m\,\frac{\mathrm{d}_i s}{\mathrm{d}t} + m\,\frac{\mathrm{d}_e s}{\mathrm{d}t}\,.$$

Such a division of the increment of the entropy, may be made, starting from the Equation (11.9), with the aid of assumption, presented by the formulae below.

The quantity $m(\mathrm{d}_e s/\mathrm{d}t)$ is defined as the increment of the entropy due to the influx of heat to the volume ΔV of the medium from the external objects, interacting with the given medium, $\mathrm{d}Q^{(e)}_{\text{mass}}$ and the external influx of entropy through the boundary Σ of a particle. Therefore, on the basis of the formula (10.9) we put[1)]

$$\begin{aligned} m\,\frac{\mathrm{d}_e s}{\mathrm{d}t} &= \frac{\mathrm{d}Q^{(e)}}{T\,\mathrm{d}t} + \left[\sum_{k=1}^{N} \operatorname{div}\frac{\mu_k \boldsymbol{I}_k}{T} - \operatorname{div}\frac{\boldsymbol{q}}{T}\right]\Delta V \\ &= \frac{\mathrm{d}Q^{(e)}_{\text{mass}}}{T\,\mathrm{d}t} + \int_{\Sigma} \frac{\sum_{k=1}^{N} \mu_k \boldsymbol{I}_k - \boldsymbol{q}}{T}\cdot \boldsymbol{n}\,\mathrm{d}\sigma\,. \end{aligned} \tag{11.10}$$

Correspondingly, after fixing $m(\mathrm{d}_e s/\mathrm{d}t)$, for the production of entropy per unit time $m(\mathrm{d}_i s/\mathrm{d}t)$ due to the internal irreversible effects we obtain

1) Note, that the presence of the term $(1/T)(\mathrm{d}Q^{(e)}_{\text{mass}}/\mathrm{d}t)$ in the formula (11.10), is linked with the assumption of the reversible nature of this heat flux.

$$m\,\frac{\mathrm{d}_i s}{\mathrm{d}t} = \left\{ -\frac{\operatorname{grad} T}{T_2} \cdot \boldsymbol{q} + \frac{\tau^{ij} e_{ij}}{T} + \sum_{k=1}^{N} \left[\frac{F_k}{T} - \operatorname{grad}\left(\frac{\mu_k}{T}\right) \right] \right.$$
$$\left. \cdot\, I_k - \sum_{\alpha=1}^{r} \sum_{k=1}^{N} \frac{\mu_k M_k \nu_{k\alpha}}{T}\, w_\alpha \right\} \Delta V . \tag{11.11}$$

The function of dissipation. The formula for the increment of entropy can be rewritten in the following form:

$$\rho\,\frac{\mathrm{d}_i s}{\mathrm{d}t} = \sigma = \sum_\beta X_\beta\, \chi^\beta > 0\,, \tag{11.12}$$

where the quantities χ^β are the components of the generalized fluxes q^i, I_k^i; components of tensor of viscous stresses $\tau^{ij}(i, j = 1, 2, 3;\ k = 1, 2, \ldots, N)$ and scalar quantities $\omega_\alpha(\alpha = 1, 2, \ldots, r)$, X_β denote the coefficients at corresponding χ^β.

Further construction of the models is based on the assumption that between the quantities X_β and χ^β exist the relations. Therefore, the quantity σ may be considered as a function of X_β or χ^β, and it contains as the parameters p, T and the concentrations of the components of mixture $c_i = \rho_i/\rho$ i.e.,

$$\sum_\beta X_\beta\, \chi^\beta = \sigma(p, T, c_i, X_j) = \widetilde{\sigma}(p, T, c_i, \chi^j)\,. \tag{11.13}$$

In some cases the problem of determining the kinetic equation can be regarded as one of establishing functional relations

$$\chi^\beta = \chi^\beta(g_{ij}, p, T, c_1, c_2, \ldots, c_N, X_1, X_2, \ldots)\,, \tag{11.14}$$

where g_{ij} are the components of metric tensor and the quantities ρ, T, c_i are scalars. In the general case the relations (11.14) for I_k^i must satisfy the conditions

$$\sum_{k=1}^{N} I_k^i = 0\,.$$

The corresponding systems of the conjugate functions X_β and χ^β can be presented with the aid of the following designations

$$X_\beta \to \chi^\beta ,$$

$$-\frac{\operatorname{grad} T}{T^2} = \omega = \omega^i \mathbf{э}_i \to \boldsymbol{q} ,$$

$$\frac{1}{T} \boldsymbol{F}_k - \operatorname{grad} \frac{\mu_k}{T} = \omega_k = \omega_k^i \mathbf{э}_i \to \boldsymbol{I}_k ,$$

$$-\sum_{k=1}^{N} \frac{\mu_k M_k \nu_{k\alpha}}{T} = \gamma_\alpha \to \omega_\alpha ,$$

$$e_{ij} \to \frac{1}{T} \tau^{ij} .$$

Between the arguments in relations (11.14) besides scalars and components of vectors, denoted by X_α, there appears also only one parametric tensor — the metric tensor with components g_{ij}. This assumption is equivalent to the physical condition of isotropy of considered model of mixture. Just as σ — the scalar quantity, then in general case as the arguments of σ in (11.13) may appear the scalars p, T, c_k, γ_α and invariant combinations only, composed of the following components of vectors and tensors: $\omega^i, \omega_k^i, e_{ij}, g_{ij}$.

The general form of the quadratic function of dissipation for the isotropic medium; the Onsager's relations for the isotropic medium. When investigating irreversible processes, which differ slightly from reversible equilibrium processes, considered in §10, in many cases we can assume that the functions (11.14) are linear with respect to X_β, i.e., that

$$\chi^\beta = \sum_\delta L^{\beta\delta} X_\delta , \qquad (11.15)$$

where the matrix of the coefficients $L^{\beta\delta}$ consists of scalars, components of second-order tensors and of fourth-order tensors.

For isotropy the corresponding tensors, described by the component $L^{\beta\delta}$ depend on scalars and the metric tensor[1)] g^{ij} only. The dependence of $L^{\beta\delta}$ on g^{ij} is obtained below, and dependence on the scalar parameters

[1)] The detailed basis of this direct and rather obvious statement is in Appendix I.

of state must be assumed additionally. In the presence of linear relations (11.15) the quantity $\sigma(X_\beta)$ represents itself the quadratic form of X_β.

It is easy to observe, that for the isotropic medium the quadratic form for σ can be written in the following manner:

$$\begin{aligned}\sigma &= a_0 T^2 g_{ij}\omega^i\omega^j + 2a^k g_{ij}\omega^i\omega^j_k + a^{kl} g_{ij}\omega^i_k\omega^j_l + b^{sm} T\gamma_s\gamma_m \\ &+ 2b^s\gamma_s g^{ij} e_{ij} + \frac{\lambda}{T}(g^{ij}e_{ij})^2 + \frac{2\mu}{T} g^{ij}g^{mn} e_{im}e_{jn}\,.\end{aligned} \tag{11.16}$$

From this it follows, that for a complete determination of the function σ it is necessary to fix the components $a_0, a^k, a^{kl} = a^{lk}$, $b^{sm} = b^{ms}$, b^s, λ and μ where $k, l = 1, 2, \ldots, N$; $m, s = 1, 2, \ldots, r$. The number of these independent coefficients in (11.16) is equal

$$\begin{aligned}&1 + N + \frac{N(N+1)}{2} + \frac{r(r+1)}{2} + r + 2 \\ &= 3 + \frac{N(N+3)}{2} + \frac{r(r+3)}{2}\,.\end{aligned}$$

The coefficients in (11.16) should also obey the condition $\sigma \geqslant 0$.

In accordance with the general theory of irreversible processes, stated in §8 with the use of Onsager's principle, the relations (11.15) take the form

$$\begin{aligned}q_i &= \frac{1}{2}\frac{\partial\sigma}{\partial\omega^i} = a_0 T^2\omega_i + \sum_k a^k\omega_{ki} \\ &= -a_0\frac{\partial T}{\partial x^i} + \sum_k a^k\left[\frac{F_{ki}}{T} - \frac{\partial}{\partial x^i}\left(\frac{\mu_k}{T}\right)\right],\end{aligned} \tag{11.17}$$

$$\begin{aligned}I_{ki} &= \frac{1}{2}\frac{\partial\sigma}{\partial\omega^i_k} = a^k\omega_i + a^{kl}\omega_{li} = -a^k\frac{1}{T^2}\frac{\partial T}{\partial x^i} \\ &+ \sum_l a^{kl}\left[\frac{F_{li}}{T} - \frac{\partial}{\partial x^i}\left(\frac{\mu_l}{T}\right)\right],\end{aligned} \tag{11.18}$$

$$\omega_\alpha = \frac{1}{2}\frac{\partial \sigma}{\partial \gamma_\alpha} = b^\alpha \operatorname{div} v + \sum_s b^{s\alpha} T \gamma_s$$
$$= - b^\alpha \frac{1}{\rho}\frac{d\rho}{dt} - \sum_{k,s} b^{s\alpha} \mu_k M_k \nu_{ks} , \tag{11.19}$$

$$\frac{\tau^{ij}}{T} = \frac{1}{2}\frac{\partial \sigma}{\partial e_{ij}} = g^{ij} \sum_s b^s \gamma_s + \frac{\lambda}{T} g^{ij} \operatorname{div} v + \frac{2\mu}{T} e^{ij} . \tag{11.20}$$

The last relation differs from the usual Navier-Stokes law for the viscous stresses, by the presence of the first term influencing the magnitude of the normal to the considered plane component of the force vector on account of the irreversible chemical and phase transitions.

The relations $\sum_{k=1}^{N} I_{ki} = 0$ imply $N + 1$ relations for the coefficients a^k and a^{ki}, namely

$$\sum_{k=1}^{N} a^k = 0 \quad \text{and} \quad \sum_{k=1}^{N} a^{kl} = 0, \qquad l = 1, 2, \dots, N . \tag{11.21}$$

Thus, the formulae (11.16)–(11.20) contain only

$$n = 2 + \frac{N(N+1)}{2} + \frac{r(r+3)}{2} ,$$

independent coefficients, which are, generally speaking, the functions of p, T, c_k. Note, that the primary formulae of general form (the linear relations (11.15)) contain the $n_0 = (3 + 3N + r + 6)^2 = (9 + 3N + r)^3$ coefficients $L^{\beta\delta}$.

The previous general theory allows to reduce the number of coefficients, which is necessary to give, to[1])

$$n_0 - n = 79 + \frac{17}{2} N^2 + \frac{107}{2} N + \frac{r^2}{2} + \frac{33r}{2} + 6Nr .$$

[1]) In practice, it takes additionally, that $b^s = 0$ $(s = 1, 2, \dots, r)$.

For example, in the case of the three-component mixture ($N = 3$) and one reaction ($r = 1$), we have $n_0 = 361$, and the number of independent coefficients $n = 10$, i.e., the number of coefficients which must be determined is cancelled by $n_0 - n = 351$.

Some simplifications of the formulae (11.17) ÷ (11.18) may be made by rewriting out fully the expression $\operatorname{grad}(\mu_k(p, T, c_k)/T)$ in its and performing calculation of coefficients of $(\partial T/\partial x^i)$, $(\partial p/\partial x^i)$ and $(\partial c_k/\partial x^i)$.

The coefficients $a_0, a^k, a^{kl}, b^s, b^{s\alpha}$, λ and μ or its combinations, can be obtained by the methods of statistical theories and the corresponding experimental data.

The previous theory is applicable not only to the mixture of perfect gases, but also to other models of real media. The further simplifications and applications of this theory can be found in the special monographs and in particular research publications.

The connection of the phenomenological coefficients a_0, a^k, a^{kl} *with the transfer coefficients for the mixture of two perfect gases.* If for the mixture of the two perfect gases one uses the formula (9.19) for the chemical potentials, then it is possible to represent the flows q_i and I_{ki}, given by (11.17) and (11.18) through the gradients of gasodynamical parameters T, p, c_i $(i = 1, 2, \ldots, N)$. In this case we write (9.19) in the form

$$\mu_k = \mu_k^0(p, T) + R_k T \ln z_k \,, \tag{11.22}$$

where

$$\mu_k^0(p, T) = U_{ok} + R_k T_0 - T s_{ok} + \int_{T_0}^{T} c_{p_k} \, dT - \int_{T_0}^{T} \frac{c_{p_k}}{T} \, dT + RT \ln \frac{p}{p_{ok}} \,,$$

and z_k is the molar concentration in kth component

$$z_k = \frac{p_k}{p} = \frac{n_k}{n} = \frac{M}{M_k} c_k \,.$$

On the basis (11.22) we have

$$T\,\mathrm{d}\,\frac{\mu_k}{T} = T\,\mathrm{d}\left(\frac{\mu_k^0}{T} + R_k \ln z_k\right) = \frac{R_k T}{z_k}\,\mathrm{d}z_k$$
$$+ \left(\frac{\partial \mu_k^0}{\partial p}\right)\mathrm{d}p + T\,\frac{\partial}{\partial T}\left(\frac{\mu_k^0}{T}\right)_p \mathrm{d}T\,, \tag{11.23}$$

$$\left(\frac{\partial \mu_k^0}{\partial p}\right) = \frac{R_k T}{p} = \frac{R_0 T}{M_k p} = \frac{1}{nM_k}, \tag{11.24}$$

$$T\,\frac{\partial}{\partial T}\left(\frac{\mu_k^0}{T}\right) = -\frac{1}{T}\left(U_{0k} + R_k T_0 + \int_{T_0}^{T} c_{p_k}\,\mathrm{d}T\right) = -\frac{i_k}{T}\,, \tag{11.25}$$

where i_k — the specific enthalpy of kth component. Using (11.23), (11.24) and (11.25), we obtain for ω_k the following expressions:

$$\omega_k = \frac{1}{T}\left[F_k - T\nabla\frac{\mu_k}{T}\right] = \frac{1}{T}\left(F_k - \frac{p}{\rho_k}\,\nabla z_k - \frac{1}{nM_k}\nabla p + \frac{i_k}{T}\nabla T\right). \tag{11.26}$$

From formula (11.18), taking into account (11.26), we arrive at

$$I_k = -\frac{a^k}{T^2}\nabla T + \sum_l \frac{a^{kl}}{T}\left[F_l - T\nabla\left(\frac{\mu_l}{T}\right)\right] = -p\sum_l \frac{a^{kl}}{T\rho_{li}}\left(\nabla z_l\right.$$
$$\left. + \frac{z_l}{p}\,\nabla p - \frac{\rho_t}{p}\,F_l\right) - \left(\frac{a^k}{T} - \sum_l \frac{a^{kl}}{T} i_l\right)\frac{1}{T}\,\nabla T \tag{11.27}$$
$$= \rho\sum_l \frac{M_k M_l}{M^2} D^{kl}\,d_l - D_k^T\,\frac{1}{T}\,\nabla T\,,$$

where

$$d_l = \nabla z_l + (z_l - c_l)\,\frac{1}{p}\,\nabla p + \frac{c_l}{p}\left(\sum_k \rho_k F_k - \rho F_l\right),$$

and the introduced multicomponents coefficients of diffusion D^{kl} and thermodiffusion D_k^T are in accordance with the formulae

$$D^{kl} = -\frac{p}{\rho T}\frac{M^2}{M_k M_l}\left(\frac{a^{kl}}{\rho_l} - \frac{a^{kk}}{\rho_k}\right) = -\frac{R_0 M}{M_k M_l}\left(\frac{a^{kl}}{\rho_l} - \frac{a^{kk}}{\rho_k}\right), \tag{11.28}$$

$$D_k^T = \frac{a^k}{T} - \sum_l \frac{a^{kl}}{T} i_l \quad (k, l = 1, 2, \ldots, N). \tag{11.29}$$

The correctness of formulae (11.27) and equalities

$$\sum_k D_k^T = 0, \quad D^{kk} = 0 \quad \text{and} \quad \sum_{k=1}^{n} \mathrm{d}_k = 0,$$

follows from properties of (11.21). From (11.28) and (11.29) it is easy to obtain the reverse formulae for a^k and a^{kl} in terms of D_k^T and D^{kl}.

In the analogous way, the formulae (11.17) can be rewritten for the mixture of perfect gases in the form:

$$q = -a_0 \nabla T + \sum_k \frac{a^k}{T}\left[F_k - T\nabla\left(\frac{\mu_k}{T}\right)\right] = -k\nabla T$$
$$+ \sum_l i_l I_l - p\sum_l \frac{D_l^T}{\rho_l} d_l, \tag{11.30}$$

where

$$k = a_0 - 2\sum_k \frac{a^k i_k}{T^2} + \sum_{k,l} \frac{a^{kl} i_k i_l}{T^2}. \tag{11.31}$$

In the kinetic theory of gases, there usually appears the coefficients D^{kl}, D_k^T and k, which are determined here by the phenomenological coefficients a_0, a^k and a^{kl}, from formulae (11.28), (11.29) and (11.31).

If the medium is anisotropic, for example due to the interaction with the electromagnetic field, then the previous theory becomes more complicated. In this case, in addition to the metric tensor, it is necessary to introduce also other tensors as the arguments of σ.

Chapter VI

Basic concepts and equations of electrodynamics

§1. Basic concepts of electrodynamics. Electromagnetic field. Maxwell's equations in vacuum

In recent times, problems of the study of motion of continuous media with inclusion of electromagnetic effects have acquired increasing significance. In this context, in continuum mechanics, one must present an account of the basic results of electrodynamics.

It will be assumed that the simplest experimental facts and laws of electrodynamics are known from an elementary general course of physics. The basic laws of electrodynamics will be formulated axiomatically in the form of Maxwell's equations, as a consequence of the results of theoretical treatments and generalizations of experience and observations.

The formulation of Maxwell's equations is based on a utilization of a number of abstract mathematical concepts on electromagnetic characteristics introduced for the description of electromagnetic properties of bodies and fields. The introduction and analysis of these concepts and equations occurred as a result of extensive work and large historical experience during the intensive development of scientific studies of electric effects over more than one hundred years.

The usefulness of the theoretical methods, resting on Maxwell's equations, is confirmed by all practical descriptions of electromagnetic effects and all known macroscopic experimental data.

The following account of electrodynamics does not pursue historical aspects, but presents a concise description of its foundations, keeping in mind applications of electrodynamics to problems of continuum mechanics. A note will only be made here of the fact that the theory of the electromagnetic field, as a theory of a real object, is due to Faraday and was tested convincingly by experiments only at the end of the last century when Hertz succeeded in obtaining electromagnetic waves and in demonstrating in this way the reality of the existence of the electromagnetic field.

First, consider the reasons why electromagnetic effects may have significance in continuum mechanics. In physics, four basic types of interaction between material objects are known: gravitational, electromagnetic, nuclear and weak interactions in the theory of elementary particles.

The internal force interactions studied earlier are linked, fundamentally speaking, to electromagnetic effects; they cause the macroscopic stresses in rigid, liquid and gaseous bodies. In particular, during collisions of molecules and atoms, the role played by electromagnetic forces is basic.

Further, under certain conditions, a body may have electric charges, and currents may flow through it. One must take into account the forces of interaction, which arise under these conditions, in the general balance of the forces acting on a volume of a continuum. Such effects are especially strong in the motion of a plasma, i.e., of a gas which contains a large number of free electrons and ions. Therefore a plasma interacts appreciably with an electromagnetic field.

Several basic definitions and concepts of electrodynamics will now be recalled.

Coulomb's law for charges at rest. Experiments show that two particles of a medium at rest which have electric charges e_1 and e_2 interact with a force, analogous to the gravitational force between two masses:

$$\boldsymbol{F} = \pm k \frac{e_1 e_2}{r^2} \operatorname{grad} r\,, \tag{1.1}$$

where r is the distance between the charged particles and k is a constant coefficient. This relation is referred to as Coulomb's law.

In contrast to the gravitational force which is always a force of attraction, the force $\boldsymbol{F}$ is attractive, if the charges e_1 and e_2 have different signs, and repulsive, if they have the same sign.

The influence of the force of gravitational attraction is significant when the interacting masses m_1 and m_2 are very large. The electric forces of interaction are many times larger. The ratio of the forces of electric repulsion between two electrons to the force of their gravitational attraction is of the order 10^{39} , a very big number which is difficult to represent with the aid of normal numbers, gravitational attraction between two electrons to the forces of their electric repulsion is of the order of 10^{-39}, a very small number which it is difficult to observe with the aid of normal apparatus.

Charge density. In continuum mechanics, it is natural to introduce a continuous distribution e of charge, in a similar manner as continuous distributions of mass have been introduced earlier. The charge density is defined as follows

$$\rho_e = \lim_{\Delta V \to 0} \frac{\Delta e}{\Delta V},$$

where Δe is the total charge in the volume ΔV.

For electrically neutral bodies, one has $\Delta e = 0$ and $\rho_e = 0$. The charge density ρ_e, depending on accumulations of ions or electrons in a definite place, may be positive as well as negative. In reality, for all bodies, the density ρ_e is almost equal to zero, because equal signed charges are always repelled; therefore large accumulations of positively or negatively charged particles cannot exist for long.

Electric current. Electric current is a consequence of motion of charged particles. Thus, if there are free electrons in a body which move in a directed manner with respect to ions, this means that an electric current flows in the body. Under these conditions, in spite of the presence of macroscopic flows of ions and electrons, the macroscopic total velocity of the particles of the body may be equal to zero.

Electric current, by its very nature, is similar to intensive regulated diffusion. Currents may occur in resting as well as in moving media. The current density is characterized by the vector $\boldsymbol{j}$ of current density, the magnitude of which is equal to the magnitude of the total charge which passes in unit time through a unit area, perpendicular to the vector $\boldsymbol{j}$.

Polarization of atoms. Under ordinary conditions an atom is electrically neutral, i.e., its total charge is equal to zero; however, if a substance contains a group of atoms which are close to each other, then the charges in each of them are displaced, i.e., neutral atoms are similar to dipoles or multipoles. This phenomenon is referred to as polarization. A large accumulation of polarized molecules or atoms, in their ordered distributions under the influence of an external electric field, leads to the macroscopic effect of electric polarization of macroscopic parts of a medium. Obviously, generally speaking, chaotic thermal motion impedes the phenomenon of macroscopic polarization.

Magnetic interaction. In addition to the interactions of resting charged particles, described above, there exist still magnetic interactions. For example, it is known that magnetized iron attracts filings, that the magnetic needle of a compass aligns itself along Earth's meridians. Consequently, there exist on Earth always forces which swing the needle of a compass into a definite direction.

In connection with the presence of magnetic interactions, experiments lead to the consideration of magnetic charges and the description of their law which is analogous to Coulomb's law for electric charges. However, it must be realized that magnetic charges do not exist in nature, but that magnetic interactions are due to interactions of electric currents (i.e., moving charges). In particular, magnetic properties of different substances indicate the presence of microscopic currents, flowing inside atoms as a consequence of the motion of electrons around kernels and of characteristic "rotations" of electrons and kernels (spins). Under ordinary conditions, in the majority of bodies, atoms are orientated in a random manner and the phenomenon of interatomic currents does not manifest itself. However, if there occurs ordered orientation of the elementary particles constituting a body, then a corresponding macroscopic effect expresses itself in the magnetic properties of the bodies which exhibit the phenomenon of magnetization.

Electric and magnetic stress vectors. Let there be given in space a set of charges which are at rest with respect to some inertial coordinate system K. In physics, one adopts a general principle with the aid of which one can study the electrical forces which are exerted by the system of charges on an elementary test charge e located at the point x^1, x^2, x^3 of space. As is known, the force, acting on a fixed test charge, depends only on its position and magnitude. It is given by

$$\boldsymbol{F} = e\boldsymbol{E}\,, \tag{1.2}$$

where $\boldsymbol{E}=\boldsymbol{E}(x^1, x^2, x^3)$ is a certain vector, referred to as electric field vector. The field of $\boldsymbol{E}$ is a result of the addition of the fields of individual charges.

It has been assumed initially that this field represents a mathematical abstraction which is convenient for the evaluation of the forces acting on a test charge. Recent studies have shown that one may consider the electric field $\boldsymbol{E}$ as an object which exists in space independently of the existence of a test charge; in other words, one may speak of an electric field as a material object which differs from a material body.

Two points of view may be adopted with respect to the relationship between charges and fields. It may be assumed that a field generates charges or that charges are singular points (physically small objects) of an existing electric field.

Analogous to the electric field vector, summation of the effects of elementary currents leads to the magnetic field vector $\boldsymbol{H}$ as a characteristic of a magnetic field with the aid of which one may assess the forces of magnetic interaction. In order to characterize an elementary test magnet or current, one introduces a small magnetic dipole moment $\boldsymbol{d}$ such that the moment of the couple, exerted by the field on the elementary magnetic needle with moment $\boldsymbol{d}$, is given by

$$\mathfrak{M} = \boldsymbol{d} \times \boldsymbol{H} \,. \tag{1.3}$$

If a magnetic field is homogeneous, one has $\boldsymbol{H} = \text{const.}$, and the general force exerted by the field on the elementary magnetic moment vanishes. Only the moment $\mathfrak{M}$ differs from zero.

If a field $\boldsymbol{H}$ is non-homogeneous, then one has, in addition to the moment $\mathfrak{M}$, a force $\mathbf{F}'$ given by

$$\mathbf{F}' = (\boldsymbol{d} \cdot \nabla)\boldsymbol{H} = (\mathrm{d}^i \nabla_i)\boldsymbol{H} = \mathrm{d}\,\frac{\partial \boldsymbol{H}}{\partial s}\,,$$

where $\partial/\partial s$ denotes differentiation in the direction of the moment of the dipole at the location of the elementary current.

Under static conditions, the stresses of an electric and magnetic field at a given point may be defined by the force (1.2) and the moment (1.3) which act on a fixed electric test charge located at this point and on an orientated elementary magnet (elementary electric current), respectively.

These simple experiments with test elements may be made more complicated, generalized and extended to the case of moving test charges and currents and to fields $\boldsymbol{E}$ and $\boldsymbol{H}$ which vary in time.

Thus, an electromagnetic field is characterized at every point of empty space and at every instant of time by two vectors: the electric field $\boldsymbol{E}$ and the magnetic field $\boldsymbol{H}$. The vectors $\boldsymbol{E}$ and $\boldsymbol{H}$, the charge density ρ_e and the vector of electric current density $\boldsymbol{j}$ are fundamental concepts of electrodynamics.

Maxwell equations in electrostatics. It is readily verified that Coulomb's law, determining the field $\boldsymbol{E}$ of a system of concentrated and distributed charges which are fixed with respect to an inertial coordinate system, may be written in the differential form

$$\operatorname{div} \boldsymbol{E} = 4\pi\rho_e\,, \quad \operatorname{curl} \boldsymbol{E} = 0\,. \tag{1.4}$$

The general solution of (1.4) in infinite space, under the condition that the vector $\boldsymbol{E}$ vanishes at infinity, leads to Coulomb's law.

The transition from the simple experimental Coulomb law to Equations (1.4), representing Maxwell's equations for a system of fixed charges, is a simple reformulation of the fundamental law of electrodynamics in terms of differential equations. This transition is in many respects analogous to that from the law of universal Newtonian gravitation to the differential equations in the theory of the Newtonian potential. In view of the importance of this transition, the above remarks will be explained in more detail.

The differential equations of the potential field in Newtonian mechanics. Let there be given two material points with masses m and m_k. They attract each other in accordance with Newton's law. The point mass m_k acts on the mass m with the force

$$\boldsymbol{F}_k = -f\frac{mm_k}{r_k^2}\,\boldsymbol{r}_k^0\,,$$

where r_k is the distance between m and m_k, f is the gravitational constant and $\boldsymbol{r}_k^0$ is the unit vector directed from m_k to m. This force is known to have a potential U_k, so that $\boldsymbol{F} = \operatorname{grad} U_k$ and

$$U_k = f\frac{mm_k}{r_k}\,.$$

If there exist in space n point masses m_k $(k = 1, 2, \ldots, n)$ and one considers their effect on a point mass $m = 1$ which may be placed at different points of space (test mass), then the test mass $m=1$ is subject, on account of all point masses m_k, to the force $\boldsymbol{F}=\Sigma \boldsymbol{F}_k$ with the potential

$$U=\sum_k U_k = f\sum_k \frac{m_k}{r_k}, \qquad \boldsymbol{F}_k = \text{grad } U_k . \tag{1.5}$$

The distribution of the masses m_k generates in space a gravitational field with the potential U which may be detected with the aid of a test mass placed at any space point under consideration.

Consider the differential equation to be fulfilled by the potential U of the gravitational force.

The function $1/r_k$, where

$$r_k=\sqrt{(x-x_k)^2+(y-y_k)^2+(z-z_k)^2}$$

is the distance between the point x, y, z of the test mass and the point x_k, y_k, z_k, the location of the kth mass generating the gravitational field, is harmonic. At all points x, y, z for which $r_k \neq 0$, the function $1/r_k$ satisfies Laplace's equation

$$\Delta\frac{1}{r_k}=\frac{\partial^2(1/r_k)}{\partial x^2}+\frac{\partial^2(1/r_k)}{\partial y^2}+\frac{\partial^2(1/r_k)}{\partial z^2}=0 .$$

Consequently, the potential U_k of the gravitational field of a single point mass satisfies the equation

$$\Delta U_k = 0$$

or

$$\text{div } \boldsymbol{F}_k = 0 , \quad \boldsymbol{F}_k = \text{grad } U_k , \quad \text{i.e., curl } \boldsymbol{F}_k = 0 . \tag{1.6}$$

Laplace's equation is linear. On the basis of (1.5), the potential $U(x,y,z)$ of the gravitational field, formed by a continuous distribution of mass over some volume V, may be written in the form

$$U(x, y, z)=f\int_V \frac{\rho\, d\tau}{r} \qquad (\rho\, d\tau = dm) . \tag{1.7}$$

Obviously, this function $U(x, y, z)$ satisfies at all points, where there is no mass, Laplace's equation

$$\Delta U = 0$$

which is equivalent to the equations

$$\operatorname{div} \boldsymbol{F} = 0, \quad \operatorname{curl} \boldsymbol{F} = 0, \quad \boldsymbol{F} = \operatorname{grad} U. \tag{1.8}$$

It may be shown that under very general, practically acceptable assumptions with respect to the distribution of the density ρ, the potential U of the gravitational field (1.7) for points x, y, z inside V satisfies Poisson's equation

$$\Delta U = -4\pi\rho, \tag{1.9}$$

which is equivalent to the equations

$$\operatorname{div} \boldsymbol{F} = 4\pi\rho, \quad \operatorname{curl} \boldsymbol{F} = 0, \quad \boldsymbol{F} = \operatorname{grad} U. \tag{1.10}$$

Thus, the problem of the determination of the potential of a gravitational field and of the forces, exerted by the field on a unit test mass, may be formulated as a problem of the determination of $U(x, y, z)$, vanishing at infinity and satisfying Laplace's equation everywhere outside V and Poisson's equation everywhere inside V, or as a problem of the determination of the force $\boldsymbol{F}$, satisfying Equations (1.8) and (1.10).

A formulation of the problems of electrostatics on the basis of Maxwell's equations (1.4) is completely analogous to the above type of formulation of problems in the theory of the Newtonian potential. It may be shown that the solution in infinite space of the problem of a function U which vanishes at infinity reduces to (1.7), expressing the law of gravitational attraction.[1)]

In the presence of unsteady electric and magnetic fields, one may obtain, on the basis of the experimental laws of induction, a generalization of Equations (1.4) by analogous means.

Maxwell's equations for an electromagnetic field in vacuum. In empty space, unoccupied by material bodies (where $\rho_e = 0$, since only such bodies can hold charges), these equations, referred to an inertial coordinate system, assume the form

$$\operatorname{curl} \boldsymbol{E} = -\frac{1}{c}\frac{\partial \boldsymbol{H}}{\partial t}, \qquad \operatorname{div} \boldsymbol{H} = 0, \tag{1.11}$$

$$\operatorname{curl} \boldsymbol{H} = \frac{1}{c}\frac{\partial \boldsymbol{E}}{\partial t}, \qquad \operatorname{div} \boldsymbol{E} = 0. \tag{1.12}$$

1) For more details see § 26, Chapter VIII.

Maxwell's equations (1.11) and (1.12) are conveniently introduced as an initial fundamental mathematical abstract formulation of experimental observations in place of Coulomb's law and of the other laws of electrodynamics, linked historically and practically more closely to ingenious experiments. In these equations, the constant c with the dimension of a velocity must be considered as the velocity of propagation of electromagnetic disturbances, i.e., as the velocity of light.

Equations (1.11) and (1.12), forming the basis of physics, are fundamental to optics and wireless engineering. They describe the propagation of light and, in general, of electromagnetic waves in vacuum as well as many other phenomena.

On dimensional units of electromagnetic quantities. Many electromagnetic characteristics are dimensional quantities. The writing down of concrete formulae and equations imposes definite links between dimensional units for the quantities involved. In particular, the use of Maxwell's equations in the form (1.11) and (1.12) assumes that $\boldsymbol{E}$ and $\boldsymbol{H}$ are measured in identical units.

In order to arrive at an independent choice of dimensional units for force, mass and acceleration, one must introduce into Newton's basic equation $\boldsymbol{F} = km\boldsymbol{a}$ the dimensional constant k; for k equal to the non-dimensional unit, the dimensional units depend on each other.

One may set the dimensional constant c in (1.11) and (1.12) or the constant f in the law of gravitation equal to unity; this is easily done and several authors have adopted this approach. However, if $c = 1$ in (1.11) and (1.12), then this means that either the velocity of light is accepted in the capacity of a unit for the measurement of velocities or the units for the measurement of $\boldsymbol{E}$ and $\boldsymbol{H}$ depend on the units of length and time. Analogously, for $f=1$ in the law of gravitation and simultaneously $k=1$ in Newton's law $\boldsymbol{F} = km\boldsymbol{a}$, the dimensional unit of mass is seen to depend on the units of distance and time, etc. Such types of supplementary conditions in many areas of experimentation, where these quantities are encountered, are not linked essentially to physical quantities and processes. Therefore such conditions, in general, are inconvenient, although logically acceptable. For example, the velocity of trains may, of course, be measured as fractions of the velocity of light; however, this is inconvenient, in spite of the fact that light is used in traffic controls along the railway tracks.

In astronomy and geography, it is nonsensical to measure distances between celestial bodies or towns exactly within centimetres; hence the use of the centimetre as unit of the length of astronomical or geographical distances is possible, but from a practical point of view senseless. It must be recalled here that the derivation of a number which characterizes a given length in an experiment involves the sequential placement of a measuring rod (unit of length) against the distance to be measured.

Naturally, physical scales for different quantities must be comparable with the quantities requiring measurement.

In general, an attainment of complete standardization and unification of units does not justify itself in many cases, since there arise various inconveniences and, in particular, a loss of feeling for the objects under study and a rupture with established living practice and tradition. In practice, one encounters difficulties even in the presence of the necessity, indisputable from a scientific point of view, of applying only a single dimensional unit, for example, the centimetre or the inch.

In general, it is inconvenient to fix physical constants by a choice of units. The velocity of light has a fundamental significance in physics. However, in many problems, it is quite inessential or may be set equal to infinity. Such problems are only made unnaturally complicated by the introduction of limitations connected with the magnitude of the velocity of light.

The form of Maxwell's equations (1.11) and (1.12) is "respectable"; it has been accepted in many texts, monographs and scientific publications, by many outstanding authors of the past and present.

No distraction will be introduced here by consideration of different systems of units for electromagnetic quantities, many of which are in existence. The question of units in electrodynamics is studied in detail in elementary and general physics texts. For numerical solution of concrete problems, a meaning of units is absolutely unnecessary. After fixing equations (1.11) and (1.12) and other laws, further developments of the general theory of the mechanics of continuous media are not linked to a concrete choice of a system of units.

The closed nature and lack of contradiction of the system of Maxwell's equations in vacuum. The system (1.11) and (1.12) contains eight equations for the determination of only six unknown characteristics of an electromagnetic field: E_1, E_2, E_3, H_1, H_2, H_3. However, this system is unique, because the relations

$$\operatorname{div} \boldsymbol{H} = 0 , \quad \operatorname{div} \boldsymbol{E} = 0$$

may be conceived as consequences of the first equations (1.11) and (1.12), if the initial conditions are given in a corresponding manner.

In fact, if $\boldsymbol{A}(x, y, z, t)$ is an arbitrary vector field which is differentiable with respect to the coordinates, then div curl $\boldsymbol{A} = 0$. Hence, it follows from the first equation (1.11) that

$$\frac{\partial}{\partial t} \operatorname{div} \boldsymbol{H} = 0 ,$$

i.e., that div $\boldsymbol{H}$ does not depend on the time t. If div $\boldsymbol{H} = 0$ at some initial instant of time, then div $\boldsymbol{H} = 0$ at all other times. This condition that the magnetic field is to be solenoidal is always fulfilled, if this condition is satisfied by the initial conditions. Thus, the equation div $\boldsymbol{H} = 0$ may be conceived as a limitation imposed on the initial conditions.

Analogously, it follows from the first equation (1.12) that

$$\frac{\partial}{\partial t} \operatorname{div} \boldsymbol{E} = 0 ,$$

i.e., if div $\boldsymbol{E} = 0$ at some definite instant of time, then it vanishes at all other times.

The equation div $\boldsymbol{E} = 0$ may be considered as a condition by which it is impossible for charges to appear in empty space, if they were not there at some "initial" instant of time; similarly, the equation div $\boldsymbol{H} = 0$ may be interpreted as the condition that magnetic charges are absent.

Although the equations div $\boldsymbol{H} = 0$ and div $\boldsymbol{E} = 0$ are not entirely consequences of the first equations, they do not contradict them. These conditions are essential limitations on the supplementary data for which solutions of Maxwell's equations have physical significance.

On the applicability of Maxwell's equations (1.11) *and* (1.12) *inside material bodies.* Maxwell's equations in the form (1.11) and (1.12) do not only serve for the description of electromagnetic fields in vacuum, but also for the study of such a field inside a body, if that body is electrically and magnetically neutral, i.e., when it does not contain macroscopic charges and under the influence of an external electromagnetic field there do not arise inside it electric currents, macroscopic polarization and magnetization.

Each molecule constituting a body has always around it an electromagnetic field. The interaction of these fields determines the internal stresses. However, charges of the particles which constitute a body may be distributed in such a manner that the mean of the proper electromagnetic field at each point of the body as well as the macroscopic charge in the body vanish:

$$\boldsymbol{E}_{\text{mean}} = 0\,, \quad \boldsymbol{H}_{\text{mean}} = 0\,;$$

at times, however, the quantities $\boldsymbol{E}_{\text{mean}}$ and $\boldsymbol{H}_{\text{mean}}$ remain zero also in the presence of an external electromagnetic field.

Thus, although microscopic fields always exist in atoms and molecules, due to the random motion of the atoms and other causes, they do not become apparent macroscopically.

In addition, note yet the very important experimental fact that Maxwell's equations (1.11) and (1.12) are also applicable to the description of microscopic fields right down to atomic scales.

On the significance of Maxwell's equations. The transition from elementary experimental facts and laws, describing in physics the first acquaintance with electric theory, to Maxwell's equations, equivalent to them, represents a simple reformulation.

In every day scientific life, one finds the opinion ingrained that reformulations of already established propositions and representations may, in essence, only be trivial. However, the example of the transition to the concentrated formulation of the laws of electrodynamics in the form of Maxwell's equations disproves this opinion. This is a brilliant achievement which serves as the basis for the development of all of contemporary physics. In the wake of the formulation of Maxwell's equations, the analysis of the nature of the electromagnetic field and of the properties of the system of Maxwell's equations has become the source of the theory of relativity and the corresponding fundamental revision of old concepts on inertial frames of reference, on space and time.

§2. Maxwell's equations in Minkowski space

Formulation of Maxwell's equations in four-dimensional space. For a more complete elucidation of the physical essence of Maxwell's equations, they will be rewritten in a new notation. Firstly, simply as notation, introduce the anti-symmetric matrix $F_{ij} = -F_{ji}$ in accordance with

$$\|F_{ij}\| = \begin{Vmatrix} 0 & F_{12} & F_{13} & F_{14} \\ F_{21} & 0 & F_{23} & F_{24} \\ F_{31} & F_{32} & 0 & F_{34} \\ F_{41} & F_{42} & F_{43} & 0 \end{Vmatrix} = \begin{Vmatrix} 0 & H^3 & -H^2 & cE_1 \\ -H^3 & 0 & H^1 & cE_2 \\ H^2 & -H^1 & 0 & cE_3 \\ -cE_1 & -cE_2 & -cE_3 & 0 \end{Vmatrix}, \tag{2.1}$$

and consider the four-dimensional manifold, i.e., the space with the coordinates $x^1, x^2, x^3, x^4 = t$, where x^1, x^2, x^3 will be accepted as the usual orthogonal Cartesian coordinates in three-dimensional geometrical space.

It is readily verified that the four equations (1.11) in projections on to Cartesian coordinate axes may then be written in the form

$$\frac{\partial F_{jk}}{\partial x^i} + \frac{\partial F_{ki}}{\partial x^j} + \frac{\partial F_{ij}}{\partial x^k} = 0 \qquad (i, j, k = 1, 2, 3, 4). \tag{2.2}$$

If one introduces into this system side by side with the matrix F_{ij} the matrix F^{ij}, defined by[1)]

$$\|F^{ij}\| = \begin{Vmatrix} 0 & H_3 & -H_2 & -\frac{1}{c}E^1 \\ -H_3 & 0 & H_1 & -\frac{1}{c}E^2 \\ H_2 & -H_1 & 0 & -\frac{1}{c}E^3 \\ \frac{1}{c}E^1 & \frac{1}{c}E^2 & \frac{1}{c}E^3 & 0 \end{Vmatrix}, \tag{2.3}$$

then the four equations (1.12) assume the form

$$\frac{\partial F^{ij}}{\partial x^j} = 0 \qquad (i = 1, 2, 3, 4). \tag{2.4}$$

[1)] The arrangement of the indices in (2.1) and (2.3) is essential in the case of transformation to the curvilinear coordinates only (For more details see pp. 205–207, Chapter IV).

Interpretation of the quantity c as the velocity of light. It is not difficult to show that the general solution of (2.2) may be given the form

$$F_{ij} = \frac{\partial A_i}{\partial x^j} - \frac{\partial A_j}{\partial x^i}, \tag{2.5}$$

where A_i, $i = 1, 2, \ldots, 4$ are four arbitrary functions of x^1, x^2, x^3. Obviously, the asymmetry of F_{ij} will be satisfied in this manner.

Note that the values of F_{ij} do not change, if one adds to A_i the components of the four-dimensional gradient vector $\partial\Psi/\partial x_i$, respectively. Using this arbitrariness, the choice of the functions A_i will be normalized further by the supplementary condition

$$\frac{\partial A_1}{\partial x^1} + \frac{\partial A_2}{\partial x^2} + \frac{\partial A_3}{\partial x^3} - \frac{1}{c^2}\frac{\partial A_4}{\partial x^4} = 0\,. \tag{2.6}$$

For any given system of four functions A_i, one may determine a function Ψ such that (2.6) is satisfied. Equation (2.6) may be interpreted as a condition which removes the arbitrariness in the choice of the function Ψ.

After replacing F^{ij} in (2.4) by

$$F_{ij} = \frac{\partial A_i}{\partial x^j} - \frac{\partial A_j}{\partial x^i},$$

one obtains on the basis of (2.6) for the four functions A_i the four identical equations

$$\frac{\partial^2 A_i}{\partial (x^1)^2} + \frac{\partial^2 A_i}{\partial (x^2)^2} + \frac{\partial^2 A_i}{\partial (x^3)^2} - \frac{1}{c^2}\frac{\partial^2 A_i}{\partial t^2} = 0 \quad (i = 1, 2, 3, 4)\,. \tag{2.7}$$

Thus, the solution of Maxwell's equations may be reduced to the problem of the determination of the four functions A_i, satisfying the relations (2.6), each of which fulfills Equation (2.7), known as the wave equation.

The general properties of the solutions of the wave equation will be studied in greater detail in Chapter VIII, Vol. II. However, one typical property of the solutions of this equation will be stated immediately. Let $f(\xi)$ be an arbitrary function which is twice differentiable with respect to its argument. It is easily seen that the function

$$A = f(x^1 - ct)\,, \tag{2.8}$$

is a solution of the wave equation. According to this solution, a given value $A = f(\xi)$, corresponding to some fixed value $\xi = x^1 - ct$, propagates along the x^1-axis with velocity c. Hence the meaning of c as the velocity of light is explained.

Minkowski space. Consider the four-dimensional, metric, pseudo-Euclidean[1)] Minkowski space corresponding to the coordinates $x^1, x^2, x^3, x^4 = t$ in which, by definition, the metric is given by[2)]

$$ds^2 = -dx^{1^2} - dx^{2^2} - dx^{3^2} + c^2 dt^2 = g_{ij} dx^i dx^j . \tag{2.9}$$

For this metric, one finds

$$\|g_{ij}\| = \left\| \begin{matrix} -1 & 0 & 0 & 0 \\ 0 & -1 & 0 & 0 \\ 0 & 0 & -1 & 0 \\ 0 & 0 & 0 & c^2 \end{matrix} \right\| , \qquad \|g^{ij}\| = \left\| \begin{matrix} -1 & 0 & 0 & 0 \\ 0 & -1 & 0 & 0 \\ 0 & 0 & -1 & 0 \\ 0 & 0 & 0 & 1/c^2 \end{matrix} \right\|$$

It is seen from Definitions (2.1) and (2.3) that F^{ij} and F_{ij} are interrelated by

$$F^{ij} = F_{pq} g^{pi} g^{qj} .$$

Transformation of Maxwell's equations to arbitrary curvilinear coordinates. If one introduces side by side with the coordinates $x^1, x^2, x^3, x^4 = t$ an arbitrary curvilinear coordinate system y^1, y^2, y^3, y^4, linked to the $x^1, x^2, x^3, x^4 = t$ by the transformation

$$y^i = f^i(x^1, x^2, x^3, x^4) \qquad (i = 1, 2, 3, 4) , \tag{2.10}$$

then the transformed formula for ds^2 assumes the form

1) This space is said to be pseudo-Euclidean, because the metric (2.9) is not positive definite, but g_{ij} may be constant throughout space.

2) Here and farther on the following conditions are assumed: for the time component $g_{44} > 0$, for the space components $g_{\alpha\alpha} < 1$. In addition, the four-dimensional indices are denoted by the Latin letters while the three-dimensional space components, introduced separately are denoted by the Greek letters. These conditions have been used by H. Weyl (⟨⟨ Space-Time-Matter ⟩⟩, 1922) and nowadays the same conditions are accepted by a lot of authors and used in widely known textbooks. On the other hand, in the number of books and in some scientific publications just the opposite conditions, with respect to the sign of g_{44} and $g_{\alpha\alpha}$ and the denotation of indices, are used.

$$ds^2 = g'_{ij}dy^i dy^j\,, \quad \text{for} \quad g'_{ij} = g_{pq}\frac{\partial x^p}{\partial y^i}\frac{\partial x^q}{\partial y^j}\,.$$

The transformed Maxwell equations (2.2) and (2.4) and Equation (2.5) are readily written down, if one considers F_{ij} and A_i as components of a tensor and vector in Minkowski space, i.e., F'_{ij} and A'_i in the new system are determined by

$$F'_{ij} = F_{pq}\frac{\partial x^p}{\partial y^i}\frac{\partial x^q}{\partial y^j}\,, \qquad A'_i = A_p\frac{\partial x^p}{\partial y^i} \tag{2.11}$$

The four-dimensional tensor $F = F_{ij}\mathbf{з}^i\mathbf{з}^j$ is known as the tensor of the electromagnetic field, and the four-dimensional vector $A = A_i\mathbf{з}^i$ as the vector potential. With the aid of the formulae of tensor calculus one obtains

$$\nabla'_k F'_{ij} + \nabla'_j F'_{ki} + \nabla'_i F'_{jk} = 0\,, \tag{2.12}$$

$$\nabla'_j F'^{ij} = 0\,, \tag{2.13}$$

$$F'_{ij} = \nabla'_j A'_i - \nabla'_i A'_j = \frac{\partial A'_i}{\partial y^j} - \frac{\partial A'_j}{\partial y^i}\,. \tag{2.14}$$

The wave equation (2.7), when the condition

$$\nabla'^{i\cdot} A'_i = 0 \tag{2.15}$$

is fulfilled, transforms to

$$\nabla'^q \nabla'_q A'_i = 0\,. \tag{2.16}$$

These equations follow from the tensor form of (2.2), (2.4), (2.5) and (2.7) in Minkowski space for which, obviously, derivatives with respect to the coordinates and time in the system x^1, x^2, x^3, t coincide with the covariant derivatives, since in this system all Christoffel symbols are equal to zero.

Note that in view of the asymmetry of F_{ij} in an arbitrary coordinate system the terms with Christoffel symbols in (2.12) cancel, and therefore (2.12) may be written in arbitrary curvilinear coordinates also in the form

$$\frac{\partial F'_{ij}}{\partial y^k} + \frac{\partial F'_{ki}}{\partial y^j} + \frac{\partial F'_{jk}}{\partial y^i} = 0\,. \tag{2.17}$$

Equation (2.13), in expanded form, becomes

$$\nabla'_j F'^{ij} = \frac{\partial F'^{ij}}{\partial y^j} + F'^{pj}\Gamma^i_{pj} + F'^{ip}\Gamma^j_{jp} = 0 .$$

It follows from the antisymmetry of F^{pj} and the symmetry with respect to the subscripts of Γ^i_{pj} that $F'^{pj}\Gamma^i_{pj}=0$; in addition, by (3.6),

$$\Gamma^j_{pj} = \frac{1}{\sqrt{-g}}\,\frac{\partial\sqrt{-g}}{\partial y^p} ,$$

where $g=|g_{ij}|$. Hence the following form of the second pair of Maxwell's equations (1.12) is also valid:

$$\nabla'_j F'^{ij} = \frac{1}{\sqrt{-g}}\,\frac{\partial\sqrt{-g}\,F'^{ij}}{\partial y^j} = 0 . \tag{2.18}$$

Thus, Maxwell's equations may be written in tensor form with the aid of the tensor F_{ij} in four-dimensional Minkowski space which has been especially introduced by definition.

With the aid of the tensor equations derived, one may consider in different reference frames the question of the form of Maxwell's equations and of the components of the magnetic and electric stresses H^1, H^2, H^3 and E^1, E^2, E^3, starting from (2.1) and (2.3) and the transformation formulae (2.11).

Metric Minkowski space has been introduced as an auxiliary mathematical image. This has been done almost solely in connection with the fact that one may consider for coordinate transformations in Minkowski space the tensor F_{ij} and that Maxwell's equations in this space may be studied as tensor equations.

An interpretation of the Minkowski space, thus introduced, as a physical space and, in this connection, even the interpretation of the tensor of the electromagnetic field as a physical object arise only after the acceptance of supplementary physical postulates the existence and meaning of which will be considered next.

Transformation restricted to space coordinates. The following derivatives will still be noted which are a consequence of the mathematical definitions. Select a particular transformation of the form

$$\left.\begin{aligned} y^\alpha &= f^\alpha(x^1, x^2, x^3)\,, \\ t' &= y^4 = x^4 = t\,, \end{aligned}\right\} \quad \alpha = 1, 2, 3, \tag{2.19}$$

in which only the space coordinates are transformed and the time coordinate is conserved.

It is readily verified, on the basis of (2.11) and the matrix formula[1] (2.1), that the corresponding transformation formulae for E_k and H^α represent transformation formulae for the covariant and contravariant components of polar and axial vectors, respectively, in three-dimensional space. For the transformation (2.10) in general form, one may likewise consider in different coordinate systems the quantities E_i and H^i; however, the corresponding transformations will not be transformations of the components of any vectors.

Lorentz transformation. Consider the transformation $y^i = f^i(x^k)$ of the most general form which conserves the appearance of the quadratic form (2.9) for the metric Minkowski space, i.e., a transformation for which

$$\begin{aligned} ds^2 &= -dx^{1^2} - dx^{2^2} - dx^{3^2} + c^2 dt^2 \\ &= -dy^{1^2} - dy^{2^2} - dy^{3^2} + c^2 dt^2 . \end{aligned} \tag{2.20}$$

As for any coordinate transformation, also for such transformations, which are referred to as Lorentz transformations, the expanded forms of the tensor equations (2.12), (2.13) and (2.17), (2.18), respectively, differ only in the notation of the coordinates and the components F_{ij}.

Invariance of Maxwell's vector equations with respect to the Lorentz transformation. The Lorentz transformation preserves also the form of Maxwell's vector equations (1.11) and (1.12). However, one finds from the transformation formulae (2.11) that the electric and magnetic stress vectors $\boldsymbol{E}$ and $\boldsymbol{H}$ in the system x^i and the vectors $\boldsymbol{E}'$ and $\boldsymbol{H}'$ in the system y^i are different. The following section deals with the transition formulae from $\boldsymbol{E}$, $\boldsymbol{H}$ to $\boldsymbol{E}'$, $\boldsymbol{H}'$ in the case of Lorentz transformations.

Thus, Maxwell's equations under corresponding conditions with respect to the transformation of the vectors $\boldsymbol{E}$ and $\boldsymbol{H}$ are invariant to Lorentz transformations. It can be sensed from (2.17) and (2.18) that, beside the Lorentz transformation, a class of transformations wider than

[1] In a curvilinear space coordinate system one must understand by H^i in (2.1) the quantity $H^i \sqrt{\operatorname{Det} \| g_{\alpha\beta} \|}$, $\alpha, \beta = 1, 2, 3$ (*cf.*, § 3, Chapter IV, Formula (3.26)).

the Lorentz transformations may be stated for which the invariance of Maxwell's equations holds true. However, as will be shown below, the Lorentz transformation has special physical significance.

Galilei transformations. Transformations of the form

$$\left.\begin{aligned} y^\alpha &= x^\alpha + a_0^\alpha - v^\alpha t\,, \quad \alpha = 1, 2, 3\,, \\ y^4 &= t' = t + t_0 = x^4 + t_0\,, \end{aligned}\right\} \tag{2.21}$$

where t_0, a_0^α and v^α are constants, are referred to as Galilei transformations. In Newtonian mechanics, these transformations correspond to uniform rectilinear translation of the reference frame y^α with respect to the reference frame x^α, where v^α are the velocity components of this motion in the system x^α.

Obviously, Galilei transformations are not Lorentz transformations, since they do not fulfill the equality (2.20). Galilei transformations and Formulae (2.21) may be made more complicated by supplementary rotation of the system y^α by a fixed finite angle about some arbitrarily selected axis and reflections in the coordinate planes. Incidentally, it should be noted that any rotation may be replaced by a set of reflections with respect to certain planes.

Tensor and vector characteristics in Minkowski space; energy-impulse tensor. In connection with Maxwell's equations, the tensor $F = F_{ij}\mathbf{з}^i\mathbf{з}^j$ of the electromagnetic field and the vector potential $A = A_i\mathbf{з}^i$ have been introduced in four-dimensional Minkowski space. In order to develop the theory further, many other vectors and tensors will be introduced. Examples are the four-dimensional electric current vector $\boldsymbol{J} = J^i\mathbf{з}_i$ (§4), the four-dimensional force vector $\boldsymbol{F}$ (§5), the four-dimensional velocity vector $\boldsymbol{u} = u^i\mathbf{з}_i = \mathrm{d}\boldsymbol{r}/\mathrm{d}s$, where $\mathrm{d}\boldsymbol{r} = \mathrm{d}x^i\mathbf{з}_i$ is the four-dimensional displacement of an individual point and $\mathrm{d}s = |\mathrm{d}\boldsymbol{r}|$, the energy-impulse tensor of the electromagnetic field

$$S = S_i^k\mathbf{з}^i\mathbf{з}_k\,, \tag{2.22}$$

where

$$S_{i\cdot}^{\cdot k} = -\frac{1}{4\pi}\left[F_{mi}F^{mk} - \tfrac{1}{4}\delta_i^k F_{mn}F^{mn}\right],$$

and many other vectors and tensors.

From (2.12) and (2.13) follow the important relations

$$\nabla_k S_{i.}^{\cdot k} = 0 \ , \quad i = 1, 2, 3, 4 \tag{2.23}$$

which may be treated as the impulse and energy equations for the electromagnetic field in vacuum.

§3. Lorentz transformations and inertial reference frames

The Principle of relativity in physics. A basic physical premise, the universal principle of relativity of Galilei, consists of the assertion that all physical laws of interactions in material media and of fields are formulated equally and that all physical processes and phenomena proceed equally, from the point of view of an observer, in any inertial reference system. The existence of a set of inertial coordinate systems, which may move with respect to each other, is formulated. However, the set of inertial reference systems may be defined in different theories by different means.

Inertial systems in Newtonian mechanics. In Newtonian physics, it is assumed that physical space is three-dimensional (Euclidean) and time is absolute; there may be defined coordinate systems which translate with constant velocity in universal time with respect to each other in all possible manners and which form the entire set of inertial reference systems. The set of inertial systems is determined by the condition that in these systems an isolated material point is at rest or moves with constant velocity.

The principle of relativity of Galilei-Newton states that all physical equations and laws must be invariant with respect to Galilei transformations (2.21), determining in Cartesian coordinate systems the transition from a Cartesian coordinate system x^1, x^2, x^3, t to another Cartesian inertial coordinate system $y^1, y^2, y^3, t + t_0$.

In Newtonian physics, Equations (2.21) and the kinematic definition of velocity lead to the law of addition of velocities

$$\boldsymbol{v}_x = \boldsymbol{v}_y + \boldsymbol{v}\,, \tag{3.1}$$

where $\boldsymbol{v}_y$ is the velocity of an object, a point, with respect to the system y, $\boldsymbol{v}_x$ is the velocity of the same object with respect to the system x and $\boldsymbol{v}$ is the velocity of translation of the inertial system y with respect to the system x.

The velocity of light in empty space can be defined as the velocity of the front of electromagnetic disturbances or, more simply, as the velocity of motion in empty space of electromagnetic particles, i.e., of photons. By (3.1), in Newtonian physics, one must find that the velocity of light differs for different observers, performing measurements in their respective inertial coordinate systems. Besides, it follows from the absolute nature of time that in Newtonian physics signals may be transmitted at infinite speeds.

Postulate on the constancy of the speed of light. However, these deductions are found to be in radical contradiction with experiments which show that light propagates through empty space with one and the same velocity with respect to any observers who move with respect to each other with constant velocity and that it is isotropic with respect to any observer, i.e., it travels at the same speed in all directions. Michelson's significant experiment and a multitude of other experiments have shown that the speed of light does not depend on the choice of an inertial coordinate system.

A deeper study of physical processes likewise displays that motion of material objects and propagation of energy with a velocity in excess of the speed of light are impossible; the speed of light may be considered to be the limiting possible velocity of any relative motion of material objects.

Therefore, contemporary physics has been based on the postulate that the speed of light is constant in all inertial reference frames. In order to preserve Galilei's basic physical principle of relativity, the postulate on the constancy of the speed of light serves as a base for a change of the ideas on inertial systems and for a search for new transformations, to replace the Galilei transformations (2.21), in order to determine the transition from one inertial system to another.

In this case, Galilei's transformations (2.21) become inapplicable; these transformations become more complicated and the existence of absolute time must be questioned.

On inertial coordinate systems in the special theory of relativity. Consider now the conditions imposed by the constancy of the speed of light on the coordinate transformations for transitions from one inertial system to another.

The set of inertial systems may be obtained with the aid of a system of transformations from a unique system which is made distinct by a condition based on experimental data.

In order to find the corresponding transformation formulae, use will be made of the basic condition on the constancy of the speed of light in all inertial systems as well as of the natural conditions of the equivalence of any two inertial systems and of their reversible symmetry, i.e., isotropy and homogeneity. The concept of homogeneity, connected with the geometric and kinematic equivalence of all points of space, will be explained further in a more concrete mathematical formulation.

Transformation of coordinates for transition from one inertial system to another. As the initial inertial system K, the choice of which is linked to experimental data, select the coordinate system $x^1, x^2, x^3, x^4 = t$ in which the space coordinates x^1, x^2, x^3 are considered to be Cartesian coordinates in three-dimensional Euclidean space and the variable t the time.

Let $y^1, y^2, y^3, y^4 = t'$ be another inertial coordinate system K' in which y^1, y^2, y^3 are likewise Cartesian coordinates and t' its time. Consider the problem of establishing the properties and of studying the conditions for the determination of the transformation formulae

$$y^i = f^i(x^1, x^2, x^3, x^4), \qquad i = 1, 2, 3, 4, \tag{3.2}$$

which must replace the transformation formulae (2.21) of Galilei-Newton.

Let $\mathrm{d}x^1, \mathrm{d}x^2, \mathrm{d}x^3$ be the displacement components of some movable point M over the time $\mathrm{d}x^4 = \mathrm{d}t$ in the system K and, correspondingly, for this same point $\mathrm{d}y^1, \mathrm{d}y^2, \mathrm{d}y^3$ the displacements and $\mathrm{d}y^4 = \mathrm{d}t'$ the time interval in the system K'.

The corresponding three-dimensional velocities $\boldsymbol{v}$ and $\boldsymbol{v}'$ of the point M in the systems K and K' are given by

$$v^2 = \frac{\mathrm{d}x^{1^2} + \mathrm{d}x^{2^2} + \mathrm{d}x^{3^2}}{\mathrm{d}t^2}, \qquad v'^2 = \frac{\mathrm{d}y^{1^2} + \mathrm{d}y^{2^2} + \mathrm{d}y^{3^2}}{\mathrm{d}t'^2}.$$

In these formulae, the quantities y^α and x^β ($\alpha, \beta=1, 2, 3$) correspond to each other by (3.2).

Introduce the quantities

$$ds_x^2 = (c^2 - v^2)dt^2 = -dx^{1^2} - dx^{2^2} - dx^{3^2} + c^2 dt^2 ,$$

$$ds_y^2 = (c^2 - v'^2)dt'^2 = -dy^{1^2} - dy^{2^2} - dy^{3^2} + c^2 dt'^2 .$$

Consider the motion of a photon, which is known to move in vacuum with the speed of light, in the system K as well as in the system K',

If one has simultaneously $v = v' = c$, then

$$ds_y^2 = ds_x^2 = 0 .$$

If one sets $ds_x^2=0$, then the equality

$$ds_y^2 = \left(\frac{\partial f^1}{\partial x^p} dx^p\right)^2 + \left(\frac{\partial f^2}{\partial x^p} dx^p\right)^2 + \left(\frac{\partial f^3}{\partial x^p} dx^p\right)^2 - c^2\left(\frac{\partial f^4}{\partial x^p} dx^p\right)^2 = 0$$

must be fulfilled as a consequence of the special form of the transformation (3.2); hence, in the general case $v \neq c$, one must satisfy

$$ds_y^2 = \varkappa(x^1, x^2, x^3, x^4) ds_x^2 ,$$

where $\varkappa$ may be an arbitrary function of its arguments.

It follows from the symmetry of the transitions $K \to K'$ and $K' \to K$ that

$$ds_x^2 = \varkappa(y^1, y^2, y^3, y^4) ds_y^2 = \varkappa(y^i)\varkappa(x^i) ds_x^2 ,$$

whence

$$\varkappa(y^i)\varkappa(x^i) = 1 . \tag{3.3}$$

It follows from the homogeneity of the space that $\varkappa$ must be independent of the points of space, i.e., of the coordinates. It follows from the constancy of $\varkappa$ and (3.3) that $\varkappa = 1$.

Consequently, it is seen that for transformations of transition from one inertial system K to another inertial system K' the quantity

$$ds^2 = -dx^{1^2} - dx^{2^2} - dx^{3^2} + c^2 dt^2 \tag{3.4}$$

must be invariant. This formula may be considered as the definition of four-dimensional metric space.

Thus, physical space may be considered as a Minkowski space and the transformations of transition from one inertial coordinate system to another as Lorentz transformations.

It has now been demonstrated that Minkowski space and Lorentz transformations, introduced earlier as auxiliary mathematical devices for the study of transformations of Maxwell's equations, have fundamental physical significance.

Obviously, great importance in the development of the theory attaches to the invariance of Maxwell's vector equations with respect to Lorentz transformations. The assumptions and deductions made above constitute the foundations of contemporary physics.

The acceptance of such a system of postulates in global systems for finite bodies represents the special theory of relativity. The restricted employment of these postulates in local, small elements of material bodies or fields establishes the base for the construction of the general theory of relativity.

Properties of Lorentz transformations. The properties of Lorentz transformations will now be studied in greater detail. First of all, it will be shown that the functions $f^i(x^k)$ in (3.2), which define the Lorentz transformations, are linear.

A more general proposition will be proved. Let there be given the transformation

$$y^i = f^i(x^1, x^2, x^3, x^4)\,, \qquad \frac{\partial f^i}{\partial x^k} = f^i_k\,, \quad |f^i_k| \neq 0\,,$$

and let

$$ds^2 = g'_{ij} dy^i dy^j = g_{pq} dx^p dx^q\,, \; g'_{ij} f^j_p f^i_q = g_{pq}\,, \tag{3.5}$$

where $g'_{ij}=\text{const.}$, $g_{pq}=\text{const.}$ and $|g'_{ij}| \neq 0$. It will be shown that then the functions $f^i(x^1, x^2, x^3, x^4)$ are linear in all their arguments.

Differentiating (3.5) with respect to x^s, one finds

$$g'_{ij} f^i_{ps} f^j_q + g'_{ij} f^i_p f^j_{qs} = 0\,, \tag{3.6}$$

where p, s, q are arbitrary subscripts assuming the values 1, 2, 3, 4. Obviously, by their symmetry, one has $g'_{ij}=g'_{ji}$ and $f^k_{sp}=f^k_{ps}$. After transpositions $p \to q \to s \to p$ in (3.6), one may write

$$g'_{ij} f^i_{qp} f^j_s + g'_{ij} f^j_q f^i_{ps} = 0\,. \tag{3.7}$$

Subtraction of (3.7) from (3.6) yields

$$g'_{ij}f^i_p f^j_{qs} - g'_{ij}f^l_{qp}f^j_s = 0\,. \tag{3.8}$$

Now transpose in (3.8) $p \to s \to q \to p$ and subtract the result from (3.7) to obtain

$$g'_{ij}f^j_q f^i_{ps} = 0 \quad \text{or} \quad A_{iq}f^i_{ps} = 0\,, \tag{3.9}$$

where p, q, s are any subscripts 1, 2, 3, 4. The determinant $|A_{iq}|$ is non-zero, since one has the matrix equation

$$\|A_{iq}\| = \|g'_{ij}\|\,\|f^j_q\| \quad \text{and} \quad |g'_{ij}| \neq 0\,, \qquad |f^j_q| \neq 0\,.$$

Since $|A_{iq}| \neq 0$, it follows from the system of homogeneous equations (3.9) that

$$f^i_{ps} = \frac{\partial^2 f^i}{\partial x^p \partial x^s} = 0\,. \tag{3.10}$$

Equations (3.10) are true for arbitrary i, p and s; hence the general solution of the system (3.10) has the form

$$y^i = f^i(x^k) = f^i_0 + c^i_k x^k\,, \tag{3.11}$$

with f^i_0 and c^i_k constants.

Obviously, the proposition proved is true for any n-dimensional space. Transformation (3.11) corresponds to a homogeneous deformation of four-dimensional pseudo-Euclidean space; it is more general than the Lorentz transformation.

In order to isolate among the transformations (3.11) the Lorentz transformations, one must replace (3.5) by stronger conditions following from (2.20):

$$\left.\begin{aligned} &g'_{11} = g_{11} = -1\,, \qquad g'_{22} = g_{22} = -1\,,\\ &g'_{33} = g_{33} = -1\,, \qquad g'_{44} = g_{44} = c^2\,,\\ &g'_{ij} = g_{ij} = 0 \quad \text{for} \quad i \neq j\,. \end{aligned}\right\} \tag{3.12}$$

Conditions (3.12) impose on the 16 coefficients $c^{i}_{.k}$ the ten algebraic relations

$$g_{ij}c^i{}_p c^j{}_q = g_{pq} \quad \text{or} \quad c_{ip}c^{iq} = \delta^{q\cdot}_{\cdot p}\,. \tag{3.13}$$

Thus, the general Lorentz transformation depends on the four constants f^i_0, determining a simple translation, and on six independent parameters in terms of which one may express by means of (3.13) the sixteen quantities c^i_k.

The transformations of Galilei and Lorentz contain the same number of independent parameters, namely ten. Galilei transformations form a group which depends on ten parameters; four of these determine the translation, three the rotation and mirror reflection of three-dimensional space and three the translatory velocity components of the coordinate system.

Lorentz transformations likewise form a group which depends on ten parameters; from the four of these, which determine the translation, and from the six, which determine the mirror reflection and rotation of the four-dimensional Minkowski space, one obtains independent parameters in terms of which the coefficients $c^i_{.k}$ are expressed.

Transformations which satisfy (3.13) for an arbitrary metric and in spaces of any dimensionality are said to be orthogonal. Orthogonal transformations form a group. Orthogonal transformations with the determinant $\Delta = |c^{i}_{.k}| > 0$ form a semi-group, the so-called characteristic group of rotations.

Infinitesimal Lorentz transformations. For the identity transformation in the group of orthogonal transformations, one has

$$c^i{}_k = \delta^i{}_k .$$

Instead of $c^i_{.k}$, introduce the quantities $\Omega^i_{.k}$, defined by

$$c^i{}_k = \delta^i{}_k + \Omega^i{}_k .$$

For the identity transformation, one has $\Omega^i_{.k} = 0$. If $\Omega^i_{.k}$ is infinitesimal, then the coefficients $c^i{}_k$ define a transformation which belongs to the characteristic group of rotations, since $\Delta \approx 1$. The quantities $c^i_{.k}$ and $\Omega^i_{.k}$ may be interpreted as components of tensors on the base of the system K or on the base of the system K' or as vectors with respect to the indices i and k in K and K', respectively.

The conditions of orthogonality for infinitesimal rotations after elimination of small second order quantities have the form

$$g_{iq}\Omega^i{}_p + g_{pj}\Omega^j{}_q = 0 \tag{3.14}$$

or

$$\Omega_{qp} = -\Omega_{pq} .$$

The asymmetric matrix elements Ω_{pq} may be considered as components of a corresponding, four-dimensional, anti-symmetric tensor in Minkowski space. Six independent components of this tensor may be considered as independent parameters which determine infinitesimal rotations in four-dimensional Minkowski space. Two three-dimensional vectors $\boldsymbol{\Omega}$ and $\boldsymbol{U}$ may be brought into correspondence with the four-dimensional tensor Ω_{pq} in inertial Cartesian coordinate systems by the relationship

$$\|\Omega_{pq}\| = \begin{Vmatrix} 0 & -\Omega^3 & \Omega^2 & -U_1 \\ \Omega^3 & 0 & -\Omega^1 & -U_2 \\ -\Omega^2 & \Omega^1 & 0 & -U_3 \\ U_1 & U_2 & U_3 & 0 \end{Vmatrix}, \qquad \Omega^i{}_q = g^{ip}\Omega_{pq} . \tag{3.15}$$

The axial vector $\boldsymbol{\Omega}(\Omega^1, \Omega^2, \Omega^3)$ is readily seen to characterize an infinitesimal, three-dimensional space rotation, and the polar vector $U(U_1, U_2, U_3)$ the infinitesimal translatory velocity of the inertial system K' with respect to the system K.

Lorentz transformations for translation of the system K' with respect to the system K. Consider an important finite particular Lorentz transformation (3.11) of the form

$$y^1 = c^1{}_1 x^1 + c^1{}_4 t , \quad y^2 = x^2 , \quad y^3 = x^3 , \quad t' = c^4_{\ 1} x^1 + c^4{}_4 t .$$

It corresponds to a translation along the x^1-axis of the system K' with respect to the system K with constant velocity $U = -c^1_4 / c^1_1$. In this case, the conditions (3.13) are readily solved; one obtains, under the condition $\Delta > 0$ which is linked to the absence of an additional reflection,

$$\left.\begin{aligned} y^1 &= \frac{x^1 - Ut}{\sqrt{1 - \dfrac{U^2}{c^2}}}, & t' &= \frac{t - \dfrac{U}{c^2} x^1}{\sqrt{1 - \dfrac{U^2}{c^2}}}, \\ x^1 &= \frac{y^1 + Ut'}{\sqrt{1 - \dfrac{U^2}{c^2}}}, & t &= \frac{t' + \dfrac{U}{c^2} y^1}{\sqrt{1 - \dfrac{U^2}{c^2}}}. \end{aligned}\right\} \tag{3.16}$$

The Lorentz transformation (3.16) becomes a Galilei transformation when U/c is small, after neglecting quantities of the order $(U/c)^2$.

The relativity of the concept of time. By the general formulae (3.11) and, in particular, by (3.16) for the transition from K to K', the time t' in the "mobile system" differs from the time in the original "immobile" system. In addition, the geometrical coordinates and time remain to a certain degree equivalent. However, one cannot speak of a complete elimination of the differences between geometric distances and time intervals. This difference becomes apparent, in particular, through the fact that the elements $dx^{\alpha 2}$ and dt^2 enter into the definition of the metric for ds^2 with opposite signs. It is known from the theory of quadratic forms that for any material transformations of variables, which preserve the canonical appearance of the quadratic form for ds^2, the signs stated are invariant.

Different time intervals. Consider the link, which follows from (3.16), between corresponding intervals of time $dt'$and dt at the point M, immobile in System K' ($y^1 = \text{const.}$) and, consequently, moving with velocity U in System K. One has

$$dt = \frac{dt'}{\sqrt{1 - \frac{U^2}{c^2}}}, \quad \text{i.e.,} \quad dt > dt'. \tag{3.17}$$

Analogously, let the point N be immobile in System K ($x^1 = \text{const.}$); then

$$\widetilde{dt}' = \frac{\widetilde{dt}}{\sqrt{1 - \frac{U^2}{c^2}}}, \quad \text{i.e.,} \quad \widetilde{dt}' > \widetilde{dt}. \tag{3.18}$$

Thus, from the point of view of an observer in K, the special time to the observer in K' flows more slowly. On the other hand, for the observer in K', the characteristic time in K flows more slowly. This is the result of the complete equivalence of the inertial systems K and K'.

Similar relations will also apply to the length of corresponding segments, lying in the direction of the velocity $\boldsymbol{U}$. The lengths of segments, perpendicular to the direction of $\boldsymbol{U}$, are the same in K and K'.

Characteristic reference system. In applications, special significance attaches to the characteristic reference system; this inertial reference system K^* is chosen at a given point M of a medium in such a manner that the velocity of the point M with respect to K^* at a given instant of time vanishes; the velocities of adjoining points and of the point M at other instants of time in this system may differ from zero.

With the use of the characteristic reference system and of characteristic time, one deals with personal sensations. Characteristic time is this invariant characteristic of ageing and of all possible internal processes and internal interactions.

For motions of elementary particles, atoms and molecules, the laws for all internal interactions and all typical times are the same only in characteristic reference systems; for example, the periods of radiated light or radio waves, the half-times of radioactive substances, the times of existence of unstable "elementary" particles, etc.

Characteristic time. For a study of the motion of a continuum, one may consider at each point and at each instant of time its characteristic coordinate system and use in this system values of the components of different tensors, vectors and their characteristic time interval $d\tau$, defined by

$$d\tau = \frac{1}{c}\,ds = dt\sqrt{1 - \frac{v^2}{c^2}} \qquad (v^2 < c^2,\ ds > 0)\,, \tag{3.19}$$

where dt is the corresponding time interval and v is the velocity in the fixed reference frame of an observer.

Characteristic and concomitant coordinate systems. Obviously, characteristic coordinate systems do not, in general, coincide with concomitant coordinate systems in which the velocities of all particles always vanish; a characteristic system is inertial, while, in general, a concomitant system is, of course, not inertial.

Time paradox. In order to consider various motions of one and the same point or various motions of different points, one may compute finite intervals of characteristic time with the aid of integrals, along the path (along the world line), of the form

$$\Delta\tau = \tau - \tau_0 = \frac{1}{c}\int ds = \int_{t_0}^{t}\sqrt{\left(1 - \frac{v^2}{c^2}\right)}\,dt\,. \tag{3.20}$$

The characteristic time interval $\Delta\tau$ is invariant, and therefore (3.20) yields the same result in calculations for use of any observer's coordinate system, whether inertial or not.

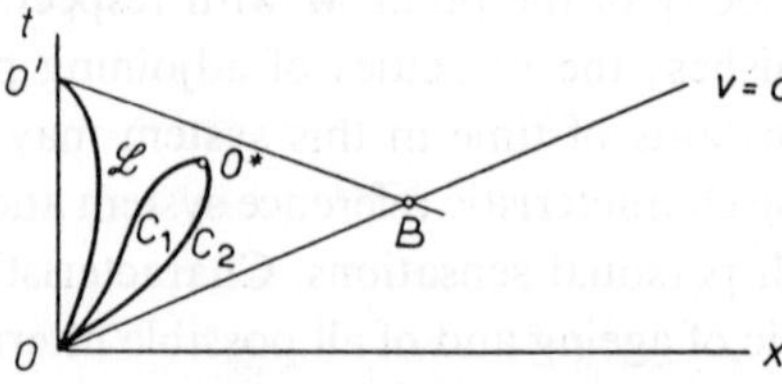

Fig. 40. The interval of the characteristic time depends on the law of motion.

In a given inertial coordinate system of an observer, the integral (3.20) depends on the curve $x^i = x^i(t)$ in Minkowski space. In particular, in the case of rectilinear motion, the interval $\Delta\tau$ is a functional of the path of integration in the plane x^1, t (Fig. 40). The values of $\Delta\tau$, computed along different paths C_1 and C_2 going from O to O^*, are different.

For a body at rest on Earth, applicable as inertial coordinate system, the path of integration coincides with the time axis; for a cosmonaut, flying in a rocket along the x-axis and then returning to Earth, the law of motion is illustrated by a curve of the form $\mathscr{L}$.

The integrand in the last integral (3.20) for $v \neq 0$ is less than for $v = 0$; therefore, on the return of the cosmonaut to the point O', the intervals of characteristic time for an observer on Earth $\Delta\tau_{\text{Earth}} = OO'$ and for the cosmonaut

$$\Delta\tau_{\text{cosm}} = \int_{t(O)}^{t(O')} \sqrt{1 - \frac{v^2}{c^2}}\, dt$$

will differ, so that

$$\Delta\tau_{\text{cosm}} < \Delta\tau_{\text{Earth}} \tag{3.21}$$

In the case of a cosmonaut and a resident on Earth who are twins, after the return of the cosmonaut to Earth, the cosmonaut will turn out to be younger than his Earth brother.

Let the segments OB and BO' correspond to the motions of a photon $x^1 = ct$ and $x^1 = -ct + x_B$, moving with the speed of light. Obviously, one has for the photon always $\Delta\tau = 0$, i.e., the characteristic time of a photon does not flow.

Transformation of **E** *and* **H** *for transition from K to K′.* It is readily explained how the vectors **E** and **H** change for the transition from one inertial coordinate system K to another "mobile" inertial coordinate system K', and, in particular, for the transition to the characteristic coordinate system $K' = K^*$.

From (2.11), (3.16) and (2.1) one finds readily

$$\left.\mathbf{E}'_{\parallel} = \left(\mathbf{E} + \frac{\mathbf{v}}{c} \times \mathbf{H}\right)_{\parallel}, \quad \mathbf{E}'_{\perp} = \left(\mathbf{E} + \frac{\mathbf{v}}{c} \times \mathbf{H}\right)_{\perp} \frac{1}{\sqrt{1 - \frac{v^2}{c^2}}}, \right\} \tag{3.22}$$

$$\left.\mathbf{H}'_{\parallel} = \left(\mathbf{H} - \frac{\mathbf{v}}{c} \times \mathbf{E}\right)_{\parallel}, \quad \mathbf{H}'_{\perp} = \left(\mathbf{H} - \frac{\mathbf{v}}{c} \times \mathbf{E}\right)_{\perp} \frac{1}{\sqrt{1 - \frac{v^2}{c^2}}} \right\} \tag{3.23}$$

where the symbol $\parallel$ denotes the components of the vectors parallel to the velocity $\mathbf{v}$ of the mobile system K' and the symbol $\perp$ the component perpendicular to $\mathbf{v}$. It is easily seen that the terms $(\mathbf{v}/c) \times \mathbf{H}$ and $(\mathbf{v}/c) \times \mathbf{E}$ do not contribute to $\mathbf{E}'_{\parallel}$ and $\mathbf{H}'_{\parallel}$, so that $\mathbf{E}'_{\parallel} = \mathbf{E}_{\parallel}$ and $\mathbf{H}'_{\parallel} = \mathbf{H}_{\parallel}$. However, in (3.22) and (3.23), these terms have been retained for the sake of the symmetry of the formulae which is very convenient for the writing down of approximate expressions for small $(v/c)^2$.

Formulae (3.22) and (3.23) show that the fundamental, normally employed characteristics of an electromagnetic field, the vectors **E** and **H**, depend essentially on the choice of the inertial coordinate system, and therefore their physical meanings have a very limited character.

In Newtonian mechanics, when one employs an accelerating coordinate system, one must introduce inertial forces. Thus, the field of external forces in non-relativistic mechanics depends on the choice of the mobile coordinate system. However, it is essential that the force field is the same in inertial systems and changes only for transition to a coordinate system which accelerates with respect to the original system. The electromagnetic field changes even for transition from one inertial coordinate system to another.

In addition, note that, if in some coordinate system K there were only an electric field, then for the transition to a system, moving with respect to K, unfailingly a magnetic field would be present, and conversely.

For example, consider an electron in its characteristic coordinate system K^*. In this coordinate system, there is no magnetic field: $\boldsymbol{H}^*=0$, and the electric field $\boldsymbol{E}^*$ will be a Coulomb field. In a system of coordinates which moves uniformly and rectilinearly with velocity $\boldsymbol{v}$ relative to the coordinate system in which the electron is at rest, the vectors $\boldsymbol{E}'$ and $\boldsymbol{H}'$ are determined as solutions of Maxwell's equations with the aid of (3.22) and (3.23) in which one must set $\boldsymbol{H}=0$ and the vector $\boldsymbol{E}$ is expressed in terms of r by the formula

$$\boldsymbol{E}=-\frac{e}{r^2}\operatorname{grad} r\,.$$

A more complete study of these fields $\boldsymbol{E}$ and $\boldsymbol{H}$ may be based on (3.22) and (3.23).

On invariant characteristics of an electromagnetic field. Each of the vectors **E** and **H** separately depends on the choice of the inertial coordinate system. The tensor $F=F_{ij}\mathfrak{z}^i\mathfrak{z}^j$ of the electromagnetic field, defined in terms of E and H by the matrix (2.1), is an invariant physical characteristic of the electromagnetic field, similar to the temperature, force vector, strain tensor, etc.

The tensor F has six independent components and only the two independent invariants

$$\begin{aligned}&\tfrac{1}{2}F_{ij}F^{ij}=H^2-E^2\,,\\&\tfrac{1}{4}\left[F^{ik}F^{lm}F_{il}F_{km}-\tfrac{1}{2}(F^{ik}F_{ik})^2\right]=(\boldsymbol{E}\cdot\boldsymbol{H})^2\,.\end{aligned}\tag{3.24}$$

A Lorentz transformation may completely destroy the magnetic or electric field only if $\boldsymbol{E}\cdot\boldsymbol{H}=0$, i.e., when the vectors **E** and **H** are perpendicular to each other.

If $H^2=E^2$ in anyone coordinate system, then this equality is fulfilled in any other inertial coordinate system. If $H^2-E^2=0$ and $\boldsymbol{E}\cdot\boldsymbol{H}=0$, then the vectors **E** and **H** are equal in magnitude and perpendicular to each other.

On the vectors E and H in the reference systems normally used in practical applications. In practice, one encounters, as a rule, different inertial coordinate systems for which the relative velocity is small in comparison with the speed of light. Therefore often the difference between the vectors $\boldsymbol{E}$, $\boldsymbol{H}$ and $\boldsymbol{E}'$, $\boldsymbol{H}'$, respectively, is small. It is seen from (3.22) and (3.23) that the vectors $\boldsymbol{E}$ and $\boldsymbol{H}$ may be considered to be invariant physical characteristics, if one neglects quantities of the order v/c.

Approximate transformation formulae for **E** *and* $\boldsymbol{H}$ *(including terms of the order* v/c*).* In the sequel, consideration will be given to non-relativistic mechanics of continuous media with inclusion of electromagnetic effects. Therefore use will be made only of such coordinate systems for which the relative velocity v is small compared with the velocity of light c ($v^2/c^2 \ll 1$). Then, if small second order quantities of v/c are neglected, but small first order quantities are included, the transition from the coordinate system K to a coordinate system K', moving with constant velocity with respect to K, will be given by the Galilei transformation

$$y^\alpha = x^\alpha - v^\alpha t\,, \qquad t' = t\,,$$

and the transformation formulae for $\boldsymbol{E}$ and $\boldsymbol{H}$ simplify to

$$\boldsymbol{E}' = \boldsymbol{E} + \frac{1}{c}(\boldsymbol{v} \times \boldsymbol{H})\,, \qquad \boldsymbol{H}' = \boldsymbol{H} - \frac{1}{c}(\boldsymbol{v} \times \boldsymbol{E}), \tag{3.25}$$

where dashed quantities relate to K', undashed to K.

In particular problems, for example when the vector $\boldsymbol{E}$ is small, one may introduce in (3.25) additional simplifications.

§4. Interaction of electromagnetic field with conductors

A conductor is a body in which there arises an electric current under the influence of an electromagnetic field. In this section, phenomena linked to polarization and magnetization will not be considered. As examples of conductors may serve metallic bodies: copper, iron, etc.; an important example of a conducting medium is a plasma, i.e., an ionized gas.

Three-dimensional and four-dimensional current density vectors. Conductance current. Electric currents, arising in conductors, represent the motion of charged particles (electrons and ions). If the microscopic velocities of electrons and ions in some small volume of a medium are denoted by $\boldsymbol{v}_k$, and the charges by e_k, then the current density $\boldsymbol{j}$ may be introduced in the form $\Sigma_i e_i \boldsymbol{v}_i$, averaged over a small volume ΔV:

$$\boldsymbol{j} = \frac{1}{\Delta V} \sum_i e_i \boldsymbol{v}_i = \boldsymbol{j}^* + \rho_e \boldsymbol{v} , \tag{4.1}$$

where $\boldsymbol{v}$ is the macroscopic velocity of the medium. The vector $\boldsymbol{j}^*$ represents the usual "technical current". Such currents arise in fixed as well as in moving conductors under the influence of an electromagnetic field and are referred to as *conduction currents*. The vector $\rho_e \boldsymbol{v}$ represents the current which is linked to the transfer of the macroscopic charge.
Since

$$\sum_i e_i \boldsymbol{v}_i = \sum_i e_i (\boldsymbol{v}_i - \boldsymbol{v}) + \Delta e \boldsymbol{v} , \quad (\Delta e = \sum_i e_i) ,$$

the conduction current vector $\boldsymbol{j}^*$ may be expressed in terms of the diffusion flow vector $\boldsymbol{I}_i$ of Chapter III by the formula

$$\boldsymbol{j}^* = \sum_i \frac{e_i}{m_i} \boldsymbol{I}_i ;$$

the ratios e_i/m_i depend only on the type of ions which transfer the charges.

Side by side with the three-dimensional vector $\boldsymbol{j}$, defined in geometrical space, introduce yet the four-dimensional current density vector in Minkowski space which in the characteristic Cartesian coordinate system is defined by

$$\left.\begin{aligned} &\boldsymbol{J} = J^i \boldsymbol{\mathfrak{z}}_i , \\ &J^1 = j^1 , \quad J^2 = j^2 , \quad J^3 = j^3 , \quad J^4 = \rho_e . \end{aligned}\right\} \tag{4.2}$$

The components of the vector $\boldsymbol{J}$ in any vector coordinate system are determined in terms of the components in the characteristic system by the general transformation formulae for a four-dimensional vector in Minkowski space.

Maxwell equations in conductors. Maxwell's equations in the presence of currents and charges and in the absence of polarization and magnetization in bodies have the form

$$\operatorname{curl} \boldsymbol{E} = -\frac{1}{c}\frac{\partial \boldsymbol{H}}{\partial t}, \quad \operatorname{curl} \boldsymbol{H} = \frac{4\pi}{c}\boldsymbol{j} + \frac{1}{c}\frac{\partial \boldsymbol{E}}{\partial t} \tag{4.3}$$

$$\operatorname{div} \boldsymbol{H} = 0, \quad \operatorname{div} \boldsymbol{E} = 4\pi\rho_e . \tag{4.4}$$

In particular, in the case of a stationary electromagnetic field,

$$\operatorname{curl} \boldsymbol{E} = 0, \tag{4.5}$$

whence $\boldsymbol{E}$ is a potential vector and

$$\operatorname{curl} \boldsymbol{H} = \frac{4\pi}{c}\boldsymbol{j}, \tag{4.6}$$

i.e., the electric current leads always to the appearance of a vortex field $\boldsymbol{H}$ of magnetic stress.

Displacement current. The quantity $(1/c)\partial \boldsymbol{E}/\partial t$ also causes appearance of a magnetic field and is referred to as displacement current. In many practical cases, this displacement current is very small. The displacement current was introduced into Maxwell's equations by Maxwell in accordance with experiments and in addition to the laws of electrodynamics of Coulomb, Ampère and Faraday which were known previously.

Law of conservation of total charge. If one takes the divergence of both sides of Equations (4.3) and uses (4.4), one obtains the important consequence of Maxwell's equations

$$\frac{\partial \rho_e}{\partial t} + \operatorname{div} \boldsymbol{j} = 0, \tag{4.6$'$}$$

which may be considered as the equation of continuity for charges or as the condition of conservation of charge.

In fact, integrating (4.6′) over some fixed geometric volume in a continuum which is a conductor, one finds

$$\int_V \frac{\partial \rho_e}{\partial t}\mathrm{d}\tau = \frac{\partial}{\partial t}\int_V \rho_e \mathrm{d}\tau = -\int_V \operatorname{div} \boldsymbol{j}\,\mathrm{d}\tau = -\int_\Sigma j_n \mathrm{d}\sigma, \tag{4.7}$$

where Σ is the surface bounding V and $\boldsymbol{n}$ is the external normal to Σ. The current vector $\boldsymbol{j}$ transfers the charges through the surface Σ, the quantity $-\int_\Sigma j_n \mathrm{d}\sigma$ is the total charge entering volume V through Σ in unit time. This quantity is equal to the change of the charge in V in unit time, i.e., to

$$\frac{\partial}{\partial t}\int_V \rho_e \mathrm{d}\tau = \frac{\partial e}{\partial t},$$

where e is the total charge inside V. If $j_n=0$ on Σ, then $\partial e/\partial t=0$ and the charge inside V is conserved.

The condition of conservation of total charge is an exact consequence of Maxwell's equations (4.3) and (4.4). Note that the law of conservation of charge, in contrast to the law of conservation of mass, is at the present time the fundamental law of physics which is always fulfilled.

The system of Maxwell's equations in conductors is an incomplete system. The system of Maxwell's equations (4.3) and (4.4) is incomplete. The number of equations in it is equal to seven, since the equation $\operatorname{div} \boldsymbol{H}=0$ is a direct consequence of the first equation (4.3) for corresponding initial data. However, the number of unknowns in it is ten: $\boldsymbol{E}, \boldsymbol{H}, \boldsymbol{j}, \rho_e$.

On a link between problems of electrodynamics and continuum mechanics. In addition, since in moving conductors $\boldsymbol{j}=\boldsymbol{j}^*+\rho_e\boldsymbol{v}$, the charge density distribution depends on the motion of the medium; therefore the problems of electrodynamics turn out to be linked to the problems of continuum mechanics.

Ohm's law for immobile conductors. The system of Equations (4.3) and (4.4) may be closed for fixed conducting media by a vector relation by means of Ohm's law, which establishes a link between the conduction current density $\boldsymbol{j}^*$ and the characteristics of the electromagnetic field. This link depends on a property of the conductor. In many important cases, Ohm's experimental law for immobile conductors has the form

$$\boldsymbol{j}^* = \sigma \boldsymbol{E}\,. \tag{4.8}$$

Conductivity. The coefficient σ is called the conductivity. For isotropic conductors, the conductivity σ is a scalar such that

$$\sigma = \frac{1}{R},$$

where R is the resistance of the conductor. For anisotropic conductors, for example, crystals, the conductivity σ is a second order tensor; in (4.8), the tensor σ contracts with the vector $\boldsymbol{E}$.

The conductivity σ is different for different conductors, and for a given conductor it may depend on its temperature T and other thermodynamic parameters. As the temperature increases, the conductivity of a gas grows. For example, air under normal conditions is almost not ionized and is thus a bad conductor; however, as the temperature increases or for intensive irradiation, the degree of ionization of air grows, the number of free electrons in air is magnified, and air becomes a good conductor; for rigid bodies, as the temperature increases, the conductivity σ may decrease. In many cases, the conductivity is considered to be a physical constant of a material, analogous to the viscosities μ and ζ or the heat conductivity $\varkappa$.

Ohm's law in the characteristic coordinate system. For mobile conductors, it is postulated that Ohm's law (4.8) is fulfilled at every point of a medium in the characteristic coordinate system (for the definition, *cf.*, §3).

In the characteristic coordinate system, Ohm's law has the form

$$\tilde{\boldsymbol{j}}^* = \sigma \tilde{\boldsymbol{E}}\,, \tag{4.9}$$

where $\sim$ denotes the fact that the corresponding quantities are considered in the characteristic coordinate system. After a transition from the characteristic system to a basic inertial reference system, with respect to which the motion of the medium is being studied, one obtains, by the transformation formulae (3.25) for the stress vector $\boldsymbol{E}$ of the electric field Ohm's law for movable conductors in the form

$$\boldsymbol{j}^* = \sigma\left(\boldsymbol{E} + \frac{\boldsymbol{v}}{c} \times \boldsymbol{H}\right). \tag{4.10}$$

Experiment and more detailed theoretical analysis show that Ohm's law in the form (4.10) may not always be employed. For example, in the case of a strong magnetic field, Ohm's law must take the form

$$\boldsymbol{j}^* = \sigma\left(\boldsymbol{E} + \frac{\boldsymbol{v}}{c} \times \boldsymbol{H}\right) + k(\boldsymbol{j}^* \times \boldsymbol{H})\,,$$

where k is some constant or a function of the thermodynamic parameters of a medium. The additional term $k(\boldsymbol{j}^* \times \boldsymbol{H})$ is referred to as Hall current.

Media with infinite conductivity. There exist media the conductivity of which is very large (for example, iron or strongly ionized plasma). In this connection, in practice, one considers often media for which σ is infinite, and the resistance R is zero. An introduction of such media, in a certain sense, is analogous to the introduction of ideal fluids instead of viscous fluids.

It follows from Ohm's law in the form (4.8) and (4.10), since, by definition, the current density $\boldsymbol{j}^*$ must be finite, that inside a medium with infinite conductivity at rest one must have

$$\boldsymbol{E} = 0\,, \tag{4.11}$$

and in a medium with infinite conductivity moving with velocity $\boldsymbol{v}$

$$\boldsymbol{E} = -\frac{\boldsymbol{v}}{c} \times \boldsymbol{H}\,. \tag{4.12}$$

Thus, the field of the electric stress vector $\boldsymbol{E}$ in a medium with infinite conductivity is determined in terms of the magnetic stress field $\boldsymbol{H}$ and the field of the macroscopic velocity $\boldsymbol{v}$ of a medium. In this case, Maxwell's two equations (4.3) may serve for the determination of $\boldsymbol{H}$ and $\boldsymbol{j}$.

Lorentz force. Forces which are exerted on a medium by an electromagnetic field are referred to as ponderomotive forces. If a medium is at rest, then there acts on its infinitesimal element $\mathrm{d}\tau$ with charge $\mathrm{d}e$, by Coulomb's law, the ponderomotive force

$$\boldsymbol{F}\,\mathrm{d}\tau = \frac{\mathrm{d}e}{\mathrm{d}\tau}\boldsymbol{E}\,\mathrm{d}\tau = \rho_e \boldsymbol{E}\,\mathrm{d}\tau\,.$$

If there is present in the element $\mathrm{d}\tau$ beside the charge $\mathrm{d}e$ a current $\boldsymbol{j}$, then a unit volume of the medium is subject to the ponderomotive force

$$\boldsymbol{F} = \rho_e \boldsymbol{E} + \frac{1}{c}(\boldsymbol{j} \times \boldsymbol{H})\,, \tag{4.13}$$

which is referred to as Lorentz force.

If one has a moving medium, then one may assume that in the characteristic coordinate system one has for the ponderomotive force

$$\tilde{\boldsymbol{F}} = \rho_e \tilde{\boldsymbol{E}} + \frac{1}{c}(\tilde{\boldsymbol{j}} \times \tilde{\boldsymbol{H}})\,, \tag{4.14}$$

where all vectors refer to the characteristic coordinate system.

If approximate formulae are used for the transition from the characteristic coordinate system K', moving with velocity $\boldsymbol{v}$ with respect to the inertial coordinate system K (*cf.*, (3.25)), then after neglecting small second order terms $(v/c)^2$ in the inertial system K one obtains

$$\boldsymbol{F} = \rho_e\left(\boldsymbol{E} + \frac{\boldsymbol{v}}{c} \times \boldsymbol{H}\right) + \frac{1}{c}(\boldsymbol{j}^* \times \boldsymbol{H}), \tag{4.15}$$

where it has been assumed that $\widetilde{\boldsymbol{j}^*} = \boldsymbol{j}^*$. Hence, comparing (4.15) with (4.13), one finds

$$\tilde{\boldsymbol{F}} = \boldsymbol{F}$$

i.e., that the Lorentz force in a non-relativistic approximation in the reference system K is given by (4.13), just as in the case of an immobile medium. Equality (4.13), determining the ponderomotive force acting on unit volume of a moving medium, has been established on the basis of experimental facts and considered to be one of the basic postulates of electrodynamics or to be one of the foundations for the evaluation of the electrodynamic characteristics of a field and current.

Impulse equations with inclusion of ponderomotive forces. The ponderomotive forces, exerted by the electromagnetic field on the particles of a conducting fluid, is a body force such as, for example, gravity; thus it must be introduced into the impulse equation for the material medium

$$\rho \boldsymbol{a} = \nabla_i p^i + \boldsymbol{F}_{\text{Lorentz}} + \rho \boldsymbol{F}_{\text{gravitation}} \,. \tag{4.16}$$

The problem of the determination of the motion of a conducting continuum is in the general case quite complex; it requires the simultaneous solution of the equations of continuum mechanics and electrodynamics. Note that the expression (4.13) for the ponderomotive force only makes allowance for the presence of charges and currents in the medium and that it must be expanded for polarized and magnetized media.

Energy interactions between a field and conducting media. Consider now the energy interaction between an electromagnetic field and a conducting medium. For example, it is known that an immobile conductor, through which passes a current, is heated; this heating is connected with an exchange of energy between the electric field and the conductor.

Next, the equation of Umov–Poynting will be obtained from Maxwell's equations; it determines the change of energy in an electromagnetic field. Subtracting from the first Maxwell equation (4.3), multiplied scalarly by $\boldsymbol{H}$, the second equation (4.3), multiplied scalarly by $\boldsymbol{E}$, one finds

$$\boldsymbol{H}\cdot\operatorname{curl}\boldsymbol{E}-\boldsymbol{E}\cdot\operatorname{curl}\boldsymbol{H}=-\frac{1}{c}\left(\boldsymbol{H}\cdot\frac{\partial \boldsymbol{H}}{\partial t}+\boldsymbol{E}\cdot\frac{\partial \boldsymbol{E}}{\partial t}\right)-\frac{4\pi}{c}(\boldsymbol{j}\cdot\boldsymbol{E}). \tag{4.17}$$

In Cartesian coordinates, the left-hand side of this relation may be written in the form of the difference of two determinants:

$$\begin{vmatrix} H_1 & H_2 & H_3 \\ \partial/\partial x_1 & \partial/\partial x_2 & \partial/\partial x_3 \\ E_1 & E_2 & E_3 \end{vmatrix} - \begin{vmatrix} E_1 & E_2 & E_3 \\ \partial/\partial x_1 & \partial/\partial x_2 & \partial/\partial x_3 \\ H_1 & H_2 & H_3 \end{vmatrix}.$$

After transpositions in the rows of these determinants, one may write

$$\begin{vmatrix} \partial/\partial x_1 & \partial/\partial x_2 & \partial/\partial x_3 \\ E_1 & E_2 & E_3 \\ \boxed{H_1} & \boxed{H_2} & \boxed{H_3} \end{vmatrix} + \begin{vmatrix} \partial/\partial x_1 & \partial/\partial x_2 & \partial/\partial x_3 \\ \boxed{E_1} & \boxed{E_2} & \boxed{E_3} \\ H_1 & H_2 & H_3 \end{vmatrix},$$

where the symbol $\square$ denotes quantities which must not be differentiated as the determinant is expanded.

It is now seen that the left-hand side of (4.17) may be rewritten in the form

$$\begin{vmatrix} \dfrac{\partial}{\partial x^1} & \dfrac{\partial}{\partial x^2} & \dfrac{\partial}{\partial x^3} \\ E_1 & E_2 & E_3 \\ H_1 & H_2 & H_3 \end{vmatrix}$$

Umov–Poynting vector and Umov–Poynting equation. If one introduces the vector

$$\boldsymbol{S}=\frac{c}{4\pi}(\boldsymbol{E}\times\boldsymbol{H}), \tag{4.18}$$

the so-called Umov–Poynting vector, then (4.17) assumes the form

$$\operatorname{div} \boldsymbol{S} + \frac{\partial}{\partial t}\frac{1}{8\pi}(H^2 + E^2) + \boldsymbol{j}\cdot\boldsymbol{E} = 0 . \tag{4.19}$$

This is the Umov–Poynting equation. Integration of (4.19) over a fixed finite volume V yields

$$\int_\Sigma S_n \mathrm{d}\sigma + \frac{\mathrm{d}}{\mathrm{d}t}\frac{1}{8\pi}\int_V (H^2 + E^2)\mathrm{d}\tau + \int_V (\boldsymbol{j}\cdot\boldsymbol{E})\mathrm{d}\tau = 0 , \tag{4.20}$$

where Σ is the surface bounding V and $\boldsymbol{n}$ is the external normal to Σ.

Energy of an electromagnetic field. Each of the terms of the integral relationship (4.20) has physical significance. The three-dimensional scalar

$$\frac{1}{8\pi}(H^2 + E^2)$$

is introduced, by definition, as the volume density $\mathscr{E}$ of the energy of the electromagnetic field $\int_V \boldsymbol{j}\cdot\boldsymbol{E}\,\mathrm{d}\tau\,\mathrm{d}t$ may be considered as the element of work done by the electric field $\boldsymbol{E}$ on moving charged particles, where for microscopic internal motion one takes account of the current $\boldsymbol{j}^*$ and for macroscopic motion of $\rho_e\boldsymbol{v}$.

Joule heat. In the case of a conductor at rest, the term

$$\int_V (\boldsymbol{j}\cdot\boldsymbol{E})\mathrm{d}\tau\,\mathrm{d}t = \int_V (\boldsymbol{j}^*\cdot\boldsymbol{E})\mathrm{d}\tau\,\mathrm{d}t = \mathrm{d}Q_\mathrm{J} \tag{4.21}$$

represents the Joule heat. For an immobile conductor, Equation (4.20) assumes the form

$$\frac{\mathrm{d}\mathscr{E}}{\mathrm{d}t} = -\int_\Sigma S_n \mathrm{d}\sigma - \frac{\mathrm{d}Q_\mathrm{J}}{\mathrm{d}t} . \tag{4.22}$$

The total energy $\mathscr{E}$ of the electromagnetic field in the volume V of a fixed conductor changes on account of the Umov–Poynting current vector through the surface Σ bounding a volume V, and due to the passage of Joule heat into the medium.

Inflow of heat energy from a field into a medium at rest. The inflow of heat energy $\mathrm{d}q_{\mathrm{el}}^{(e)}$ from a field into unit mass of a conducting medium at rest is given by

$$dq_{\text{el}}^{(e)} = \frac{dQ^{(e)}}{dm} = \frac{1}{\rho}(j \cdot E)dt\,. \tag{4.23}$$

Change of energy of an electromagnetic field in vacuum. Note that the energy of the electromagnetic field in a volume V does not only change on account of interactions of the field with a medium. The Umov–Poynting equation is also valid in the presence of Maxwell's equations (1.11) and (1.12) in vacuum. The Relation (4.22) may be rewritten

$$\frac{d\mathscr{E}}{dt} = -\int_{\Sigma} S_n d\sigma\,,$$

i.e., the total energy of the electromagnetic field in volume V changes in this case only due to the Umov–Poynting flow vector. However, this change is non-zero only for an unsteady electromagnetic field. For a stationary field ($\partial \boldsymbol{H}/\partial t = \partial \boldsymbol{E}/\partial t = 0$), the Umov–Poynting flow vector through a closed surface in vacuum is always zero, by Maxwell's equations. In conductors, the Umov–Poynting flow vector is non-zero, also in the case of a stationary electromagnetic field. Through an open part of Σ, the Umov–Poynting flow vector generally is non-zero, if $\boldsymbol{E} \times \boldsymbol{H} \neq 0$. The Umov–Poynting vector characterizes the energy exchange between different parts of an electromagnetic field, i.e., the flow of energy through parts of the field subdivided by a surface Σ.

In the presence of macroscopic motion of a medium, all the preceding interpretations may apply to infinitesimal elements of the medium, when the Umov–Poynting equation is written in terms of the corresponding characteristic inertial coordinate system.

In inertial coordinate systems, moving relative to the characteristic inertial coordinate system,

$$\int_{\Delta V} (j \cdot E) d\tau dt \neq \int_{\Delta V} (\tilde{j} \cdot \tilde{E}) d\tau dt\,, \tag{4.23'}$$

i.e., the integral on the left-hand side is not equal to the Joule heat. The difference is equal to the work done by the ponderomotive force (4.13).

Equation of heat flow for conducting media. The heat flow equation for moving particles of a conducting medium in the characteristic coordinate system may be rewritten in the general case in the form

$$dU = \frac{p^{\alpha\beta}\,d\varepsilon_{\alpha\beta}}{\rho} + dq^{(e)}_{\text{nonel}} + \frac{1}{\rho}(\tilde{j}\cdot\tilde{E})dt + dq^{**}\,, \tag{4.24}$$

where the $\sim$ indicates the fact that the corresponding quantities are referred to the characteristic coordinate system. The inflow of energy $\rho^{-1}(\tilde{j}\cdot\tilde{E})dt = dq^{(e)}_{el}$ represents the Joule heat, i.e., the influx to a particle of a medium of heat energy which is contained in the electromagnetic field. In the case under consideration, the contribution of the electromagnetic field to dq^{**} has been taken to be equal to zero. In the next section, it will be shown that dq^{**} must, in general, be taken into consideration, if effects of polarization and magnetization are essential.

Within the framework of the relations introduced in this section for the ponderomotive forces and interchange of energy between a field and a medium constructed in magneto-hydrodynamics the basic conditions of which will be considered below, attention will now be given to the problem of the ponderomotive forces and energy fluxes in the medium for the case when the effects of polarization and magnetization of a medium are essential.

§5. Interaction of an electromagnetic field with bodies in the presence of polarization and magnetization

In certain bodies, there occurs polarization and magnetization under the influence of an external electromagnetic field. An external field generates inside a body a macroscopic electromagnetic field which deforms and modifies the external field.

Maxwell's equations in the presence of polarization and magnetization. Vectors of electric and magnetic induction, magnetization and polarization. In such bodies, Maxwell's equations have the form

$$\operatorname{curl} \boldsymbol{E} = -\frac{1}{c}\frac{\partial \boldsymbol{B}}{\partial t}, \qquad \operatorname{div} \boldsymbol{B} = 0\,, \tag{5.1}$$

$$\operatorname{curl} \boldsymbol{H} = \frac{4\pi}{c}\boldsymbol{j} + \frac{1}{c}\frac{\partial \boldsymbol{D}}{\partial t}\,, \qquad \operatorname{div} \boldsymbol{D} = 4\pi\rho_e\,, \tag{5.2}$$

where $\boldsymbol{D}$ and $\boldsymbol{B}$ are the vectors of electric and magnetic induction, respectively. Side by side with the vector of magnetic induction $\boldsymbol{B}$, one may consider the vector of magnetization $\boldsymbol{M}$ which is linked to $\boldsymbol{B}$ by

$$\boldsymbol{B} = \boldsymbol{H} + 4\pi\boldsymbol{M}\,. \tag{5.3}$$

The vector $\boldsymbol{M}$ characterizes, from a macroscopic point of view, the orderly distribution of magnetic dipoles in a body. Analogously, one may consider with the vector of electric induction $\boldsymbol{D}$ the polarization vector $\boldsymbol{P}$ which is related to $\boldsymbol{D}$ by

$$\boldsymbol{D} = \boldsymbol{E} + 4\pi\boldsymbol{P}\,. \tag{5.4}$$

The polarization vector $\boldsymbol{P}$ characterizes the distribution of electric dipoles in a body.

The second equation (5.1), as before, is a consequence of the first equation (5.1), when initially $\operatorname{div} \boldsymbol{B} = 0$.

Maxwell's equations in integral form. The system of equations (5.1) and (5.2) may be written in integral form:

$$\begin{aligned}
&\oint_{\mathscr{L}} \boldsymbol{E}\cdot \mathrm{d}\boldsymbol{s} = -\frac{1}{c}\int_{\Sigma_1} \frac{\partial B_n}{\partial t}\,\mathrm{d}\sigma\,, &&\int_{\Sigma} B_n \mathrm{d}\sigma = 0\,,\\
&\oint_{\mathscr{L}} \boldsymbol{H}\cdot \mathrm{d}\boldsymbol{s} = \frac{4\pi}{c}\int_{\Sigma_1} j_n \mathrm{d}\sigma + \frac{1}{c}\int_{\Sigma_1} \frac{\partial D_n}{\partial t}\,\mathrm{d}\sigma\,, &&\int_{\Sigma} D_n \mathrm{d}\sigma = 4\pi\int_V \rho_e \mathrm{d}\tau\,,
\end{aligned} \tag{5.5}$$

and their consequence, i.e., the condition of conservation of total charge, in the form

$$-\frac{\mathrm{d}}{\mathrm{d}t}\int_V \rho_e \mathrm{d}\tau = \int_{\Sigma} j_n \mathrm{d}\sigma\,,$$

where $\mathscr{L}$ is a closed contour which is at rest in the selected inertial coordinate system Σ_1 is a surface stretched over this contour, the subscript n denotes normal components to surfaces Σ_1 and Σ, the direction of the normal $\boldsymbol{n}$ to Σ_1 is chosen such that the direction of the circuit during the integration along $\mathscr{L}$ and the direction $\boldsymbol{n}$ form a right-handed system, Σ is a closed surface which is at rest in the coordinate system under consideration and bounds the volume V.

Maxwell's tensor equations in the presence of electric currents, polarization and magnetization. Just as the equations for an electromagnetic field in vacuum, Maxwell's equations in material media may be written in tensor form in Minkowski space. For this purpose, one must introduce the four-dimensional flow vector $\boldsymbol{J} = J_i\, \mathbf{3}_i$ from (4.2) and instead of the covariant and contravariant components F_{ij} and F^{ij} of the tensor of the electromagnetic field, introduced in §2 by the matrices (2.1) and (2.3), one must specify two antisymmetric tensors F and H the components of which in a "Cartesian" coordinate system ($ds^2 = -(dx^1)^2 - (dx^2)^2 - (dx^3)^2 + c^2 dt^2$) are defined by

$$\|F_{ij}\| = \left\| \begin{matrix} 0 & B^3 & -B^2 & cE_1 \\ -B^3 & 0 & B^1 & cE_2 \\ B^2 & -B^1 & 0 & cE_3 \\ -cE_1 & -cE_2 & -cE_3 & 0 \end{matrix} \right\|, \tag{5.6}$$

$$\|H^{ij}\| = \left\| \begin{matrix} 0 & H_3 & -H_2 & -\frac{1}{c}D^1 \\ -H_3 & 0 & H_1 & -\frac{1}{c}D^2 \\ H_2 & -H_1 & 0 & -\frac{1}{c}D^3 \\ \frac{1}{c}D^1 & \frac{1}{c}D^2 & \frac{1}{c}D^3 & 0 \end{matrix} \right\|. \tag{5.7}$$

Obviously, in the absence of polarization and magnetization, i.e., when $\boldsymbol{P} = 0$ and $\boldsymbol{M} = 0$, the matrix of F_{ij} in (2.1) coincides with the matrix of F_{ij} in (5.6), and the matrix of H^{ij} in (5.7) with the matrix of F^{ij} in (2.3).

Instead of the tensor H, one may introduce the polarization tensor $\mathscr{P}$ with the components $\mathscr{P}_{ij}$, defined by

$$\mathscr{P}_{ij} = \frac{1}{4\pi}(F_{ij} - H_{ij})\,.$$

It is easily verified directly that the tensor form of Equations (5.1) and (5.2) in four-dimensional Minkowski space in any curvilinear coordinate system is

$$\nabla_i F_{jk} + \nabla_j F_{ki} + \nabla_k F_{ij} = 0\,, \tag{5.8}$$

$$\nabla_k H^{ik} = \frac{4\pi}{c} J^i\,. \tag{5.9}$$

It is seen directly from the tensor form of these equations that they are invariant with respect to Lorentz transformations.

Transformations of the basic vectors of the electromagnetic field during transition from one inertial reference system to another. During transition from one "immobile" inertial system K to another "mobile" inertial system K', the vectors $\boldsymbol{E}, \boldsymbol{D}, \boldsymbol{M}$ and $\boldsymbol{H}, \boldsymbol{B}, \boldsymbol{P}$ transform in accordance with (3.22) and (3.23), respectively. The components of the four-dimensional electric current density vector J^i transform in accordance with the usual transformation formulae for vector components. In particular, if the system K' translates with respect to K along the x^1-axis with velocity v, then, on the basis of (3.16), one has

$$\left.\begin{aligned} &J'^1 = j'^1 = \frac{j^1 - v\rho_e}{\sqrt{1-\dfrac{v^2}{c^2}}}\,, \qquad J'^2 = j^2\,, \\ &J'^3 = j^3\,, \qquad\qquad J'^4 = \rho'_e = \frac{\rho_e - \dfrac{v}{c^2} j^1}{\sqrt{1-\dfrac{v^2}{c^2}}}\,. \end{aligned}\right\} \tag{5.10}$$

The components and the magnitude of the three-dimensional current density vector and likewise the charge density depend on the choice of the inertial coordinate system.

Vector potential. As for the electromagnetic field in vacuum, one may introduce for the electromagnetic field in a medium the vector potential $\boldsymbol{A} = A_i \boldsymbol{3}^i$, setting

$$F_{ij} = \nabla_j A_i - \nabla_i A_j\,,$$

when Equations (5.8) will be satisfied identically. In the absence of polarization and magnetization equation (5.9) reduces for $J^i \neq 0$ and $\nabla^\alpha A_\alpha = 0$ to the non-homogeneous wave equation with in Cartesian coordinates has the form

$$\frac{\partial^2 \boldsymbol{A}}{\partial x^{1^2}} + \frac{\partial^2 \boldsymbol{A}}{\partial x^{2^2}} + \frac{\partial^2 \boldsymbol{A}}{\partial x^{3^2}} - \frac{1}{c^2}\frac{\partial^2 \boldsymbol{A}}{\partial t^2} = \frac{4\pi}{c}\boldsymbol{J}.$$

Minkowski tensor. In the capacity of the energy-impulse tensor of an electromagnetic field, in the absence of polarization and magnetization one may take the Minkowski tensor which reduces, by definition, to

$$S_i^k = -\frac{1}{4\pi}\left[F_{mi}H^{mk} - \tfrac{1}{4}\delta_i^k F_{mn}H^{mn}\right]. \tag{5.11}$$

This formula is a direct generalization of (2.22) for the energy-impulse tensor of the electromagnetic field in vacuum. It is easily verified that, in general, the Minkowski tensor is non-symmetric:

$$S^{ij} \neq S^{ji}.$$

Four-dimensional vector of the ponderomotive force. In this case the equations of impulse and energy assume the form

$$\nabla_k S^{ik} = -F^i, \tag{5.12}$$

where F^i is the four-dimensional vector of the ponderomotive force.[1)]

[1)] In the Cartesian coordinate system employed with the four-dimensional matrix

$$ds^2 = -dx^{1^2} - dx^{2^2} - dx^{3^2} + c^2 dt^2$$

the spatial components of the four-dimensional force satisfy the following equality.

$$F^\alpha_{\text{four}} = -F_{\alpha\,\text{four}} \quad (\alpha = 1, 2, 3).$$

The components of the usual three-dimensional force satisfy the relations:

$$F^\alpha_{\text{four}} = F^\alpha_{\text{three}} = F_{\alpha\,\text{three}}$$

These relations are likewise valid in the case of transition from arbitrary four-dimensional vectors in pseudo-Euclidean space to the corresponding three-dimensional vectors in the three-dimensional Euclidean space.

The general formulae introduced above for the three-dimensional ponderomotive force and the energy exchange between a field and a medium in the presence of polarization and magnetization are obtained with the aid of (5.12). The expressions for F^1, F^2, F^3 contain the components of the Lorentz force. In the absence of polarization and magnetization of a medium, but in the presence of currents, the three-dimensional part of the four-dimensional ponderomotive force $\boldsymbol{F}$ simply yields the Lorentz force and the fourth component the quantity $\boldsymbol{j} \cdot \boldsymbol{E}$ which in the characteristic coordinate system is equal to the produced Joule heat, referred to unit time and unit volume of the material medium.

After the introduction of the energy-impulse tensor $S = S^{ij}\mathbf{з}_i\mathbf{з}_j$ of the electromagnetic field, Formula (5.12) determines the four-dimensional vector $\boldsymbol{F} = F^i\mathbf{з}_i$, the volume density of the forces which are external with respect to the body. The volume density of the four-dimensional forces exerted by a field on a body, the ponderomotive forces, are determined by the distribution of the characteristics of the field and are introduced in this way for the general case, when the material medium (body) moves in any manner.[1)]

In order to establish the dependence of the components F^i on the usual three-dimensional vector characteristics of a field, one must employ an inertial coordinate system in which Maxwell's equations (5.1) and (5.2) were written down. For this purpose, one may select different coordinate systems as inertial systems. In particular, one may take the fixed reference system of an observer in which the motion of the medium is determined, or one may use a set of characteristic inertial coordinate systems at each point of the material medium and at each instant of time.

The components of the three-dimensional vector characteristics of a field, defined by the method stated above in characteristic coordinate systems, may be evaluated in a concomitant coordinate system which generally is not inertial. In particular, if the concomitant system is inertial, then at each point the concomitant system and the characteristic coordinate system coincide; therefore, in this case, the interaction

[1)] Formulae (5.12) and (5.34) represent a generalization of the experimental fundamental laws of electrodynamics, due to Coulomb, Lorentz, Joule, etc., to the general case of motion of magnetized and polarized material bodies. It follows from the later work that this generalization is linked to the conditions of the choice of the energy-impulse tensor and the characteristic movement of the field which may differ.

between the field and the material medium may be considered as an interaction between the field and an immobile body in a concomitant coordinate system.

In each coordinate system, the three components F^α ($\alpha=1, 2, 3$) and the fourth component F^4 for the space coordinate transformations

$$y^\alpha = y^\alpha(x^\beta), \quad y^4 = x^4 \; ; \quad \alpha, \beta = 1, 2, 3$$

form, respectively, the components of a three-dimensional vector and a three-dimensional scalar.

As has been stated above, this vector and scalar depend on the choice of the initial coordinate system x^i. Within the framework of the special theory of relativity, even for inertial systems translating with respect to each other with constant velocity, this three-dimensional vector and this three-dimensional scalar are not invariant.[1)]

Nevertheless, in inertial reference systems which are mobile with respect to each other, these characteristics differ only little, when the relative velocity of motion is small compared with the speed of light. In applications within the framework of non-relativistic mechanics, these differences may be neglected.

In the general case, one may use as natural and convenient characteristics of internal physical processes in a material medium the three-dimensional density vectors of the ponderomotive force and the three-dimensional scalar $F_4 = c^2 F^4$ in the characteristic coordinate system.

On additional relations of electrodynamics which close Maxwell's equations (5.1) *and* (5.2). The system of Maxwell's equations (5.1) and (5.2) is not closed. It involves seven independent equations and the sixteen unknowns $\boldsymbol{E}, \boldsymbol{H}, \boldsymbol{B}, \boldsymbol{D}, \boldsymbol{j}, \rho_e$. These equations are insufficient for the determination of the enumerated characteristics of an electromagnetic field.

In order to close the system, one requires at least three additional vector relations linking the vectors $\boldsymbol{E}$, $\boldsymbol{H}$, $\boldsymbol{j}^* = \boldsymbol{j} - \rho_e \boldsymbol{v}$, $\boldsymbol{B}$ and $\boldsymbol{D}$.

[1)] This is a general property of any three-dimensional vector in the special theory of relativity. For example, this is so in the case of the current density vector (*cf.*, (3.16) and (5.10)).

One may use as such relations Ohm's law and the laws of polarization and magnetization of a medium. These additional relations do not bear a general character; in essence, they differ for different bodies and processes. In many cases, one uses Ohm's law and the laws of magnetization and polarization of a body in the form

$$j^* = \sigma\left(\boldsymbol{E} + \frac{\boldsymbol{v}}{c} \times \boldsymbol{H}\right), \tag{5.13}$$

$$\widetilde{\boldsymbol{B}} = \mu\widetilde{\boldsymbol{H}}, \tag{5.14}$$

$$\widetilde{\boldsymbol{D}} = \varepsilon\widetilde{\boldsymbol{E}}, \tag{5.15}$$

where σ denotes, as before, the conductivity, μ the magnetic permeability and ε the dielectric constant. In practice, one may often assume that σ, μ and ε are constants. In vacuum, one has $\sigma=0$, $\mu=\varepsilon=1$. The quantities σ, μ and ε, similar to the viscosity and thermal conductivity, may be considered to be physical characteristics of a medium; they may depend on the temperature (for low temperatures, polarization and magnetization of bodies are stronger) and have a tensorial character, as, for example, in the case of anisotropic bodies.

Laws of magnetization and polarization in tensorial form. The three-dimensional vector relations (5.14) and (5.15), referred to the characteristic coordinate system, are equivalent to a single four-dimensional tensor relation which is invariant with respect to the choice of coordinate systems. In component form, it becomes

$$F_{ij} = C_{ijkl}H^{kl}, \quad C_{jikl} = -C_{ijkl} = C_{ijlk}, \tag{5.16}$$

where the components of the tensor C_{ijkl} depend on the components of the metric tensor g_{ij}, the components of the three-dimensional velocity vector of points of the medium and, in general, on other physical parameters, characterizing the physical states of the particles of a medium, for example, on the temperature, the components of the strain tensor, etc.

In the isotropic case, when the coefficients μ and ε in (5.14) and (5.15) are scalars, it may be verified that one has for the components of the tensor C_{ijkl} the following formulas

$$C_{ijkl} = \tfrac{1}{2}\{\mu(\gamma_{ik}\gamma_{jl} - \gamma_{il}\gamma_{jk})$$
$$+\varepsilon^{-1}(g_{ik}u_j u_l - g_{jk}u_i u_l + g_{jl}u_i u_k - g_{il}u_j u_k)\},$$

where $\gamma_{ij} = g_{ij} - u_i u_j$ and u_i are the covariant components of the four-dimensional velocity vector. This verification is conveniently performed for the immobile body in a Cartesian inertial coordinate system, linked to the body. In this system,

$$u^\alpha = u_\alpha = 0 \,, \quad \alpha = 1, 2, 3 \;; \quad u^4 = c^{-1} \,, \qquad u_4 = c \,.$$

Note that the law of magnetization (5.14) is not fulfilled for very strong magnetic fields, when the magnetization vector

$$\boldsymbol{M} = \frac{\mu - 1}{4\pi} \boldsymbol{H}$$

attains maximal values for complete saturation, whereas the magnitude of the vector $\boldsymbol{H}$ may increase on account of external currents. There also exist materials in which dipole charges (so-called polarized bodies) remain unchanged in the absence of an external electric field $\boldsymbol{E}$. For such materials, the law of polarization (5.15) is untrue. The laws of polarization and magnetization of bodies (5.14) and (5.15) may be employed for the solution of a large class of problems in the case of not too strong electromagnetic fields.

Ponderomotive forces. Next, an expression will be derived for the ponderomotive forces, i.e., for the forces which an electromagnetic field exerts on a body in which polarization and magnetization take place. This expression is linked to Maxwell's equations (5.1) and (5.2); it is true for any Ohm's law and any laws of polarization and magnetization.

The three-dimensional ponderomotive force per unit volume, due to polarization and magnetization, differs from the Lorentz force; in any curvilinear inertial coordinate system, it has, by (5.12), the form

$$\begin{aligned} \boldsymbol{F} = \rho_e \boldsymbol{E} + c^{-1} (\boldsymbol{j} \times \boldsymbol{B}) \\ + (8\pi)^{-1} (D_\alpha \nabla E^\alpha - E^\alpha \nabla D_\alpha + B_\alpha \nabla H^\alpha - H^\alpha \nabla B_\alpha) \,, \end{aligned} \tag{5.17}$$

where $\nabla = \mathbf{3}^k \nabla_k$ is the three-dimensional gradient vector which in an orthogonal Cartesian coordinate system becomes

$$\nabla = \frac{\partial}{\partial x} \boldsymbol{i} + \frac{\partial}{\partial y} \boldsymbol{j} + \frac{\partial}{\partial z} \boldsymbol{k} \,.$$

It is easily seen that in the last term of Formula (5.17) for the ponderomotive force the vectors of electric and magnetic induction $\boldsymbol{D}$ and $\boldsymbol{B}$ may be replaced by the vectors of polarization and magnetization $\boldsymbol{P}$ and $\boldsymbol{M}$, respectively. By (5.17), the force $\boldsymbol{F}$ differs from the Lorentz force

$$\boldsymbol{F}_{\text{Lorentz}} = \rho_e \boldsymbol{E} + c^{-1}(\boldsymbol{j} \times \boldsymbol{H}) .$$

It is easily seen that the last term in brackets in (5.17) vanishes, when the laws (5.14) and (5.15) apply, if μ and ε are scalars and identical for different particles of a medium (i.e., independent of the coordinates). Generally speaking, this term differs from zero in anisotropic media, when μ and ε are tensors, and in isotropic media, when the laws (5.14) and (5.15) are not satisfied or when μ and ε are functions of parameters which depend on the coordinates (for example, the temperature and the density).

Let us consider the three-dimensional equations of law of change of quantity of motion of the material medium. The three-dimensional ponderomotive volume force, external with respect to the medium, is defined by (5.17) in any inertial coordinate system. The invariance of a conception of a force as the four-dimensional vector and non-invariance of the vector of three-dimensional ponderomotive force in a different, moving one with respect to the other inertial coordinate systems (like any other three-dimensional vectors in the special theory of relativity) represents a relativistic effect. In practice, as it has been shown in §4, in the case of the Lorentz force, when the relative velocity of the inertial systems is small in comparison with the speed of light, the difference of the components of three-dimensional ponderomotive force vectors, due to their non-invariancy is small in comparison with their magnitude.[1)]

During formation of the equations of motion in the non-inertial coordinate systems and during the study of the accelerated motion of particles in the inertial coordinate systems, as it has been shown above, one can locally introduce the characteristic inertial coordinate system. In these systems, in accordance with the fundamental physical assumption, the four-dimensional forces of interaction between a field and a medium, taking into account the law of action and reaction, can be determined with the aid of (5.12), and the three-dimensional forces — with the aid of (5.17). In

[1)] The vectors $\boldsymbol{B}$ and $\boldsymbol{D}$ during the transition from one inertial coordinate system to another "mobile" inertial coordinate system transform in accordance with formulas, analogous to the formulas of transformation of the vectors $\boldsymbol{H}$ and $\boldsymbol{E}$.

such a manner, the laws introduced for interaction of an electromagnetic field and a material body in rest, are extended to the case of interaction of an electromagnetic field and a material continuum, moving arbitrarily with respect to the inertial observer.

In connection with the fundamental three-dimensional equation of impulses, one has to have in mind the following remark; in accordance with the definition of the energy-impulse tensor of the electromagnetic field, the field possesses the impulse and energy, expressed by the characteristics of the field and, in particular, by the vectors of polarization $\boldsymbol{P}$ and magnetization $\boldsymbol{M}$.

In the Newtonian mechanics, an impulse and a kinetic energy of a material point or a finite body in translational motion, are expressed by the mass and the velocity of the body. Generally, if the body (material continuum) is polarized and magnetized, then in accordance with the theory of relativity, the impulse and the kinetic energy in translational motion of the body, depends not on its mass velocity but on "internal parameters", as the vectors of polarization and magnetization $\boldsymbol{P}$ and $\boldsymbol{M}$.

The influence of the vectors $\boldsymbol{P}$ and $\boldsymbol{M}$ on the motion of the particles of the body represents the relativistic effect which, in the frames of the Newtonian mechanics, can be considered with the aid of comparison of the corresponding parts of the energy-impulse tensor of the medium with the energy-impulse tensor of the electromagnetic field. Thus, instead of the Minkowski tensor, one can introduce other tensors for electromagnetic field and obtain, in connection with it, modification of the form of formula for the ponderomotive forces, which will contain the additional terms. If for this equation of motion one uses the Minkowski tensor for the field, then the additional terms appear as the consequence of complication of the dynamical properties of polarized and magnetized continuum.

The consideration, stated above, can serve to clarify that for the foundation of the theory of the model of polarized and magnetized continuum it is convenient and, in essence, it is necessary to use the relativistic mechanics.

Flux of energy from a field to a body. Besides the interaction of forces, there occurs also an energy exchange between a field and a material body. From (5.12) and the formulas for components of the Minkowski energy-impulse tensor, one obtains for $i = 4$ in a Cartesian inertial coordinate system

$$F^4c^2 = F_4 = \boldsymbol{E}\cdot\boldsymbol{j} + E_\beta \frac{\partial P^\beta}{\partial t} + H_\beta \frac{\partial M^\beta}{\partial t} - \frac{\partial}{\partial t}\left(\frac{E_\beta P^\beta + H_\beta M^\beta}{2}\right). \tag{5.18}$$

In fact, on the basis of (5.6), (5.7) and (5.1), one has

$$S_{4.}^{\cdot 4} = -\frac{1}{4\pi}(F_{m4}H^{m4} - \tfrac{1}{4}F_{mn}H^{mn}) = \frac{\boldsymbol{B}\cdot\boldsymbol{H}+\boldsymbol{E}\cdot\boldsymbol{D}}{8\pi} \tag{5.19}$$

and, in addition,

$$\boldsymbol{S} = S_{4.}^{\cdot\alpha}\boldsymbol{\mathfrak{z}}_\alpha = \frac{c}{4\pi}(\boldsymbol{E}\times\boldsymbol{H}).$$

By definition,

$$F_4 = -\left(\frac{\partial S_{4.}^{\cdot 4}}{\partial t} + \frac{\partial S_{4.}^{\cdot 1}}{\partial x^1} + \frac{\partial S_{4.}^{\cdot 2}}{\partial x^2} + \frac{\partial S_{4.}^{\cdot 3}}{\partial x^3}\right) = -\frac{\partial S_{4.}^{\cdot 4}}{\partial t} - \operatorname{div}\boldsymbol{S}. \tag{5.20}$$

In the case under consideration, in the presence of magnetization and polarization, the Umov–Poynting equation assumes, on the basis of Maxwell's equations (5.1) and (5.2), the somewhat modified form

$$\operatorname{div}\boldsymbol{S} + \frac{1}{4\pi}\left(\boldsymbol{H}\cdot\frac{\partial \boldsymbol{B}}{\partial t} + \boldsymbol{E}\cdot\frac{\partial \boldsymbol{D}}{\partial t}\right) + \boldsymbol{j}\cdot\boldsymbol{E} = 0. \tag{5.21}$$

Substituting for $S_{4.}^{\cdot 4}$ from (5.19) and for div $\boldsymbol{S}$ from (5.21) into (5.20), one arrives after some simple manipulations at (5.18). It is directly obvious (*cf.*, (5.3)) that one may replace the component H_β of the vector $\boldsymbol{H}$ by the component B_β of $\boldsymbol{B}$ in (5.18) without changing the quantity F_4.

The flux of energy from a field to a volume element $d\tau$ of a material body during time dt due to an electric current, polarization and magnetization is now given by

$$F_4 d\tau\, dt. \tag{5.22}$$

For the continuous motions this quantity appears on the right-hand side of the energy equation (2.18) Chapter V, as the part of the external macroscopic flow of energy $dQ_{el} = dq_{el}\, dm$, which depends on interaction between the material medium and the electromagnetic field, dq_{el} — flow of energy referred to the unit of mass. In accordance with (5.18) and (5.22), in any inertial system of coordinates one can write

$$
\begin{aligned}
F_4\, \mathrm{d}t\, \mathrm{d}\tau = \mathrm{d}Q_{el} = \mathrm{d}q_{el}\, \mathrm{d}m = \Big[& \boldsymbol{E}\cdot\boldsymbol{j}\, \mathrm{d}t + E_\alpha\, \mathrm{d}P^\alpha \\
& + H_\alpha \mathrm{d}M^\alpha + \frac{1}{2}\, \mathrm{d}(E_\alpha P^\alpha + H_\alpha M^\alpha) \Big] \mathrm{d}\tau = \Big[\frac{1}{\rho} \boldsymbol{E}\cdot\boldsymbol{j}\, \mathrm{d}t \\
& + E_\alpha\, \mathrm{d}\pi^\alpha + H_\alpha\, \mathrm{d}m^\alpha - \frac{1}{2}\, \mathrm{d}(E_\alpha \pi^\alpha + H_\alpha m^\alpha) \\
& - \frac{1}{2} (E_\alpha \pi^\alpha + H_\alpha m^\alpha)\, \nabla_\beta v^\beta\, \mathrm{d}t \Big]\, \mathrm{d}m\,,
\end{aligned}
\tag{5.23}
$$

where $\mathrm{d}m = \rho \mathrm{d}t$, element of mass, $\pi^\alpha = (P^\alpha/\rho) = (P^\alpha\, \mathrm{d}\tau/\mathrm{d}m)$, $m^\alpha = (M^\alpha/\rho) = (M^\alpha\, \mathrm{d}\tau/\mathrm{d}m)$ are the components of the vectors of the density of polarization and magnetization, respectively, referred to as the mass of particle. The increment of components of the vectors π^α and $\boldsymbol{m}^\alpha$ during the time $\mathrm{d}t$, $\mathrm{d}\pi^\alpha$ and $\mathrm{d}m^\alpha$, taken with respect to the inertial system of coordinates, and in particular, for a given point it can be the characteristic system of coordinates. In formula (5.23) the derivative of density of particle with respect to the time $(\mathrm{d}\rho/\mathrm{d}t)$ is in accordance with the equation of continuity (1.3) Chapter III, replacing $\rho\, \nabla_\beta v^\beta = -\rho \operatorname{div} v$, where v^β denotes the components of the velocity vector of the material points.

Formula (5.23) for the flux of the electromagnetic energy is closely linked with the Umov–Poynting equation (5.21). Using (5.19) and (5.20), the formula (5.23) may be written in the form:

$$
\mathrm{d}q_{el} = -\frac{1}{\rho}\left(\mathrm{d}\, \frac{\boldsymbol{E}\cdot\boldsymbol{D} + \boldsymbol{H}\cdot\boldsymbol{B}}{8\pi} + \operatorname{div} S\, \mathrm{d}t \right). \tag{5.24}
$$

For the medium at rest or for the mobile medium in characteristic system of coordinates one has $\mathrm{d}q_{el} = \mathrm{d}q^*_{el} = \mathrm{d}q^{(e)}_{el} + \mathrm{d}q^{**}_{el}$. The value of $\mathrm{d}q_{el}$, calculated in any inertial system of coordinates differs from $\mathrm{d}q^*_{el}$ by the value of the elementary work done by the ponderomotive electromagnetic forces, external with respect to the medium. If the work done by the external forces is not included into the heat flow equation, then in the heat flow equation appears $\mathrm{d}q^*_{el}$.

On the basis of (5.23), the heat flow equation (2.20) Chapter V for the material medium in the presence of the polarization and the magnetization becomes

$$dU = \left(\frac{1}{\rho} p^{\alpha\beta} + g^{\alpha\beta} \frac{E_\gamma \pi\gamma + H_\gamma m\gamma}{2}\right) \nabla_\beta v_\alpha \, dt + \frac{1}{\rho} \boldsymbol{E} \cdot \boldsymbol{j} \, dt$$

$$+ E_\alpha d\pi^\alpha + H_\alpha dm^\alpha - \frac{1}{2} d(E_\alpha \pi^\alpha + H_\alpha m^\alpha) + dq_1^* . \tag{5.25}$$

Here, all the vectors of the electromagnetic nature and their increments are taken in the characteristic system of coordinates, and $dq_1 = dq_1^{(e)} + dq_1^{**}$ — the total flux of energy to the particle, referred to the unit of mass, supplementary to the work done by the macroscopic forces and the flux of energy from the electromagnetic field. In particular, the $dq_1^{(e)}$ can be the energy flow due to the heat conductivity, and in the models of continuum with complicated properties, dq_1^{**} can depend on interaction of the considered particle with other particles of material medium.

The theoretical construction of models of a material medium is connected with the definition of the internal energy U, entering into the equation (5.25), referred to the unit of mass of the medium. Thus, one can take the total differential

$$d\left[\frac{E_\alpha \pi^\alpha + H_\alpha m^\alpha}{2}\right]$$

to the left-hand side of (5.25) and include it into the transformed formula for the interaction energy of a medium.

$$U_1 = U + \frac{1}{2}(E_\alpha \pi^\alpha + H_\alpha m^\alpha) . \tag{5.26}$$

The transition from U to U_1 is analogous to the transition from the internal energy to the free energy or to the Gibbs' thermodynamic potential in the energy equation or in the heat flux equation (*cf.*, §6 of Chapter V). Omitting the index "one" in the notation of the function U_1, the heat flux equation (5.25) has the form:

$$dU = \left(\frac{1}{\rho} p^{\alpha\beta} + \frac{E_\gamma \pi\gamma + H_\gamma m\gamma}{2} g^{\alpha\beta}\right) \nabla_\beta v_\alpha dt$$

$$+ \frac{1}{\rho} \boldsymbol{E} \cdot \boldsymbol{j} \, dt + E_\alpha \, d\pi^\alpha + H_\alpha dm^\alpha + dq_1^* . \tag{5.27}$$

If the medium is in rest, then $v^\alpha = 0$, and in this case the equation (5.25) takes the simpler form:

$$d(\rho U) = \boldsymbol{E} \cdot \boldsymbol{j}\, dt + E_\alpha dP^\alpha + H_\alpha dM^\alpha + \rho\, dq_1^* . \tag{5.28}$$

This form of the heat flux equation frequently appears in physics in the static problems. Essentially, taking into account the motion of a medium, one has (5.27) instead of (5.28). The equation (5.27) is valid in the mobile characteristic coordinate system, in which the Maxwell's equations can be written down. The physical meaning of different terms of the formulae (5.23) is especially perceptible in the considered point of medium, when the characteristic system of coordinates is used.

In a characteristic coordinate system,

$$\rho^{-1}(\boldsymbol{j} \cdot \boldsymbol{E})dt = dq_{el}^{(e)} , \tag{5.29}$$

where $dq_{el}^{(e)}$ is the Joule heat and

$$\rho^{-1}(\boldsymbol{E} \cdot d\boldsymbol{P} + \boldsymbol{H} \cdot d\boldsymbol{M}) = dq_{el}^{**} , \tag{5.30}$$

if the processes of polarization and magnetization are reversible.

Note that such a decomposition of dq^* into $dq_{el}^{(e)}$ and dq_{el}^{**} is valid in the case of immobile media. However, if a medium moves relatively to a reference frame, then (5.29) and (5.30) hold true at every point only in the characteristic reference system (*cf.*, (4.2)).[1)]

In the presence of the laws of magnetization and polarization (5.14) and (5.15), the processes of polarization and magnetization are reversible, and the electromagnetic field spends on the polarization and magnetization of a body the non-zero amount of macroscopic energy

$$\boldsymbol{E} \cdot d\boldsymbol{P} + \boldsymbol{H} \cdot d\boldsymbol{M} \neq 0 .$$

The quantities $dq_{el}^{(e)}$, dq_{el}^{**} and the components of $\widetilde{\boldsymbol{E}}$, $\widetilde{\boldsymbol{H}}$, $\widetilde{\boldsymbol{P}}$ and $\widetilde{\boldsymbol{M}}$ and their increments in a characteristic coordinate system may be expressed in terms of the corresponding components and increments in a concomitant coordinate system with due consideration of the motion of the concomitant coordinate system with respect to a given inertial reference

[1)] *Cf.*, p. 380.

frame.[1)]

Ponderomotive forces and forces which depend on the characteristics of the electromagnetic field. Previously, we introduced the theory of the ponderomotive forces, external with respect to the material medium, and the inflow of energy from the electromagnetic field to the medium, which does not depend on the properties of a specific model of a material continuous medium. In general case, the equation of state for the internal stresses and for the internal energy of the elements of the medium depends also on the vectors $\boldsymbol{\pi}$ and $\boldsymbol{m}$. The term $\nabla_\beta p^{\alpha\beta}$ in the impulse equation and the term $(1/\rho) p^{\alpha\beta} \nabla_\beta v_\alpha$ in the heat flux equation do not depend only on the microscopic mechanical and thermodynamical characteristics of a motion and a state of the elements of the medium, but also on the components of the vectors $\boldsymbol{\pi}$ and $\boldsymbol{m}$. But in a given case, these terms result from the internal interactions in the material medium, for which the vectors $\boldsymbol{\pi}$ and $\boldsymbol{m}$, appearing also in a non-closed system of Maxwell's equations, are the characteristics of a state of the medium.

Observer systems, characteristic systems, concomitant systems. For an arbitrary motion of the medium, the characteristic system of coordinates can be determined at any point and at any instant of time. Obviously, the totality of the characteristic coordinate systems in the four-dimensional space-time is the nonholonomic set of the system of coordinates. In other words, there does not exist one transformation of coordinates from the observer system immediately to all characteristic systems of coordinates. On the other hand, there exists one transformation (this is the law of motion of the medium) from the observer system to the concomitant system of coordinates, but it is non-inertial. In connection with it, for the natural physical laws it is convenient to use one immobile concomitant system of coordinates, such that the velocities of all points of the medium with respect to it vanish, just as the velocity of the considered point at a given instant of time is equal to zero with respect to each characteristic system of coordinates.

[1)] The corresponding theory is developed in L. I. Sedov, On the ponderomotive forces of interaction of an electromagnetic field and the acceleration of a material continuum moving with inclusion of finite deformation, *PPM*, 29, No. 1, 1965, pp. 4 – 17.

Because of the arbitrariness in the choice of a spatial system of coordinates for the concomitant and for the characteristic coordinate system, one can assume, without losing the generality in the general theory (in Euclidean and non-Euclidean spaces), that at any given fixed instant of time t' the coordinate lines and the system of the basic vectors $\hat{\mathbf{э}}_i$ and $\tilde{\mathbf{э}}_i$ in the concomitant and in the characteristic system, coincide in each point of the medium, i.e., that $\hat{\mathbf{э}}_\alpha = \tilde{\mathbf{э}}_\alpha$ at the fixed instant of time t' (in inertial coordinate system of an observer).

Note, that in a concomitant system it is possible to choose as a coordinate, individualizing the points of an arbitrary introduced "ideal medium", the value of its coordinates; for example, in the inertial system of observer at some fixed instant of time, which has the global meaning.[1)] From the observer point of view, one can always present the law of motion of the medium in the three-dimensional form

$$x^\alpha = x^\alpha(\xi^\beta, t), \quad \alpha, \beta = 1, 2, 3 \ ,$$

where ξ^β are the Lagrangian coordinates, and for $t = t'$ one can put

$$x^\alpha = x^\alpha_0 = \xi^\alpha \ .$$

Obviously, with such a definition of the Lagrangian coordinates and with the spatial system of coordinates, determined by any method, one has

[1)] The above considerations are valid also in the frame of the general theory of relativity, just as for any reference system with nonisotropic world line (for example, for the observer system or for the concomitant system. The matrix of general form

$$ds^2 = g_{ij}\, dy^i\, dy^j \ ,$$

corresponding to the transformation

$$y^4 = f(\xi^\alpha, t) \ \text{ and } \ y^\alpha = g^\alpha(\xi^1, \xi^2, \xi^3), \quad \alpha = 1, 2, 3 \ ,$$

for which the family of world lines and, consequently, the reference frame (*cf.*, pp. 19–21) are preserved, can be globally reduced to the form

$$ds^2 = c^2 dt^2 + 2\hat{g}_{\alpha 4} d\xi^\alpha dt + \hat{g}_{\alpha\beta} d\xi^\alpha d\xi^\beta \ ,$$

where c — the velocity of light. After that, one comes to the conclusion that the variable t, determined in each reference system, globally plays the role of the variable time t in the inertial reference frame, defined in STR. In different reference systems, just as in STR, the variable t differs in one and the same point. See, Sedov L. I., "Global time in the general theory of relativity". — Sov. Phys. Dokl. 28 (9), September 1983, pp. 727 – 729.

$$\hat{\mathbf{3}}_\alpha = \tilde{\mathbf{3}}_\alpha ,$$

when $t = t'$. Choosing the system of coordinates in such a manner, one obtains that at the instant $t + \mathrm{d}t$ the separation of the coordinate line will occur. The characteristic systems of coordinates displace as the rigid bodies with the different translational velocity v, equal to the velocity of the considered point of the medium, for the constant basis vectors $\tilde{\mathbf{3}}_\alpha$, with respect to the inertial reference frame of an observer. During the time $\mathrm{d}t$, the concomitant coordinate system displaces in such a manner, that in the general case the tetrahedron of the basis vectors $\hat{\mathbf{3}}_\alpha$ at each point of the medium, is deformed and rotated and in addition, displaced translationally with the velocity $\boldsymbol{v}$. Generally, at instant $t' + \mathrm{d}t$ one obtains $\hat{\mathbf{3}}_\alpha \neq \tilde{\mathbf{3}}_\alpha$ for each point of the medium on the space.

The different definitions of the time derivatives of vectors and tensors. The tensorial and vectorial characteristics of the field and of the medium and its derivatives and increments with respect to the coordinates and the time, determined with respect to the characteristic system of coordinates, can be calculated in the concomitant coordinate system. If the spatial tetrahedrons of the concomitant system and of the characteristic coordinate system at each point of the medium and at the given instant of the time coincides, what always can be implied, then the components of all three-dimensional tensors and vectors also coincide. Just as the characteristic and concomitant coordinate systems move with respect to the observer frame in the different manner, the increments and the derivatives of the components of tensors, taken with respect to the time in characteristic and concomitant system are different but linked by the simple formulae. For example, for any vector $A = \tilde{A}^\alpha \tilde{\mathbf{3}}_\alpha = \hat{A}^\alpha \hat{\mathbf{3}}_\alpha = \tilde{A}_\alpha \tilde{\mathbf{3}}^\alpha = \hat{A}_\alpha \hat{\mathbf{3}}^\alpha$ the following formulae are valid:

$$\frac{\mathrm{d}A}{\mathrm{d}t} = \frac{\mathrm{d}\tilde{A}^\alpha}{\mathrm{d}t} \tilde{\mathbf{3}}_\alpha = \frac{\mathrm{d}\hat{A}^\alpha}{\mathrm{d}t} \hat{\mathbf{3}}_\alpha + \hat{A}^\beta \frac{\mathrm{d}\hat{\mathbf{3}}_\beta}{\mathrm{d}t} = \left(\frac{\mathrm{d}\hat{A}^\alpha}{\mathrm{d}t} + \hat{A}^\beta \hat{\nabla}_\beta \hat{v}^\alpha \right) \hat{\mathbf{3}}_\alpha ,$$

and

$$\frac{\mathrm{d}A}{\mathrm{d}t} = \frac{\mathrm{d}\tilde{A}_\alpha}{\mathrm{d}t} \tilde{\mathbf{3}}^\alpha = \left(\frac{\mathrm{d}\hat{A}_\alpha}{\mathrm{d}t} - \hat{A}_\beta \hat{\nabla}_\alpha \hat{v}^\beta \right) \hat{\mathbf{3}}^\alpha ,$$

just as

$$\frac{d\hat{\mathbf{\jmath}}_\beta}{dt} = \hat{\nabla}_\beta \hat{v}^\alpha \hat{\mathbf{\jmath}}_\alpha \quad \text{and} \quad \frac{d\hat{\mathbf{\jmath}}^\beta}{dt} = - \hat{\nabla}_\alpha \hat{v}^\beta \hat{\mathbf{\jmath}}^\alpha .$$

If at the considered instant of the time $\widetilde{\mathbf{\jmath}}_\alpha = \hat{\mathbf{\jmath}}_\alpha$ and $\widetilde{\mathbf{\jmath}}^\alpha = \hat{\mathbf{\jmath}}^\alpha$, then the following equalities are valid

$$\frac{d\widetilde{A}^\alpha}{dt} = \frac{d\hat{A}^\alpha}{dt} + \hat{A}^\beta \hat{\nabla}_\beta \hat{v}^\alpha \quad \text{and} \quad \frac{d\widetilde{A}_\alpha}{dt} = \frac{d\hat{A}_\alpha}{dt} - \hat{A}_\beta \hat{\nabla}_\alpha \hat{v}^\beta .$$

These formulae in the curvilinear coordinate system are the generalization of the well known in the kinematics of rigid bodies, the Euler's formulae, to the case of the deforming moving coordinate system. This Euler's formulae describe the link between the derivatives of the components of a vector with respect to the time, for the absolutely rigid, immobile and rotating systems of coordinates. In the similar way it is easy to write the formulas connecting the derivatives of any degree with respect to the time of components of tensors of any order in the characteristic and the concomitant systems of coordinates. The formula (5.23) and its consequences can be rewritten in the global concomitant system.

Moment equation in four-dimensional form. Experiments with a magnetized needle in a magnetic field show that the interaction of a body and a field does not reduce only to ponderomotive forces. Experience indicates that this interaction becomes apparent likewise on account of the action of couples distributed through a volume, given by their moments. For this purpose, the differential equation for the moments of a field will be derived and analyzed. For a field, one may consider the usual three-dimensional moment equation into which enter only the usual moments of three-dimensional forces. However, within the framework of the special theory of relativity, i.e., within the framework of the theory of the electromagnetic field, described by Maxwell's equations, a force represents a four-dimensional vector. In this context, one must consider the moment equation in four-dimensional form; this step is connected with a generalization of the concept of moments of forces and of the moment of an impulse for material media as well as for fields.

It follows from the four-dimensional expressions for the components

S^{ij} of the energy-impulse tensor written in a Cartesian inertial coordinate system, that

$$x^j \nabla_k S^{ik} - x^i \nabla_k S^{jk} = -(x^j F^i - x^i F^j), \qquad i, j = 1, 2, 3, 4, \tag{5.31}$$

where x^i are the coordinates of a point in the volume occupied by a body. The anti-symmetric tensor with the six independent components $x^j F^i - x^i F^j$ may be considered as a generalization to the four-dimensional case of the normal concept of the three-dimensional anti-symmetric tensor, corresponding to a single axial vector, the moment of the three-dimensional ponderomotive force about the origin of coordinates.

It must be emphasized that in the special theory of relativity a moment, a second order antisymmetric tensor with six independent components, may be reduced in the general case in a Cartesian coordinate system to two three-dimensional vectors (one of which is axial, the other polar).

It is easily seen that (5.31) may be rewritten in the form

$$\nabla_k (S^{ik} x^j - S^{jk} x^i) + S^{ij} - S^{ji} = -(x^j F^i - x^i F^j). \tag{5.32}$$

The components of the third order tensor $S^{ik} x^j - S^{jk} x^i$ may be considered as the components of a generalized tensor of the moment density of the energy-impulse of a field. Relation (5.32) (in analogy to the kinetic energy theorem) represents a simple consequence of the definition of the ponderomotive forces and is fulfilled identically.

The energy equation in thermodynamics represents in the general case an equation which is independent of the kinetic energy theorem. Analogously, one may introduce new characteristics of a field: the tensors of the volume densities of the characteristic internal moment with the components $Q^{ijk} = -Q^{jik}$ and of the ponderomotive moment acting from the field into the body, with the components $\mathscr{H}^{ij} = h^{ij} - (x^i F^j - x^j F^i)$ which may be determined with the aid of the moment equation for the field and the moment equation for the medium, independent of (5.32).

Such independent moment equations for an electromagnetic field may be taken in the form

$$\nabla_k (S^{ik} x^j - S^{jk} x^i + Q^{ijk}) = -\mathscr{H}^{ij} = -h^{ij} + (x^j F^i - x^i F^j). \tag{5.33}$$

The components of the tensor of the ponderomotive moment $\mathscr{H}^{ij}$, external with respect to the body, include, besides the components of the moment of the ponderomotive forces still the components of an additional volume ponderomotive moment, namely of the second order antisymmetric tensor h^{ij}, due to the field.

In writing down (5.33), it has been assumed that the externally distributed volume moments, acting on an electromagnetic field, are only caused by material bodies.

Moment equations, analogous to (5.33), may be written down separately also for material media. In the equations for material media, there will appear on the right-hand side with a positive sign the components $\mathscr{H}^{il}$ of the tensor of the ponderomotive moment due to the action of the field on the medium. In addition, there may be present on the right-hand side of the moment equation for media also moments caused by internal interactions between particles of the medium itself and other external objects (bodies and fields which are not electromagnetic). From (5.33) and (5.32), there follows

$$h^{ij} = S^{ij} - S^{ji} - \nabla_k Q^{ijk} . \tag{5.34}$$

This relation may be considered as an equation for characteristic moments (four-dimensional analogues of equation (3.6), Chapter III). Equation (5.34) and the corresponding equation for material media together with experimental results, as foundation for the construction of a model of a medium, may serve as a source for the definition of h^{ij}, S^{ij} and Q^{ijk} which must be introduced for modelling of an electromagnetic field as a macroscopic physical object.

If the model of a material medium is fixed and the components of the ponderomotive moments h^{ij} have been determined by experiment (for example, the moment exerted on an elementary magnetic needle by the field), then (5.34) may be conceived as a link between S^{ij} and Q^{ijk}. If by an additional condition, entering into the definition of an electromagnetic field, these quantities do not depend on any new essential characteristics of the field, then (5.34) and (5.33), respectively, must be fulfilled identically together with (5.32).

In correspondence with experimental results, Relation (5.34) may be turned into an identity with the aid of different conditions. In particular, one introduces as the simplest and most natural conditions:

1. The energy-impulse tensor is the Minkowski tensor the components of which are defined by (5.11);

2. The components of the internal moment tensor Q^{ijk} for points inside a given volume of a field vanish, or, in a more general form, one has

$$\nabla_k Q^{ijk} = 0 , \tag{5.35}$$

which is satisfied identically or by the strength of the impulse equation.

These two basis conditions which may be included in the definition of the model of an electromagnetic field lead to agreement with experiment (*cf.*, (5.39)).

The tensor of the ponderomotive moment of an electromagnetic field. On the basis of (5.11) and (5.35), one obtains from (5.34) and the two conditions above

$$h^{ij} = \frac{1}{4\pi}\left[F^{i\cdot}_{\cdot m}H^{mj} - F^{j\cdot}_{\cdot m}H^{mi}\right]. \tag{5.36}$$

The manipulations in (5.36) will now be performed. In a Cartesian inertial coordinate system, the matrix $\|H^{ij}\|$ is defined by (5.7) and the matrix $\|F_{ij}\|$ by (5.6). Since $F^{\cdot l}_{m.} = g^{lk}F_{mk}$ and $g^{11} = g^{22} = g^{33} = -1$, $g^{44} = 1/c^2$, $g^{ij} = 0$ for $i \neq j$, one has for the matrix $\|F^{\cdot l}_{m.}\|$

$$\|F^{\cdot l}_{m.}\| = \left\|\begin{matrix} 0 & -B^3 & B^2 & E_1/c \\ B^3 & 0 & -B^1 & E_2/c \\ -B^2 & B^1 & 0 & E_3/c \\ cE_1 & cE_2 & cE_3 & 0 \end{matrix}\right\|. \tag{5.37}$$

With the aid of (5.7) and (5.37), one can easily write down the matrix $\|h^{ij}\|$ on the basis of (5.36).

Three-dimensional space vectors for the ponderomotive moment. For the four-dimensional antisymmetric ponderomotive moment tensor with components h^{ij}, one may introduce in a Cartesian coordinate system the two three-dimensional vectors $\mathscr{M}(\mathscr{M}_1, \mathscr{M}_2, \mathscr{M}_3)$ and $\vec{\mathscr{L}}(\mathscr{L}^1$,

$\mathscr{L}^2, \mathscr{L}^3$) entering into $\|h^{ij}\|$ (*cf.*, (3.26) Chapter IV) which may be written in the form

$$\|h^{ij}\| = \left\| \begin{matrix} 0 & \mathscr{M}_3 & -\mathscr{M}_2 & -\mathscr{L}^1/c \\ -\mathscr{M}_3 & 0 & \mathscr{M}_1 & -\mathscr{L}^2/c \\ \mathscr{M}_2 & -\mathscr{M}_1 & 0 & -\mathscr{L}^3/c \\ \mathscr{L}^1/c & \mathscr{L}^2/c & \mathscr{L}^3/c & 0 \end{matrix} \right\| . \tag{5.38}$$

With the aid (5.7) and (5.37), after some simple manipulations, one finds the formulae[1)]

$$\boldsymbol{\mathscr{M}} = (4\pi)^{-1}(\boldsymbol{B}\times\boldsymbol{H}) + (4\pi)^{-1}(\boldsymbol{D}\times\boldsymbol{H}) = \boldsymbol{M}\times\boldsymbol{H} + \boldsymbol{P}\times\boldsymbol{E} \tag{5.39}$$

$$\vec{\mathscr{L}} = (4\pi)^{-1}(\boldsymbol{E}\times\boldsymbol{H} - \boldsymbol{D}\times\boldsymbol{B}) = \boldsymbol{S} - \boldsymbol{S}^* \ , \tag{5.40}$$

where $\boldsymbol{S}$ is the Poynting vector and $\boldsymbol{S}^* = (4\pi)^{-1}(\boldsymbol{D}\times\boldsymbol{B})$ is of an analogous nature. In the matrix of the Minkowski tensor, the components of the vector $(1/c^2)\boldsymbol{S}$ form a fourth row and the components of $(1/c^2)\boldsymbol{S}^*$ a fourth column.

The three-dimensional vector $\mathscr{M}$ is nothing else but the ordinary ponderomotive moment. Formula (5.39) corresponds well to experimental measurements within the framework of models of material media which are satisfactory in practice. It is a natural generalization of (1.3) to the case of polarization. Obviously, one has $\mathscr{M} = 0$, when the equations of state (5.12) and (5.13) hold true with ε and μ scalars. In the general case of anisotropic media, with the same equations of state, one finds $\mathscr{M} \neq 0$.

For material media, in the special theory of relativity, the moment equation is presented in 6 component equations: three "ordinary" moment equations as projections in a three-dimensional formulation corresponding to the vector $\boldsymbol{\mathscr{M}}$, and three equations as projections corresponding to the vector $\vec{\mathscr{L}}$. The last three moment equations, on the strength of the definitions of the models of magnetized and polarized material media, may be satisfied identically or may turn out to be essential

1) *Cf.*, footnote p. 385.

relations for the determination of certain characteristics of a medium.

Formulae (5.39) and (5.40) have been established in an inertial coordinate system (the three-dimensional space part of the coordinate system may be an arbitrary curvilinear system). Their confirmation by experiment relates to immobile bodies.

On different definitions of the energy-impulse tensor and the moment tensor of an electromagnetic field. For a study of a system, consisting of an electromagnetic field and a material body, as a single object, when different mechanical and electromagnetic characteristics are involved, one may introduce the energy-impulse and moment tensors for the entire system as sums of the corresponding tensors for the field and the medium. One may write for the components of these tensors

$$S^{ij} + T^{ij}, \qquad Q^{ijk}_{\text{field}} + Q^{ijk}_{\text{medium}},$$

where T^{ij} and Q^{ijk}_{medium} are the components of the corresponding tensors relating to the medium. Obviously, the impulse and moment equations for the system as a whole will not contain ponderomotive forces and moments. For a medium and for a field the equations of impulses, energy and moments can be written in the form

$$\nabla_j (S^{ij} + T^{ij}) = \mathscr{F}^i, \quad \nabla_k (Q^{ijk}_{\text{medium}} + Q^{ijk}_{\text{field}}) + S^{ij} + T^{ij}$$
$$- S^{ji} - T^{ji} = M^{ij} \qquad (i, j, k = 1, 2, 3, 4),$$

where $\mathscr{F}_i$ and M^{ij} are the components of the four-dimensional mass forces and moments, external with respect to the medium and field. In many cases one can assume $\mathscr{F}^i = 0$ and $M^{ij} = 0$. The problems of foundation of the models of the field and specific medium are linked to the definition of here enumerated tensors by given or found characteristic values.

The subdivision of the general resultant energy-impulse and moment tensors with a general electromagnetic character into corresponding tensors for the field and medium separately may, in general, be effected according to different conditions. For a definite sum, these quantities for the medium are determined uniquely by their values for the field, and conversely. Hence it is expedient for operations with different material media to determine the impulse, energy and moment of the electromagnetic field always by the same method.

Nevertheless, these definitions of the dynamic properties of a field, introduced for these cases in accordance with the fundamental condition, may be arrived at in a different manner. A description of the dynamic properties of an electromagnetic field has been given above with the aid only of the single non-symmetric Minkowski energy-impulse tensor with components S^{ij} defined by (5.11), in correspondence with formulae for the ponderomotive forces and moments also established above.

Abraham energy-impulse tensor of the electromagnetic field. However, many authors introduce instead of the Minkowski tensor as energy-impulse tensor of the electromagnetic field the symmetric Abraham tensor with components A^{ij} for which in the characteristic coordinate system the space part is obtained simply by symmetrization of the components of the Minkowski tensor

$$A^{\alpha\beta} = \frac{1}{2}(S^{\alpha\beta} + S^{\beta\alpha}) \ ,$$

and the contravariant components of the vectors of the time part coincide with the corresponding components of the Poynting vector, divided by c^2, i.e.,

$$A^{4\alpha} = A^{\alpha 4} = \frac{1}{4\pi c}\,(\boldsymbol{E} \times \boldsymbol{H})^{\alpha} \ ,$$

and, finally by definition,

$$A^{44} = S^{44} \ .$$

Besides the Minkowski and Abraham energy impulse tensor of the electromagnetic field other tensors of electromagnetic fields are proposed.

It is characteristic that the difference between the Minkowski, Abraham and other tensors, appears only in the case, when at the points of volume, occupied by the field exists the material medium and the polarization tensor differs from zero, i.e., $P^{ij} \neq 0$. In the vacuum or conducting medium, where the electrical polarization and magnetization vanish, the energy-impulse tensors of the electromagnetic field, defined by various methods, coincide.

The previous formulas, on account of the definition of the Minkowski tensor, can be written in the characteristic coordinate system, in the form

$$A^{ij} - S^{ij} = \Omega^{ij} \ ,$$

in addition

$$\Omega^{\alpha\beta} = \frac{1}{2} h^{\alpha\beta}, \quad \Omega^{\alpha 4} = h^{\alpha 4} \quad \text{and} \quad \Omega^{4i} = 0 , \tag{5.41}$$

for

$$\alpha, \beta = 1, 2, 3; \quad i = 1, 2, 3, 4 ,$$

where the components of the tensor h^{ij} are determined by (5.38), (5.39) and (5.40).

The difference of the ponderomotive forces in Minkowski's and Abraham's approach. For components of the ponderomotive four-dimensional force in Minkowski's and Abraham's approach, one has

$$F^i_{\text{Abr}} = - \nabla_j A^{ij} \quad \text{and} \quad F^i_{\text{Min}} = - \nabla_j S^{ij} .$$

From (5.41) it is easy to obtain that in the characteristic system of coordinates

$$F^\alpha_{\text{Abr}} = - \nabla_j A^{ij} \quad \text{and} \quad F^i_{\text{Min}} = \nabla_j S^{ij} , \tag{5.42}$$

and on the basis of (5.38) these equalities in the three-dimensional vectorial form are

$$F_{\text{Abr}} = F_{\text{Min}} - \frac{1}{2} \operatorname{curl} \mathscr{M} + \frac{1}{c} \frac{\partial \vec{\mathscr{L}}}{\partial t} . \tag{5.43}$$

For the specific influx of energy

$$F^4_{\text{Abr}} = F^4_{\text{Min}} . \tag{5.44}$$

If for the isotropic body the laws of magnetization and polarization (5.14) and (5.15) are valid, then $\mathscr{M} = 0$, and consequently in this case the equality (5.43) has the form:

$$F_{\text{Abr}} = F_{\text{Min}} + \frac{1}{4\pi c} \frac{\partial}{\partial t} (\boldsymbol{D} \times \boldsymbol{B} - \boldsymbol{E} \times \boldsymbol{H}) .$$

If the energy-impulse tensor of the medium is fixed, independently of the definition of the energy-impulse of the electromagnetic field, then from (5.42) follows that the equations of motion (impulses) for a medium are different, just as $F_{\text{Abr}} \neq F_{\text{Min}}$. However, if one defines the energy-impulse tensor of the medium in different manner, then the form of the equation of motion can be preserved.

From (5.44) follows, that the fluxes of the electromagnetic energy in a medium, appearing in the heat-flux equation, described with the aid of the Minkowski energy-impulse tensor as well as the Abraham tensor, have the same form:

Ponderomotive moments and tensors of energy-impulse of the field. As it follows from the considered four-dimensional equations of motion, if one accepts as an experimental fact, that in the magnetized and polarized medium acts due to the field the distributed four-dimensional moment, characterized by the tensor, components h^{ij} of which are determined by (5.38), (5.39) and (5.40), on the basis of the law of action and reaction, following from the law of conservation of impulse for the system (medium plus field), follows that when using for the energy-impulse tensor of the field, the Minkowski tensor for the components of the tensor of the density of the characteristic moment of the field, then the equalities are valid

$$\nabla_k Q^{ijk}_{\text{Min}} = 0 ,$$

and, in particular, one can assume $Q^{ijk}_{\text{Min}} = 0$. On the other hand, using the Abraham tensor as the energy-impulse tensor of the field, since $A^{ij} = A^{ji}$, one has

$$h_{ij} = - \nabla_k Q^{ijk}_{\text{Abr}} \neq 0 . \qquad (5.45)$$

Thus, for utilization of the Abraham's energy-impulse tensor of the electromagnetic field it is necessary to attach to it the complementary characteristics — the tensor $Q^{ijk} = - Q^{jik}$ of the internal moments of motion (for $i = 4$ and $j = \alpha$) and of the surface couples (for $i = \alpha, j = \beta$). The divergence of this tensor $\nabla_k Q^{ijk}$ is defined by the equality (5.45) if for the distributed mass ponderomotive moments, acting on the medium due to the field, the formulas (5.38)–(5.40) are valid.

Thus, if one does not employ Minkowski's definition of the energy-impulse tensor, then one must introduce on three-dimensional surface sections of four-dimensional volumes in the electromagnetic field not only surface forces of interaction, but also interactions due to distributed surface moments (couples). With the utilization of Minkowski's energy-impulse tensor, moments distributed over sections inside an electromagnetic field need not be introduced.

In connection with these general, essentially equivalent possibilities of establishing different conditions for the definition of the dynamic properties of an electromagnetic field, and likewise in connection with the analysis of the moment equation for the electromagnetic field and of the formulae for the ponderomotive moment, one may select in the capacity of a universal and natural condition the Minkowski tensor as the energy impulse tensor of the electromagnetic field. If this condition is accepted, then one must not only follow it in the description of the dynamic properties of the electromagnetic field, but also in the construction of models of material media.

The energy-impulse tensors of the medium for the various energy-impulse tensors of the field. Let us observe, that if one takes for the resultant energy-impulse tensor N^{ij} of the field and medium

$$N^{ij} = T^{ij}_{\text{Abr}} + A^{ij} = T^{ij}_{\text{Min}} + S^{ij} ,$$

then it turns out, that for the contravariant components of the energy-impulse tensor of the medium in general, the following inequalities take place

$$T^{ij}_{\text{Abr}} \neq T^{ij}_{\text{Min}} \quad \text{and} \quad T^{(ij)}_{\text{Abr}} \neq T^{(ij)}_{\text{Min}} , \tag{5.46}$$

where $T^{(ij)} = \frac{1}{2}(T^{ij} + T^{ji})$. It follows, that not only components of the energy-impulse tensors of the medium are not equal between themselves, but also their symmetric parts. It is easy to check that the above considerations are valid also in the case when the resultant tensor N^{ij} is symmetric.

Motives for the introduction of a symmetric energy-impulse tensor. The following considerations lead to an attempt of introducing a symmetric energy-impulse tensor for an electromagnetic field. The energy-impulse tensor, figuring in the general theory of relativity or derivable by averaging of the microscopic energy-impulse tensor for the field in vacuum, is automatically symmetrical; however, it either relates to the field and the medium as a whole (since microscopic interactions inside a medium have an electromagnetic character) or to the electromagnetic field in vacuum or to absence of magnetized and polarized material media. In all these cases, either the Minkowski tensor is symmetric, since $\boldsymbol{P}=\boldsymbol{M}=0$, or only the resultant energy-impulse tensor for the medium and the field is symmetric; in fact, only for this tensor one obtains automatically symmetry in the general theory of relativity also for averages.

On the other hand, it is obvious that the values of the components of the four-dimensional ponderomotive force (a three-dimensional force and a flux of energy) remain unchanged, if one takes instead of any energy-impulse tensor with components S^{ij} another tensor with components

$$S^{*ij} = S^{ij} + \Omega^{ij},$$

when the components of the supplementary tensor Ω^{ij} fulfill identically

$$\nabla_j \Omega^{ij} = 0 . \tag{5.47}$$

Addition of the tensor with components Ω^{ij} under the condition (5.47) does not influence the formula for the ponderomotive force. In many cases, the components Ω^{ij} subject to (5.47) may be chosen in such a manner that the tensor S^{*ij} will be symmetric. By such an operation, the energy-impulse tensor S^{ij} may even be symmetrized, while preserving the physical four-dimensional vector of the ponderomotive force. However, in the result of such a symmetrization, in order to obtain acceptable formulae for the volume ponderomotive moment acting on a material body, one must introduce for the field the internal moment $\nabla_k Q^{ijk} \neq 0$.

Finally, in many classical models, the energy-impulse tensor for material media is symmetric; hence it is natural to strive to preserve this property in the general case. However, a non-symmetric energy-impulse tensor for a material medium may occur only in complicated models for which theoretical and experimental studies have not yet been performed in a sufficiently detailed and wide manner.

§6. Hydrodynamics of the conducting fluid

As an example of a model of a continuous medium, in which electromagnetic effects are taken into account, consider the model of the conducting fluid or gas in which no polarization and magnetization occur, while an electric current may flow, i.e., $\boldsymbol{M} = \boldsymbol{P} = 0$, $\boldsymbol{j}^* \neq 0$. Setting $dq^{**} = 0$, it will be assumed that in the models under consideration individual particles of the continuum may exchange with adjoining particles and other external objects only mechanical and thermal energy.

Further, for the sake of simplicity, ideal media will be considered: $p^{ij} = -pg^{ij}$. In more general cases of magnetohydrodynamics, one takes account of the viscosity of the medium.

The following system of equations of magnetohydrodynamics is approximate within the framework of Newtonian mechanics: the scalar equation of continuity:

$$\frac{\partial \rho}{\partial t} + \operatorname{div}(\rho \boldsymbol{v}) = 0\ ; \tag{6.1}$$

the vectorial momentum equation

$$\rho\left[\frac{\partial \boldsymbol{v}}{\partial t} + (\boldsymbol{v}\cdot\nabla)\boldsymbol{v}\right] = -\operatorname{grad} p + \rho_e \boldsymbol{E} + \frac{1}{c}(\boldsymbol{j}\times\boldsymbol{H}) + \rho \boldsymbol{F}_{\text{sup}}\ , \tag{6.2}$$

where $\boldsymbol{F}_{\text{sup}}$ denotes the density of ordinary mass forces which do not arise from the interaction of the fluid with an electromagnetic field, for example, gravity; the scalar heat flow equation

$$\mathrm{d}U + p\mathrm{d}\frac{1}{\rho} = \frac{1}{\rho}(\tilde{\jmath}\cdot\tilde{E})\mathrm{d}t + \mathrm{d}q^{(e)}_{\text{sup}}\ , \tag{6.3}$$

where $\mathrm{d}q^{(e)}_{\text{sup}}$ denotes the flux of heat to unit mass of fluid from outside due to heat conduction or radiation, etc.; the scalar relations, expressing the second law of thermodynamics,

$$T\mathrm{d}s = \frac{1}{\rho}(\tilde{\jmath}\cdot\tilde{E})\mathrm{d}t + \mathrm{d}q^{(e)}_{\text{sup}}\ , \tag{6.4}$$

where it has been assumed that $\mathrm{d}q' = 0$.

If the medium is determined by specification of the internal energy U as a function of ρ and s, then one obtains from

$$\mathrm{d}U = T\mathrm{d}s - p\mathrm{d}\frac{1}{\rho}$$

the two additional scalar equations of state

$$T = \left(\frac{\partial U}{\partial s}\right)_\rho, \qquad -p = \left(\frac{\partial U}{\partial \frac{1}{\rho}}\right)_s . \tag{6.5}$$

The dynamic and thermodynamic equations (6.1)–(6.5) must yet be augmented by electromagnetic equations, namely Maxwell's equations

$$\left.\begin{aligned} \operatorname{curl} \boldsymbol{E} &= -\frac{1}{c}\frac{\partial \boldsymbol{H}}{\partial t}, \qquad & \operatorname{div} \boldsymbol{H} &= 0, \\ \operatorname{curl} \boldsymbol{H} &= \frac{4\pi}{c}\boldsymbol{j} + \frac{1}{c}\frac{\partial \boldsymbol{E}}{\partial t}, \qquad & \operatorname{div} \boldsymbol{E} &= 4\pi\rho_e, \end{aligned}\right\} \tag{6.6}$$

and Ohm's law which in the usual presentation of magnetohydrodynamics assumes the form

$$\boldsymbol{j} = \sigma\left(\boldsymbol{E} + \frac{1}{c}\boldsymbol{v}\times\boldsymbol{H}\right) + \rho_e \boldsymbol{v}. \tag{6.7}$$

The system of equations (6.1)–(6.7) is closed, if expressions for $dq^{(e)}_{\text{sup}}$ and F_{sup} are given.

The simplification of this system for the case of large and small conductivity of the medium (more precisely for the large and small values of the parameter $(\sigma L/c)$, where L is characteristic linear dimension of the specimen t) leads to the equations of magnetohydrodynamics (MHD) and electrohydrodynamics (EHD). To obtain the equations of MHD and EHD one must evaluate the order of values in the Maxwell's equations also in the expression for the Lorentz force and the Joule heat, and neglect the terms, small as compared with others.

Let us denote the characteristic values of the electric and magnetic field, velocity, linear and time parameters by E, H, v, L, T. One obtains the MHD and EHD equations in non-relativistic approximations, when

$$\left(\frac{v}{c}\right)^2 \ll 1, \quad \left(\frac{L}{cT}\right)^2 \ll 1. \tag{6.8}$$

If the characteristic time is of the order (L/v), then the second condition (6.8) is the direct consequence of the first. It is usually satisfied in the problems of mechanics, v represents itself the characteristic velocity of the medium or the characteristic velocity of the propagation of waves.

Let us consider the equation

$$\operatorname{curl} \boldsymbol{H} = \frac{4\pi}{c}\boldsymbol{j} + \frac{1}{c}\frac{\partial \boldsymbol{E}}{\partial t}.$$

In the above equation, after substituting the expression for j from the Ohm's law (6.7), one can evaluate the order of magnitude of the terms of obtained equation.

In addition, one can use the evaluation $\rho_e \sim (E/L)$ which follows from the equation $\operatorname{div} E = 4\pi\rho_e$. One has (the maximal order of magnitude is written down under the relevant term)

$$\begin{array}{ccccccccc} \operatorname{curl} \boldsymbol{H} = & \dfrac{4\pi}{c}\rho_e v & + & \dfrac{4\pi\sigma}{c} E & + & \dfrac{4\pi\sigma}{c^2} v \times H & + & \dfrac{1}{c}\dfrac{\partial E}{\partial t} & . \\ \dfrac{H}{L} & \dfrac{v}{c}\dfrac{E}{L} & & \dfrac{\sigma L}{c}\dfrac{E}{L} & & \dfrac{v}{c}\dfrac{\sigma L}{c}\dfrac{H}{L} & & \dfrac{L}{cT}\dfrac{E}{L} & . \end{array} \tag{6.9}$$

Using this evaluation, one can show that in the general case, from

$$\left(\frac{\sigma L}{c}\right)^2 \gg 1 \ , \tag{6.10}$$

follows that $H^2 \gg E^2$ and the corresponding equations are called the equations of magnetohydrodynamics, and when

$$\left(\frac{\sigma L}{c}\right)^2 \ll 1 \ , \tag{6.11}$$

the solutions with $E^2 \gg H^2$ are possible. Such problems are considered in electrohydrodynamics.

The equations of magnetohydrodynamics. The condition $(\sigma L/c) \gg 1$ is satisfied for a number of motion of the liquid metals (for mercury $\sigma \sim 6 \cdot 10^{15}$ 1/sec, strongly ionized gases — plasma ($\sigma \sim 10^{13} - 10^{14}$), for majority of phenomena, considered in astrophysics (not only for large σ, but also for large L). Because of this, the domain of application of MHD is very large (MHD – generators of the electric current, tokamaks, the instruments for realization of the thermonuclear reactions etc.).

From evaluations of (6.9) it follows that under the conditions (6.10) the terms $(4\pi/c)\,\rho_e\, v$ and $(1/c)\,(\partial E/\partial t)$ can be neglected comparing with $(4\pi\sigma/c)\boldsymbol{E}$.

Then (6.9) takes the form

$$\text{curl}\,\boldsymbol{H} = \frac{4\pi\sigma}{c}\left(\boldsymbol{E} + \frac{v}{c} \times \boldsymbol{H}\right)$$

or

$$\boldsymbol{E} = \frac{c}{4\pi c}\ \text{curl}\,\boldsymbol{H} - \frac{v}{c} \times \boldsymbol{H}\,. \tag{6.12}$$

From it follows that under the conditions (6.8), (6.10)

$$E^2 \ll H^2\,.$$

Next, from evaluation of (6.9) and the conditions (6.8) and (6.10) follows, that in the general case, in the expression for the current, the term ρ_e^v can be neglected in comparison with the conduction current

$$j = j^* = \tilde{j}\,,$$

where

$$j = \frac{c}{4\pi}\,\text{curl}\,\boldsymbol{H}\,.$$

In addition, the magnetic field intensity can be considered as the same in all systems of coordinates, moving one with respect to the other with velocities, much smaller than the velocity of light

$$\boldsymbol{H}' = \boldsymbol{H} - \frac{v}{c} \times \boldsymbol{E} \approx \boldsymbol{H}\,,$$

just as $\boldsymbol{E}^2 \ll \boldsymbol{H}^2$.

Consider now the expression for the Lorentz force in MHD

$$F_1 = \rho_e \boldsymbol{E} + \frac{1}{c}\boldsymbol{j} \times \boldsymbol{H} = \rho_e \boldsymbol{E} + \frac{1}{4\pi}\ \text{curl}\,\boldsymbol{H} \times \boldsymbol{H}\,.$$

Since the order of magnitude of the first term is E^2/L, of the second is H^2/L^2, and $\boldsymbol{E}^2 \ll \boldsymbol{H}^2$, then

$$F_L = \frac{1}{4\pi}\ \text{curl}\,\boldsymbol{H} \times \boldsymbol{H}\,.$$

Finally, the expression for Joule heat has the form:

$$\tilde{j} \cdot \tilde{E} = \frac{1}{\sigma} \tilde{j}^2 = \frac{1}{\sigma} j^2 = \frac{c^2}{16\pi^2 \sigma} (\operatorname{curl} \boldsymbol{H})^2 \ ,$$

because $\tilde{j} = j$.

Thus, in the system of equations for the medium appears only the magnetic field intensity $\boldsymbol{H}$ while the magnitudes $\boldsymbol{E}, \rho_e, \boldsymbol{j}$ does not.

From the Maxwell's system of equation one can obtain the equation in which only $\boldsymbol{H}$ appears. Consider the equation

$$\operatorname{curl} \boldsymbol{E} = -\frac{1}{c} \frac{\partial \boldsymbol{H}}{\partial t} . \tag{6.13}$$

In accordance with (6.12) one has

$$\boldsymbol{E} = \frac{1}{c} \left(\frac{c^2}{4\pi\sigma} \operatorname{curl} \boldsymbol{H} - \boldsymbol{v} \times \boldsymbol{H} \right) . \tag{6.14}$$

The parameter $c^2/4\pi\sigma$ has the dimension of the coefficient of kinematic viscosity and is referred to as the coefficient of magnetic viscosity ν_m

$$\nu_m = \frac{c^2}{4\pi\sigma} .$$

Substituting (6.14) into (6.13), one obtains the equation for $\boldsymbol{H}$, referred to as the equation of induction:

$$\frac{\partial \boldsymbol{H}}{\partial t} = \operatorname{curl}(\boldsymbol{v} \times \boldsymbol{H}) - \boldsymbol{v}_m \operatorname{curl} \operatorname{curl} \boldsymbol{H} . \tag{6.15}$$

This equation can be supplemented by the equation $\operatorname{div} \boldsymbol{H} = 0$ (*cf.*, §1). If $(vH/L)^2 \ll (\nu_m H/L^2)^2$, i.e., if $(vL/\nu_m) \ll 1$ then the first term in the RHS of the equation of induction can be neglected, comparing with the second one. In such a case, the equations for $\boldsymbol{H}$ do not depend on the motion of a medium (if σ does not depend on the temperature determined by the motion of a medium). But the motion of a medium obviously depends on $\boldsymbol{H}$. Non-dimensional parameter vL/ν_m is referred to as the magnetic Reynolds number, $\mathrm{Re}_m = vL/\nu_m$. If, on the contrary, $\mathrm{Re}_m \gg 1$, then in the conduction equation (6.15) one can neglect the second term of the RHS and the magnetic field obeys the equations

$$\frac{\partial \boldsymbol{H}}{\partial t} = \operatorname{curl}(\boldsymbol{v} \times \boldsymbol{H}), \quad \operatorname{div} \boldsymbol{H} = 0\,, \tag{6.16}$$

and appears to be frozen in the medium (see the following section).

Hence, the system of equations of magnetohydrodynamics in the simpler version (without taking into account the properties of viscosity and heat conduction of a medium) has the form

$$\begin{aligned}
&\frac{d\rho}{dt} + \rho \operatorname{div} \boldsymbol{v} = 0\,,\\
&\rho \frac{d\boldsymbol{v}}{dt} = \rho \boldsymbol{F} - \operatorname{grad} \boldsymbol{p} + \frac{1}{4\pi} \operatorname{curl} \boldsymbol{H} \times \boldsymbol{H}\,,\\
&\rho T \frac{ds}{dt} = \frac{c^2}{16\pi^2 \sigma} (\operatorname{curl} \boldsymbol{H})^2\,,\\
&U = U(\rho, s), \quad p = \rho^2 \frac{\partial U}{\partial \rho}\,, \quad T = \frac{\partial U}{\partial s}\,,\\
&\frac{\partial \boldsymbol{H}}{\partial t} = \operatorname{curl}(\boldsymbol{v} \times \boldsymbol{H}) - \operatorname{curl} v_m \operatorname{curl} \boldsymbol{H}\,,\\
&\operatorname{div} \boldsymbol{H} = 0\,.
\end{aligned} \tag{6.17}$$

$\boldsymbol{E}$, ρ_e, $\boldsymbol{j}$ do not appear in this system. If it is necessary, one can calculate them, having the solution of the system, with the aid of the formulae

$$\begin{aligned}
&\boldsymbol{j} = \frac{c}{4\pi} \operatorname{curl} \boldsymbol{H}\,,\\
&\boldsymbol{E} = \frac{1}{c}(\nu_m \operatorname{curl} \boldsymbol{H} - \boldsymbol{v} \times \boldsymbol{H})\,,\\
&\rho_e = \frac{1}{4\pi} \operatorname{div} \boldsymbol{E}\,.
\end{aligned} \tag{6.18}$$

Equations of electrohydrodynamics. Consider now the phenomena in which $(\sigma L/c) \ll 1$, for example, the fluxes of a low conducting medium with some (small) number of charged particles. The charged particles frequently appear in the fluxes as the result of an interaction with streamlined bodies, or they can be added artificially to obtain the possibility to control the flux with the aid of the electromagnetic field.

Consider now the simplest case, when in the medium there exists only the charged particles of one type, with the charge e (its number for unit of volume n_1) and the neutral molecules (its number for unit of volume n). If one denotes by $\boldsymbol{v}_{\mathrm{rel}}$ the average velocity of the charged particles with respect to the medium

$$\boldsymbol{v}_{\mathrm{rel}} = \frac{\sum \boldsymbol{v}_{\mathrm{rel}}}{n_1} ,$$

then the conductivity current (the current in the coordinate system, with respect to which the velocity of the medium vanishes) can be, obviously, described by

$$j' = \rho_e \boldsymbol{v}_{\mathrm{rel}} ,$$

and for the current in the coordinate system at rest, one has

$$j = \rho_e \boldsymbol{v} + \rho_e \boldsymbol{v}_{\mathrm{rel}} .$$

Usually, one introduces the coefficient of mobility b of the charged particles by the formula

$$\boldsymbol{v}_{\mathrm{rel}} = b\boldsymbol{E} ,$$

then

$$j = \rho_e b \boldsymbol{E} . \tag{6.19}$$

Thus, in the considered case, one can introduce the notation

$$\sigma = \rho_e b . \tag{6.20}$$

The coeffficient of mobility is proportional to the number of the charged particles and inversely proportional to the density of the medium. Considering the equation (6.9), one can observe, that if $(\sigma L/c) \ll 1, (v^2/c^2) \ll 1$, $(L/cT) \ll 1$, then this equation permits a class of solutions, for which $\boldsymbol{H}^2 \ll \boldsymbol{E}^2$. Such a case is considered in EHD. The Maxwell's system of equations for $\boldsymbol{E}$ can be simplified, namely, in equation

$$\operatorname{curl} \boldsymbol{E} + \frac{1}{c}\frac{\partial \boldsymbol{H}}{\partial t} = 0\,,$$

the second term is of the small magnitude of the higher order, comparing with the first

$$\left(\frac{E}{L} \ll \frac{L}{cT}\frac{H}{L}\right).$$

Therefore, this equation can be replaced by the equation curl $\boldsymbol{E} = 0$.

Thus, the electric field $\boldsymbol{E}$ in EHD is described by the equations of electrostatics.

$$\begin{aligned} &\operatorname{curl} \boldsymbol{E} = 0 \\ &\operatorname{div} \boldsymbol{E} = 4\pi\rho_e\,. \end{aligned} \qquad (6.21)$$

Note, that $\boldsymbol{E}$ can depend on time.

In addition, in EHD, the intensity of an electric field can be considered as the same in all inertial system, moving one with respect to the other with non-relativistic velocities

$$\boldsymbol{E}' = \boldsymbol{E} + \frac{1}{c}\,\boldsymbol{v} \times \boldsymbol{H} \approx \boldsymbol{E}\,,$$

and linked to it, that $\boldsymbol{H}^2 \ll \boldsymbol{E}^2$. Then the Ohm's law can be written in the form

$$\boldsymbol{j} = \rho_e \boldsymbol{v} + \sigma \boldsymbol{E}\,,$$

or in the considered simple case in the form

$$\boldsymbol{j} = \rho_e \boldsymbol{v} + \rho_e b \boldsymbol{E}\,. \qquad (6.22)$$

Consider the expression for the Lorentz force

$$\boldsymbol{F}_L = \rho_e \boldsymbol{E} + \frac{1}{c}\,\boldsymbol{j} \times \boldsymbol{H} = \rho_e \boldsymbol{E} + \frac{1}{c}\,\rho_e \boldsymbol{v} \times \boldsymbol{H} + \frac{\sigma}{c}\,\boldsymbol{E} \times \boldsymbol{H}\,.$$

Since the order of the first term is E^2/L, the second — $(v/c)\,(E/L)\,H$, the third — $(\sigma L/c)\,(E/L)\,H$, and $H^2 \ll E^2$, $(v^2/c^2) \ll 1$, $(\sigma L/c) \ll 1$, then the second and third term can be neglected comparing with the first, i.e.,

$$F_L = \rho_e E . \tag{6.23}$$

The expression for the Joule heat has the form

$$j' \cdot E' = \sigma E \cdot E = \sigma E^2 ,$$

since $E' = E$ in EHD. For the medium containing the charged particles of one type only

$$j' \cdot E' = \rho_e b E^2 . \tag{6.24}$$

Now, the complete system of equations of EHD, in simpler version (without taking into account the viscosity and the heat conductivity of the medium) has the form

$$\begin{aligned}
&\frac{d\rho}{dt} + \rho \operatorname{div} v = 0 , \\
&\rho \frac{dv}{dt} = \rho F - \operatorname{grad} p + \rho_e E , \\
&\rho T \frac{ds}{dt} = \sigma E^2 = \rho_e b E^2 , \\
&T = \frac{\partial U}{\partial s}, \quad p = \rho^2 \frac{\partial U}{\partial \rho}, \quad U = U(\rho, s) , \\
&\operatorname{curl} E = 0 , \\
&\operatorname{div} E = 4\pi\rho_e , \\
&\frac{\partial \rho_e}{\partial t} + \operatorname{div} j = 0 , \\
&j = \rho_e b E + \rho_e v .
\end{aligned} \tag{6.25}$$

The equation

$$\frac{\partial \rho_e}{\partial t} + \operatorname{div} \boldsymbol{j} = 0 ,$$

as has been shown in §1, is the direct consequence of the Maxwell equations. In this system, the intensity of magnetic field is not included. If it is necessary, one can calculate the intensity of magnetic field (having the solution of a system) from the formula

$$\operatorname{curl} \boldsymbol{H} = \frac{4\pi}{c} \boldsymbol{j} + \frac{1}{c} \frac{\partial \boldsymbol{E}}{\partial t} , \qquad \operatorname{div} \boldsymbol{H} = 0 . \tag{6.26}$$

In EHD one introduces the so-called electric Reynolds number $\mathrm{Re}_{el} = (v/bE)$; if $\mathrm{Re}_{el} \to 0$ (the coefficient of mobility of charges is big) then the equations which describe the electric field, do not depend on the parameters of motion (if b does not depend on these parameters). Thus, for the small electric Reynolds number, the motion of medium does not influence the electric field. If $\mathrm{Re}_{el} \to 0, b \to 0$ then the charges are moving together with the medium — they are frozen in the medium.

Besides the magnetohydrodynamics and electrohydrodynamics, other branches of sciences, in which the subject of a study is a motion of continuous media interacting with the electromagnetic field, are developed. For example, the growing applications have in practice the so-called ferroelectric fluids (the fluids with the small iron particles suspended in it), due to this the ferrohydrodynamics has appeared and is developed. There exists the hydrodynamics of polarized fluids, magnetoelasticity, piezoelectricity etc.

§7. The laws of freezing in of magnetic and vortex flux lines

The general properties of a vector field $\boldsymbol{A}(x, y, z, t)$ which obeys relations of the form

$$\frac{\partial \boldsymbol{A}}{\partial t} = \operatorname{curl}(\boldsymbol{v} \times \boldsymbol{A}) , \quad \operatorname{div} \boldsymbol{A} = 0 \tag{7.1}$$

will now be explained.

It has been shown above (*cf.*, (6.16)) that the field $\boldsymbol{H}$ of the magnetic stress satisfies these conditions in the case when $\mathrm{Re}_m \gg 1$ and in particular in the case of a medium with infinite conductivity. However, these conditions are also satisfied by the field of the vorticity vector $\boldsymbol{\omega} = \frac{1}{2} \operatorname{curl} \boldsymbol{v}$ of the velocity in the case of barotropic processes in an ideal fluid in a field of potential mass forces. In fact, consider the momentum equation for an ideal non-conducting fluid in the form of Lamb–Gromeka:

$$\frac{\partial \boldsymbol{v}}{\partial t} + \operatorname{grad} \frac{\boldsymbol{v}^2}{2} + 2\boldsymbol{\omega} \times \boldsymbol{v} = -\frac{1}{\rho} \operatorname{grad} p + \boldsymbol{F} . \tag{7.2}$$

Assume that the external mass forces $\boldsymbol{F} = \operatorname{grad} U$ and $p = f(\rho)$. Then one may introduce the pressure function

$$\mathscr{P} = \int_{p_0}^{p} \frac{dp}{\rho(p)} , \tag{7.3}$$

the gradient of which is readily seen to be given by

$$\operatorname{grad} \mathscr{P} = \frac{1}{\rho} \operatorname{grad} p .$$

Under these assumptions, the momentum equation (7.2) may be presented in the form

$$\frac{\partial \boldsymbol{v}}{\partial t} + \operatorname{grad} \frac{\boldsymbol{v}^2}{2} + 2\boldsymbol{\omega} \times \boldsymbol{v} = -\operatorname{grad} \mathscr{P} + \operatorname{grad} U .$$

Taking the curl on both sides of this equation, one obtains

$$\frac{\partial \boldsymbol{\omega}}{\partial t} = \operatorname{curl}(\boldsymbol{v} \times \boldsymbol{\omega}) , \tag{7.4}$$

where

$$\operatorname{div} \boldsymbol{\omega} = 0 .$$

These equations actually coincide with Equations (7.1) and (6.16). Note that (7.4) is kinematic by its nature; however, it was obtained as a consequence of the dynamic momentum equation.

Thus, in establishing the general properties of a vector field $\boldsymbol{A}$ satisfying the conditions (7.1), one arrives, in particular, at very important properties of the magnetic field $\boldsymbol{H}$ in the case of a medium with infinite conductivity and of the vortex field ω for barotropic processes in an ideal fluid in a field of potential mass forces.

Derivation of the formula for the time derivative of the flow of a solenoidal vector through a fluid surface. First of all, a formula will be derived for the derivative with respect to time of the flow of the solenoidal vector $\boldsymbol{A}(x, y, z, t)$ through some open fluid surface Σ, bounded by a curve C, i.e., through a surface moving together with the particles of a continuous medium.

As is known, the flow of a vector through a surface Σ is defined by the integral

$$\int_{\Sigma}(\boldsymbol{A}\cdot\boldsymbol{n})\mathrm{d}\sigma = \int_{\Sigma} A_n \mathrm{d}\sigma ,$$

where $\boldsymbol{n}$ is the unit normal to the element $\mathrm{d}\sigma$ of the surface Σ.

As before, the positive direction of the normal $\boldsymbol{n}$ and the circuit of C will be linked in such a manner, that seen from the end of $\boldsymbol{n}$ the circuit of C is counter-clockwise.

If at time t the surface occupies the position Σ, and at time $t+\Delta t$ the position Σ_1 (Fig. 41) then, by definition, the time derivative is given by

$$\frac{\mathrm{d}}{\mathrm{d}t}\int_{\Sigma} A_n \mathrm{d}\sigma = \lim_{\Delta t\to 0}\frac{\int_{\Sigma_1} A_n(x, y, z, t+\Delta t)\mathrm{d}\sigma - \int_{\Sigma} A_n(x, y, z, t)\mathrm{d}\sigma}{\Delta t}. \tag{7.5}$$

It will be shown that, by (7.1), this derivative vanishes and that hence for such a vector

$$\int_{\Sigma} A_n \mathrm{d}\sigma = \mathrm{const}.$$

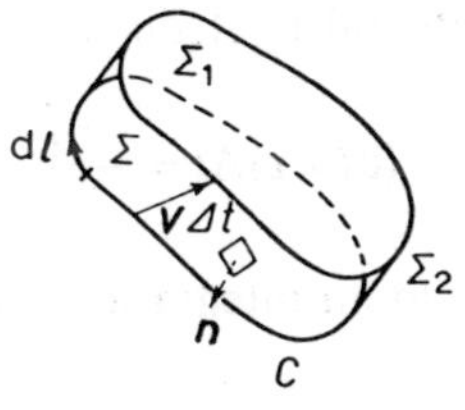

Fig. 41. On the derivation of the formula for the time derivative of the flow of a solenoidal vector through a fluid surface.

In fact, let Σ_2 denote the surface formed by the trajectories of the points of the contour C as it moves during the time Δt and consider the closed surface formed by Σ, Σ_1 and Σ_2.

By the theorem of Gauss–Ostrogradskii, using the second condition (7.1), one finds

$$\int_{\Sigma+\Sigma_1+\Sigma_2} A_n(x, y, z, t)\mathrm{d}\sigma = \int_V \operatorname{div} A\,\mathrm{d}\tau = 0\,, \tag{7.6}$$

where V is the volume bounded by the closed surface $\Sigma+\Sigma_1+\Sigma_2$ and $\boldsymbol{n}$ is the outward normal with respect to V. Adding and subtracting $\int_{\Sigma_1} A_n(x, y, z, t+\Delta t)\mathrm{d}\sigma$ to (7.6) and reversing the direction of the normal $\boldsymbol{n}$ to the elements of Σ_1, one obtains readily

$$\begin{aligned}\int_{\Sigma_1} A_n(x, y, z, t+\Delta t)\mathrm{d}\sigma - \int_{\Sigma} A_n(x, y, z, t)\mathrm{d}\sigma \\ = \int_{\Sigma_1} A_n(x, y, z, t+\Delta t)\mathrm{d}\sigma - \int_{\Sigma_1} A_n(x, y, z, t)\mathrm{d}\sigma \\ + \int_{\Sigma_2} A_n(x, y, z, t)\mathrm{d}\sigma\,.\end{aligned}$$

Hence one finds for the time derivative (7.5) the expression

$$\begin{aligned}\frac{\mathrm{d}}{\mathrm{d}t}\int_{\Sigma} A_n(x, y, z, t)\mathrm{d}\sigma \\ = \int_{\Sigma} \frac{\partial A_n}{\partial t}\mathrm{d}\sigma + \lim_{\Delta t\to 0}\frac{1}{\Delta t}\int_{\Sigma_2} A_n(x, y, z, t)\mathrm{d}\sigma\,.\end{aligned} \tag{7.7}$$

Obviously, the vectorial element of area $\boldsymbol{n}\mathrm{d}\sigma$ of the side surface Σ_2 is

$$\boldsymbol{n}\,\mathrm{d}\sigma = \boldsymbol{v}\Delta t \times \mathrm{d}\boldsymbol{l}\,,$$

where $\mathrm{d}\boldsymbol{l}$ is the element of the curve C, bounding Σ (Fig. 41). Therefore, one may go over in the integral $\int_{\Sigma_2} A_n\mathrm{d}\sigma$ to an integration along C

$$\int_{\Sigma_2} (\boldsymbol{A}\cdot\boldsymbol{n})\mathrm{d}\sigma = -\int_C \boldsymbol{A}\cdot(\mathrm{d}\boldsymbol{l}\times\boldsymbol{v})\Delta t = -\int_C \mathrm{d}\boldsymbol{l}\cdot(\boldsymbol{v}\times\boldsymbol{A})\Delta t\,,$$

or, using Stokes' theorem, to an integration over the original surface Σ, stretched over C:

$$\frac{1}{\Delta t}\int_{\Sigma_2} A_n \mathrm{d}\sigma = -\int_C (\boldsymbol{v}\times\boldsymbol{A})\cdot \mathrm{d}\boldsymbol{l} = -\int_\Sigma [\mathrm{curl}(\boldsymbol{v}\times\boldsymbol{A})]_n \mathrm{d}\sigma .$$

The expression (7.7) for the unknown derivative is now reduced to the form

$$\frac{\mathrm{d}}{\mathrm{d}t}\int_\Sigma A_n \mathrm{d}\sigma = \int_\Sigma \left[\frac{\partial \boldsymbol{A}}{\partial t} - \mathrm{curl}(\boldsymbol{v}\times\boldsymbol{A})\right]_n \mathrm{d}\sigma . \tag{7.8}$$

This is the general formula of vector analysis for the derivative of the flow of a solenoidal vector $\boldsymbol{A}$ through a liquid surface Σ.

Obviously, in order for (7.8) to be true, the values of the velocity vector $\boldsymbol{v}$ must be defined only at points of Σ and of the contour C bounding it.

Conservation of vector surfaces, tubes and lines. If the vector field $\boldsymbol{A}$ satisfies (7.1), then one obtains immediately from (7.8)

$$\frac{\mathrm{d}}{\mathrm{d}t}\int_\Sigma A_n \mathrm{d}\sigma = 0 .$$

Thus, the assertion above has been proved that

$$\int_\Sigma A_n \mathrm{d}\sigma = \mathrm{const} , \tag{7.9}$$

where Σ is a surface which moves together with the particles of a continuous medium in a region where the vector field $\boldsymbol{A}(x, y, z, t)$ satisfies (7.1). A number of very important consequences follow from the property (7.9) of the vector field $\boldsymbol{A}$ under consideration.

First consequence

Vector surfaces of such a vector field during motion always become vector surfaces. In fact, consider at some instant of time t a certain vector surface Π of such a vector field $\boldsymbol{A}$, i.e., a surface at each point of which the vector $\boldsymbol{A}$ is tangential to it. By the strength of the continuity of the motion, the surface Π becomes at time $t+\Delta t$ some other surface Π' which will again be a vector surface. Since Π is, by assumption, a vector surface, then the flux of the vector $\boldsymbol{A}$ through any surface Σ belonging to Π vanishes. By (7.9), also the flux of the vector $\boldsymbol{A}$ through Σ', occupied by Σ after the time interval Δt and belonging, as a consequence of the continuity of the motion, to Π', i.e., of the vector $\boldsymbol{A}$ at time $t+\Delta t$ will lie in a plane tangential to Π': the surface Σ, and consequently also the surface Σ' may then be assumed to be infinitesimally close and conveniently to lie on Π; then, without fail, at each point of Π' one has $A_n=0$, whence Π' will again be a vector surface.

In particular, the side surface of a vector tube moves in time into the side surface of a vector tube, i.e., vector tubes become vector tubes.

Second consequence

Vector lines of a vector field A, satisfying (7.1), in time always become vector lines. In fact, one may draw through a vector line l at time t two vector surfaces Π_1 and Π_2, when l will be the line of intersection of these surfaces. By the continuity of the motion, at time $t+\Delta t$, the line l becomes the line l' of intersection of the surfaces Π_1' and Π_2' into which the surfaces Π_1 and Π_2, respectively, move and each of which, by strength of the first consequence of (7.9), remains a vector surface. Consequently, it is seen that l' is again a vector line.

Third consequence

The intensity of any vector tube of the vector field A under consideration remains constant at all times of a motion.

In fact, the intensity of a vector tube is known (§8, Chapter II) to be defined as the flux of the vector A through a cross-section of a tube (in a solenoidal vector field, it is constant along the tube). Therefore the assertion formulated above is a direct consequence of the property (7.9) of the vector field A under consideration. For the vector field A, satisfying the condition (7.1) one has

$$\int_{\Sigma} A_n d\sigma = \int_{\Sigma'} A_n d\sigma ,$$

where Σ and Σ' are arbitrary cross-sections of the vector tube at times t and t', respectively.

Thus, in a vector field A, satisfying (7.1), vector surfaces, tubes and lines are conserved during the time of motion in the sense that they move in space together with the particles of a continuous medium. Vector surfaces, lines and tubes are frozen in the medium.

As has been studied above, the vector field H of magnetic stress in the case of a medium with infinite conductance and the vorticity vector $\omega=\frac{1}{2}$ curl v in the case of barotropic processes in an ideal fluid subject to potential external mass forces satisfy (7.1). Therefore the fields of magnetic stress $\boldsymbol{H}$ and vorticity $\boldsymbol{\omega}$ are frozen in the medium in these cases in the sense stated above.

Thus, for example, if there were not in some region $\mathscr{D}$ occupied by a continuous medium with infinite conductivity at some initial instant t_0 a magnetic field $\boldsymbol{H}$, then there will be no such field in the region $\mathscr{D}'$ into which $\mathscr{D}$ moves at some arbitrary instant of time t.

The magnetic field moves with the particles of the continuous medium. If there occurs on the sun an eruption of plasma, representable by a cloud of burning gas of infinite conductivity, then the magnetic field moves together with the plasma and extends from the sun into interplanetary space.

In view of the practical importance of the properties of the vortex field ω in the above cases, these problems will now be considered in greater detail.

Thomson's theorem. If Σ is a fluid surface, Property (7.9) of the vector field $\omega = \frac{1}{2}$ curl v may be written in the form

$$\int_\Sigma \omega_n \mathrm{d}\sigma = \text{const.}, \tag{7.9'}$$

i.e., the flux of the vorticity vector $\boldsymbol{\omega}$ through any surface Σ, moving with the particles of a continuous medium, is constant.

With the aid of Stokes' theorem, Property (7.9′) may be rewritten in the form

$$\int_C (\boldsymbol{v}, \mathrm{d}\boldsymbol{l}) = \text{const.}, \tag{7.10}$$

(where C is an arbitrary closed contour which consists all the time of the same fluid particles) or in the form

$$\Gamma = \text{const.} \tag{7.11}$$

Thus, if the external mass forces have a potential and the motion of an ideal fluid is barotropic, then the circulation of the velocity around any closed "fluid" contour at all times of motion remains unchanged. This statement is known as Thomson's theorem.

The possibility of an appearance of closed contours with $\Gamma \neq 0$ in a potential flow with surfaces of velocity discontinuity. It must be recalled that the fact that the circulation is constant, which follows from Thomson's theorem, only holds true along contours which are obtained from each other by continuous deformation.

If a fluid lies at rest at some instant of time, then the circulation at this instant is zero along all contours. However, during further motion, in the presence of discontinuous changes in velocity in the flow, there may appear closed contours of the form $\mathscr{L}$ along which the circulation will be non-zero (Fig. 42).

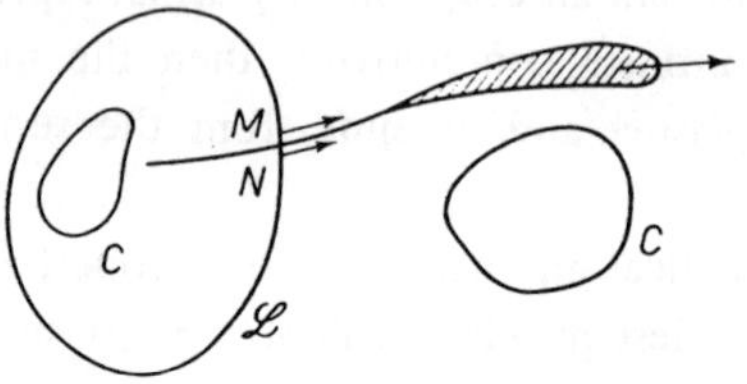

Fig. 42. The circulation along the contour C vanishes, along $\mathscr{L}$ it may be non-zero.

In fact, if the contour $\mathscr{L}$ intersects a line of discontinuity of the velocity, then the circulation along $\mathscr{L}$, in general, differs from zero, since the velocities through M and N differ on a surface of discontinuity; therefore the contour $\mathscr{L}$, closed at the instant of time under consideration, becomes open at subsequent instants of time and was, in general, open at preceding times and, in particular, at the initial instant, when the fluid was at rest. Thomson's theorem cannot be applied to open fluid contours and the circulation is not conserved along such contours.[1] A direct consequence of Thomson's theorem is the following theorem of Lagrange.

Lagrange's theorem. If under the condition of Thomson's theorem there are no vortices in some part of a liquid at some given instant of time, then they did not occur in this part of the liquid for $t < t_0$ and will not occur for $t > t_0$. Note that in the formulation of Lagrange's theorem one is not speaking of a definite part of space, but of a definite mass of fluid which moves in a continuous manner.

Lagrange's theorem will now be proved. By assumption, there are no vortices in a region $\mathscr{D}$ at $t = t_0$, and, consequently, at each point of this

[1] In many cases the experiments and the theory show that for the description of a real motion of fluid, it is necessary to consider the velocity field with discontinuity surfaces.

region $\boldsymbol{\omega}=0$. Therefore, by Stokes theorem, the circulation $\Gamma=2\int_{\Sigma}\omega_n\,\mathrm{d}\sigma$ vanishes along any closed contour belonging to $\mathscr{D}$:

$$\oint_C \boldsymbol{v}\cdot\mathrm{d}\boldsymbol{l} = 0\,.$$

By Thomson's theorem, the circulation will also vanish along any closed fluid contour at all times. Then it follows again from Stokes' theorem that for any surface Σ, entirely lying in the region $\mathscr{D}$ of the fluid,

$$\oint_{\Sigma} \omega_n\,\mathrm{d}\sigma = 0\,.$$

However, this can only be true when at any point of $\mathscr{D}$ and in any direction $\boldsymbol{n}$ one has $\omega_n=0$, i.e., when $\boldsymbol{\omega}=0$ at any instant of time in any particle of the fluid.

An absence of vortices is known to be equivalent to the existence of a velocity potential. Therefore Lagrange's theorem may likewise be formulated in the following manner.

If at some instant of time t_0 the motion of a fluid or gas has a potential, then under the conditions of Thomson's theorem the flow was and will be potential at all times. (In the general case, the surfaces of discontinuity for the potential and for the corresponding field of velocity can form and develop, descending from boundaries of fluid, in the potential fluxes.)

It may be said that in a fluid vortices may form and vanish, generally speaking, only when one of the assumptions of Thomson's theorem is violated, i.e., either when the fluid is viscous or when the motion is not barotropic or when the external mass forces are not potential or when the continuity of the velocity field is disturbed.

The properties of conservation of vector surfaces and lines proved above, and likewise of conservation of the intensity of the vector tubes for fields of the vorticity vector of the velocity during barotropic motions of an ideal fluid or gas are referred to as Helmholtz' dynamic theorems. They will be formulated next.

Helmholtz' dynamic vortex theorems. First theorem: Under the assumptions of Thomson's theorem, particles of a fluid, which form at some instant of time a vortex surface, tube or line, form at all times of the motion a vortex surface, tube or line, respectively.

Second theorem: Under the assumptions of Thomson's theorem, the intensity of a vortex tube at all times of a motion remains constant, i.e.,

$$\Gamma = \int_C \boldsymbol{v} \cdot \mathrm{d}\boldsymbol{l} = \text{const.},$$

where C is a closed contour embracing the vortex tube.

Helmholtz' theorem will be utilized in Chapter VIII when dealing with problems of hydrodynamics.

Chapter VII

On the formulation of problems in continuum mechanics

§1. General foundations of the formulation of concrete problems

Models and reference systems. For theoretical studies of concrete problems in continuum mechanics, one must choose without fail explicitly or implicitly[1)] the observer's reference system in which the motion and state of the system under consideration is described. In Newtonian mechanics, one may select as observer's reference system any "immobile" or moving system. However, one must always point out the inertial reference system, since only on this basis one can make use of information on inertial forces.

In cases of need, one likewise must have in mind the existence of a Lagrangian coordinate system which effectively employs individualization of points of a continuum and, in essence, always rests on the basis of a definition of the characteristics of a motion and of the states of the particles of a medium.

In the preceding chapters, the universal equations of mechanics, thermodynamics and electrodynamics have been established and described. These equations are the fundamental relations within the framework of which theories of any concrete models of continuous media are

[1)] For the sake of distinctness and clarity, it is best of all in the formulation of problems to state all assumptions and conditions in a clear manner.

constructed. They are accepted for all models and within the framework of definite models for all possible separate cases of motion and physical processes. As has been explained above, the universal equations for continuous smooth distributions may be written down in the form of partial differential equations.

Side by side with the differential equations, the same physical laws have been stated in integral form: the integral form of the equation of continuity ((1.2), Chapter III), the impulse equation ((2.2), Chapter III), the first law of thermodynamics and energy equation ((8.1), Chapter V), the second law of thermodynamics ((8.2), Chapter V) and the general Maxwell equations ((5.5), Chapter VI).

For smooth continuous distributions of the characteristic of a motion, the differential and integral formulations are equivalent. However, one has also to consider discontinuous distributions of the characteristics of phenomena in space and in time. In the presence of jumps, the integral formulation appears to have a more general character as it retains its meaning also in this case. The differential formulations retain their significance in regions of continuous phenomena, but require supplementary conditions at discontinuities. For integral formulations, such supplementary conditions are already inherent. In the following sections, the corresponding conditions at discontinuities will be derived from the universal integral relations.

As has been stated earlier, derivation of systems of equations, which permit a detailed study of the motion of a given continuous medium, always demands introduction of supplementary hypotheses, i.e., of assumptions which fix particular properties and the physical nature of a model under consideration. It is necessary to choose a model. The problem of this choice may form the object of special and extended investigations; in many cases, this is a basic problem of physics linked to idealization, schematization and introduction of different types of concepts and characteristics. A choice and construction of new models are necessary for the description of newly discovered effects and phenomena the essence of which is already known or only is starting to show itself in developing areas of science and engineering.

In the preceding chapters, a number of important classical models have already been encountered: the models of the ideal fluid and gas, the model of the elastic body, the model of the viscous fluid, the model of the conducting fluid in magnetohydrodynamics, etc. This list by no means exhausts the range of known models; there exist a number of other models several of which will be introduced below. At the present time, in connection with applications of new materials, a widening of the range of utilization of already employed materials, the necessity for taking electromagnetic properties into account in mechanics, the application of conditions of high vacuum or, conversely, of very large pressures, of very low temperatures or, conversely, of very high temperatures, in connection with studies of complex phenomena in living organisms, etc., the problem of the construction of new models has become acute. The theory of the construction of new models in physics and mechanics is undergoing at the present time vigorous development.

One may introduce additional assumptions on external effects, for example, on external mass forces $\boldsymbol{F}^{(e)}$ and on elementary fluxes of energy $dq^{(e)}$ and dq^{**}, within the framework of each fixed model. Within the framework of a given model, one may likewise pose the problem of the determination of these quantities during a motion, fixed in a hypothetical manner, or following results of experimental observations.

The necessity for supplementary conditions, separating different motions. After choice of a model, one must still establish additional conditions in order to subdivide a definite phenomenon or class of phenomena and motions. This is obvious! In fact, within the framework of the theoretical model of an incompressible ideal fluid, one may consider a wide spread of manifold motions of water, oil and many other fluids and even of air and other gases, when compressibility may be neglected; for example, the various flows and wave motions of water in oceans and lakes, the motion of water in jets escaping from vessels, the motion of water over spillways at weirs, the motion of water due to the motion of ships and submarines in the sea, the motion of air caused by the motion of dirigeables travelling at small subsonic velocities, and many other problems. In all these cases, one may employ one and the same closed system of differential equations. Consequently, the system of differential equations, which is satisfied at every point of a volume occupied by a fluid and which is closed, is quite insufficient for the solution of mathematical problems on the determination of the fields of velocity and pressure in a liquid volume. As is known, the general solution of the differential equations of motion contains arbitrary functions and constants which must be determined from special conditions. The different typical supplementary conditions which subdivide the various motions will be studied next.

The region occupied by a continuous body and the time interval of a motion. Solutions of mathematical problems are presented in the form of functions, defined at points of a volume, occupied by a medium, and over a time interval, during which a motion is being considered. The time interval may be finite, it may begin at some characteristic "initial" instant $t=t_0$ or be "attached" to it; studies of motions and states of continuous media may take place for any $t>t_0$ or $t<t_0$ or, generally, for all $t \gtreqless t_0$. In the sense of a problem, it may be required to search for different characteristic instants of time.

The region of a volume $\mathscr{D}$ occupied by a moving medium may in some cases be given, in others be unknown at the start. For example, the region may be assumed to be known when a liquid moves in a given vessel and completely fills the vessel or when a liquid occupies all space during motion around different fixed and previously given obstacles inside a liquid. In many cases, the region $\mathscr{D}$ is unknown beforehand; it must be determined, for example, in the process of solution of the problems of the escape of water from a vessel, of the deformation of an elastic body under the influence of a given system of external loads, etc.

In certain cases, the boundary of a region may consist of known parts, for example, the surface of the ocean floor, the wall of a vessel or, generally, moving surfaces of bodies inside a fluid, and of parts which are unknown beforehand and which must be found in the process of solution of a problem, for example, the free surface of the sea in wave motion or the boundary of a jet leaving a vessel, etc.

A region $\mathscr{D}$ occupied by a continuous medium may be finite, for example, the inside of a vessel or tube containing liquid or the volume of a rod, beam, part of a machine, etc. made of a deformable "rigid" body. In order to schematicize a problem, one must often deal with a region $\mathscr{D}$ which contains the point at infinity, when the fluid or "rigid" deformable body or electromagnetic field occupies the entire space or the outside of some system of given bodies or when one has jets or strips, i.e., films which extend to infinity.

Conditions at infinity. In order to solve a problem for a region which includes the point at infinity, one must, on the basis of assumptions of a physical nature, give supplementary conditions at infinity.

In the capacity of such conditions, in many cases, it is assumed that a phenomenon under study bears the character of a local perturbation and that for large distances the state and motion of the medium are given. For example, in order to investigate the absolute motion of an unbounded volume of fluid due to the motion inside it of a rigid body of finite dimensions, one may assume that at infinity the velocity of the fluid tends to zero and the pressure to a specified value and that the electromagnetic field has properties, given beforehand; analogously, one may assume that a deformable body finds itself in its natural state at infinity, which is determined by supplementary physical conditions, etc.

In the problem of gas or liquid flow around immobile bodies, one must assume that upstream towards infinity, and in certain cases in any direction of the flow, certain properties are specified: velocity, pressure and temperature, etc.

However, the conditions at infinity may also bear a more complex character; for example, when one has at infinity in different directions certain regular, may be, periodic processes, there may, generally speaking, appear differently directed waves of a given type, etc.

Not infrequently, the problem of formulation of the supplementary conditions at infinity is linked to subtle effects arising from the absence or presence of energy flux from infinity or from radiation.

Singular points inside a medium. The conditions at infinity may be considered to be supplementary conditions at a singular point at infinity. In a study of different effects in a medium or field, one may also introduce singular points in finite parts of a region $\mathscr{D}$. For example, one may construct point sources or sinks of fluid mass, dipoles and multipoles of mass transport, charges, dipoles and multipoles in electromagnetic fields, singular points of various, but definite types to take account of effects due to concentrated external forces and sources of energy, etc. Such singularities may modify the influence and presence of other bodies at large distances from the region of motion which are of interest in the sense of a problem under consideration.

Thus, one may introduce in the region $\mathscr{D}$ singular points at which the asymptotic behaviour of certain functions is prescribed; the presence of such points may also be called into existence by certain external phenomena which are not allowed for in the equations of motion.

Initial conditions. In the theory of ordinary and partial differential equations, special significance attaches to Cauchy's problem which, for example, for ordinary second order differential equations

$$\frac{d^2 x}{dt^2} = f\left(t, x, \frac{dx}{dt}\right) \tag{1.1}$$

may be formulated in the following manner. Find a solution $x(t)$ of Equation (1.1) such that for $t = t_0$

$$(x)_{t=t_0} = x_0 , \quad \left(\frac{dx}{dt}\right)_{t=t_0} = x'_0 , \tag{1.2}$$

where t_0, x_0 and x'_0 are given. It is known from the theory of differential equations that Cauchy's problem has in many cases a unique solution. The supplementary conditions (1.2) are referred to as initial data or Cauchy data.

In an analogous manner, one may introduce Cauchy's problem for partial differential equations and specify, as initial data *for unsteady motions*, depending on the form of the equations, the unknown functions and certain of their derivatives at time $t = t_0$. For example, for the solution of a dynamic problem for an elastic body with the aid of Lamé's equations (2.26), Chapter IV, one must state the initial displacements and velocities at all points of the body.

In order to solve the problem of the unsteady motion of an ideal homogeneous incompressible fluid, i.e., the system of equations (1.10), Chapter IV, it is sufficient to specify the velocity distribution over the entire region $\mathscr{D}$ at the initial instant of time.

If the region $\mathscr{D}$ is unbounded, then in a number of cases initial conditions are sufficient to separate out a definite solution. The number and form of the initial data depend on the order of the system of equations. However, the question of the formulation of Cauchy data and the problems of the existence and uniqueness of solutions must be solved separately in each concrete case.

Boundary conditions. If a region $\mathscr{D}$ is finite or infinite, but has a boundary S, then, in addition to initial conditions, the establishment of definite solutions demands specification and exploitation of special conditions at the boundary S. These conditions are referred to as edge or boundary conditions. Boundary conditions may have very different forms. They are established on the basis of additional physical reasoning. Certain typical and important types of such conditions will now be stated.

Conditions of adhesion at boundaries for displacements and velocities. Let the position and motion of an entire bounding surface S, or of any part S_1 of it be known. As the boundary surface S_1 is approached from the side of the medium, one has, by definition, contact between the medium and its boundary S_1; therefore the displacements of individual points of the medium near S_1 and of the very surface S_1 must be linked by a condition of maintenance of contact. In the absence of slipping of points of a medium along the tangent to the surface S_1, the displacement vector $\boldsymbol{w}_{\text{medium}}$ of the medium and $\boldsymbol{w}_{\text{boundary}}$ of points of the surface S_1, respectively, will be identical.

For example, this may happen when an elastic body is fixed on supports of a given type, when external objects penetrate into a "rigid" deformable medium or when a viscous fluid flows around a rigid body of given shape, etc.

Obviously, in such a case, one has on the surface S_1 the conditions

$$\boldsymbol{w}_{\text{medium}} = \boldsymbol{w}_{\text{boundary}}, \qquad \boldsymbol{v}_{\text{medium}} = \boldsymbol{v}_{\text{boundary}}. \tag{1.2'}$$

If the motion of the boundary S_1 is specified, then, in the absence of slipping along S_1 on this boundary, Conditions (1.2′) apply in which the displacement and velocity vectors

$$\boldsymbol{w}_{\text{boundary}}(\xi^1, \xi^2, \xi^3, t) \quad \text{and} \quad \boldsymbol{v}_{\text{boundary}} = \left(\frac{d\boldsymbol{w}_{\text{boundary}}}{dt}\right),$$

respectively, are known functions of the Lagrangian coordinates ξ^1, ξ^2, ξ^3. Such a condition is often encountered in the mechanics of "rigid" deformable bodies. Conditions (1.2′) likewise apply in the theory of motions of viscous fluids and are called the conditions of adhesion.

In the theory of elasticity, fundamental significance attaches to conditions in terms of displacements, since the displacements determine the strain and stress tensors.

In the theory of motion of liquid and gaseous bodies, the displacements of particles do not enter directly into the equations of motion which instead involve the velocity components; therefore, the boundary condition (1.2′) for the velocity plays a fundamental role in hydrodynamics.

Obviously, for continuous motions, the velocity condition (1.2′) is fulfilled automatically, if that for the displacements is satisfied.

If a boundary S_1 is moved from some initial position to a given position, then $w_{\text{boundary}}(\xi^1, \xi^2, \xi^3) \neq 0$; if the originally displaced boundary S_1 remains immobile, then, after displacement of S_1, the velocity v_{boundary} will vanish.

Each of the vectorial conditions (1.2′) for the displacements or velocities is equivalent to three scalar equalities.

The number of initial and boundary conditions depends on the order of the differential equations; therefore the boundary conditions and their numbers differ from model to model.

Condition of stream line flow around a body for an ideal fluid. For example, Euler's dynamic equations of motion for an ideal fluid contain only first order derivatives of the velocity components.

Navier–Stokes' equations for a viscous fluid contain second order partial derivatives of the velocity components with respect to the coordinates. In both cases, it is natural and convenient to consider the boundary condition (1.2′) for the velocities.

However, for an ideal fluid, the three adhesion conditions (1.2′) are too strong. Under conditions of complete adhesion to a wall, there does not exist a solution of Euler's equations; hence one must admit for ideal fluids and gases the possibility of slippage of particles at boundaries with external rigid or deformable bodies.

For an ideal fluid, Condition (1.2′) on S_1 is replaced by the single scalar condition

$$v_{n\,\text{fluid}} = v_{n\,\text{boundary}} \quad \text{on} \quad S_1 , \tag{1.3}$$

where $v_{n\,\text{fluid}}$ and $v_{n\,\text{boundary}}$ are the velocity components normal to S_1 of fluid particles and points of the bounding surface, respectively. Condition (1.3) expresses the maintenance of contact between a liquid and a given surface S_1; by this condition, the fluid cannot intrude into the body, which is contiguous with it along S, and it cannot tear itself loose from S_1.

On a surface S_1, bounding some body contiguous with an ideal fluid, one has, as a rule, the condition

$$v_{\tau\,\text{fluid}} \neq v_{\tau\,\text{body}} , \tag{1.4}$$

where the subscript τ refers to the velocity component which is tangential to S_1. By (1.4), there occurs slippage of the ideal fluid along S_1, moving in a specified manner.

If the motion of an ideal fluid has a potential: $v = \text{grad}\ \varphi$, then (1.3) may be written in the form

$$v_{n\,\text{fluid}} = \frac{\partial \varphi}{\partial n} = v_{n\,\text{boundary}} \quad \text{on} \quad S_1 .$$

Consequently, the condition of flow at S_1 specifies the value of the normal derivative of the potential φ. If S_1 is fixed, then (1.3) assumes the form

$$v_{n\,\text{fluid}} = 0 \quad \text{on} \quad S_1$$

or

$$\frac{\partial \varphi}{\partial n} = 0 \quad \text{on} \quad S_1 .$$

At given boundaries, one may impose, besides (1.2′) or (1.3), for different models a number of other conditions. For example, the temperature or the flux of heat may be specified on S_1.

If a closed system of equations includes electromagnetic characteristics, there must be given on S conditions, for example, with regard to the vectors $\boldsymbol{E}$, $\boldsymbol{H}$ and $\boldsymbol{j}$.

In the formulation of boundary conditions, one must be guided by the general conditions at surfaces of discontinuity which will be considered later on. However, already at this stage, a wide range of boundary conditions will be studied in detail which may be formulated with the aid of the conditions at surfaces of discontinuity.

Conditions at a free surface. In many cases, the surface S, or some part S_2 of it, of the region of continuous motion of a continuous medium is unknown beforehand and must be determined during the solution of a problem. As a rule, the external loading is given on the unknown boundary S_2. In the theory of elasticity and in other theories, there may be known on the surface S_2 the density of the surface forces

$$\boldsymbol{p}_n = p_{nn}\,\boldsymbol{n} + p_{n\tau}\,\boldsymbol{\tau} = \boldsymbol{f}(M, t)\,, \tag{1.5}$$

where M is a point of S_2. Condition (1.5) yields three relations on S_2. Such a boundary condition is typical for a practical engineering analysis of some component. In the study of problems of the propagation of elastic or seismic waves, one may consider surfaces, so called free surfaces, which are subject to surface stresses, for example, to atmospheric pressure p_0, acting along the normal $\boldsymbol{n}$ to S_2. In this case, one has

$$p_{nn} = -p_0\,, \quad p_{n\tau} = 0\,. \tag{1.6}$$

Boundary conditions of the form (1.5) or (1.6) are also encountered in problems of motions of viscous fluids, when the free surface is the surface of contact with another viscous or ideal fluid, respectively, where all characteristics of the motion may be assumed known.

Conditions at free surfaces in an ideal fluid. In problems of the motion of ideal fluids or gases, one likewise often considers free surfaces. The condition that a medium is ideal always means, by definition, that

$$p_n \cdot \tau = 0\,, \tag{1.7}$$

therefore, in an ideal medium, the conditions (1.6) reduce at a free boundary to the single equality

$$p = p_0\,,$$

where p_0 is the given value of the pressure in the external medium which, for example, at the surface of water, where the perturbation is not known beforehand, may be assumed to be equal to the atmospheric pressure. In practice, one often meets the case when the pressure given at the surface, differs from the atmospheric pressure. For example at free boundaries of a fluid which partially fills a closed tank, the pressure p_0 may have any value.

In order to determine the motion and shape of the boundary S_2 in terms of the unknown components of the velocity or displacement fields, one must also make use of equalities of the form (1.2′). For this purpose, one must keep in mind that for free surfaces S_2 Equations (1.2′) link only velocity components which are normal with respect to S_2, since the velocity of the free boundary, as a rule, is determined as the velocity component of this boundary normal to itself.

In the general case, one must solve problems with mixed boundary conditions, when the volume of a moving medium is partly bounded by given external, immobile walls or walls with given conditions of support, and partly by free surfaces. In addition, parts of a boundary may be subject to other types of conditions, when a boundary is only partly given or there are specified boundaries with more complicated edge conditions in the form of linear or non-linear, finite or differential relations between the unknown functions. Several such conditions will be encountered below and in the second volume.

At this stage, remarks will be limited to stressing the necessity of the formulation and explanation of the meaning of certain boundary conditions. The formulation of concrete problems and their solution will be considered further when dealing with hydrodynamics, elasticity and plasticity (*cf.* Volume II).

§ 2. Typical simplifications in the formulations of certain problems, leading to a reduction of the number of independent variables

The mathematical problems of the determination of solutions of the equations describing the motion and other physical processes from the Eulerian point of view reduce to the search for unknown functions of the four variables x^1, x^2, x^3, t, for example, velocity, pressure, temperature, density, electric and magnetic stresses, etc.

In many cases, such a mathematical problem is very difficult and its solution demands the introduction of additional schematization, linked to the formulation of concrete physical problems, and of permissible simplifications in their mathematical formulation.

The presence of a large number (four in the present case) of independent variables greatly complicates a problem. In a number of problems, important simplifying assumptions, leading to successful solutions, are linked to the question of reduction of the number of independent variables or to assumptions on the completely definite dependence of the unknown functions on several of the variables, taken in a corresponding, specially selected coordinate system of particular form.

Steady motions. Obviously, not always, but in certain practical acceptable cases, in a corresponding coordinate system, one may assume many motions and processes to be stationary. As a consequence, when using an Eulerian approach, one can reduce the number of independent variables by one, since the time t is excluded. For steady motions, there is no need for initial conditions with respect to the time t, since the time derivatives drop out of all the equations for steady motion. As a rule, this circumstance simplifies the solution of a mathematical problem.

Two-dimensional motions. The motion of a continuum is said to be two-dimensional or plane-parallel, if one may select a Cartesian system x, y, z such that the velocities of all particles of a medium are parallel to the planes (x, y), so that all characteristics of a motion and all states depend only on the two variables x, y and, may be, on the time t.

In this case, the motion and state of the particles of a medium do not depend on the coordinate z; the motion of the medium takes place parallel to the plane (x, y) and the motions in all planes, parallel to the plane (x, y), are identical.

The assumption that a motion is two-dimensional is only admissible in particular cases, for example, in the aerodynamic problem of motion perpendicular to its generators of an infinite cylindrical wing in gas or water and in certain problems of surface gravity waves on a fluid; among a number of problems of the theory of elasticity, one has the problem of the equilibrium of a long cylindrical beam the cross-section of which is subject to external statical loading with zero resultant in its plane, when the loads do not depend on the longitudinal coordinates and the displacements in the longitudinal direction are restrained by clamping conditions, etc.

The mathematical theory of the effective solution of problems of two-dimensional motions has been strongly developed and established as the basis of many approximate methods for the solution of three-dimensional problems, when the assumption of plane-parallel patterns for an unknown field is unrealistic.

Two-dimensional potential motions of an incompressible fluid. The great success of the theory of two-dimensional motions of an incompressible fluid is linked to the fact that for potential motions the velocity potential $\varphi(x, y, t)$ is a harmonic function:

$$\frac{\partial^2 \varphi}{\partial x^2} + \frac{\partial^2 \varphi}{\partial y^2} = 0 . \tag{2.1}$$

For each harmonic function $\varphi(x, y, t)$, one may define a conjugate harmonic function $\psi(x, y, t)$ in accordance with the Cauchy–Riemann equations

$$\frac{\partial \psi}{\partial x} = -\frac{\partial \varphi}{\partial y}, \quad \frac{\partial \psi}{\partial y} = \frac{\partial \varphi}{\partial x}, \tag{2.2}$$

i.e.,

$$d\psi(x, y, t) = -\frac{\partial \varphi}{\partial y} dx + \frac{\partial \varphi}{\partial x} dy . \tag{2.3}$$

The condition of integrability (2.3) ensures Equation (2.1). The function $\psi(x, y, t)$ is referred to as *stream function.* By (2.3) and the equation of the stream line, one has the $\psi = \text{const.}$ along any such line.

As is known, one may introduce on the basis of the Cauchy–Riemann equations (2.2) an analytic function w of the complex variable $z = x + iy$, where $i=\sqrt{-1}$:

$$w(z, t) = \varphi(x, y, t) + i\psi(x, y, t)\,. \tag{2.4}$$

The function $w(z, t)$ will be referred to as complex potential.

The problem of the search for the velocity potential $\varphi(x, y)$ may now be reduced to the problem of the determination of a complex potential. For the solution of this problem, one has at one's disposal the powerful apparatus and methods of the theory of functions of a complex variable. It makes possible the solution of many difficult problems and the further development of the hydrodynamics of potential two-dimensional motions of an incompressible fluid.

For steady, two-dimensional motions, one requires only one independent quantity, the complex variable $z=x+iy$. This is an additional very strong assumption of the entire theory.

In a somewhat more complicated form, analogous methods of solution of plane problems, based on applications of the theory of functions of a complex variable, have been developed in the theory of elasticity (Volume II, Chapter XI).

Axisymmetric motions. Problems involving axial symmetry form another important class. In such problems, it is assumed that one may select a cylindrical coordinate system in which the essential arguments of the unknown functions will only be the coordinates r, z and the time t, while the angular coordinate φ is unimportant. All equations and formulae, yielding solutions, will be invariant with respect to rotations by any angle about the z-axis.

In particular, many problems of the strength and motion of bodies of rotation, for example, problems on tubes, tanks, special shells, etc., or problems of translations inside fluids and gases of bodies of rotation along axes of symmetry or their rotations about axes of symmetry, and many other problems can be considered within the framework of the theory of motions of continua with axial symmetry.

Two-dimensional and axisymmetric motions are examples in which only two geometrical coordinates have essential significance.

One-dimensional unsteady motions. Motions and processes in which only the single geometrical coordinate η is essential are referred to as one-dimensional motions. When the time t is also essential, the motion is said to be *unsteady*. It may be shown[1)] that in fluids one-dimensional unsteady motions, during which displacements are directed orthogonally to the coordinate surfaces $\eta=\text{const.}$, are only possible in the following three cases.

Motion with plane waves. 1. Motions with plane waves occur when one may select a Cartesian coordinate system such that the only essential independent variable arguments will be the coordinate x (below, $x=r$ may be used) and the time t. Then, on the plane $x=\text{const.}$ (phase plane of the wave), all characteristics of a motion are identical, i.e., all derivatives of the unknown characteristics of a motion and processes with respect to y and z vanish.

Motion with cylindrical waves. 2. Motions with cylindrical waves occur when one can select a cylindrical coordinate system such that the only essential independent variables will be the distance r from an axis of symmetry and the time t. In this case, on the cylindrical surface $r=\text{const.}$ (phase surface of the wave), all characteristics of a motion are constant, i.e., all derivatives of unknown quantities with respect to z and the polar angle φ vanish.

Motion with spherical symmetry. 3. Motions with spherical waves occur when one may select a spherical coordinate system such that the only essential independent variables will be the distance r from the centre of symmetry and the time t. In this case, on the spheres $r=\text{const.}$ (phase surface of the wave), all characteristics of a motion are identical, i.e., all derivatives of unknown quantities with respect to the longitude θ and latitude φ vanish.

Many important theoretical and practical problems can be studied within the framework of one-dimensional motions, for example, problems of the theories of propagation of light and sound waves, of the theory of shock waves, of the theory of detonations, etc.

1) *Cf.*, Lipschitz, *Z.f. reine und angew. Math.*, Vol. 100, 1887, p. 89. Also, G. A. Liubimov, *On possible forms of one-dimensional unsteady motions of a viscous gas*, sb. No. 19, "Teor. Gidromekh.", No. 7, p. 132, Oborongiz, 1956.

Thus, the simplifications stated above lead to the elimination of one, two or even three independent variables; in steady one-dimensional motions, only the single variable r is essential, in the zero-dimensional case of unsteady motion only the single variable t.

Self-similar motions. Solutions in which a reduction of the number of arguments in the unknown functions is achieved by taking into consideration, in essence, only certain combinations of the independent variables have great significance. An example of such solutions are self-similar motions when one may introduce instead of the four variables x, y, z, t only the three essential independent arguments

$$\frac{x}{t^\alpha}, \quad \frac{y}{t^\alpha}, \quad \frac{z}{t^\alpha},$$

where α is some constant.

For one-dimensional unsteady motions, one may introduce instead of the two variables r and t, in the case of self-similar motions, the single variable

$$\lambda = r/t^\alpha .$$

Obviously, in this case, the partial differential equations in r and t become ordinary differential equations in the single independent variable λ.

As will be seen below, the presence of self-similarity for an unknown solution may be established directly, starting from the formulation of a problem, with the aid of the reasoning of dimensional analysis. For this purpose, one does not even require to write down the equations of motion and boundary conditions; it is sufficient to know only the parameters and characteristics which enter into these equations and conditions. Having in mind such reasoning, one may schematize in certain cases a phenomenon beforehand and pose the problem in such a manner that the simplifications described may be applied, and, in particular, that the unknown solution may be self-similar. If a solution is self-similar, then this is a very valuable property, since from the point of view of the theory of reduction of partial differential equations to ordinary differential equations, such a step is already a great achievement which permits numerical solution of problems by simpler methods.

§3. Linearization of equations and of problems of continuum mechanics

The non-linearity of problems of continuum mechanics. Generally speaking, the basic equations of continuum mechanics are non-linear. The non-linearity of the problems of continuum mechanics arises from the fact that, in the general case, the unknown functions enter into equations and boundary conditions in a non-linear manner. For example, in Euler's equations (in the expression for the acceleration), one has the products $u_i(\partial u_k/\partial x^i)$; in the case of the motion of compressible media with strong changes in density and pressure, the equations contain the non-linear terms $(1/\rho)\,\partial p/\partial x^k$.

Maxwell's equations are linear for a field in vacuum; non-linearity arises on account of the interactions between the electromagnetic field and a medium as a consequence of Ohm's law and the complicated laws for electric polarization and magnetization.

The presence of a number of special physical effects, which in the general case have great practical significance, is connected with the non-linearity of the initial equations. The property of non-linearity causes immense difficulties in the mathematical methods of investigation and the solution of the problems under consideration.

Small perturbations of states of equilibrium or of basic motions. In certain problems of continuum mechanics, the motions and processes under study bear the character of small perturbations of certain states of equilibrium or basic motion.

For example, in elastic bodies (in machine components and structures), deformations are often small and the components of the strain tensor, which in a Cartesian coordinate system are numbers, are of the order of a few percent; therefore the linear theory of infinitesimal strains in which products of small quantities are neglected has found wide application.

In the theory of waves in a heavy fluid, one considers often motions of water for which the free surface of the water differs little from the horizontal plane, the level of still water; in this case, the magnitudes of absolute velocities and the corresponding displacements of water particles are small.

In aerodynamics, one studies often the motion of various slender bodies (aerofoils, projectiles, etc.) through air in the direction of their basic dimension (Fig. 43). When the angle of inclination of the velocity of flight to an element of the surface of a body is small, then this motion causes in the basic mass of air small velocity perturbations which are proportional to the product of the velocity of flight and the small angle of inclination.

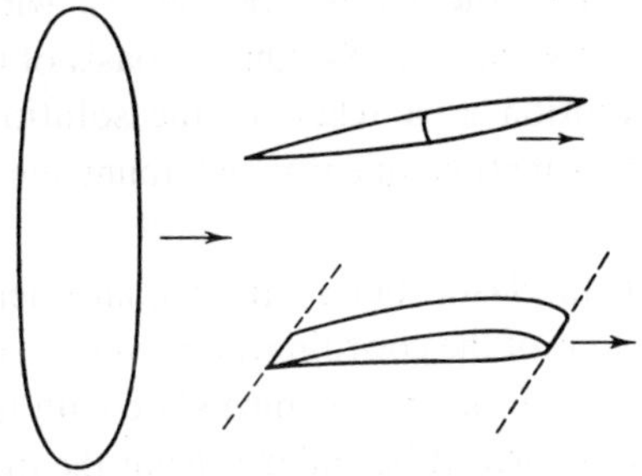

Fig. 43. Motion of slender wings and bodies of rotation.

Very many other examples of this type may be listed. Generally speaking, the conclusions stated regarding the smallness of perturbations of the unknown functions for displacements, velocities, density changes, pressures, temperatures, characteristics of electromagnetic fields, etc., are not strict, and in separate small regions of a flow they are always simply untrue. However, in many cases, these assumptions have turned out to be good in the principal, practically important areas of work.

Of course, there exist also important cases when the smallness of the perturbations of the unknown functions is quite unacceptable and essential non-linear effects must be taken into account.

Linearization of problems of continuum mechanics. In those cases when the smallness of the perturbations of the unknown functions is acceptable, one may linearize the formulations of problems and arrive at the following essential simplifications:

Linearization of the equations. 1. For unknown functions which may be assumed to be small of first order, all equations, supplementary relations of the type of equations of state, relations expressing initial and boundary conditions, etc., may be written in the form of linear equations, after small terms of higher order than the first have been neglected.

Linearization of boundary conditions. 2. Under the assumption that the deformation of the boundaries is small, the boundary conditions on a deformed surface S, bounding a region $\mathscr{D}$, may be transferred along the normal to the boundary S_0 of $\mathscr{D}_0$, corresponding to the basic unperturbed state, to S_0.

Thus, the unknown functions are determined as solutions of linear systems of equations in the known region $\mathscr{D}_0$ with linear boundary conditions on the known surface S_0. On the basis of the functions found for the perturbed motion as a result of the solution of the boundary problem in first approximation, one may determine the deformations of the boundary S.

For example, for an elastic body, the displacement of its points are determined as functions of the coordinates by means of solutions of linear equations in the region of the undeformed state with linearized boundary conditions on the undeformed boundary; from the displacements found, one may compute the small deformations and also, in first approximation, the shape of the deformed boundary.

In particular, in studying the small deformations of an elastic beam with its lower clamped face subject to distributed forces, one may write down the boundary conditions on the undeformed surface S_0 of the beam (Fig. 44a).

In the theory of waves of small amplitudes on the surface of a heavy fluid, the condition at the free surface S (condition of constancy of pressure) is transferred to the horizontal plane, coinciding with the level of the fluid at rest (Fig. 44b). After finding the linearized velocity field from the velocities defined at the horizontal level S_0, one may evaluate the displacements of the points of the free surface and in this way find exactly to first order of smallness the shape of the free wavy surface.

In linearized aerodynamics, the complicated region $\mathscr{D}$, occupied by the disturbed motion of a gas, with boundaries coinciding with the surface of a thin wing, is replaced by the outside of a flat plate close to which, by assumption, is the surface of the wing. The flow boundary conditions on the surface of the wing, with consideration of only small first order quantities, are transferred, respectively, to different sides of the flat plate. Subsequently, the motion of a fluid or gas in infinite space is considered and the bounding flat plate becomes the surface of the pressure and velocity discontinuities; the jump in pressure is balanced then by the external distributed forces exerted on the liquid or gas by the side of the

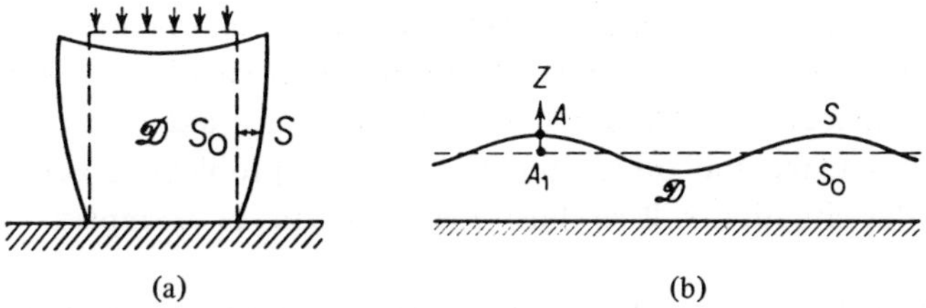

Fig. 44. On the linearization of boundary conditions. (a) Elastic beam subject to distributed forces, with lower face held rigidly. (b) On the formulation of problems in the theory of waves of small amplitude.

wing. In an approximate formulation, these forces are exerted on the fluid or gas by the side of the plate. By studying the motion of an infinite fluid with velocity discontinuities on the surface corresponding to the wing, one must derive expressions for the external distributed forces.

Replacing the inside of the wing by fluid by introduction of discontinuities and external surface forces, distributed over the surface of the wing, one may also arrive at a weakly non-linear formulation of the problem.

Linearized formulations establish the basis of the theory of elasticity within the framework of the theory of small deformations. This theory forms the base of a wide range of methods of analysis of engineering problems.

In aerodynamics, side by side with linearized theories, non-linear theories have been developed extensively, since in many cases it is impossible to assume that the perturbations in a flow are small.

The theory of waves on the surface of a heavy fluid and many problems of electrodynamics and other areas of physics have been developed within the framework of linearized formulations of problems. In the mathematically difficult theory of waves of finite amplitude on the surface of a heavy fluid, when the boundary conditions are non-linear and must be fulfilled at an unknown free surface, only a small number of problems can be studied.

Superposition of solutions. Linear differential equations have the remarkable property that solutions can be superimposed. The sum of certain particular solutions is again a solution. Obviously, for non-linear equations, the sum of particular solutions is not a solution. By forming finite sums, series or integrals of particular solutions, one may construct solutions of the linear equations of continuum mechanics which contain arbitrary functions and sets of constants; with their aid, one may satisfy initial, boundary and other conditions of the problems posed. In the sequel, this situation will be illustrated by means of examples.

Standing waves. Among the particular solutions of linear equations which contain partial derivatives of the unknown functions with respect to the time t and have coefficients independent of t, special significance have solutions of the form

$$F = \text{Real}\,[g(x, y, z, \omega)\, e^{i\omega t}]\,, \tag{3.1}$$

where $i=\sqrt{-1}$, $\omega=\omega_1+i\omega_2$ is a number which is, in general, complex and F is an unknown function. Formula (3.1) determines the dependence of F on the time; the complex function $g(x, y, z, \omega)$ characterizes the distribution of the amplitude and phase in space.

These functions depend on the "frequency" ω. The surface $\arg(g)=$ const. is called the phase surface; the surfaces where $|g|=0$ are called nodal, and where $|g|$ has maxima anti-nodal surfaces. The oscillations (3.1) are called standing waves. As a rule, when speaking of standing waves, it is understood that the region $\mathscr{D}$ contains nodes.

In applications, one considers usually standing oscillations with phases which are the same throughout space.

In certain cases, for example for rigid clamping at points of a boundary S, the boundary conditions reduce to the requirement that the boundaries are to coincide with nodes; in other cases, the conditions at the free boundaries may reduce to the demand that sections of the boundary must coincide with anti-nodes. The boundary conditions, generally speaking, impose limitations on the admissible values of ω. If $\omega=\omega_1$ is real, then the dependence (3.1) of the unknown quantities on the time represents a harmonic oscillation with different amplitudes and, generally speaking, with moving phases at different points and for different quantities. In the presence of (3.1), the problem reduces to the determination of the function $g(x, y, z, \omega)$ and the frequency ω. For $\omega_2<0$, the amplitude grows, for $\omega_2>0$, it fades.

For linear equations, one can construct more general solutions with the aid of Fourier's method which involves superposition of standing waves of the form (3.1) with different ω which in some problems may assume certain discrete values, in others may form a continuous set.

Progressive waves. Another important particular form of the solutions of linear equations are solutions of the type of undamped progressive waves:

$$F(x, y, z, k, \omega, t) = \text{Real } g(y, z, k, \omega) e^{i(kx-\omega t)}, \tag{3.2}$$

where k and ω are constant real numbers and g is some complex function. In the general case, one has a set of admissible values of ω and k, where k may depend on ω. The solution (3.2) corresponds to a periodic distribution of sinusoidals with different phases of unknown functions of x and t.

Fixed values of F propagate along the x-axis with the velocity $a = \omega/k$. The quantity a is referred to as the velocity of propagation of the progressive wave, in the given case, of a sinusoidal wave.

If for different ω or k the quantity a differs, i.e., $a(k) \neq \text{const.}$, one is dealing with dispersion of waves. Waves of different lengths propagate at different speeds.

For plane sinusoidal progressive waves, propagating in the direction of the vector $\boldsymbol{\varkappa} = \varkappa_1 \boldsymbol{i} + \varkappa_2 \boldsymbol{j} + \varkappa_3 \boldsymbol{k}$, one has

$$F = \text{Real } Ae^{i(\boldsymbol{\varkappa}\cdot\boldsymbol{r}-\omega t)}, \tag{3.3}$$

where A is a constant and $\boldsymbol{r}$ is a radius vector. The velocity of propagation of the waves in this case is given by $a = \omega/|\boldsymbol{\varkappa}|$.

In the general case of progressive waves, non-sinusoidal forms with plane surfaces of equal phases correspond to solutions of the form

$$F = f(\boldsymbol{\varkappa}\cdot\boldsymbol{r} - \omega t). \tag{3.4}$$

If $\boldsymbol{\varkappa}$ and ω are constants, then a plane wave propagates along the direction of $\boldsymbol{\varkappa}$ like a rigid body; if $\boldsymbol{\varkappa}$ and ω are functions of the state of the particles, then different states propagate with different velocities, and therefore the shape of a wave will deform in dependence on the function f of the coordinates.

On linearization with the aid of special variables. In certain particular cases, one can transform without any approximation nonlinear equations into linear equations by means of a transition to new specially selected independent variables. Such linearizations are encountered in the theory of steady barotropic two-dimensional potential gas flows.[1)]

§4. Conditions on surfaces of strong discontinuity

On surfaces of discontinuity in continuum mechanics. Hitherto, in the presentation of the basic concepts and derivation of systems of equations connected with models of continuous media, it has been assumed that in the region D, occupied by a medium, at points of which the corresponding equations must be fulfilled, given and unknown functions are continuous and have continuous derivatives which are needed.

Such an assumption is a very strong limitation which is not applicable in a number of important practical applications. In fact, for example in the study of vibrations of a system consisting of an ideal fluid and of elastic bodies submerged in it, one must consider interacting continuous media with sharply differing properties and characteristics of motion. On the interfaces of these media, such characteristics of state and motion as density, velocity, displacement, etc. may be, in general, discontinuous functions of the coordinates.

In this example, for a study of continuous motions of a fluid and elastic bodies, the interfaces may be considered as surfaces of discontinuity on which one must impose on the unknown functions special conditions which play the role of edge conditions on, in general, mobile and in advance unknown boundaries. One such simple condition has been encountered in §1 of this chapter. In the general case of this type, the conditions on surfaces of discontinuity may bear a more complicated character, for example, on interfaces with melting or forming ice supported in water, or on interfaces of air and products of combustion, of gun powder moving and burning in the atmosphere.

Besides surfaces of discontinuity, on different sides of which different

[1)] B. Riemann, *Über die Fortpflanzung endlicher Schwingungswellen,* Gesammelte math. Werke, N.Y., Dover, 1963. P. Molenbrock, *Archiv Mathem. und Physik,* Ser. 2, Vol. 9, 1890. S. A. Chaplygin, *On gas jets,* Selected Works, Vol. II, Gostekhizdat, 1948.

models are used to describe the motion of media, one may also have to consider such mobile surfaces of discontinuity of density, velocity, pressure, entropy, etc. on different sides of which the continua must be considered within the framework of a single model (for example, the model of the ideal perfect gas).

In a number of problems of gas dynamics for an ideal perfect gas and in many other cases, the requirement that the unknown functions are to be continuous in the coordinates in the region $\mathscr{D}$ occupied by the medium leads to non-existence of solutions. Removal of the requirement of continuity and admission of sectionally smooth unknown solutions ensures for corresponding formulations of problems the existence and uniqueness of solutions. The discontinuous solutions obtained may well correspond to real effects observed in practice.

It turns out that on surfaces of discontinuity, which are conserved in their capacity as isolated surfaces separating regions of continuity of processes, one must fulfill certain universal relations between the characteristics of motion and state on either side. In what follows, general methods will be discussed for the derivation of such relations and for their actual establishment in practice.

Subsequent deductions of conditions on surfaces of discontinuity are based on assumptions that such surfaces exist; the question of the real presence of such surfaces in an unknown concrete solution is a special problem which is linked to the properties of the model selected and to the mathematical features of the particular problem under consideration.

Discontinuous solutions may be used in approximate methods of solution for the simplification of problems and for the derivation of effective solutions likewise in cases when continuous solutions also exist.

On discontinuous motion as the limit of continuous motions for different complicated models. Although, in practice, one considers often motions of continuous media with surface discontinuities and admits the fruitfulness of such an approach and solutions, there exists the prevalent point of view that for the description of real phenomena within the framework of continuum mechanics one may and must, generally speaking, consider continuous motions only. In cases when a continuous solution does not exist or ceases to exist starting from some instant of time, one must, in order to derive continuous solutions, turn the attention to other, more complicated models and introduce into the equations of motion additional terms and relations to allow for dissipative effects in their layers inside or on the boundary of a region $\mathscr{D}$ which arise as a consequence of sharp gradients in the distributions of velocity, temperature, density, pressure, etc.

For example, one may quote the case when a solution is discontinuous for a problem posed in terms of Euler's equations within the framework of an ideal perfect gas, while it is continuous with sharp variations in the parameters of motion and state in thin layers for the Navier–Stokes equations within the framework of the model of a viscous gas.

Structure of discontinuities. A study of continuous solutions, corresponding to discontinuities in simplified models, for more complicated models leads to the establishment of the structure of discontinuities. The existence of a solution, its uniqueness and the effective solution of the problem of the structure of jumps is linked closely to methods of derivation of complicated models. In the case of a large number of unknown functions or non-linear equations of processes, the problem of structure, in general, is mathematically difficult.

Application of the theory of continuous transitions to a theory in which discontinuous solutions are obtained and studied is based on the assumption that it is possible to obtain discontinuous solutions within the framework of a given simple model as a limit of continuous solutions of the same problem for successively more complicated models, as the coefficients in the equations of motion of the complicated model tend to the coefficients in the equations for the simplified model. For example, as the coefficients of viscosity tend to zero, the Navier–Stokes equations for the viscous gas become Euler's equations for the ideal gas.

The problem of the existence and uniqueness of the limiting process for corresponding solutions is mathematically complicated, since an effective search for such a sequence of solutions is, as a rule, impracticable.

Complicated models are described by more involved equations with higher order terms, where in the limiting process under consideration the coefficients of the higher derivatives in the partial differential equations vanish. For example, this is the case in the transition from the equations for the viscous fluid to those for the ideal fluid.

For transitions from a given model with discontinuous solutions to complicated models with continuous solutions, the complicated models may be introduced in different ways by means of different dissipation laws, different modes of non-linear viscosity, etc. There arises the question regarding the independence of the limiting process from the derivation of the auxiliary complicated models.

Discontinuous solutions as limits of continuous solutions within the framework of an established model. Within the framework of an established model, these questions are resolved in a somewhat simple manner for those special systems of equations which admit a sequence of continuous solutions tending to the given limiting discontinuous solution. Such a situation occurs for several systems of linear partial differential equations. However, this circumstance does not even always apply in the case of systems of linear equations. This position is connected with the fact that discontinuous solutions are, in general, irreversible and characterized by finite losses of mechanical energy even when the continuous solutions are, generally speaking, reversible.

For example, in the case of the non-linear equations of gas dynamics for adiabatic processes, there do not, in general, exist sequences of continuous solutions which tend in the limit to the adiabatic irreversible discontinuous solution under consideration.

Discontinuous solutions as limits of continuous solutions with suitable external effects. Limiting discontinuous motions and the corresponding conditions at discontinuities can still be derived in the following manner. Consider a sequence of continuous motions for a given system of equations in which in thin layers, with continuous, but sharp variations in the characteristics of the motion, suitable external mass forces, external fluxes of heat and other forms of energy are introduced. Then perform limiting processes in which the total characteristics of the external effects either tend to zero or have a given value, depending on the properties of the discontinuity in the sense of the problem under study (the lifting surface in aerodynamics, surface of heat generation for combustion in a thin layer or for any other chemical reaction, etc.).

Thus, by operation on the equations of processes for given models, or on systems of complicated equations for "approximate" models or with the aid of artificial external effects, every time with assistance of some supplementary explicitly formulated or implicit assumptions, one may, in general, derive motions with discontinuities as limits of continuous motions and obtain the conditions at the discontinuities.

On the expediency of the study of discontinuous solutions. In order to illustrate the significance of questions relating to the relations between continuous and discontinuous solutions, consider the following exotic problem of the motion in the hands of a gesticulating guest of the mechanical system which consists of a goblet containing wine and pieces of ice swimming in the wine. The walls of the glass and the interfaces between the ice and the wine, obviously, are conveniently considered as surfaces of discontinuity of the density of matter. In this case, even for a very detailed mechanical study within the framework of the theory of continuous media, one would scarcely introduce thin layers with continuous density changes, although it is impossible altogether to exclude such a treatment.

In this example, as in many other "practically more important" cases (the problem of the goblet with wine may be replaced by the more acute problem of the motion of fuel and gas, filling the tanks of a rocket flying in cosmic space), one may speak of continuous solutions only from a particular theoretical point of view, such as a source for obtaining different criteria and supplementary relations on discontinuities or as basis for assessment of the realness and suitability of the actively determined discontinuous solutions. If one were to remain only within the framework of complicated theories with thin layers, sharp but continuous variations of the parameters of motion and state, then one would not speak of obtaining effective solutions of many problems, which have already been solved with utilization of surfaces of discontinuity inside and on the boundaries of continuous media.

The role of integral laws in the determination of models. The need for such a type of limiting process disappears when the basic physical equations are formulated in integral terms in which continuity of the unknown functions, in essence, is not implied. Integral formulations of physical laws are completely equivalent to the differential formulations for continuous processes. For discontinuous processes, an integral formulation possesses greater generality.

For a given system of difficult equations one can write the various systems of integral relations over arbitrary volumes of medium, which in the case of continuous motion are equivalent to each other and to a given system of differential equations. For the motion with strong discontinuities, the different systems of such integral relations can be non-equivalent. The choice of the integral laws which are valid not only for continuous processes, but also for processes with internal surfaces of strong discontinuities in the medium, is linked to the complementary physico-chemical hypothesis, validity of which should be conformed to the experiments.

Particularly, in the integral relations in the general case, the framework of reversible and irreversible must be extended, e.g. the additional effects of the growth of entropy in particles which pass through surfaces of strong discontinuity, the dissipation of energy on the surface of discontinuity.

On the additional conditions at discontinuities. The above statements must still be augmented by arguments on the properties of stability of strong discontinuities. In fact, in certain strong discontinuities, all the conditions established below may be fulfilled including conditions connected with a growth of entropy. Nevertheless, discontinuities exist which may not be realizable on account of their instability, due to the form of the jump and the properties of the system of differential equations describing the continuous motions on both sides of the jump.

Likewise, one must keep in mind that for a utilization of physically admissible discontinuities (which are stable and satisfy the universal conditions of mechanics and thermodynamics), in order to ensure uniqueness and correspondence of unknown solutions to reality, one must in certain problems impose at the jumps supplementary conditions of a physical nature.

At the present time, the conditions of stability of discontinuities and supplementary physical relations, mentioned above, are considered in magnetohydrodynamics[1)] and in the general mathematical theory of complicated models of continuous media.[2)] In the theory of shocks for the model of the perfect gas, these questions do not arise.

On the solution of problems in the class of piecewise smooth functions. In order to obtain solutions of the equations in integral form in the class of piecewise smooth functions, discontinuous solutions taking account of

[1)] A. G. Kulikovskii, G. A. Liubimov, *On magnetohydrodynamic shockwaves, ionizing gases.* DAN/SSSR; Vol. 129, No. 1, 1958, A. G. Kulikovskii, A. A. Barmin, *On shock waves, ionizing gases in an electromagnetic field,* DAN/SSSR, Vol. 178, No. 1, 1968.

[2)] A. G. Kulikovskii, *On surface discontinuities separating ideal media with different properties. Waves of recombination in magnetohydrodynamics,* PMM, No. 6, 1968.

a definite supplementary growth in entropy at discontinuities and of other additional conditions are, generally speaking, derived automatically from the formulation of a problem.

The construction of sectionally smooth solutions for systems of differential equations by means of generalizations of these equations with the aid of the corresponding integral relations leads to the theory of generalized solutions developed for linear equations in the general theory of the differential equations of mathematical physics.

A theory of piecewise smooth solutions may be constructed and naturally developed for the formulation of the problems of physics and mechanics with the aid of variational principles (*cf.* Appendix II, p. 581).

Weak and strong discontinuities. There exist surfaces of weak and strong discontinuity. Surfaces on which the unknown functions are continuous, while only several of their derivatives with respect to the coordinates and time are discontinuous, are said to be .weak discontinuities; when the unknown functions themselves undergo discontinuities during a passage through a surface, one speaks of strong discontinuities.

In what follows, consideration will be given to the conditions at surfaces of strong discontinuity. Such surfaces may be introduced as given surfaces with specified laws of motion in the form of external links, or as bearers of specified or unknown force or other external effects, or as surfaces without external effects the form of motion of which are unknown beforehand and, generally speaking, must be found during the process of solution of a problem.

Velocity of the points of a surface of discontinuity. Consider the mobile surface S with the equation

$$f(x, y, z, t) = 0\,. \tag{4.1}$$

As a consequence of its motion, let the surface S occupy at times t and $t + \Delta t$ different positions S and S' (Fig. 45). Select on S at time t some point M and assume that there exists at M a definite normal to S.[1)] The unit normal vector $\boldsymbol{n}$ to S at M will be directed to the side of the vector $\overline{MN}$,

1) In fact, such a normal exists if grad f is defined uniquely at each point of S. In what follows, the case of surfaces S with fractures and other singularities will be excluded.

where N is the point of intersection of the displaced surface S' with the normal to S at M.

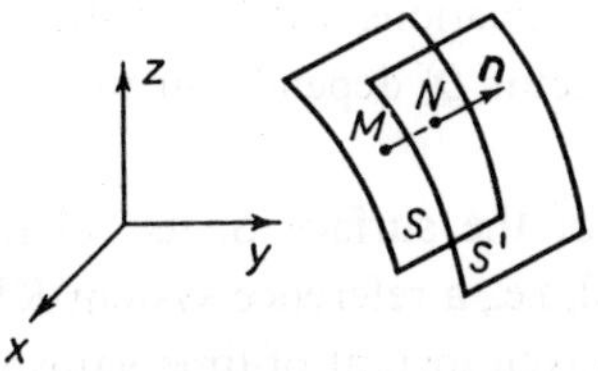

Fig. 45. On the determination of the velocity of the points of a moving surface.

The sign of $f(x, y, z, t)$ will be determined from the condition

$$f(M, t) = 0, f(N, t) > 0\,, \quad \text{whence} \quad \boldsymbol{n} = \frac{\operatorname{grad} f}{|\operatorname{grad} f|}\,. \tag{4.2}$$

The velocity of the displacement in space of the surface S at the point M is the vector $\mathscr{D}$, normal to S and defined by the limit

$$\mathscr{D} = \boldsymbol{n} \lim_{\Delta t \to 0} \frac{MN}{\Delta t}\,. \tag{4.3}$$

If the equation (4.1) of the surface S is given, then the vector $\mathscr{D}$ is readily evaluated. In fact, denoting by

$$n_x = \frac{1}{|\operatorname{grad} f|}\frac{\partial f}{\partial x}\,, \quad n_y = \frac{1}{|\operatorname{grad} f|}\frac{\partial f}{\partial y}\,, \quad n_z = \frac{1}{|\operatorname{grad} f|}\frac{\partial f}{\partial z}$$

the components of the unit vector $\boldsymbol{n}$, one may write

$$f(x + MNn_x\,, \quad y + MNn_y\,, \quad z + MNn_z\,, \quad t + \Delta t) = 0\,,$$

whence one obtains, apart from small higher order terms,

$$MN\left(\frac{\partial f}{\partial x}n_x + \frac{\partial f}{\partial y}n_y + \frac{\partial f}{\partial z}n_z\right) + \frac{\partial f}{\partial t}\Delta t = 0$$

or

$$MN\,|\operatorname{grad} f| + \frac{\partial f}{\partial t}\Delta t = 0\,.$$

The definition (4.3) now yields

$$\mathscr{D} = -\frac{\dfrac{\partial f}{\partial t}}{|\operatorname{grad} f|}\boldsymbol{n}\,. \tag{4.4}$$

Obviously, the vector $\mathscr{D} = 0$ at every point M of S, if f in (4.1) does not depend on the time.

For coordinate transformations with transitions to differently moving systems, the velocity vector $\mathscr{D}$ depends on the choice of the coordinate system.

For each point M of the surface S, its "characteristic" coordinate system may be specified, i.e., a reference system K^* in which the velocity $\mathscr{D}$ of the point M at a given instant of time vanishes. In this context, the word "characteristic" has been placed in inverted commas, since previously the characteristic coordinate system has been defined as an inertial system in which the velocity of the point of the medium considered vanishes. In this sense, for an approach to a surface of velocity discontinuity S from different sides, one obtains different characteristic systems, since the velocities of points of the medium on different sides of S are different.

Choice of coordinate systems. Consider now the conditions on a surface of strong discontinuity. For passage through a smooth surface S (with definite normal), let different characteristics of state and motion occur at the discontinuity. Select on S some (any) point M. Since all mechanical, thermodynamic, electrodynamic and general physical equations preserve their form in any inertial coordinate system, select for the derivation of the unknown conditions at M as a reference system the "characteristic" system K^* in which the velocity $\mathscr{D}$ for a given point of the surface at a given instant of time vanishes, i.e., $\mathscr{D}=0$.

Subdivision of the volume V adjoining a surface of discontinuity. In relation to a certain part of an isolated surface of discontinuity S, consider now the closed surface Σ as the boundary of a volume obtained by the following method. At each point of a separated part of S draw the normal and mark off along it on both sides of S segments of length $h/2$, where h is a very small constant length. The set of such segments, drawn at all points of the section of S under consideration, forms at a given instant of time the corresponding volume V, bounded by the surface Σ (Fig. 46).

Now, consider in the selected system K^* two volumes: firstly, the volume V defined above, a fixed volume, secondly, the volume V^*, a

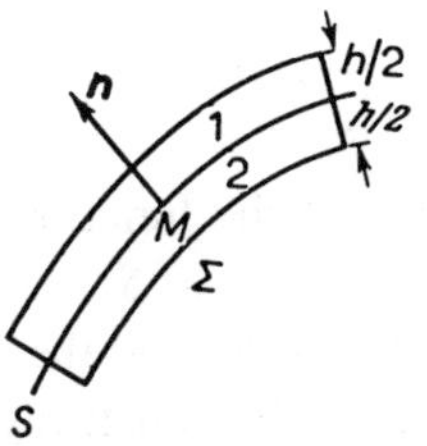

Fig. 46. The lay-out of the surface of discontinuity S and the closed surface Σ.

moving volume, i.e., a substantial volume, connected with the points of the material medium, where at the instant of time t under consideration both these volumes coincide. At the following instant of time $t+\mathrm{d}t$, the volume V^* is displaced with respect to its position V at time t. The surface of discontinuity S also moves inside the volume V; however, the point M under consideration during the infinitesimal time[1] $\mathrm{d}t$ retains its position, since the velocity of M in the system K^* at time t vanishes.

By definition, the volume V is immobile in the system K^*, the volume V^*, in general, is mobile; this volume is immobile only in the case when the points of the medium on the surface Σ in the system K^* are immobile, or have zero velocity components normal to Σ.

It follows from (8.15), Chapter II that one has for any integrable piecewise smooth function $A(x, y, z, t)$ in the system K^*

$$\frac{\mathrm{d}}{\mathrm{d}t}\int_{V^*} A\,\mathrm{d}\tau = \frac{\mathrm{d}}{\mathrm{d}t}\int_{V} A\,\mathrm{d}\tau + \int_{\Sigma} A v_n \mathrm{d}\sigma\,, \tag{4.5}$$

where v_n is the projection of the velocity of points of the medium relative to K^* on to the outward normal to Σ.

Limit of the derivative of the integral with respect to a fluid volume as it is shrunk to a point of the surface of discontinuity. If the surface of discontinuity S inside V is fixed, then one may write for the derivative of the integral with respect to the volume V fixed in the system K^*

[1] In the special theory of relativity, the time element $\mathrm{d}t$ must be taken in the system K^*, in Newtonian mechanics, it does not depend on the choice of the inertial coordinate system. In what follows, all reasoning and, in particular, the rule to be used for addition of velocities takes place within the framework of Newtonian mechanics.

$$I = \frac{\mathrm{d}}{\mathrm{d}t} \int_V A \, \mathrm{d}\tau = h \frac{\mathrm{d}}{\mathrm{d}t} \int_S A^* \, \mathrm{d}\sigma ,$$

where A^* is the mean value of the function A over the corresponding segment of length h, perpendicular to S.

Obviously, if the surface S is fixed, and the quantity A does not depend explicitly on the time, then $I = 0$. For unsteady motion, when the function A is finite and continuous together with its derivatives with respect to the coordinates and the time on both sides of the fixed surface S (where there may be discontinuities), the quantity I is a continuous function of t which vanishes as h tends to zero.

The choice of the system K^* is determined by the condition $\mathscr{D} = 0$ at M. At adjoining points of the surface S, the velocity $\mathscr{D} \neq 0$, and therefore the surface S moves, in general, inside V. For an infinitesimal element of S near the point M, the velocities of the neighbour points are infinitesimally small, then shrinking V to the point, one finds

$$\lim_{\substack{\Delta\sigma \to 0 \\ h \to 0}} \frac{1}{\Delta\sigma} \frac{\mathrm{d}}{\mathrm{d}t} \int_V A \, \mathrm{d}\tau = 0 , \tag{4.6}$$

where $\Delta\sigma$ is the element of S which shrinks into M.

The characteristics of motion and state on the two sides of S will be denoted by the subscripts 1 and 2, respectively. Select this numbering of the sides of S in such a manner that the direction of the normal corresponds to a transition from Side 2 to Side 1.

Letting $h \to 0$ and $\Delta\sigma \to 0$, one obtains from (4.5) the kinematic relationship

$$\lim_{\substack{h \to 0 \\ \Delta\sigma \to 0}} \frac{1}{\Delta\sigma} \int_\Sigma A \, v_n \, \mathrm{d}\sigma = A_1 \, v_{n1} - A_2 \, v_{n2} , \tag{4.7'}$$

where v_{n1} and v_{n2} are the projections of the velocities of the points of the medium on the two sides of S on to one and the same positive direction of the normal to S.

Thus, one has the equality

$$\lim_{\substack{\Delta\sigma \to 0 \\ h \to 0}} \frac{1}{\Delta\sigma} \frac{\mathrm{d}}{\mathrm{d}t} \int_{V^*} A \, \mathrm{d}\tau = A_1 v_{n1} - A_2 v_{n2} . \tag{4.7}$$

With the aid of (4.7), one may write down in the reference system K^* all universal dynamic and thermodynamic conditions at strong discontinuities for material media. Subsequently, attention will be given to discontinuities for electromagnetic fields.

The universal equations of mechanics and thermodynamics. For the sake of convenience, the basic equations derived earlier will be written down first in integral form.

1. The equation of continuity (*cf.* (1.1), Chapter III);

$$\frac{\mathrm{d}}{\mathrm{d}t}\int_{V^*}\rho\,\mathrm{d}\tau = 0\,. \tag{4.8}$$

2. The momentum equation (*cf.* (2.2), Chapter III);

$$\frac{\mathrm{d}}{\mathrm{d}t}\int_{V^*}\rho\boldsymbol{v}\,\mathrm{d}\tau = \int_V \boldsymbol{F}\rho\,\mathrm{d}\tau + \int_\Sigma \boldsymbol{p}_n\,\mathrm{d}\sigma\,. \tag{4.9}$$

3. The moment equation (*cf.* (3.4), Chapter III);

$$\frac{\mathrm{d}}{\mathrm{d}t}\int_{V^*}(\boldsymbol{r}\times\boldsymbol{v})\rho\,\mathrm{d}\tau + \frac{\mathrm{d}}{\mathrm{d}t}\int_{V^*}\rho\boldsymbol{k}\,\mathrm{d}\tau$$

$$= \int_V \boldsymbol{h}\rho\,\mathrm{d}\tau + \int_\Sigma \boldsymbol{Q}_n\,\mathrm{d}\sigma + \int_V (\boldsymbol{r}\times\boldsymbol{F})\rho\,\mathrm{d}\tau + \int_\Sigma (\boldsymbol{r}\times\boldsymbol{p}_n)\,\mathrm{d}\sigma\,. \tag{4.10}$$

4. The energy equation (*cf.* (8.1), Chapter V);

$$\frac{\mathrm{d}}{\mathrm{d}t}\int_{V^*}\rho\left(\frac{v^2}{2} + U\right)\mathrm{d}\tau = \int_V \rho(\boldsymbol{F}\cdot\boldsymbol{v})\,\mathrm{d}\tau$$

$$+ \int_\Sigma (\boldsymbol{p}_n\cdot\boldsymbol{v})\,\mathrm{d}\sigma - \int_\Sigma q_n^*\,\mathrm{d}\sigma + \int_V \frac{\mathrm{d}q^*_{\mathrm{mass}}}{\mathrm{d}t}\rho\,\mathrm{d}\tau\,, \tag{4.11}$$

where q_n^* denotes the total external flux of supplementary specific energy, thermal as well as non-thermal (including the work of surface moments, etc.) through the bounding surface Σ and $\mathrm{d}q^*_{\mathrm{mass}}/\mathrm{d}t$ the total specific supplementary flux of energy due to mass sources of energy in unit time. Supplementary flux of energy means additional flow of energy compared with the flux of mechanical energy allowed for in (4.11) by the first two terms which are equal to the work done by the macroscopic mass and surface forces entering into the momentum equation.

Here and below, only the standard models will be considered for which the internal energy U and the entropy S are additive functions of the mass.

5. The entropy equation which follows from the second law of thermodynamics and may be written in the form

$$\frac{dS}{dt} = \frac{d}{dt}\int_{V^*} s\rho\, d\tau = \int_V \frac{\rho}{T}\left(\frac{dq^{(e)}}{dt} + \frac{dq'}{dt}\right) d\tau\,, \qquad \frac{dq'}{dt} \geqslant 0\,. \qquad (4.12)$$

For models with reversible processes, one has in regions of continuous motion $dq' = 0$. However, as has been noted above, in a study of strong discontinuities with sharp variations of the characteristics of motion, an assumption of reversibility of this phenomenon leads to a contradiction with the second law of thermodynamics, expressed by (4.12). In the general case, it is impossible to assume beforehand that $dq' = 0$ as particles of a medium pass through the surface of discontinuity.

For given $dq^{(e)}$ and, in particular, for adiabatic processes, Equation (4.12) may be used to compute the right-hand side of (4.12) written in the form of a sum of volume integrals which has a finite value as the volume V shrinks to nothing.

Equations (4.8)–(4.12) will now be applied to the volume V^*, defined above and containing the surface of discontinuity S; at the point M of S, Formula (4.7) will be employed, subdividing beforehand the right-hand and left-hand sides of (4.8)–(4.12) by the area element $\Delta\sigma$ of S.

Surface density of external effects on a surface of discontinuity. The conditions at a surface of discontinuity will now be written down under the following assumptions.

1. All integrands under the surface integrals taken over Σ, as Σ shrinks to S, have finite values; however, generally speaking, these values are different on different sides of S.

2. As $h \to 0$, one has the limiting equalities:

$$\lim_{h\to 0}\int_V \rho \boldsymbol{F}\, d\tau = \int_S \boldsymbol{R}\, d\sigma\,,$$

$$\lim_{h\to 0}\int_V \rho \boldsymbol{h}\, d\tau = \int_S \mathfrak{M}\, d\sigma\,,$$

$$\lim_{h\to 0}\int_V \left(\rho \boldsymbol{F}\cdot \boldsymbol{v}+\rho\frac{\mathrm{d}q^*_{\mathrm{mass}}}{\mathrm{d}t}\right)\mathrm{d}\tau=\int_S W\,\mathrm{d}\sigma\,,$$

$$\lim_{h\to 0}\int_V \frac{\rho}{T}\left(\frac{\mathrm{d}q^{(e)}}{\mathrm{d}t}+\frac{\mathrm{d}q'}{\mathrm{d}t}\right)\mathrm{d}\tau=\int_S \Omega\,\mathrm{d}\sigma\,,$$

where $\boldsymbol{R}$, $\mathfrak{M}$ and W are the surface density distributions over S of the forces, moments and energy fluxes which are external to the medium, and Ω gives the density distribution on S of the change of entropy on account of external heat flux and irreversible growth of entropy, on account of irreversible processes of the passage through the discontinuities.

Obviously, if $\rho\boldsymbol{F}$, $\rho\boldsymbol{h}$ and $\rho\boldsymbol{F}\cdot\boldsymbol{v}+\rho\,(\mathrm{d}q^*_{\mathrm{mass}}/\mathrm{d}t)$ are finite in the volume V, then

$$\boldsymbol{R}=0\,,\quad \mathfrak{M}=0\,,\quad W=0\,. \tag{4.13}$$

In particular, this will be the case when the external mass forces are forces of gravity or forces of inertia for a study of relative motion and, in general, for any continuous field of mass forces, including ponderomotive forces, moments and energy fluxes acting on the medium as a result of an electromagnetic field (*cf.* (5.17), (5.39) and (5.22), Chapter IV), when the electromagnetic field is continuous on S.

In the important case when the surface S is a surface of discontinuity not only of mechanical characteristics, but also of the characteristics of an electromagnetic field, then the quantities $\boldsymbol{R}$ and W are, in general, non-zero. In the following section, the formulae for $\boldsymbol{R}$ and W in terms of the values of the components of the vectors $\boldsymbol{E}$, $\boldsymbol{H}$, $\boldsymbol{B}$ and $\boldsymbol{D}$ on different sides of the surface S will be written down.

On discontinuities, which model lifting surfaces of wings or in the case of discontinuities-active discs modelling water or air screws, providing traction, the quantities $\boldsymbol{R}$, W and, may be, $\mathfrak{M}$ are non-zero.

On discontinuities which arise inside gas flows, in aerodynamics, in the theory of explosions and in many other areas, one always has $\boldsymbol{R}=\mathfrak{M}=W=0$ (but $\Omega\neq 0$). In this context, the vanishing of the external effects on the discontinuities is a typical condition used in these applications of continuum mechanics.

Conditions on surfaces of discontinuity in the "characteristic" reference system. One obtains from the law of conservation of mass, on the basis of (4.7),

$$\rho_1 v_{n1} = \rho_2 v_{n2}\,, \tag{4.14}$$

from the momentum equation

$$\boldsymbol{R} + \boldsymbol{p}_{n1} - \rho_1 \boldsymbol{v}_1 v_{n1} = \boldsymbol{p}_{n2} - \rho_2 \boldsymbol{v}_2 v_{n2}\,, \tag{4.15}$$

from the moment equation, taking into account (4.15),

$$\mathfrak{M} + \boldsymbol{Q}_{n1} - \rho_1 \boldsymbol{k}_1 v_{n1} = \boldsymbol{Q}_{n2} - \rho_2 \boldsymbol{k}_2 v_{n2}\,, \tag{4.16}$$

from the energy equation

$$W + \boldsymbol{p}_{n1} \cdot \boldsymbol{v}_1 - \rho_1 \left(\frac{v_1^2}{2} + U_1 \right) v_{n1} - q_{n1}^* = \boldsymbol{p}_{n2} \cdot \boldsymbol{v}_2 - \rho_2 \left(\frac{v_2^2}{2} + U_2 \right) v_{n2} - q_{n2}^* \tag{4.17}$$

and, finally, from the entropy equation

$$\rho_1 v_{n1} s_1 - \rho_2 v_{n2} s_2 = \Omega\,. \tag{4.18}$$

When the process is adiabatic ($dq^{(e)} = 0$), the quantity Ω (when $\rho_1 v_{n1} = \rho_2 v_{n2} \neq 0$) is, generally speaking, non-zero. Since the process is irreversible, i.e., $dq' \geqslant 0$, then

$$\Omega = \rho_1 v_{n1} (s_1 - s_2) \geqslant 0\,. \tag{4.19}$$

For adiabatic processes, Equality (4.18) may be interpreted as definition of Ω which for really realizable processes must be non-negative.

The relations (4.15)–(4.17) for the discontinuities in all the quantities involved for values given or found from solutions of problems may serve for the evaluation of the external effects $\boldsymbol{R}$, $\mathfrak{M}$ and W.

On the disintegration of an arbitrary discontinuity. If (4.13) is fulfilled, then the conditions obtained demonstrate that the discontinuities of the different characteristics of motion and state may not be arbitrary. The initial data may be given in an arbitrary manner so that (4.14)–(4.18) may not be fulfilled. This means that at the next instant of time the given discontinuities may not exist; there occurs disintegration of the initial discontinuity into, generally speaking, several discontinuities among which one may have strong and weak jumps. An analogous position arises during collisions of two or several discontinuities.

The problem of the disintegration of an arbitrary discontinuity will not be considered here, though it is important for applications.

The form of (4.14)–(4.19) is convenient for applications when the discontinuities are fixed, in particular for steady motions of media.

Conditions on surfaces of discontinuity in an arbitrary reference system. For unsteady motions in different coordinate systems, the surfaces of discontinuity may have velocities $\mathscr{D}$ which differ in magnitude and direction. Therefore one must give the form of these relations in any reference frame which is not connected with the motion of any point of the surface of discontinuity.

In order to obtain such general conditions at different points of the surface S in one and the same coordinate system, it is sufficient to replace in all equations the vector $\boldsymbol{v}^*$ of the velocity of the motion with respect to the system K^* by the vector $\boldsymbol{v}=\boldsymbol{v}^*+\mathscr{D}$ ($\boldsymbol{v}^*=\boldsymbol{v}-\mathscr{D}$) of the velocity with respect to the fixed coordinate system K.

The corresponding conditions, after utilization of the equation of conservation of mass and of the momentum equation, may be written in the form

$$\rho_1(\mathscr{D}-v_{n1})=\rho_2(\mathscr{D}-v_{n2}), \tag{4.20}$$

$$\boldsymbol{R}+\boldsymbol{p}_{n1}+\rho_1\boldsymbol{v}_1(\mathscr{D}-v_{n1})=\boldsymbol{p}_{n2}+\rho_2\boldsymbol{v}_2(\mathscr{D}-v_{n2}), \tag{4.21}$$

$$W+\boldsymbol{p}_{n1}\cdot\boldsymbol{v}_1-q^*_{n1}+\rho_1(\mathscr{D}-v_{n1})\left(\frac{v_1^2}{2}+U_1\right)$$
$$=\boldsymbol{p}_{n2}\cdot\boldsymbol{v}_2-q^*_{n2}+\rho_2(\mathscr{D}-v_{n2})\left(\frac{v_2^2}{2}+U_2\right), \tag{4.22}$$

$$\rho_1(v_{n1}-\mathscr{D})(s_1-s_2)=\Omega, \tag{4.23}$$

where the quantity $W(v)$ in (4.22) is equal to $W(v^*)+\boldsymbol{R}\cdot\mathscr{D}$.

These relations at the discontinuities are true in any coordinate system (inertial or not inertial) and at all points of a surface of discontinuity.

The conditions for the moments have not been written down, since in what follows only such models will be considered for which $\mathfrak{M}=\boldsymbol{Q}_n=\boldsymbol{k}=0$ at all points of the region of motion.

Velocity of surface of discontinuity relative to a medium. It is easily seen that the velocities $\mathscr{D}-v_{n1}$ and $\mathscr{D}-v_{n2}$ may be interpreted as the velocities of the surface of discontinuity with respect to particles of the medium on different sides of the jump.

Tangential discontinuities. If $\mathscr{D}-v_{n1}=0$ and $\mathscr{D}-v_{n2}=0$, then the particles of the medium do not pass from one side of the discontinuity to the other and $v_{n1}=v_{n2}$. Then, generally speaking, a discontinuity in the tangential components of the velocities on the different sides of the discontinuity and an arbitrary density jump ($\rho_1 \neq \rho_2$) are possible. Such jumps are referred to as tangential discontinuities. For such a discontinuity, Conditions (4.21)–(4.23) assume the form

$$\begin{aligned} \boldsymbol{R} &= \boldsymbol{p}_{n2}-\boldsymbol{p}_{n1}\,, \\ W &= q^*_{n1}-q^*_{n2}-\boldsymbol{p}_{n1}\cdot\boldsymbol{v}_1+\boldsymbol{p}_{n2}\cdot\boldsymbol{v}_2\,, \\ \Omega &= 0\,. \end{aligned} \tag{4.24}$$

Consequently, for $\boldsymbol{R}=0$, the stress on the area element of a surface of tangential discontinuity is continuous, and the work done by the stresses over the difference in the tangential velocities (with respect to the jump) for $W=0$ is equal to the difference of the fluxes of energy q_n^* through the jump. For ideal fluids, Conditions (4.24) for $\boldsymbol{R}=0$ and $W=0$ reduce to conditions of continuity of the pressure and the normal components of the vector of energy flux on the surface of a tangential discontinuity.

Condensation and rarefaction jumps. If $v_{n2} \neq v_{n1}$, then the particles of the medium pass from one side of the surface S to the other, changing their characteristics of state and motion in the jump (shock).

It is readily verified that the difference $v_{n2}-v_{n1} \neq 0$ neither depends on the choice of the reference system nor on the method of numbering the sides of S. In fact, a change in the numbering alters the direction of the normal, i.e., it transposes the normal components of the velocity and changes their sign.

The numbering of the sides of the surface S will be done in such a manner that the medium passes from Side 1 to Side 2 through S. If a reference system is used for which $\boldsymbol{v}_1=0$, then, obviously, in this case in such a coordinate system $\mathscr{D}_n=\mathscr{D}>0$. For such a method of approach, one finds that the surface S propagates in the medium at rest (with index 1).

Obviously, if $v_{n2}-v_{n1}>0$, then in the present approach $v_{n2}>0$ and the medium behind the jump S enters a medium at rest through the jump. If at the jump the law of conservation (4.20) is fulfilled, then for such jumps $\rho_2 > \rho_1$, i.e., the density of the medium increases behind the jump. Jumps for which $v_{n2} - v_{n1} > 0$ are referred to as condensation jumps.

If $v_{n2} - v_{n1} < 0$, then the velocity component of the medium, normal with respect to S, is directed behind the jump to the side which is the reverse of the velocity of propagation of the jump in the immobile medium; therefore there arises in the medium behind the jump rarefaction, i.e., $\rho_2 < \rho_1$. Such jumps are referred to as rarefaction jumps.

The conditions established in this section for discontinuities may serve as a source for obtaining boundary conditions for solutions of differential equations in regions of continuous motions of media.

In a number of cases, one may state properties, the motion and the state of the particles of a medium on one side of a surface of discontinuity, when the corresponding characteristics on the other side must satisfy these relations.

In particular, one may obtain by such a method boundary conditions at free surfaces of liquids, on boundaries of rigid bodies, etc.

§5. Strong discontinuities in an electromagnetic field

Consider an electromagnetic field, interacting with a material medium, and assume that there is present in the field a surface of discontinuity S. The conditions will now be established which must be satisfied by the values of the electromagnetic characteristics on the two sides of the surface S. In order to obtain these relations, a start will be made from Maxwell's equations, written in integral form and extended to the case of an electromagnetic field with a surface of discontinuity.

In any inertial system, these equations are (*cf.* (5.5), Chapter V)

$$\int_\Sigma B_n \mathrm{d}\sigma = 0\,, \qquad \int_\Sigma D_n \mathrm{d}\sigma = 4\pi \int_V \rho_e \mathrm{d}\tau\,, \tag{5.1}$$

$$\begin{aligned} \int_{\mathscr{L}} \boldsymbol{E} \cdot \mathrm{d}\boldsymbol{r} &= -\frac{1}{c}\frac{\mathrm{d}}{\mathrm{d}t}\int_{\Sigma_1} B_n \mathrm{d}\sigma\,, \\ \int_{\mathscr{L}} \boldsymbol{H} \cdot \mathrm{d}\boldsymbol{r} &= \frac{4\pi}{c}\int_{\Sigma_1} j_n \mathrm{d}\sigma + \frac{1}{c}\frac{\mathrm{d}}{\mathrm{d}t}\int_{\Sigma_1} D_n \mathrm{d}\sigma \end{aligned} \tag{5.2}$$

and, as a consequence of these equations, the law of conservation of charges is given by

$$-\frac{\mathrm{d}}{\mathrm{d}t}\int_V \rho_e \mathrm{d}\tau = \int_\Sigma j_n \mathrm{d}\sigma\,, \tag{5.3}$$

where Σ is a closed surface bounding an immobile volume V and Σ_1 is an immobile open surface bounded by a contour $\mathscr{L}$.

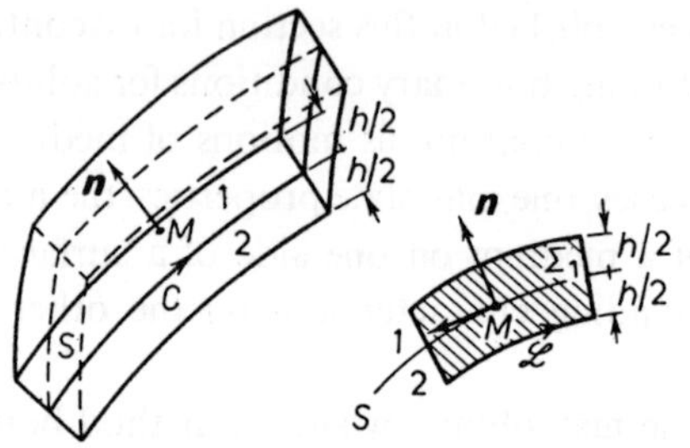

Fig. 47. Scheme of the regions of integration for the derivation of the conditions on surfaces of discontinuity.

Let M be some point on the surface of discontinuity S and K^* the inertial coordinate system in which the velocity $\mathscr{D}$ of the points M of the surface S vanishes (Fig. 47). The system K^* is the "characteristic" coordinate system for M; neighbouring points on S may have in the system K^* non-zero velocities. Assume that Equations (5.1) and (5.2) have been written down in the system K^*. In Equations (5.1) and (5.3), employ the volume V which is bounded by Σ and has been determined in the same manner as in §4. In (5.2), select as surface Σ_1 and contour $\mathscr{L}$ the intersection of V and the surface Σ with a plane which passes through the normal vector $\boldsymbol{n}$ and the tangent vector $\boldsymbol{\tau}$ to S at M. The direction of the vector $\boldsymbol{\tau}$ may be chosen arbitrarily. By assumption, the directions of $\boldsymbol{n}$, $\boldsymbol{\tau}$ and the vector of the normal $\boldsymbol{n}^*$ to Σ_1 form a right-handed system, so that $\boldsymbol{n}^* = \boldsymbol{n} \times \boldsymbol{\tau}$.

Conditions on surfaces of discontinuity. Consider a surface of discontinuity S on both sides of which the vectors $\boldsymbol{H}$, $\boldsymbol{B}$, $\boldsymbol{E}$ and $\boldsymbol{D}$ are finite and continuous, while they may exhibit dicontinuities for a passage through S. As regards the distribution of charges ρ_e and currents $\boldsymbol{j}$, assume that on S there may be surface charges γ and surface currents $\boldsymbol{i}$

(vectors lying in the tangential planes to S), which are determined at M by means of the limits

$$\lim_{V\to M}\frac{1}{\Delta\sigma}\int_{V\approx h\Delta\sigma}\rho_e\,d\tau=\gamma(M)\,, \tag{5.4}$$

$$\lim_{\substack{\Delta\sigma\to M\\ h\to 0}}\frac{h}{\Delta\sigma}\int_{\Sigma_1\approx\Delta\sigma} j_\tau\,d\sigma=\boldsymbol{i}(M)\,, \tag{5.5}$$

where $\Delta\sigma\to M$ is an element of S or Σ, shrinking into M, and $\boldsymbol{j}_\tau$ is the vector component of the volume density of the current, parallel to the tangent plane to S at M.

From (5.1) and (5.3), using (5.4) in the same manner as the corresponding equations were used in §4, one obtains

$$B_{n1}-B_{n2}=0\,,\qquad D_{n1}-D_{n2}=4\pi\gamma, \tag{5.6}$$

$$\operatorname{div}\boldsymbol{i}+j_{n1}-j_{n2}=-\frac{1}{\sqrt{G}}\frac{\partial\gamma\sqrt{G}}{\partial t}\,, \tag{5.7}$$

where G is the determinant of a matrix,

$$\operatorname{div}\boldsymbol{i}=\lim_{C\to M}\frac{1}{\Delta\sigma}\int_C i_n\,dl=\nabla_1 i^1+\nabla_2 i^2\,,$$

C is a closed contour shrinking into M on S, i_n is the component of the vector $\boldsymbol{i}$ normal to C, $\Delta\sigma$ is the area element of S, bounded by C, $\operatorname{div}\boldsymbol{i}$ is the two-dimensional divergence of the vector $\boldsymbol{i}$, determined on S, i^1 and i^2 are the components of $\boldsymbol{i}$ and ∇_k are covariant derivatives in the coordinate system on S.

In an analogous manner, one obtains from (5.2), after passing to the limit $h\to 0$ and then shrinking the contour $\mathscr{L}$ into the point M,

$$\boldsymbol{E}_{\tau 1}-\boldsymbol{E}_{\tau 2}=0\,, \tag{5.8}$$

$$H_{\tau 1}-H_{\tau 2}=\frac{4\pi}{c}\boldsymbol{i}\cdot(\boldsymbol{n}\times\boldsymbol{\tau})\,.$$

Obviously, the last equality may be written in the vectorial form

$$\boldsymbol{H}_{\tau 1}-\boldsymbol{H}_{\tau 2}=\frac{4\pi}{c}(\boldsymbol{i}\times\boldsymbol{n})\,, \tag{5.9}$$

where $\boldsymbol{H}_{\tau 1}$ and $\boldsymbol{H}_{\tau 2}$ are vector components of $\boldsymbol{H}$, parallel to a tangent plane to S at the point M.

Conditions (5.6)–(5.9) form the complete system of relations on the surface of discontinuity for the electromagnetic characteristics, including polarization, magnetization and currents. These relations have been written in the inertial system K^* which is "characteristic" for M and in which the velocity $\mathscr{D}$ of the point M of the surface of discontinuity vanishes.

On the basis of the general formulae, fulfilled for Lorentz transformations from the system K^*, moving with velocity $\mathscr{D}$ relative to the fixed "immobile" inertial reference system K, to the system K (*cf.* (3.22) and (3.23), Chapter VI), one may rewrite (5.6), (5.8) and (5.9) in the system K in the form

$$B_{n1} - B_{n2} = 0\,, \qquad D_{n1} - D_{n2} = 4\pi\gamma^*\,, \tag{5.10}$$

$$\boldsymbol{E}_{\tau 1} - \boldsymbol{E}_{\tau 2} = \frac{1}{c}[(\boldsymbol{B}_1 - \boldsymbol{B}_2) \times \mathscr{D}]_\tau\,, \tag{5.11}$$

$$\boldsymbol{H}_{\tau 1} - \boldsymbol{H}_{\tau 2} + \frac{1}{c}[(\boldsymbol{D}_1 - \boldsymbol{D}_2) \times \mathscr{D}]_\tau = \frac{4\pi}{c}\sqrt{1 - \frac{\mathscr{D}^2}{c^2}}\,(\boldsymbol{i}^* \times \boldsymbol{n})\,, \tag{5.12}$$

where $\boldsymbol{n}$ and $\boldsymbol{\tau}$ are the normal and tangential directions to the surface S. The velocity $\mathscr{D}$ is evaluated in the system K and directed along the normal to S. The quantity γ^* and the vector $\boldsymbol{i}^*$ are defined in the characteristic coordinate system. On neglecting terms of the order $\mathscr{D}^2/c^2$ for $\gamma = \gamma^*$ and $\boldsymbol{i} = \boldsymbol{i}^*$, Formula (5.7) retains its form.

Surface densities of ponderomotive forces and energy fluxes from a field to a medium on a surface of discontinuity. Consider now the formulae for the components of the vector of the surface density of the ponderomotive force $\boldsymbol{R}$ and the surface density of the energy flux W to a medium on the surface of discontinuity S of the electromagnetic quantities. In a Cartesian coordinate system, one has for the components of the body forces[1)]

$$-\rho F^\alpha = \nabla_k S^{\alpha k} = \frac{\partial S^{\alpha 1}}{\partial x^1} + \frac{\partial S^{\alpha 2}}{\partial x^2} + \frac{\partial S^{\alpha 3}}{\partial x^3} + \frac{\partial S^{\alpha 4}}{\partial t}\,, \quad \alpha = 1, 2, 3 \tag{5.13}$$

and

1) Here F^α represents the space components of the four-dimensional mass-force. In (5.12), Chapter VI there are given the components of the body force.

$$-F_4 = \nabla_k S_4^k = \frac{\partial S_4^1}{\partial x^1} + \frac{\partial S_4^2}{\partial x^2} + \frac{\partial S_4^3}{\partial x^3} + \frac{\partial S_4^4}{\partial t}, \tag{5.14}$$

where S^{ik} are the components of the energy-impulse tensor. Multiply both sides of (5.13) and (5.14) by the volume element $d\tau = dx^1 dx^2 dx^3$ and integrate over the immobile volume V in the system K^* for the given point M on S (Fig. 47). The Gauss–Ostrogradskii formula then yields

$$\int_V \rho F^\alpha d\tau = -\int_\Sigma S^{\alpha\beta} n_\beta d\sigma - \frac{d}{dt}\int_V S^{\alpha 4} d\tau, \qquad \alpha = 1, 2, 3, \tag{5.15}$$

$$\int_V F_4 d\tau = -\int_\Sigma S_4^\beta n_\beta d\sigma - \frac{d}{dt}\int_V S_4^4 d\tau, \tag{5.16}$$

where n_β are the components of the unit normal vector $\boldsymbol{n}$ to Σ in the three-dimensional Cartesian coordinate system.

For the components S^{ik} of the energy-impulse tensor, employ Minkowski's formula ((5.11), Chapter VI), viz.,

$$S^{ik} = -\frac{1}{4\pi}\left[F_{m.}^{\cdot i} H^{mk} - \tfrac{1}{4} g^{ik} F_{mn} H^{mn}\right]. \tag{5.17}$$

The matrix for S^{ik} may be written in the form

$$\|S^{ik}\| = \left\| \begin{matrix} S^{11} & S^{12} & S^{13} & S^{14} \\ S^{21} & S^{22} & S^{23} & S^{24} \\ S^{31} & S^{32} & S^{33} & S^{34} \\ S^{41} & S^{42} & S^{43} & S^{44} \end{matrix} \right\|$$

If one uses for $\|F_{ij}\|$ and $\|H^{ij}\|$ the matrix definitions (5.6) and (5.7), Chapter VI and takes into consideration that (*cf*. also (5.37))

$$\|F_{m.}^{\cdot i}\| = \|F_{mj} g^{ji}\| = \left\| \begin{matrix} 0 & -B^3 & B^2 & \frac{1}{c}E_1 \\ B^3 & 0 & -B^1 & \frac{1}{c}E_2 \\ -B^2 & B^1 & 0 & \frac{1}{c}E_3 \\ cE_1 & cE_2 & cE_3 & 0 \end{matrix} \right\|,$$

then one obtains in the Cartesian coordinate system the following expressions for S^{ik} in terms of $\boldsymbol{E}, \boldsymbol{H}, \boldsymbol{B}$ and $\boldsymbol{D}$:

$$S^{\alpha\beta} = -\frac{1}{4\pi}\left[E^\alpha D^\beta + H^\alpha B^\beta + \tfrac{1}{2} g^{\alpha\beta}(\boldsymbol{B}\cdot\boldsymbol{H}+\boldsymbol{E}\cdot\boldsymbol{D})\right], \tag{5.18}$$

where $\alpha, \beta = 1, 2, 3$ and it has been taken into consideration that in a three-dimensional Cartesian coordinate system $E_\alpha = E^\alpha$ and $D_\beta = D^\beta$, while

$$S^{\alpha 4}\mathbf{3}_\alpha = \frac{1}{4\pi c}(\boldsymbol{D}\times\boldsymbol{B}), \qquad S^{\alpha .}_{.4}\mathbf{3}_\alpha = \frac{c}{4\pi}(\boldsymbol{D}\times\boldsymbol{B}), \tag{5.19}$$

$$S^{4\alpha}\mathbf{3}_\alpha = \frac{1}{4\pi c}(\boldsymbol{E}\times\boldsymbol{H}), \qquad S^{.\alpha}_{4.}\mathbf{3}_\alpha = \frac{c}{4\pi}(\boldsymbol{E}\times\boldsymbol{H}), \tag{5.20}$$

$$S^{.4}_{4.} = \frac{1}{8\pi}(\boldsymbol{E}\cdot\boldsymbol{D}+\boldsymbol{H}\cdot\boldsymbol{B}), \tag{5.21}$$

with $\mathbf{3}_\alpha$ the base vectors in the three-dimensional coordinate system.

It should be emphasized again that in (5.18)–(5.20) the components of the tensor S^{ij} are defined in a four-dimensional coordinate system for which[1)]

$$ds^2 = g_{ij}dx^i dx^j = -dx^{1^2} - dx^{2^2} - dx^{3^2} + c^2 dt^2 .$$

This fact must be borne in mind when one compares the formulae which have been written down here with those in some other books in which other definitions of g_{ij} are used.

Thus, the integrands in (5.15) and (5.16) have been expressed in terms of $\boldsymbol{E}, \boldsymbol{H}, \boldsymbol{B}$ and $\boldsymbol{D}$ which, by assumption, are finite on S and in V.

Performing in (5.15) and (5.16) the limit transitions to the point M on the surface of discontinuity S, described in 7.4, one obtains

$$R^\alpha = (S^{\alpha\beta})_2 n_\beta - (S^{\alpha\beta})_1 n_\beta , \tag{5.22}$$

$$W = (S^{\beta}_{4})_2 n_\beta - (S^{\beta}_{4})_1 n_\beta . \tag{5.23}$$

On the basis of (5.18) and (5.20), Formulae (5.22) and (5.23) lead to the following expressions for R^α and W in terms of $\boldsymbol{E}, \boldsymbol{H}, \boldsymbol{B}$ and $\boldsymbol{D}$ in the system K^* at the point M:

1) *Cf.* pp. 383–385.

$$R^{\alpha} = -\frac{1}{4\pi}\left[E^{\alpha}D_n + H^{\alpha}B_n + \frac{n^{\alpha}}{2}(\boldsymbol{E}\cdot\boldsymbol{D}+\boldsymbol{H}\cdot\boldsymbol{B})\right]_2 + \frac{1}{4\pi}\left[E^{\alpha}D_n + H^{\alpha}B_n + \frac{n^{\alpha}}{2}(\boldsymbol{E}\cdot\boldsymbol{D}+\boldsymbol{H}\cdot\boldsymbol{B})\right]_1, \tag{5.24}$$

$$W = \frac{c}{4\pi}(\boldsymbol{E}_2\times\boldsymbol{H}_2 - \boldsymbol{E}_1\times\boldsymbol{H}_1)\cdot\boldsymbol{n}, \tag{5.25}$$

where $n^{\alpha} = -g^{\alpha\beta}n_{\beta}$ — the contravariant components of the three-dimensional vector $\boldsymbol{n}$.

These expressions may be substituted into the relations (4.14)–(4.17) at jumps for a material medium.

Since (5.39) of Chapter VI shows that the densities of the volume ponderomotive moments are finite even when $\boldsymbol{E}, \boldsymbol{D}, \boldsymbol{B}$ and $\boldsymbol{H}$ have jumps, it is obvious that in the conditions on the discontinuities for the moments the surface densities of the ponderomotive moments will vanish. The ponderomotive forces are expressed by the gradients of $\boldsymbol{E}, \boldsymbol{D}, \boldsymbol{H}$ and $\boldsymbol{B}$, then they can influence the conditions on the strong jumps.

§6. Surfaces of discontinuity inside ideal compressible media

Next, consider in greater detail the conditions (and their consequences) on surfaces of strong discontinuities in the ideal compressible media, introduced in 4.1. In these cases, by definition, internal stresses may only be pressures: $\boldsymbol{p}_n = -p\boldsymbol{n}$. Besides, this section will consider surfaces of discontinuity on which do not occur any external surface effects on a given medium, i.e., for which $\boldsymbol{R}=0$ and $W=0$; likewise, it will be assumed that $q_n^*=0$ on a surface of discontinuity, and, in particular, no attention will be given to the thermal conductivity of a medium in a jump.

Conditions on a fixed jump. In order to study motions of a medium with respect to the "characteristic" coordinate system K^* for a given element of the surface of discontinuity S, Condition (4.14) may be written in the form

$$\rho_1 v_{n1} = \rho_2 v_{n2} \,. \tag{6.1}$$

Further, let it be assumed that the velocity component v_{n1} normal to S is finite, i.e., $v_{n1} \neq 0$, so that there occurs a passage of particles from one side of the surface of discontinuity to the other. Then one obtains from (4.15) for an ideal medium

$$\boldsymbol{v}_{\tau 1} = \boldsymbol{v}_{\tau 2} \,, \tag{6.2}$$

where $\boldsymbol{v}_{\tau i}$ is the vector component of $\boldsymbol{v}$, parallel to the tangent plane to S at the point M.

Using (6.1) and (6.2), one obtains from the momentum equation (4.15) for the projection on the normal to S

$$\rho_1 v_{n1}^2 + p_1 = \rho_2 v_{n2}^2 + p_2 \,. \tag{6.3}$$

By (6.1) and (6.2), the energy equation reduces to the form

$$U_1 + \frac{v_{n1}^2}{2} + \frac{p_1}{\rho_1} = U_2 + \frac{v_{n2}^2}{2} + \frac{p_2}{\rho_2} \,. \tag{6.4}$$

Finally, by (6.1), Equation (4.18) for the entropy jump yields

$$\rho_1 v_{n1} (s_1 - s_2) = \Omega \,. \tag{6.5}$$

It must now be recalled that in (6.1)–(6.5) the velocities are taken with respect to the system K^* at which the point M of the surface of discontinuity has zero velocity.

In order to establish motions in the case of a fixed jump, it may be assumed that K^* coincides with the basic "immobile" reference frame.

Propagation of a discontinuity over the particles of a medium. It is seen from (6.1) that v_{n1} and v_{n2} have the same sign. If one introduces the reference system K in which the velocity through the jump is zero and $\mathscr{D}_n = \mathscr{D} > 0$ then one may use in K on the jump Relations (6.1)–(6.5) in which one must set

$$v_{n1} = -\mathscr{D} \,, \qquad v_{n2} = v_n - \mathscr{D} \,,$$

$$v_{\tau 1} = v_{\tau 2} = 0 \,,$$

where $v_n = v_{n2} - v_{n1}$ is the normal component to S of the velocity of the medium relative to K on Side 2. In this case, the quantity $\mathscr{D}$ may be considered as the velocity of propagation of the surface of discontinuity by the particles of the medium on Side 1. In what follows, the quantity $\mathscr{D}$ defined in this manner will be employed, and, for the sake of simplicity, the specific volume $V=1/\rho$ will be introduced instead of the density ρ.

It is easily seen that (6.1) and (6.3) are equivalent to the relations

$$-v_{n2}=V_2\sqrt{\frac{p_2-p_1}{V_1-V_2}}, \qquad -v_{n1}=\mathscr{D}=V_1\sqrt{\frac{p_2-p_1}{V_1-V_2}}, \tag{6.6}$$

$$v_{n2}-v_{n1}=v_n=\mathscr{D}\left(1-\frac{\rho_1}{\rho_2}\right)=\pm\sqrt{(p_2-p_1)(V_1-V_2)}\,. \tag{6.7}$$

Since $\mathscr{D}>0$, the positive sign in (6.7) corresponds to $\rho_1<\rho_2$, the negative sign to $\rho_1>\rho_2$. After elimination of the velocity, Relation (6.4) reduces to the form

$$U_2-U_1=\tfrac{1}{2}(p_2+p_1)(V_1-V_2)\,. \tag{6.8}$$

Equation (6.6) shows that, if $V_1>V_2$, i.e., $\rho_2>\rho_1$, then necessarily $p_2>p_1$; conversely, if $\rho_2<\rho_1$, then $p_2<p_1$.

Condensation and rarefaction jumps. Discontinuities for which

$$v_n>0\,,\quad \rho_2>\rho_1\,,\quad p_2>p_1\,, \tag{6.9}$$

are referred to as condensation jumps. Rarefunction jumps are defined by the inequalities

$$v_n<0\,,\quad \rho_2<\rho_1\,,\quad p_2<p_1\,. \tag{6.10}$$

If one of the three inequalities of (6.9) or (6.10) is fulfilled, then the other two are automatically implied.

The properties (6.9) and (6.10) have a very general character and are consequences only of the law of conservation of mass and of the energy equation, when on the surface of discontinuity the medium is not subject to external surface forces and external mass flow.

On a jump of internal energy. Relation (6.8) does not contain the velocity, it is fulfilled in any reference system and is convenient for a study of changes of density and pressure in the particles as they proceed through a discontinuity.

If a density jump is specified, then in certain important cases one may determine from (6.8) the pressure jump; subsequently, the corresponding velocities follow from (6.6) and (6.7).

In the general case, the internal energy of a homogeneous ideal material medium is a function of the specific volume V (density ρ), the pressure p and certain other parameters determining the physical and chemical properties of a medium; the vectors of polarization and magnetization may then be physical parameters which may be changed by the jump on passage through a surface of discontinuity.

For example, such a situation is encountered in the study of phenomena of propagation of fronts of combustion and detonation and of various fronts of electromagnetic waves, etc. Thus, one has for a perfect gas

$$U = c_V T + U_0 = \frac{c_V}{c_p - c_V} \frac{p}{\rho} + U_0 ,$$

where c_p and c_V are the specific heat capacities and U_0 is constant for a given gas. On passage through a jump, the components of a gas may change, and therefore c_p, c_V and U_0 may undergo jumps. If the gas itself is a mixture of perfect gases, then

$$U = \sum_i \frac{\rho_i}{\rho} \left(U_{0i} + \int_{T_0}^{T} c_{Vi}(T) \mathrm{d}T \right),$$

where ρ_i/ρ is the contribution to the mass of the i-th component of the gas mixture. On passage through the surface of discontinuity, the ratios ρ_i/ρ may suffer jumps which must be determined from additional physico-chemical laws and assumptions. As the temperature of a mixture rises, one may consider combustion (complete or incomplete), take into account pressures of dissociation or ionisation, etc.

If the physico-chemical properties of a medium do not change on passage of particles through a jump, but only the density ρ and the pressure p, then (6.8) determines for fixed p_1 and ρ_1 the link between the values of p_2 and ρ_2 beyond the jump.

Hugoniot's adiabat. If one graphs the states p, $V = 1/\rho$ in the pV-plane, then for given p_1, V_1 Equality (6.8) determines a curve in this plane, which is referred to as Hugoniot's adiabat. (Fig. 48).

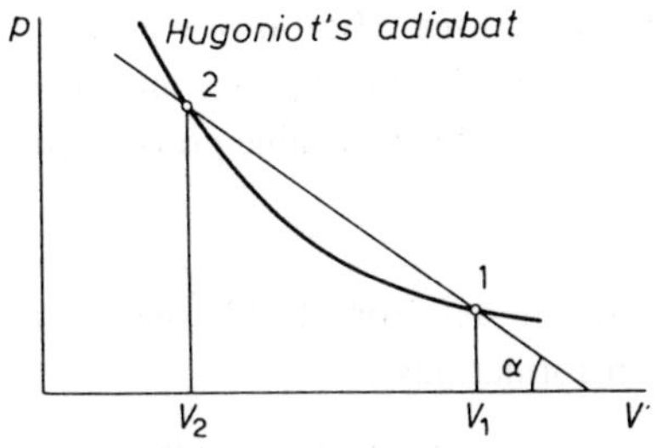

Fig. 48. Hugoniot's adiabat.

The point p_1, V_1, corresponding to the state before the jump, will lie on Hugoniot's adiabat if

$$U_2(p_1, V_1) - U_1(p_1, V_1) = 0\,, \tag{6.11}$$

i.e., if the function $U_2(p, V)$ coincides with the function $U_1(p, V)$. Equality (6.11) may be fulfilled, if all other parameters, except p and V, and constants on which the internal energy may depend, remain unchanged on passage through the discontinuity. In the presence of irreversible chemical reactions or of any other processes, Equality (6.11) may not be fulfilled; then the point p_1, V_1 may not lie on Hugoniot's adiabat.[1)]

Consider the case of jumps when (6.11) holds true. Obviously, for given p_1, V_1, it is sufficient for the determination of the point p_2, V_2 on Hugoniot's adiabat to specify only one of the quantities

$$p_2 \quad \text{or} \quad \rho_2 = 1/V_2$$

or

$$\tan\alpha = \frac{p_2 - p_1}{V_1 - V_2} = \frac{\mathscr{D}^2}{V_1^2}\,. \tag{6.12}$$

The angle α determines the secant of Hugoniot's adiabat, corresponding to States 1 and 2 before and after the jump. Obviously, it characterizes the velocity $\mathscr{D}$ of propagation of the jump over the particles of the medium in State 1.

Entropy change along Hugoniot's adiabat. Next, evaluate the

[1)] For a more complete theory of Hugoniot's adiabat see, for example, L. I. Sedov, *Plane problems of hydrodynamics and aerodynamics,* John Wiley, 1963.

change in the entropy s along Hugoniot's adiabat. For this purpose, consider between some fixed state p_1, V_1 and an arbitrary state p, V, lying on Hugoniot's adiabat, the reversible process with heat flow for which

$$T\mathrm{d}s = \mathrm{d}U + p\mathrm{d}V. \tag{6.13}$$

Substituting in (6.13) for $\mathrm{d}U$ from (6.8), one obtains for fixed p_1, V_1 after some simple transformations

$$T\mathrm{d}s = \tfrac{1}{2}(V_1 - V)\mathrm{d}(p - p_1) - \tfrac{1}{2}(p - p_1)\mathrm{d}(V_1 - V).$$

By (6.12), one finds then

$$T\mathrm{d}s = \tfrac{1}{2}(V_1 - V)^2 \mathrm{d}\frac{p - p_1}{V_1 - V} = \tfrac{1}{2}(V_1 - V)^2 \mathrm{d}\tan\alpha$$

$$= -\tfrac{1}{2}(p - p_1)^2 \mathrm{d}\frac{1}{\tan\alpha}. \tag{6.14}$$

Entropy change along Hugoniot's adiabat for a small pressure jump. On the basis of the equality

$$\mathrm{d}\frac{1}{\tan\alpha} = -\mathrm{d}\frac{V - V_1}{p - p_1}$$

$$= -\mathrm{d}\left[\left(\frac{\mathrm{d}V}{\mathrm{d}p}\right)_{p=p_1} + \tfrac{1}{2}\left(\frac{\mathrm{d}^2 V}{\mathrm{d}p^2}\right)_{p=p_1}(p - p_1) + \dots\right]$$

$$= -\tfrac{1}{2}\left(\frac{\mathrm{d}^2 V}{\mathrm{d}p^2}\right)_{p=p_1}\mathrm{d}p + O(p - p_1)\mathrm{d}p,$$

where $O(p - p_1)$ is a small quantity which vanishes together with $p - p_1$, one obtains from (6.14)

$$T\mathrm{d}s = \tfrac{1}{4}(p - p_1)^2\left(\frac{\mathrm{d}^2 V}{\mathrm{d}p^2}\right)_{p=p_1}\mathrm{d}p + (p - p_1)^2 O(p - p_1)\mathrm{d}p. \tag{6.14'}$$

Hence, for small $p - p_1$,

$$T^* \Delta s = \tfrac{1}{12}(p - p_1)^3\left(\frac{\mathrm{d}^2 V}{\mathrm{d}p^2}\right)_{p=p_1} \tag{6.15}$$

$$+ \text{ small terms of order higher than } (p - p_1)^3,$$

where T^* is some mean value of the temperature in the interval of integration.

It is seen from (6.15) that for small pressure jumps $p - p_1$ the entropy change on passage through a jump is a small quantity of order $(p - p_1)^3$.

Poisson's adiabat. In the case of continuous adiabatic motions with a change of state of particles, the entropy is conserved, i.e.,

$$s_2(p, V) - s_1(p_1, V_1) = \Delta s = 0\,. \tag{6.16}$$

Equation (6.16) determines for given p_1, V_1 the link between p and V. The corresponding curve in the pV-plane which passes through the point p_1, V_1 is known as Poisson's adiabat.

Poisson's and Hugoniot's adiabats at the point p_1, V_1 have a second order contact. On the basis of (6.15) and (6.16), for small differences $p - p_1$, the equations of Poisson's and Hugoniot's adiabats may be presented in the forms

$$V - V_1 = f(p - p_1, s = \text{const.}) \tag{6.17}$$

and

$$V - V_1 = f(p - p_1, s = \text{const.}) + k(p - p_1)^3 + \dots\,, \tag{6.18}$$

respectively; the coefficient k depends only on V_1 and p_1. Consequently, near p_1, V_1, these adiabats touch (Fig. 49).

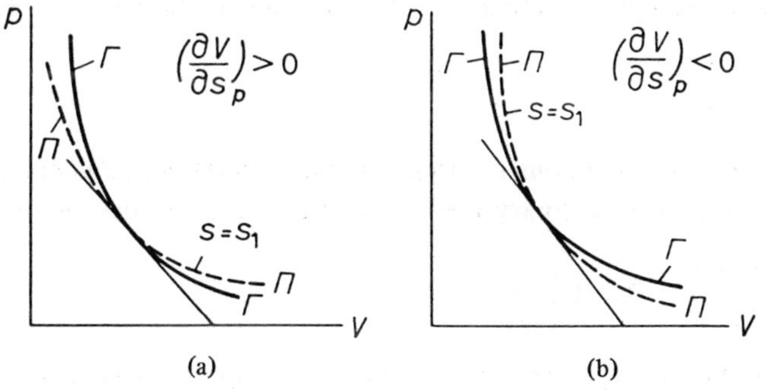

Fig. 49. Relationship between Poisson's (Π) and Hugoniot's (Γ) adiabats for $(\partial^2 V/\partial p^2)_s > 0$.

It is seen from (6.17) and (6.18) that at p_1, V_1 both adiabats have the same tangent and curvature, i.e.,

$$\left(\frac{\mathrm{d}V}{\mathrm{d}p}\right)_\Gamma = \left(\frac{\mathrm{d}V}{\mathrm{d}p}\right)_\Pi = \left(\frac{\partial V}{\partial p}\right)_s,$$

$$\left(\frac{\mathrm{d}^2 V}{\mathrm{d}p^2}\right)_\Gamma = \left(\frac{\mathrm{d}^2 V}{\mathrm{d}p^2}\right)_\Pi = \left(\frac{\partial^2 V}{\partial p^2}\right)_s. \tag{6.19}$$

Obviously, the total derivatives of the specific volume V along Poisson's adiabat are simply partial derivatives with respect to p of the specific volume $V(p, s)$, expressed with the aid of the equation of state as a function of the two independent variables p and s.

Weak jumps are spread by the particles at the velocity of sound. It is easily seen from (6.6) and (6.19) that for weak jumps, as $p_2 \to p_1$ and $V_2 \to V_1$,

$$\mathscr{D}^2 = -V^2\left(\frac{\mathrm{d}p}{\mathrm{d}V}\right)_\Gamma = \left(\frac{\mathrm{d}p}{\mathrm{d}\rho}\right)_\Gamma = \left(\frac{\mathrm{d}p}{\mathrm{d}\rho}\right)_\Pi = \left(\frac{\partial p}{\partial \rho}\right)_s. \tag{6.20}$$

The quantity

$$a = \sqrt{\left(\frac{\partial p}{\partial \rho}\right)_s} \tag{6.21}$$

is called the velocity of sound; it follows that for infinitesimal perturbations, weak jumps are spread by the particles at the velocity of sound, i.e.,

$$\mathscr{D} = a.$$

Relative locations of Hugoniot's and Poisson's adiabats near the point p_1, V_1. For ordinary media, there are satisfied the inequalities

$$\left(\frac{\partial^2 V}{\partial p^2}\right)_s > 0, \quad \left(\frac{\partial V}{\partial s}\right)_p > 0. \tag{6.22}$$

For example, for a perfect gas (*cf.* (7.11) and (4.3), Chapter V), one has

$$\left.\begin{aligned} U &= c_V T_0 \left(\frac{\rho}{\rho_0}\right)^{\gamma-1} e^{(s-s_0)/c_V} + \text{const.} \\ &= c_V T + \text{const.} = \frac{1}{\gamma-1} pV + \text{const.}, \\ \frac{p}{p_0} &= e^{(s-s_0)/c_V} \left(\frac{\rho}{\rho_0}\right)^{\gamma}, \\ \text{i.e.,} \quad \frac{\rho_0}{\rho} &= \frac{V}{V_0} = e^{(s-s_0)/c_p} \left(\frac{p_0}{p}\right)^{1/\gamma}, \end{aligned}\right\} \tag{6.23}$$

whence

$$\left.\begin{aligned} \left(\frac{\partial^2 V}{\partial p^2}\right)_s &= \frac{(1+\gamma)p_0^{1/\gamma}}{\gamma^2 \rho_0} \frac{e^{(s-s_0)/c_p}}{p^{2+1/\gamma}} > 0, \\ \left(\frac{\partial V}{\partial s}\right)_p &= \frac{V}{c_p} > 0. \end{aligned}\right\} \tag{6.23}$$

It follows from $(\partial^2 V/\partial p^2)_s > 0$ that Poisson's adiabat is concave upwards. If $(\partial V/\partial s)_p > 0$, then $s > s_1$ above and $s < s_1$ below Poisson's adiabat; if $(\partial V/\partial s)_p < 0$, then, conversely, one has $s < s_1$ above and $s > s_1$ below Poisson's adiabat (*cf.* Fig. 49a and b). Since on Hugoniot's adiabat one has Inequality (6.15), then obviously near the point p_1, V_1 Hugoniot's adiabat, which touches Poisson's adiabat, crosses for increasing V from above to below (Fig. 49a) when $(\partial V/\partial s)_p > 0$. If $(\partial V/\partial s)_p < 0$, then the relative positions of these adiabats are interchanged.

Equations of Poisson's and Hugoniot's adiabats for a perfect gas. By (6.23), the equations of Poisson's and Hugoniot's adiabats (6.8) have for a perfect gas the forms

$$pV^{\gamma} = p_1 V_1^{\gamma}, \tag{6.24}$$

$$\frac{1}{\gamma-1}(pV - p_1 V_1) = \tfrac{1}{2}(p+p_1)(V_1 - V), \tag{6.25}$$

respectively. Poisson's adiabat has the asymptotes $V=0$ and $p=0$. Hugoniot's adiabat is a hyperbola with the asymptotes

$$V = V_1 \frac{\gamma-1}{\gamma+1}, \quad p = -p_1 \frac{\gamma-1}{\gamma+1}.$$

The general form of these curves is shown in Fig. 50. Clearly, Hugoniot's

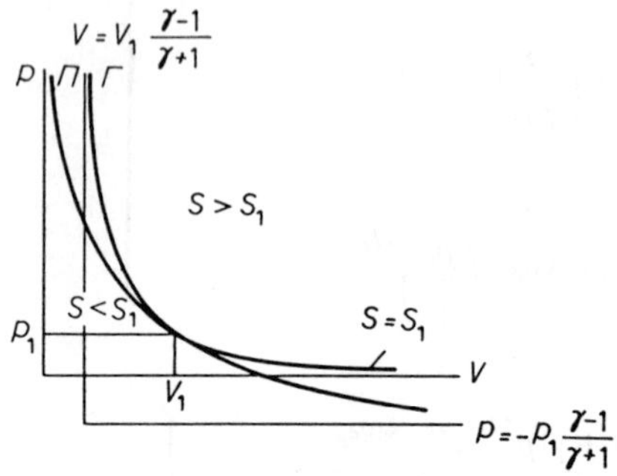

Fig. 50. Adiabats of Poisson and Hugoniot for a perfect gas.

adiabat for a perfect gas is over its entire length turned with its concave side upwards.

For the perfect gas, the entropy increases monotonically together with the pressure along Hugoniot's adiabat. For a perfect gas, it follows from (6.25) and (6.13) that along Hugoniot's adiabat

$$T ds = \frac{\gamma+1}{4 V_1} (V_1 - V)^2 dp ,$$

whence

$$\frac{\partial s}{\partial p} > 0 . \tag{6.26}$$

Hence, for a perfect gas, during upward motion along Hugoniot's adiabat, i.e., for increasing pressure P, the entropy grows monotonically.

It may be shown that also in the general case for any medium for which $(\partial^2 V/\partial p^2)_s > 0$, one has on Hugoniot's adiabat $s > s_1$ for $p > p_1$, and $s < s_1$ for $p < p_1$. In particular, for small jumps $p - p_1$, this result is seen directly from (6.14′) and (6.19).

Condensation jumps are real, rarefaction jumps cannot exist. It follows from the second law of thermodynamics that in the case under consideration only such jumps are physically admissible for which $\Omega > 0$, i.e:, for which the entropy s_2 of the particles after passage through a jump is larger than their original entropy s_1 before the jump. The preceding analysis establishes that for $(\partial^2 V/\partial p^2)_s > 0$ and $U_2(p, \rho) = U_1(p, p)$, such a situation occurs only for condensation jumps when

$$p_2 > p_1 , \quad \rho_2 > \rho_1 \quad \text{and} \quad v_n = v_{n2} - v_{n1} > 0 .$$

Thus, one arrives at the fundamental result in the theory of strong discontinuities that, if $(\partial^2 V/\partial p^2)_s > 0$, one may achieve in practice only condensation jumps. Rarefaction jumps cannot occur in reality.

This deduction is linked in an essential manner to the assumptions made: Firstly, to the inequality $(\partial^2 V/\partial p^2)_s > 0$, and, secondly, to the condition $U_2(p, \rho) = U_1(p, \rho)$.

When rarefaction jumps are feasible. If after passage of the particles of a gas through a discontinuity the chemical and physical properties of the gas mixture change in such a manner that the second of these conditions is not satisfied, then there arises the possibility of the occurrence of rarefaction jumps during real motions. In this case, Condition (6.8) determines the geometrical locus of the points p_2, V_2 of the curve which is also called Hugoniot's adiabat; this curve, as was stated above, does not pass through the point p_1, V_1.

In reality, such rarefaction jumps occur, for example, at combustion fronts.

The velocity of propagation by particles of a condensation jump through a discontinuity is supersonic. Note yet a very important property of the relative normal velocity components on the two sides of a jump. Let β_1 denote the acute angle between the tangent to Poisson's adiabat and the V-axis (Fig. 51). Since Hugoniot's adiabat is concave for arbitrary media (provided $(\partial^2 V/\partial p^2)_s > 0$) near the point p_1, V_1, and for a perfect gas over its entire length, it follows that for condensation jumps for small $p - p_1$ in any of the media under consideration and for any condensation jumps in a perfect gas one has the inequality

$$\tan \beta_1 < \tan \alpha .$$

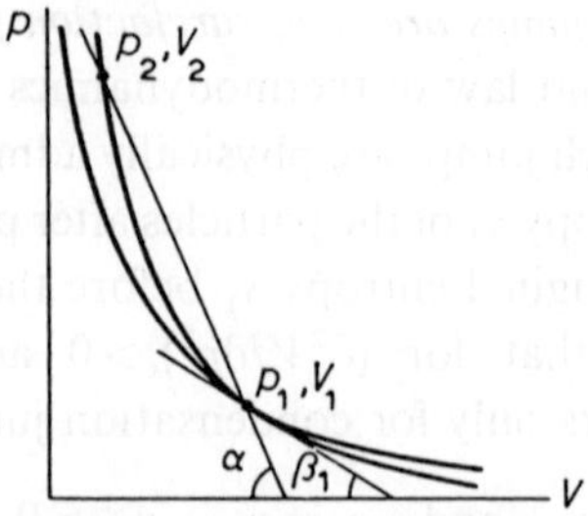

Fig. 51. β_1 is the acute angle between the tangent to Poisson's adiabat and the V-axis, α is the acute angle between the secant of Hugoniot's adiabat and the V-axis.

Since

$$a_1^2 = (\mathrm{d}p/\mathrm{d}\rho)_{p_1} = -V_1^2(\mathrm{d}p/\mathrm{d}V)_{p_1} = V_1^2 \tan\beta_1 ,$$

and

$$\mathscr{D}^2 = V_1^2 \frac{p_2 - p_1}{V_1 - V_2} = V_1^2 \tan\alpha ,$$

it follows that for condensation jumps ($p_2 > p_1$)

$$\mathscr{D}^2 = v_{n1}^2 > a_1^2 . \qquad (6.27)$$

It may likewise be shown[1)] that for any media in which $(\partial^2 V/\partial p^2)_s > 0$ the velocity of propagation by particles of the gas of a condensation jump with finite pressure increase $p_2 - p_1$ though a jump is larger than the velocity of sound.

Fixed jumps may exist only in supersonic flows. For a fixed jump, the normal component v_{n1} of the velocity of the particles, as they approach and pass through a jump, is larger than the velocity of sound. Thus, fixed jumps may only arise in supersonic flows. Obviously, moving jumps may exist in flows with any speed and, in particular, they may propagate in a medium at rest.

Different Hugoniot adiabats corresponding to the points p_1, V_1 and p_2, V_2. Previously, Hugoniot's adiabat Γ_1 corresponding to p_1, V_1 has been considered, for which

[1)] *Cf.* footnote on page 477.

$$U(p, V) - U(p_1, V_1) = \tfrac{1}{2}(p+p_1)(V_1 - V) \; ; \tag{6.28'}$$

Next, consider Hugoniot's adiabat Γ_2 corresponding to p_2, V_2 with an equation of the form

$$U(p, V) - U(p_2, V_2) = \tfrac{1}{2}(p+p_2)(V_2 - V) \,. \tag{6.28}$$

It is easily shown that the adiabats (6.28′) and (6.28) are different curves, but that under Condition (6.11) both these curves pass through the points p_1, V_1 and p_2, V_2.

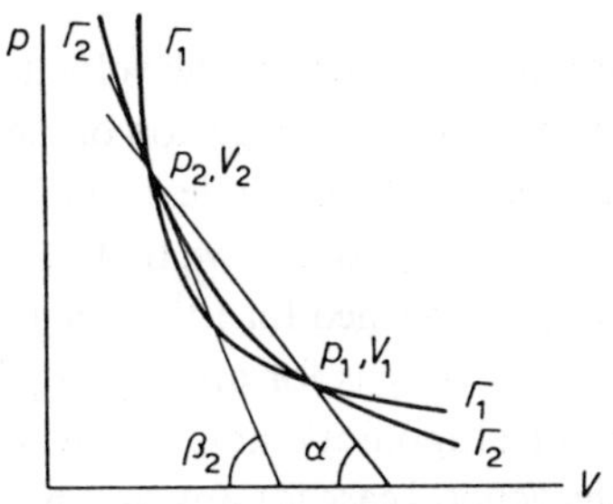

Fig. 52. Adiabat Γ_1 corresponds to (6.28′), adiabat Γ_2 to (6.28).

If for the adiabat (6.28′) the transition from p_1, V_1 to p_2, V_2 corresponds to a condensation jump, then for the adiabat (6.28) the transition from p_2, V_2 to p_1, V_1 corresponds to a rarefaction jump (Fig. 52).

Next, for the sake of simplicity, consider the case of a perfect gas. The common secant of Γ_1 and Γ_2 is the straight line which passes through the points p_1, V_1 and p_2, V_2 and which, on account of the concavity of the adiabat of a perfect gas, lies in the internal p_1, p_2 above both adiabats; therefore the acute angle β_2 between the V-axis and the tangent to the adiabat Γ_2 at the point p_2, V_2 is larger than the angle α between the V-axis and the secant of Γ_1 and Γ_2 (Fig. 52):

$$\tan \beta_2 > \tan \beta_1 \,. \tag{6.29'}$$

The velocity of particles of a medium, passing through a condensation jump, is subsonic relative to the jump. Directly from this result, there follows an important inequality for condensation jumps ($\rho_1 < \rho_2$); in fact, by (6.6) and (6.29′), one has

$$v_{n2}^2 = V_2^2 \frac{p_2 - p_1}{V_1 - V_2} = V_2^2 \tan\alpha < V_2^2 \tan\beta_2 .$$

However, since

$$V_2^2 \tan\beta_2 = \left(\frac{dp}{d\rho}\right)_{p_2} = a_2^2 ,$$

one finds

$$v_{n2}^2 = (\mathscr{D} - v_n)^2 < a_2^2 . \qquad (6.29)$$

Consequently, the normal component of the velocity of a medium relative to a jump, equal to the velocity of propagation of the jump by the particles $(\mathscr{D} - v_n)$, behind a condensation jump is less than the velocity of sound.

By a more detailed study, it may be shown that the property of condensation jumps found above, established for $(\partial^2 V/\partial p^2)_s > 0$ for any media for small $p_2 - p_1$, and for a perfect gas for any $p_2 - p_1$ (for a perfect gas, the property of concavity of Hugoniot's adiabat over its entire length has been used), is true in the general case for any $p_2 - p_1$, and for any media for which $(\partial^2 V/\partial p^2)_s > 0$.

Next, consider some examples of adiabatic motions of a perfect gas with condensation jumps.

Problem of a piston with plane waves. A beginning will be made with the problem of a piston. Let a cylindrical tube, closed on the left-hand side by a piston, contain a perfect gas (Fig. 53). At the initial instant of time $t = 0$, the piston and the weightless gas lie in a state of rest. Let ρ_1 and p_1 denote the density and pressure of the gas at rest. Consider the problem of the disturbed adiabatic motion of the gas, as the piston is displaced for $t > 0$.

Fig. 53. Problem of the motion of a gas, displaced by a piston.

Obviously, the disturbed motion of the gas essentially depends on the law of motion of the piston which may be specified with the aid of a function $v_\pi(t)$, the velocity of the piston. By the boundary condition at the piston, the velocity of the particles of the gas, which are in contact with the piston, must be equal to $v_\pi(t)$. In the general case, during continuous accelerated motion of the piston, the problem of the motion of the gas is difficult and may only be solved by means of computations on a digital computer. However, the problem is readily solved in the particular case when the initially resting piston instantly begins to move with the velocity v_π to the side of the gas and then continues its motion with the constant velocity. Obviously, all conditions of the problem are easily fulfilled, if it is assumed that at the initial instant of time a shock wave leaves the piston, moving into the gas at rest at constant supersonic speed $\mathscr{D}$ with respect to the initial state of the gas. Between the shock wave and the piston, there occurs translation with the known constant velocity v_π, with constant pressure p_2 and density ρ_2.

The velocity $\mathscr{D}$, density ρ_2 and pressure p_2 are readily computed in terms of $v_\pi = v_{n2}$, p_1 and ρ_1, from the three conditions at the discontinuities (6.6) and (6.25). The specific entropy behind the jump in all particles is the same and constant, where the discontinuity in the entropy is

$$s_2 - s_1 = c_V \log \frac{p_2}{p_1}\left(\frac{\rho_1}{\rho_2}\right)^\gamma > 0$$

If the velocity of the piston is constant, but directed away from the gas, then an analogous solution leads to a rarefaction jump which is not admissible. In reality, in this case, there arises no jump and the problem has a continuous solution.

If the velocity of the piston varies and is directed towards the gas, then a variable $\mathscr{D}$ is obtained. Small changes in the velocity of the piston propagate with velocity $v+a$ (where v is the velocity of the gas in the region behind the shock wave) and since $v+a$ behind the front of the shock wave is larger than the velocity $\mathscr{D}$ of the front, unfailingly for some time these perturbations bypass the shock wave and change the velocity of the gas behind the shock wave. Due to this the shock wave slows down or accelerates, and this, in turn, influences the magnitude of the pressure and entropy jumps. Thus, it is clear that behind the front of the wave one obtains motion of the gas with variable characteristics in terms of the coordinate (distance from the piston) and time. The entropy in the particles, thanks to the adiabatic conditions, is found to be constant, but due to the variable velocity $\mathscr{D}$ of the shock wave the entropy in different particles will be different. Therefore, in the region of continuous motion of the gas between the piston and the shock wave, one will not have barotropy, as is seen, for example, from the formula

$$\frac{p}{p_1} = e^{(s-s_2)/c_V}\left(\frac{\rho}{\rho_1}\right)^{\gamma}.$$

For a definite density ρ, the pressure at different particles may be different due to differences in entropy.

Spherical piston. The problem of the piston in a cylindrical tube may be made more complex by considering the problem of the expansion of a sphere Σ with a fixed centre O in an infinite mass of gas, initially at

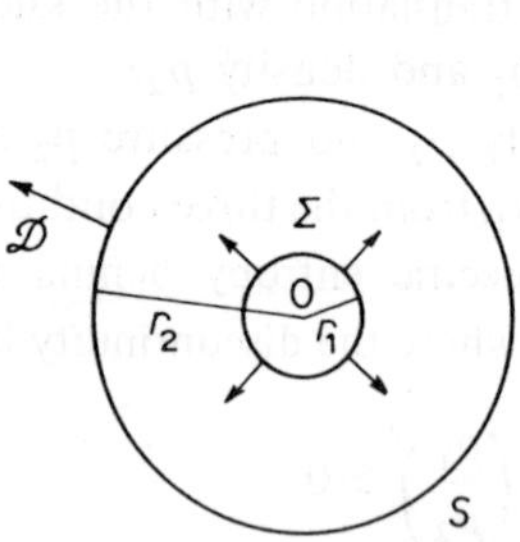

Fig. 54. On expansion of the sphere Σ in a gas, there arises the spherical shock wave S.

rest, with initial density ρ_1 and pressure p_1 (Fig. 54). The velocity of expansion of the sphere Σ is dr_1/dt. In the general case, there occurs in the gas a spherical shock wave S which advances with the velocity $\mathscr{D} = dr_2/dt$, where r_2 is the radius of the shock wave S. If Σ expands from a point (the centre of symmetry so that at $t=0$ one has $r_1=0$) and the velocity $dr_1/dt = \text{const.}$, then it may be shown that in this case even the velocity $\mathscr{D}$ of the shock wave is constant.

For the solution of this problem, it is expedient to write all equations in spherical coordinates and to utilize simplifications which arise out of the presence of spherical symmetry (the characteristics of the motion will depend on two independent variables, namely the radius r and the time t).

One can consider the problem of the spherical piston as a model of an explosion in an air atmosphere, if it is assumed that inside Σ one has products of chemical reactions, i.e., a strongly compressed gas which displaces the air, acting as a piston. In this case, there is formed in the air a shock wave which is referred to as an explosion wave. In order to determine the motion of the air between the explosion wave S and the surface Σ, in which one finds the products of the explosion, one must solve a problem of gas dynamics. For the solutions of this problem, all the equations and supplementary initial and boundary conditions have been prepared previously.

The problem of the point explosion. It is assumed in the problem of the spherical piston that the law of expansion of the sphere Σ is given; however, in the problem of an explosion, the velocity of expansion dr_1/dt of the sphere Σ which separates the products of the explosion from the air is not known beforehand.

The problem of an explosion in an atmosphere may be treated schematically, in order to take the principal effect into account which consists of the fact that a small volume contains a significant amount of energy which is then imparted to the air and that, as a consequence of this, there arises in the atmosphere quickly an expanding spherical region of moving air with sharply perturbed pressure and density fields. The problem of the point explosion may be formulated in the following manner. At the initial instant of time, in a mass of weightless perfect gas at rest under constant pressure p_1 with density ρ_1, there is released instantaneously at some point (the centre of symmetry of the subsequent motion) a given amount of energy E; it is required to determine the perturbed motion in the simplest form, when the motion of the gas particles is, by assumption, adiabatic.

In this case, just as in the case of the problem of the spherical piston, a spherical shock wave is formed which expands from the point of the explosion and separates the gas at rest from the moving gas in the region inside the shock wave (Fig. 55). All the characteristics of motion and state may be assumed to be functions of r and t alone. In order to determine the dependence of all the characteristics of state and of the velocity of the motion of the particles of the gas on the radius, one must integrate the following non-linear partial differential equations in spherical coordinates (*cf.* §3, Chapter IV):

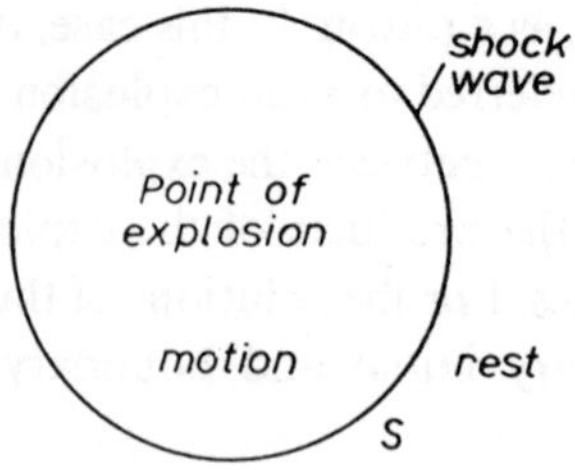

Fig. 55. Scheme for the problem of the point explosion.

$$\left.\begin{aligned}
&\text{the equation of continuity:}\\
&\frac{\partial \rho}{\partial t} + \frac{\partial \rho v}{\partial r} + \frac{2\rho v}{r} = 0\,,\\
&\text{the momentum equation:}\\
&\frac{\partial v}{\partial t} + v\frac{\partial v}{\partial r} + \frac{1}{\rho}\frac{\partial p}{\partial r} = 0\,,\\
&\text{the adiabatic condition:}\\
&\frac{\partial}{\partial t}\left(\frac{p}{\rho^{\gamma}}\right) + v\frac{\partial}{\partial r}\left(\frac{p}{\rho^{\gamma}}\right) = 0\,.
\end{aligned}\right\} \qquad (6.30)$$

In the integration of these equations, one must satisfy the initial conditions above, the boundary conditions at the jump (explosion wave) derived earlier and the condition that at each instant of time the total energy of the gas inside the spherical shock wave is equal to the sum of the explosion energy E and the original energy of the gas at rest inside the sphere S.

In §8, the general properties of the solution of this problem will be analyzed in greater detail.[1)]

On detonation and combustion waves. If during the passage through the discontinuity there occurs an exchange (gain or loss) of energy and entropy due to any physico-chemical processes (combustion, condensation, evaporation, chemical reactions, etc.), then the basic equation

[1)] For a complete solution of this problem in finite form, *cf.* L. I. Sedov, *Methods of similarity and dimension in mechanics*, Academic Press, 1959.

(6.8) changes its form. Instead of (6.11), one obtains in this case an equation of the form

$$U_2(p_1, V_1) - U_1(p_1, V_1) = q^* , \tag{6.31}$$

where q^* is the chemical or other energy, liberated on passage of the particles through the surface of discontinuity; one has $q^* > 0$ for combustion or condensation and $q^* < 0$ for evaporation. Then the equation of the corresponding Hugoniot adiabat may be written in the form[1)]

$$U_2(p_2, V_2) - U_2(p_1, V_1) = \tfrac{1}{2}(p_1 + p_2)(V_1 - V_2) + q^* . \tag{6.32}$$

A rarefaction jump for $q^* > 0$ is characteristic for a combustion front. A condensation jump for $q^* > 0$ corresponds to a detonation wave.

The problem of a piston with a detonation wave. The preceding problems of a piston with a plane and of spherical waves may be generalized, if it is assumed that a forming condensation jump represents a detonation wave. Then the gas, originally at rest, is an explosive mixture and one obtains behind the front of the wave another gas (the product of the reaction mixture). In the problem of the detonation of a mixture in front of a moving plane piston (between the piston and the detonation wave), one obtains even translation of the gas, if the piston moves with sufficiently large constant velocity towards the side of the gas. If, after the detonation wave has been formed, the velocity of the piston is small, or if the piston stops or if it begins to move away from the gas, then one will have the solution in the presence of a detonation wave which propagates in the gas, where there occurs between the piston and the shock detonation wave continuous motion of the gas with variable parameters.

Relations at fixed shock waves for a perfect gas. In many applications and, in particular, in the fundamental problems of aerodynamics of the study of motion of bodies with constant supersonic speed, one employs the model of a perfect gas and in a coordinate system which is fixed in the flying body; one deals with steady motion of a gas with fixed condensation jumps, i.e., shock waves. In these applications,

[1)] A detailed analysis of the corresponding Hugoniot adiabat may be found in L. D. Landau, E. M. Lifshitz, *Mechanics of Continuous Media*, Pergamon Press, 1958 and L. I. Sedov, *The plane problem of hydrodynamics and aerodynamics*, John Wiley, 1962.

one uses, instead of the theory based on the analysis of Hugoniot's adiabat, a theory based on the following relations.

For a perfect gas, one has

$$U = \frac{1}{\gamma - 1}\frac{p}{\rho} + \text{const.}, \qquad a^2 = \left(\frac{\partial p}{\partial \rho}\right)_s = \frac{\gamma p}{\rho}.$$

By (6.1)–(6.4), the quantities $v_{\tau 2}$, v_{n2}, ρ_2 and p_2 after the jump are readily expressed in terms of $v_{\tau 1}$, v_{n1}, ρ_1 and p_1 before the jump:

$$\boldsymbol{v}_{\tau 2} = \boldsymbol{v}_{\tau 1}, \quad v_{n2} = v_{n1}\left(\frac{\gamma-1}{\gamma+1} + \frac{2}{\gamma+1}\frac{a_1^2}{v_{n1}^2}\right), \tag{6.33}$$

$$\rho_2 = \frac{\gamma+1}{\gamma-1}\frac{\rho_1}{1 + \dfrac{2}{\gamma-1}\dfrac{a_1^2}{v_{n1}^2}},$$

$$p_2 = \frac{2}{\gamma+1}\rho_1 v_{n2}^2\left[1 - \frac{\gamma-1}{2\gamma}\frac{a_1^2}{v_{n1}^2}\right]. \tag{6.34}$$

The vectors $\boldsymbol{v}_1$ and $\boldsymbol{v}_2$ determine the plane π coinciding with the plane which passes through the normal to the surface of discontinuity and through the direction of the component $\boldsymbol{v}_{\tau 1}$. Select in the plane π Cartesian axes, denote by β the angle between an element of the shock wave in the plane π and the x-axis, by θ the angle between the velocity and the x-axis and by u, v velocity components. Obviously,

$$v_\tau = |\boldsymbol{v}| \cos(\beta - \theta) = u \cos\beta + v \sin\beta,$$
$$v_n = |\boldsymbol{v}| \sin(\beta - \theta) = u \sin\beta - v\cos\beta. \tag{6.35}$$

Select a coordinate system such that for the point under consideration on the shock wave the direction of the x-axis coincides with the direction of the velocity v_1 ($|v_1| = u_1$, $v_1 = 0$). In (6.33), replace v_τ and v_n by u, v and β; eliminating β with the aid of (6.35), one obtains

$$v_2^2 = (u_1 - u_2)^2\,\frac{\dfrac{2}{\gamma+1}\left(u_1 - \dfrac{a_1^2}{u_1}\right) - (u_1 - u_2)}{u_1 - u_2 + \dfrac{2}{\gamma+1}\dfrac{a_1^2}{u_1^2}}. \tag{6.36}$$

The shock polar is a hypercissoid. In the hodograph plane of the velocity $\boldsymbol{v}_2(u_2, v_2)$ in which u_2 and v_2 are used as coordinates (this plane

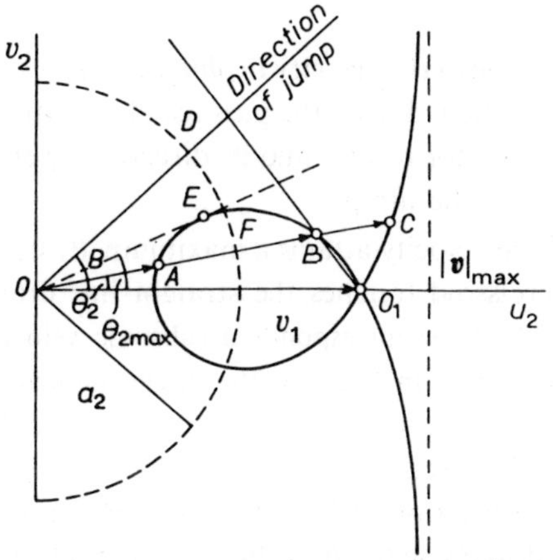

Fig. 56. Shock polar-hypercissoid. OD is the direction of the shock wave, OB is the vector of the velocity after the jump, OO_1 is the velocity vector of the gas before the jump.

corresponds to the plane π), Equation (6.36) determines a curve which is referred to as shock polar. This curve is a hypercissoid (Fig. 56).

There correspond three points on the shock polar, e.g., three magnitudes of the velocity to every value of the angle θ_2 of inclination of the velocity after the jump: The points A, B and C. It follows from the condition of the continuity of the velocity component tangent to the jump that the direction of the jump, determined by β, is obtained as the direction of the perpendicular, drawn from the origin O to the straight line through O_1 and the point on the hypercissoid, corresponding to the velocity behind the jump; in Fig. 56, the point B is such a point. Obviously, for the jump corresponding to the point C, one has the inequality

$$v_n = v_{n2} - v_{n1} < 0 .$$

Therefore, by (6.10), this value of the velocity corresponds to a rarefaction jump. Jumps, corresponding to points of the branches of the hypercissoid which go to infinity, do not have physical significance; these branches of the curve must be disregarded.

The directions of jumps of infinitesimal intensity are obtained in the limit when B tends to O and the straight line O_1B becomes the tangent to the hypercissoid at O_1.

Straight and oblique condensation jumps. Jumps for which the velocity $\mathbf{v}_1$ is perpendicular to the jump are called straight; otherwise, a condensation jump is said to be oblique.

On the rotation of the velocity in oblique condensation jumps. If a jump is not straight, then, during the passage of the particles through the jump, their velocity vector rotates and its direction approaches the direction of the tangent to the jump.

The rotation of the velocity attains a maximum, $\theta_2 = \theta_{2\,\max}$, at the point E where the hypercissoid touches the straight line drawn from O (Fig. 56). To the point E there corresponds a subsonic velocity which is close to the velocity of sound. In Fig. 56, the circle marked by a broken line corresponds to velocities equal to the velocity of sound[1] $|v_2| = v_{\text{crit}} = a_2$. This circle cuts the hypercissoid at F. The magnitude $|v_2|$ of the velocity after the jump is supersonic, when the point B lies to the right of F, subsonic for A and likewise for B in the not large interval EF. It follows from (6.33) that

$$\tan(\beta - \theta_2) = \tan\beta \left(\frac{\gamma - 1}{\gamma + 1} + \frac{1}{(\gamma+1)\mathsf{M}_1^2(1 - \cos^2\beta)} \right), \tag{6.37}$$

where $\mathsf{M} = v/a$ and $\mathsf{M}_1 = v_1/a_1$ $(v_1 = |\mathbf{v}_1|)$.

The quantity M is referred to as Mach number; obviously, for supersonic motion, $\mathsf{M} > 1$, for subsonic motion, $\mathsf{M} < 1$. It follows from (6.37) that the magnitude of the angle of rotation of the velocity $\delta = \theta_2 - \theta_1$ (on Fig. 56, it is simply θ_2) depends on M_1 and β. The largest possible angle of rotation $\delta_{\max}$ depends only on M_1 in the flow running up to the jump. In Fig. 57, the dependence of $\delta_{\max}$ on M_1 is shown.

Supersonic flow around an angle and a wedge. The conditions at skew condensation jumps make it possible to construct immediately the solution of the problem of the flow around a supersonically translating angle and wedge (Fig. 58).

It is directly obvious that the patterns of the flows shown in Fig. 58 satisfy all the conditions of the problem of the flows around an angle and wedge.

[1] It may be shown that for given parameters through a jump the critical velocity (which is equal to the local velocity of sound) is obtained independently of the angle of inclination of the jump.

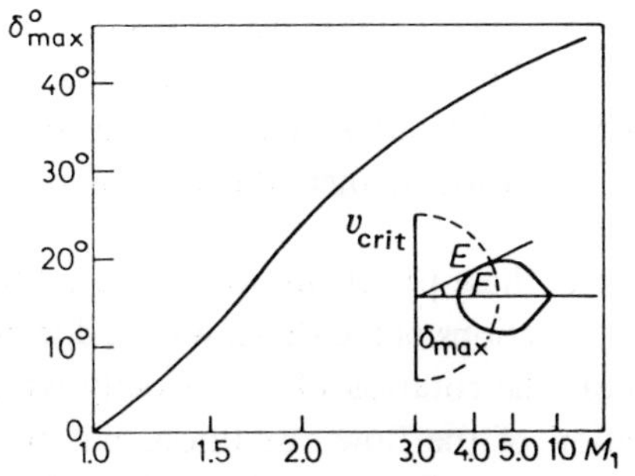

Fig. 57. Dependence on M of the largest possible angle of rotation of the velocity during the passage through a jump. The graph has been drawn for $\gamma = 1.4$.

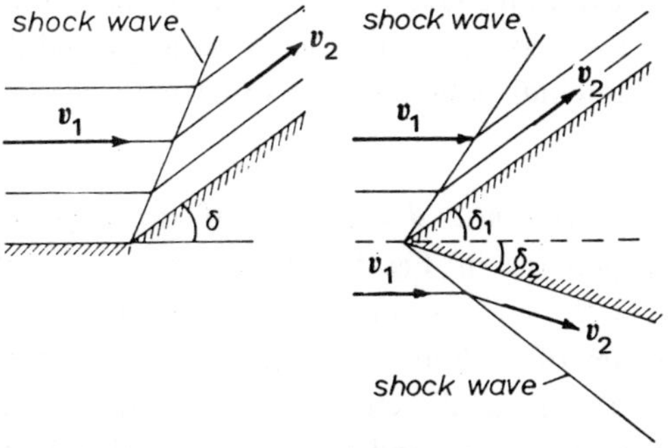

Fig. 58. Flow around an angle and wedge inclined unsymmetrically to the velocity of the oncoming flow.

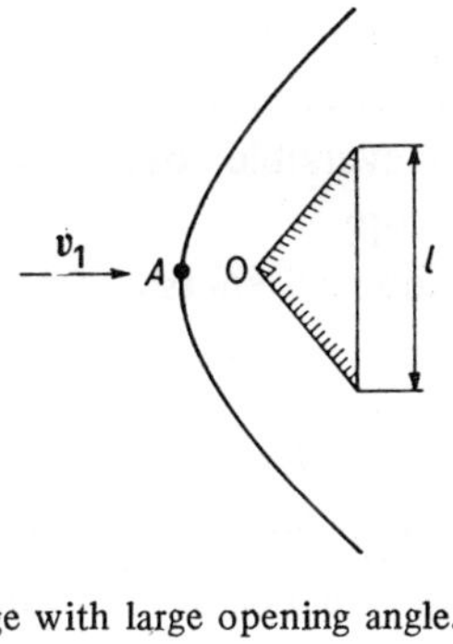

Fig. 59. Flow around a wedge with large opening angle. The shock wave is formed at some distance ahead of the wedge.

However, it follows from the preceding results that the solutions of these problems can only be obtained, if angles δ, δ_1 and δ_2 for a given Mach number of the oncoming flow are smaller or equal to the angle $\delta_{max}(\mathsf{M}_1)$.

If this condition is satisfied, then the pattern of the flow shown is admissible.

However, for any of these problems, for $\delta < \delta_{max}$, there exist two solutions with different directions of the shock waves, during passage through each of which occurs the rotation of the velocity which is necessary to guarantee satisfaction of the flow condition, but they attain different velocities behind the wave. For example, in Fig. 56 these two solutions correspond to the points B and A.

In practice, for thin bodies, there occurs motion which corresponds to a gentle shock wave with a large velocity behind the jump, corresponding to the point B. This theoretical solution for the angle and the wedge agrees well with experimental data on supersonic flow around thin pointed bodies.

Flow with detached shock wave. If the angle $\delta > \delta_{max}$, then flows with the patterns of Fig. 58 are impossible. In this case, there forms in front of the body a curved shock wave (Fig. 59). The distance AO from the vertex of the body to the closest point of the shock wave is proportional to the linear dimensions of the body and depends on the opening angle of the wedge and M_1. As $l \to \infty$, the shock wave moves forward to infinity and subsonic motion of the gas occurs in front of the wedge.

Loss of mechanical energy in jumps. In conclusion of this section, it will be noted that the presence of jumps in adiabatic flows is linked to the phenomenon of an irreversible loss of usable mechanical energy caused by a growth of entropy.

One may write for the internal energy of a perfect gas

$$U = c_V T + \text{const.} = c_V T_0 \left(\frac{p}{p_0}\right)^{(\gamma-1)/\gamma} e^{(s-s_0)/c_p} + \text{const.}$$

Consider the internal energy for certain characteristic states of rest before and after a jump (states which are obtained if the gas adiabatically brakes down to $v = 0$).[1] The pressures and temperatures in these states

[1] *Cf.* §5, Chapter VIII, Volume II.

before and after a jump (stagnation pressure and temperature) will be denoted by p_1^*, p_2^*, T_1^* and T_2^*, respectively.

If the passage of the particles through a jump is not accompanied by chemical reactions or phase transitions (i.e., in (6.31), $q^* = 0$), then it follows from (6.4) for a perfect gas that $T_1^* = T_2^*$. Hence

$$e^{(s_1 - s_0)/c_p} p_1^{*(\gamma-1)/\gamma} = e^{(s_2 - s_0)/c_p} p_2^{*(\gamma-1)/\gamma}, \quad \text{or} \quad \left(\frac{p_1^*}{p_2^*}\right)^{(\gamma-1)/\gamma} = e^{(s_2 - s_1)/c_p}$$

If the loss is irreversible, then $s_2 - s_1 > 0$ whence $p_2^* < p_1^*$. Consequently, a gas passing through a jump has less pressure in the state of rest and therefore its capacity to do work (technical usefulness) drops.

It is seen from a study of Hugoniot's adiabat that the growth of entropy and, consequently, the losses increase with increasing intensity of the condensation jump, characterized by the pressure drop $p_2 - p_1$.

For a given Mach number M_1 in the oncoming flow, the intensity of a straight jump attains a maximum. The losses in an oblique jump are always smaller than in a straight jump.[1)]

§7. Dimensionality of physical quantities and Π-theorem

Physical relations are invariant with respect to the choice of coordinate systems. During an actual writing down of equations and in concrete numerical calculations, one has to introduce and employ different coordinate systems. In the general case, these coordinate systems may be arbitrary; however, in many situations, the systems are chosen from considerations of simplicity and convenience of computations or of the analysis of numerical results. The possible arbitrariness of choice of coordinate systems, introduced by accepted methods of description of phenomena under study, is not essentially connected with these very phenomena; therefore the various equations, expressing certain physical

[1)] A more complete theory of the losses in systems of condensation jumps is contained in L. I. Sedov, *Plane problems of hydrodynamics and aerodynamics,* Wiley, 1962 and in §3, Chapter VIII, Volume II.

phenomena and facts, must have the property of invariance with respect to a choice of coordinates.

Tensorial nature of physical equations. As a consequence of this fact, all the equations and supplementary relations introduced above have the form of scalar, vector or, in general, tensor equations. In fact, it is the reason why the set of the characteristics of motion and state has an invariant tensorial nature. It is the reason for the necessity to develop and employ tensor calculus and to formulate all physical relations in invariant tensor form.

Dimensional quantities. On the other hand, the definition and specification of different quantities, i.e., the characteristics of media, fields and processes, for example, of density, energy, velocity, charge, etc., with the aid of numbers is closely linked to the utilization of definite dimensional units the choice of which likewise depends on the investigator.

Quantities, the numerical values of which in problems under consideration depend on the choice of dimensional units are said to be dimensional. For example, energy may be measured in kilogramme-metres, in Joules, in calories, in tons of coal or kilogramme of uranium, in roubles, and yet in many other dimensional units. The corresponding numbers determining an amount of energy depend essentially on the choice of dimensional units.

Operations with a large number of various characteristics interconnected by different definitions and different equations show that the dimensional units for various characteristics are, generally speaking, interrelated. For example, the dimensional unit for the velocity $v = \mathrm{d}s/\mathrm{d}t$ is linked to the dimensional units of the length $\mathrm{d}s$ and the time $\mathrm{d}t$.

Primary (fundamental) dimensional units. Side by side with dependent dimensional units, one must consider primary independent dimensional units which are introduced by experimental means, generally speaking, with the aid of special, in theory, arbitrary conditions. Thus, for example, different primary, or fundamental dimensional units are introduced by known methods for the measurement of length, time and mass.

Quantities for which dimensional units are introduced by experiment with the aid of standards are called primary or fundamental, and the corresponding dimensional units are referred to as primary or fundamental units.

Secondary (derived) dimensional units. The dimensional units for other quantities which are obtained from the definitions of these quantities in terms of primary units are known as derived or secondary dimensional units.

Various systems of dimensional units. One may use various quantities as origin of primary dimensional units. In different areas of measurement, it is advantageous and convenient to select as primary dimensional units local primary dimensional units which are different in different cases. Thus, there arose different systems of dimensional units and the problem of the transition from one system to another.

The following systems of dimensional units are widely used: CGS with the primary units centimetre, gram, second; MKS with the primary units metre, kilogram-force, second;[1)] SI with metre, kilogram-mass, second, ampère, degree Kelvin, candle power; and other systems of units. In mechanics (and, generally speaking, in practice), one employs often only three primary dimensional units (for example, this is the case in the systems CGS and MKS); the dimensional units for all remaining quantities, including the characteristics of electromagnetic fields, are considered to be derived units.

Once the fundamental dimensional units have been established, the dimensional units for other dimensional quantities with derived units are obtained automatically from their definitions.

Formula of dimensionality. The expression for a derived dimensional unit in terms of the basic units is called its dimensionality. In the CGS system, the formula of dimensionality contains three arguments: the symbol L of length, the symbol T of time and the symbol M of mass. For example, the symbol for the dimensional unit of force is written in the form

[1)] It has been assumed here that the actual conditions which determine the primary dimensional units are known from a general course of physics — it is known what is a second etc.

$$K = \frac{ML}{T^2} = MLT^{-2}.$$

In the system CGS, the formulae of dimensionality for all physical quantities are products of powers of L, T and M of the form

$$N = L^l T^t M^m, \tag{7.1}$$

where exponents l, t and m are integers or rational fractions.[1)]

In the system MKS, the dimensionality formulae have the form

$$N = L^{l_1} T^{t_1} K^{k_1}, \tag{7.2}$$

where l_1, t_1 and k_1 are the corresponding dimensionality exponents.

The transition from (7.1) to the corresponding formula (7.2) is achieved by replacing the symbol M in (7.1) by the symbol K using the formula $M = KT^2L^{-1}$.

Dimensionality formulae make it possible to compute numerical scale factors for the evaluation of corresponding characteristics in a change of the magnitudes of the primary dimensional units.

If one wishes to step over from given dimensional units of length L, time T and mass M to new dimensional units, changing L to αL, T to βT and M to γM, then the new dimensional unit for N in (7.1) will be

$$\alpha^l \beta^t \gamma^m \tag{7.3}$$

times the old unit. In this manner one can establish easily the transition scale factors of secondary dimensional units for a change of the magnitudes of the primary dimensional units.

Number of primary dimensional units. The number of primary dimensional units may be and is sometimes larger than three. For example, in many problems of gas dynamics and thermodynamics, besides the primary dimensional units metre, kilogram and second, determined by experimental means, one uses, as a rule, dimensional units for temperature: degree Celsius or degree Fahrenheit, etc., and for amounts of heat energy: gram calorie or kilogram calorie. Thus, one may consider and employ systems of dimensional units with five primary units. Systems are con-

1) For a full presentation of dimensional theory, see L. I. Sedov, *Methods of Similarity and Dimensionality in Mechanics,* Academic Press, 1959.

sidered and used with primary units for electromagnetic quantities, etc. In these cases, the dimensionality formulae will be monomials of the form (7.1) with a larger number of arguments.

On the possibility of increasing the number of primary dimensional units. Generally speaking, systems of dimensional units with any number of primary units may be introduced. In particular, one may choose for length, time and velocity independent dimensional units on the basis of experience; however, in this case, the formula for the velocity must be written in the form

$$v = k\frac{\mathrm{d}s}{\mathrm{d}t}, \tag{7.4}$$

where k is a dimensional constant which depends on the choice of the dimensional units for v, $\mathrm{d}s$ and $\mathrm{d}t$. If it is assumed that k is an absolute numerical factor (i.e., the same in all systems of dimensional units employed), equal to or not equal to unity, then one finds that in any systems of units the dimensional units for v, $\mathrm{d}s$ and $\mathrm{d}t$ are interdependent.

In practice, such a condition is usually always applied.

If one considers the relation

$$I \cdot \text{heat energy} = \text{mechanical energy}\,,$$

then one may choose separate independent dimensional units for heat energy, for example, the calorie, or for mechanical energy, for example, kilogram metre. In this case, there appears in the formulae and equations the dimensional constant I, i.e., the mechanical equivalent of heat with the dimensionality

$$[I] = \frac{\text{kilogram metre}}{\text{calorie}}; \tag{7.5}$$

the "physical" constant I is analogous to the constant k in (7.4).

On the possibility of reducing the number of primary dimensional units. The constant I may be considered as a dimensional constant for independent dimensional units for heat and mechanical energy or as a non-dimensional scale factor, if the dimensional units of heat and mechanical energy differ only like dimensional units for one and the same quantity, for example, as foot and metre. In an analogous manner, one may consider the fundamental equation of physics

$$E = mc^2 , \tag{7.6}$$

where E is energy, m is mass and c is the velocity of light in vacuum. If it is assumed in (7.6) that c is a universal constant, which may be set equal to unity in a class of admissible systems of dimensional units, analogously to k in (7.4), then the possible different dimensional units for energy and mass are found to be interdependent (just as the dimensional units for v, ds and dt in (7.4) when $k = 1$).

This type of reasoning may serve as a basis for a reduction of the number of primary dimensional units. If one fixes the dimensional universal constant in the law of universal gravitation

$$F = f\frac{m_1 m_2}{r^2} , \tag{7.7}$$

one obtains an additional link between the primary dimensional units for mass, time and distance.

In this manner, one may obtain systems of dimensional units with different numbers of primary dimensional units. In particular, one may arrive at a system of universal dimensional units, in which the dimensional units of all quantities are fixed for ever and therefore all quantities may be considered to be non-dimensional.

Relativity of the concepts of dimensional and non-dimensional quantities. The concepts of dimensional and non-dimensional (abstract) quantities are relative. This statement must be interpreted in the following manner. In order to study a given phenomenon or some set of phenomena, one introduces different variables or constant characteristics of processes and objects. In order to obtain numerical values, one utilizes in reality explicitly or implicitly or admits potentially a certain set of systems of dimensional units.

Quantities the numerical values of which depend on the choice of a concrete system of dimensional units from an admissible set of systems of dimensional units, are called dimensional. Quantities, the numerical values of which are the same in all systems of dimensional units of this set of systems are called non-dimensional or abstract.

For example, the geometrical object angle may be presented correspondingly by different numbers: radians, degree, part of a right angle, etc.; therefore, in order to include in the set of admissible systems of dimensional units a system with different dimensional units for angle, one must consider the quantity angle to be dimensional. However, if one limits consideration to such systems of dimensional units in which angles are measured only in radians, then angles may be considered as abstract non-dimensional quantities. Such a condition relating to the measurement of angles is often employed in practice.

By such means one may consider any other quantity (for example, time or length) as a non-dimensional quantity for a corresponding subdivision of the set of systems of dimensional units in which the dimensional unit for this quantity is the same in all systems of dimensional units employed in a given problem. In applications, the condition of non-dimensionality of angles is convenient, the condition of non-dimensionality of lengths inconvenient. This circumstance is linked to the fact that for geometrically similar systems corresponding angles are identical, while corresponding lengths are different.

Appearance of dimensional physical constants as arguments of functions under study. It follows from the further consideration that in certain cases it is, generally speaking, advantageous to increase the number of primary dimensional units. However, this advantage is, as a rule, outweighed by the fact that for an increase of the number of primary dimensional units there appear additional physical constants, which are essential for a problem under consideration, of the type of the mechanical equivalent of heat I or the velocity of light c or the gravitational constant f. Similar constants will enter into the equations of processes and other conditions of a problem, and they must be added to the determining parameters and included in the arguments of the functional relations to be studied. It will be shown below that therefore, in the general case, one does not obtain additional simplifications as a result of an increase in the number of primary dimensional units. However, an essential benefit is reaped from such an increase when it is known, from additional physical reasoning, that physical constants of such a type (the type I or f) are not essential in given functional relations.

This is also the reason why the standard system of dimensional units with fixed constant values of I, c and f is inconvenient and is not employed in practice in many technical and physical problems.

On the possible benefit of an application of different systems of dimensional units. In certain scientific problems, there appears clearly a tendency towards standardization and towards an administrative introduction of a universal system of dimensional units. Obviously, in a number of cases, this is very convenient and useful; however, the tying in of the universal system of dimensional units into definite physical constants or conditions is in many cases artificial. In contrast, the very possibility of utilization of arbitrary dimensional units and the independence of the laws under study from the choice of systems of dimensional units may serve as a source of useful derivations. Therefore, side by side with the introduction of common dimensional units, one must develop a theory of arbitrary and diverse classes of dimensional units as a basis for fruitful methods of experimentation and theory.

In these questions, one has complete analogy between the choice of a system of dimensional units and the choice of a coordinate system and a reference system, in general. Naturally, one may fix completely a definite universal reference system once and for all and study all phenomena only in this system. However, the very possibility of utilization of different reference systems and the possibility of employing in concrete problems special characteristic reference systems is the basis of a valuable method of investigation in physics. And what is more, the fundamental position of physics with regard to the invariance of physical laws with respect to choice of coordinate systems and reference frames (for example, inertial reference systems) is, in a known sense, another presentation of such universal laws of physics as the law of conservation of momentum, the law of conservation of energy and the law of conservation of moment of momentum.

The universal invariance of physical laws with respect to the groups of the transformations of Galilei or Lorentz in certain problems (however, it is essential to emphasize that it is not in all problems) supplements the property of invariance of the functional relations under investigation with respect to the group of similarity transformations, determined by the possibility of conservation of all equations and all supplementary conditions for similarity transformations, coinciding with a transition from one system of dimensional units to another.

Π-Theorem. Consider now the structure of the functional relations between, in general, dimensional quantities which express physical regularities, invariant with respect to a choice of systems of dimensional units.

Let there be given some dimensional or non-dimensional quantity a which is, in general, a function of dimensional, not interdependent quantities $a_1, a_2, \ldots, a_n$:

$$a = f(a_1, a_2, \ldots, a_k, a_{k+1}, \ldots, a_n). \tag{7.8}$$

In the process under consideration, certain of the quantities a_i may be constants, others may be variables. Assume with respect to the variable quantities that either their values are non-zero or finite or that the function (7.8) is continuous as the corresponding arguments tend to zero or infinity.

Further, assume that the arguments in the functional relation (7.8) include all the dimensional and non-dimensional constants or variables which determine the value of the quantity a under consideration.

The structure of the function in (7.8) will now be found under the assumption that this function expresses itself some physical law which does not depend on the choice of the system of dimensional units.

Among the dimensional quantities $a_1, a_2, \ldots, a_n$, let the first k quantities ($k \leqslant n$) have independent dimensionalities (the number of basic dimensional units must be larger or equal to k).

Independence of dimensionality means that the formula, expressing the dimensionality of one of the quantities, may not be presented as a combination in the form of a power of the monomial of the dimensionality formulae for the other quantities. For example, the dimensionalities of length L, velocity L/T and energy ML^2/T^2 are independent; the dimensionalities of length L, velocity L/T and acceleration L/T^2 are dependent.

As a rule, one has among mechanical quantities no more than three with independent dimensionalities. Assume that k is equal to the largest number of parameters with independent dimensionalities; therefore the dimensionalities of the quantities $a, a_{k+1}, \ldots, a_n$ may be expressed in terms of the dimensionalities of the parameters $a_1, a_2, \ldots, a_k$.

Select the k quantities $a_1, a_2, \ldots, a_k$ with independent dimensionalities as basic quantities and introduce for them the dimensionality notation

$$[a_1] = A_1, \quad [a_2] = A_2, \ldots, [a_k] = A_k.$$

The dimensionalities of the remaining quantities will then have the form

$$[a] \quad = A_1^{m_1} A_2^{m_2} \dots A_k^{m_k} \,,$$

$$[a_{k+1}] = A_1^{p_1} A_2^{p_2} \dots A_k^{p_k} \,,$$

$$\dots\dots\dots\dots\dots\dots$$

$$[a_n] \quad = A_1^{q_1} A_2^{q_2} \dots A_k^{q_k} \,.$$

Change now the dimensional units of the quantities $a_1, a_2, \dots, a_k$, respectively, by the factors $\alpha_1, \alpha_2, \dots, \alpha_k$; the numerical values of these quantities and of the quantities $a, a_{k+1}, \dots, a_n$ in the new system of units will then be

$$a_1' = \alpha_1 a_1 \,, \qquad a' \quad = \alpha_1^{m_1} \alpha_2^{m_2} \dots \alpha_k^{m_k} a \,,$$

$$a_2' = \alpha_2 a_2 \,, \qquad a_{k+1}' = \alpha_1^{p_1} \alpha_2^{p_2} \dots \alpha_k^{p_k} a_{k+1} \,,$$

$$\dots\dots\dots\dots\dots\dots\dots\dots$$

$$a_k' = \alpha_k a_k \,, \qquad a_n' \quad = \alpha_1^{q_1} \alpha_2^{q_2} \dots \alpha_k^{q_k} a_n \,.$$

In this new system of dimensional units, Relation (7.8) assumes the form

$$\begin{aligned} \alpha' &= a' \alpha_1^{m_1} \alpha_2^{m_2} \dots \alpha_k^{m_k} a = \alpha_1^{m_1} \alpha_2^{m_2} \dots \alpha_k^{m_k} f(a_1, a_2, \dots, a_n) \\ &= f(\alpha_1 a_1, \alpha_2 a_2, \dots, \alpha_k a_k, \alpha_1^{p_1} \alpha_2^{p_2} \dots \alpha_k^{p_k} a_{k+1}, \dots, \alpha_1^{q_1} \alpha_2^{q_2} \dots \alpha_k^{q_k} a_n) \,. \end{aligned} \tag{7.9}$$

This equality shows that the function f possesses the property of homogeneity with respect to the dimensional units of the quantities $a_1, a_2, \dots, a_k$.

The set α_i will now be used to reduce the number of independent variables in f. Let

$$\alpha_1 = \frac{1}{a_1} \,, \qquad \alpha_2 = \frac{1}{a_2} \,, \dots, \alpha_k = \frac{1}{a_k} \,.$$

For such a choice of scales $\alpha_1, \alpha_2, \dots, \alpha_k$, the values of the first k arguments on the right-hand side of (7.9) will be equal to unity, independently of the numerical values of $a_1, a_2, \dots, a_k$.[1)] Thus, using the fact that (7.8), by assumption, does not depend on the choice of the system of dimensional units, a system of dimensional units will be selected for which k of the arguments of the function f have fixed constant values, equal to unity.

[1)] It has been assumed here, for the sake of simplicity, that the parameters a_1, $a_2, \dots, a_k$ are finite and non-zero; obviously, the following results may be extended to the case when $a_1, a_2, \dots, a_k$ may become zero or infinite, if the function f is continuous for these values of the arguments.

In this relative system of dimensional units, the numerical values of the parameters $a, a_{k+1}, \ldots, a_n$ are determined by

$$\Pi = \frac{a}{a_1^{m_1} a_2^{m_2} \ldots a_k^{m_k}},$$

$$\Pi_1 = \frac{a_{k+1}}{a_1^{p_1} a_2^{p_2} \ldots a_k^{p_k}},$$

$$\cdots\cdots\cdots\cdots$$

$$\Pi_{n-k} = \frac{a_n}{a_1^{q_1} a_2^{q_2} \ldots a_k^{q_k}},$$

where $a, a_1, \ldots, a_n$ are the numerical values of the quantities under consideration in the original system of dimensional units.

It is readily seen that the values of $\Pi, \Pi_1, \ldots, \Pi_{n-k}$ do not depend on the choice of the original system of dimensional units, since they have zero dimensionality with respect to the dimensional units $A_1, A_2, \ldots, A_k$. Likewise, it is obvious that the values $\Pi, \Pi_1, \ldots, \Pi_{n-k}$ do not, in general, depend on the choice of the system of those dimensional units in terms of which the k dimensional units are expressed for the dimensionally independent quantities $a_1, a_2, \ldots, a_k$. Consequently, the quantities $\Pi, \Pi_1, \ldots, \Pi_{n-k}$ may be considered to be non-dimensional. Therefore, in any system of dimensional units, Relation (7.8) may be presented in the form

$$\Pi = f(\underbrace{1, 1, \ldots, 1}_{k}, \Pi_1, \ldots, \Pi_{n-k}), \tag{7.10}$$

where Π and all arguments of f are non-dimensional.

Thus, the link, which does not depend on the choice of the system of dimensional units, between the $n+1$ dimensional quantities $a, a_1, \ldots, a_n$ with k independent dimensionalities may be presented in the form of relations between the $n+1-k$ quantities $\Pi, \Pi_1, \ldots, \Pi_{n-k}$, which are themselves non-dimensional combinations of the $n+1$ dimensional quantities. This general derivation of dimensional theory is known as the Π-theorem.

If some non-dimensional quantity is a function of a number of dimensional quantities, then this function may depend only on non-dimensional combinations consisting of all the dimensional quantities which determine it.

Obviously, one may replace in (7.10) the system of non-dimensional parameters $\Pi_1, \Pi_2, \ldots, \Pi_{n-k}$, by changing the form of f, by another system of non-dimensional parameters which are functions of the $n-k$ non-dimensional parameters $\Pi_1, \Pi_2, \ldots, \Pi_{n-k}$.

It is readily seen that it is impossible to construct from the n parameters $a_1, a_2, \ldots, a_n$, not more than k of which are parameters with independent dimensionality, more than $n-k$ independent non-dimensional combinations. This fact follows directly from the derivation of the relations (7.10), if one takes as a any selected non-dimensional combination determined by the quantities $a_1, a_2, \ldots, a_n$.

Every physical relation between dimensional quantities may be formulated as a relation between non-dimensional quantities. As a matter of fact, this circumstance is a source of useful applications of dimensional theory in the study of physical problems.

The smaller the number of parameters, determining a quantity under study, the more limited is the form of the functional dependence and the simpler is the conduct of an investigation. In particular, if the number of usable basic dimensional units is equal to the number of determining parameters, which in this case have independent dimensionalities, then this dependence is determined completely with the aid of dimensional theory apart from a constant multiplier.

In fact, if $n=k$, then it is impossible to form from the parameters $a_1, a_2, \ldots, a_n$ non-dimensional combinations; therefore, the functional dependence (7.10) may be presented in the form

$$a = C a_1^{m_1} a_2^{m_2} \ldots a_n^{m_n} ,$$

where C is a non-dimensional constant and the exponents $m_1, m_2, \ldots, m_n$ are readily determined with the aid of the dimensionality formula for a. The non-dimensional constant C in this case may be determined either experimentally or theoretically by solving a corresponding mathematical problem.

Obviously, dimensional theory must prove the more useful the larger is the number of basic dimensional units which may be selected.

It has been seen above that the number of basic dimensional units may be chosen arbitrarily; however, an increase of the number of basic dimensional units is linked to an introduction of additional physical constants which likewise must figure among the determining parameters. When the number of basic dimensional units is increased, the number of determining parameters is also increased; in the general case, the difference $n+1-k$, equal to the number of non-dimensional parameters in terms of which a physical relation is formed, remains constant.

An increase in the number of basic dimensional units may prove useful only in the case when it is clear from additional physical reasoning that the physical constants which arise for the introduction of the new fundamental dimensional units are not essential. For example, if one considers a phenomenon in which mechanical and thermal processes take place, then one may introduce for the measurement of amounts of heat and mechanical energy two different dimensional units, i.e., calorie and Joule; however, then one must also introduce the dimensional constant I, i.e., the mechanical heat equivalent. Let it now be assumed that one is dealing with a phenomenon of heat transfer in a moving incompressible ideal fluid. In that case, there does not occur a conversion of heat energy into mechanical energy and conversely; therefore the thermal and mechanical processes will proceed independently of the value of the mechanical heat equivalent. If one were to arrange the possibility of a change of the magnitude of the mechanical heat equivalent, then this would have no effect on the values of the characteristic quantities. Consequently, in the case under consideration, the constant I does not enter into the physical relations and an increase in the number of basic dimensional units makes it possible to obtain with the aid of dimensional theory additional important deductions.

§8. Parameters determining a class of phenomena and typical examples of the application of the methods of dimensional analysis

Elucidation of the system of determining parameters on the basis of the mathematical formulation of a problem. Applications of dimensional analysis are based on applications of the Π-theorem to problems under study. In this connection, there arises the problem of the enumeration of the arguments, i.e., of the determining parameters, in functions of the form (7.8).

In thermodynamics, in Chapter 5, the concept of the system of determining parameters characterizing the state and motion of a small particle of a medium has been introduced. Now it is necessary to introduce the system of determining parameters which follows from the formulation of a separate class of problems and characterizes completely for a given problem every global problem, taken separately.

A basic and initial stage in the formulation of a problem is the choice of a model or system of models of continuous media and a schematization of the properties of the unknown solutions. At this stage there takes place a study of the conditions of symmetry and the choice of a suitable co-ordinate system. Then the system of equations, the system and classes of unknown functions and independent variables are fixed.

Independent variables (for example, x, y, z, t) and physical constants of the type of coefficients of thermal conductivity, viscosity, modulus of elasticity, etc., must be included in the list of the system of determining parameters. In addition, for a given class of problems, one must include among the number of determining parameters specific dimensional and non-dimensional characteristics of the region $\mathscr{D}$, occupied by the moving medium. Then it is necessary to characterize and include in the number of determining parameters quantities which determine the given functions for the formulation of the boundary and initial conditions.

If a problem under consideration has been formulated as a mathematical problem, then it is always easy to write down the complete table of arguments in functions of the form (7.8) for unknown physical laws. This is a common process for obtaining the system of determining parameters when a problem has been posed mathematically.

The determining parameters are all the data which must be given for the computation of the unknown functions by different means, including the analysis of engines. In order to obtain the required answers with the aid of experiments, it is also necessary to state clearly and list all determining parameters. Only under such conditions, an experiment may be repeated and a comparison of different experiments executed.

For the isolation of the table of determining parameters, a mathematical formulation of a problem is not required. The above method for constructing the table of determining parameters on the basis of a mathematical formulation of a problem is, in general, not obligatory.

One may write down the system of determining parameters also in such cases when detailed properties of a model and system of equations are generally unknown. It is sufficient to rely on preliminary data or hypotheses in the form of functions and on constants which enter or may be introduced into the definition of a model, into initial, boundary and other conditions specifying a concrete problem.

Finite system of determining parameters. In studying a set of different mechanical systems undergoing certain motions, one may always restrict the class of admissible systems and motions in such a manner that a concrete system and its specific motion, which is of interest, is determined by a finite number of dimensional and non-dimensional parameters. These limitations may be imposed by specification of a number of abstract functions or constants given by the conditions of a problem.

The system of determining parameters must be complete. Dimensional analysis, based on application of the Π-theorem, allows deductions which follow from the possibility of application of arbitrary special dimensional units for the description of physical laws. In this context, for the listing of the parameters determining a class of phenomena, one must state all dimensional parameters, connected with the essence of a problem, independently of the fact whether they are constant or variable. It is important that the parameters may take different numerical values in different systems of dimensional units.

For example, the gravitational acceleration g must always be included in the system of determining parameters, when gravity is essential, in spite of the fact that g is identical for many real motions.

After g has been introduced as a determining parameter, one may, without all kinds of complications, take into account an artificial widening of the class of motions by varying the magnitude of the gravitational acceleration g. Such a method permits at times to become aware of and obtain valuable particular deductions on the influence of parameters which, by the Π-theorem, may only be introduced in combination with the gravitational acceleration g.

The system of determining parameters must be complete. Among the determining parameters, certain of which may be physical dimensional constants, there must be, without fail, quantities with dimensionalities in terms of which the dimensionalities of all unknown quantities of interest can be expressed.

If a system of determining parameters is incomplete from the point of view of their dimensionality, and its extension, in essence, rules out a formulation of the problem, this means that a determining quantity is either zero or infinity. Such cases are often encountered when initial conditions of the source type are specified with the aid of the Dirac δ-function.

Insufficiency of dimensional theory for the solution of problems. In the general case, the methods of study of functional relations with the aid of the Π-theorem are, in essence, restricted and insufficient, since it is impossible to establish with their help links between non-dimensional quantities. All derivations of dimensional theory retain their validity for any changes in the equations of motion, provided only one does not introduce for this purpose any new specified dimensional quantities. For example, one may multiply in the equations of motion different terms by certain positive or negative non-dimensional numbers or functions which depend on the accepted system of determining parameters.

Similar modifications, which do not affect the deductions of dimensional analysis, may influence the character of physical relationships in an essential manner.

The basic utilisation of dimensional analysis for theoretical and experimental studies is linked to the possibility of a description and study of physical laws in non-dimensional form, which is invariant with respect to the choice of a system of dimensional units.

The problem of the spherical piston in a gas. Consider the problem of a spherical piston which begins to expand at time $t=0$ inside a gas at rest with initial density ρ_1 and pressure p_1 from some initial radius r_0 in accordance with the given law $v_n(t)=dr_1/dt$.

In order to determine the perturbed, spherically symmetric, adiabatic motion of an ideal perfect gas, it is unnecessary to integrate the system (6.30) of three equations in the three unknowns ρ, p and v. Obviously, one may take as independent arguments of the unknown functions the quantities

$$\gamma, \rho_1, p_1, \rho_0, r, t$$

and the law of expansion of the piston

$$v_p(t)\,, \tag{8.1}$$

where $\gamma = c_p/c_V$ is a non-dimensional constant coefficient, the index of the adiabat. The coefficient γ enters into the equation of the adiabatic condition and into the condition on the shock wave. In order to obtain a table with a finite number of determining parameters, one may select a definite form of the law of expansion $v_n(t)$ of the piston which depends on a finite number of parameters. For example, if the piston expands with uniform acceleration or, in a more general case, if $v_n(t)$ is a polynomial, then one must take as determining parameters the coefficients in the corresponding polynomial.

Consider the simplest important case when the piston begins to expand suddenly at time $t=0$ with constant velocity

$$v_n = \text{const.} = v_0\,.$$

In this case, the system of determining parameters may be written in the form

$$\rho_1,\ p_1,\ r_0,\ \gamma,\ v_0,\ r,\ t$$

or

$$\gamma,\ v_0/a_1,\ r_0/r,\ r/a_1 t, \tag{8.2}$$

where $a_1 = \sqrt{\gamma p_1/\rho_1}$ is the velocity of sound in the undisturbed medium.

On the basis of the Π-theorem, one may write down for the unknown solution formulae of the form

$$\rho = \rho_1 f_1\left(\gamma\,, \frac{v_0}{a_1}\,, \frac{r_0}{r}\,, \frac{r}{a_1 t}\right),$$

$$p = p_1 f_2\left(\gamma\,, \frac{v_0}{a_1}\,, \frac{r_0}{r}\,, \frac{r}{a_1 t}\right),$$

$$v = v_0 f_3\left(\gamma\,, \frac{v_0}{a_1}\,, \frac{r_0}{r}\,, \frac{r}{a_1 t}\right), \tag{8.3}$$

where the non-dimensional functions f_1, f_2, f_3 involve four non-dimensional parameters two of which, namely r_0/r and $r/a_1 t$, are variable.

Equation (6.30) may now be rewritten in non-dimensional form in terms of the functions f_1, f_2, f_3. The equations remain partial differential equations in terms of the two independent variables

$$\xi = r_0/r\,, \quad \eta = r/a_1 t\,.$$

On the basis of (8.3), the problem under consideration may be essentially simplified, if one introduces the additional assumption that at the initial instant of time the expansion of the piston starts from a point, i.e., $r_0 = 0$. Then the characteristic linear dimension ($r_0 = 0$) drops out in Table (8.2); therefore one of the variable arguments ($\xi = r_0/r = 0$) disappears in (8.3). Consequently, in this case, the Π-theorem leads to the result that the unknown solution has the form

$$\rho = \rho_1 f_1\left(\gamma\,, \frac{v_0}{a_1}\,, \eta\right),$$

$$p = p_1 f_2\left(\gamma\,, \frac{v_0}{a_1}\,, \eta\right), \tag{8.4}$$

$$v = v_0 f_3\left(\gamma\,, \frac{v_0}{a_1}\,, \eta\right),$$

i.e., that the unknown functions may only depend on the variables r and t in the combination

$$\eta = r/a_1 t\,.$$

Thus, on the basis of the *Π*-theorem, it has been established that the solution of the problem of the spherical piston starting from a point is self-similar.

After substituting (8.4) into (6.30), one obtains ordinary differential equations for the functions f_1, f_2, f_3. The original system of non-linear partial differential equations has been greatly simplified, since one need now only find solutions of ordinary, although even non-linear equations.

Obviously, all the preceding deductions remain valid also for other problems of one-dimensional unsteady motions of a gas in which the system of determining parameters is given by Table (8.2) or an extended table with addition, for example, of the dimensional constant parameter q^*, corresponding to the specific energy released at the front of the detonation or combustion wave. Since q^* has the dimension of the square of a velocity, in the problem of the piston expanding from a point, i.e., for $r_0 = 0$ with a detonation or combustion wave, the formulae for the self-similar solutions in the region of the disturbed motion have the form

$$\rho = \rho_1 f_1\left(\gamma, \frac{v_0}{a_1}, \frac{q^*}{a_1^2}, \frac{r}{a_1 t}\right),$$

$$p = p_1 f_2\left(\gamma, \frac{v_0}{a_1}, \frac{q^*}{a_1^2}, \frac{r}{a_1 t}\right), \tag{8.5}$$

$$v = v_1 f_3\left(\gamma, \frac{v_0}{a_1}, \frac{q^*}{a_1^2}, \frac{r}{a_1 t}\right).$$

In this case, the number of constant parameters is increased; however, the motion is self-similar, since, in essence, it has only one variable parameter.

For a more detailed study of the self-similar problem of a piston, it has been shown that the piston causes a spherical shock wave which propagates into the undisturbed gas. If one considers the characteristics of the motion at the shock wave, one finds that the radius r_2 of the spherical shock wave for the self-similar problem is determined by the parameters

$$\gamma, v_0, a_1, q^*, t\,.$$

Since it is impossible to form from these quantities a non-dimensional combination, involving a variable quantity, i.e., the time t, one obtains for r_2 the formula

$$r_2 = a_1 t f\left(\gamma, \frac{v_0}{a_1}, \frac{q^*}{a_1^2}\right). \tag{8.6}$$

It follows from (8.6) that the front of the shock in these problems, in the absence as well as in the presence of energy releases at the jump, expands uniformly with constant velocity. This is a very important result which has been obtained without effectively finding a complete solution of the problem.

The preceding reasoning may be extended and consideration may be given to all possible formulations of problems leading to self-similar solutions. Note that self-similarity of a solution provides for absence of the characteristic linear dimension r_0 and for absence of constants from which a constant with the dimensionality of time could be formed.

If one considers instead of the problem of a piston with constant velocity the problem of a uniformly accelerated expansion of a piston, then the solution ceases to be self-similar and the corresponding gas-dynamical problem will belong to a class of more difficult problems.

The problem of the point explosion. Next, the reasoning of dimensional theory will be applied to the problem of the point explosion, posed in §6. It is easily seen that the system of determining parameters in this problem is

$$\gamma,\ \rho_1,\ p_1,\ E,\ r,\ t\,. \tag{8.7}$$

It is readily verified directly that the corresponding three non-dimensional independent parameters ($n=6$, $k=3$) may be taken in the form

$$\gamma, \qquad \lambda = \frac{\rho_1^{\frac{1}{5}} r}{E^{\frac{1}{5}} t^{\frac{2}{5}}}, \qquad \tau = \frac{p_1^{\frac{5}{6}} t}{E^{\frac{1}{3}} \rho_1^{\frac{1}{2}}}\,. \tag{8.8}$$

For the pressure distribution behind the front of the explosion shock wave, one has the form

$$p = \frac{E}{r^3} f(\gamma, \lambda, \tau)\,. \tag{8.9}$$

This formula contains the two variable parameters λ and τ; therefore the problem of the point explosion and its solution are not self-similar. The form of f in (8.9) may be sought, in particular, by means of numerical calculations on the basis of (6.30) and supplementary initial and boundary conditions. One may also pose the problem of an experimental determination of this function. In both cases, a result on the structure of the functional relation (8.9) has great practical and theoretical significance.

Naturally, these results will correspond to reality only in those cases when the assumptions are fulfilled on which the schematical treatment and formulation of the problem was based (disregarding radiation, homogeneity and assuming initial state of rest of the atmosphere, etc.). Note that fortunately in many important cases (however, not always) a given formulation of a problem corresponds, in practice, to reality and, consequently, is useful.

In order to establish the form of the function (8.9) (for fixed γ), it is sufficient to perform one single computation for certain concrete data. For example, if one computes for the case of an electric flash discharge in air, then this result permits determination of dimensional quantities for an atomic bomb explosion. It is sufficient to execute one single experimental explosion with known energy at a definite height, i.e., for known ρ_1 and p_1. Measurements during such an experiment make possible the computation and prediction of results for all other experiments with other energies E and other known data relating to a homogeneous gas atmosphere.

Each such computation or experiment allows to obtain information on the function $f(\gamma, \lambda, \tau)$ which, obviously, is sufficient for the description of a point explosion under any other conditions.

Formulation of the problem of a strong explosion. For concentrated charges, when one wishes to study the effect of an explosion at distances which are large compared with the dimensions of the charge, but still sufficiently close to the centre of the charge where the pressure due to the explosion is still very large, one may introduce into the formulation of the problem and into (8.9) further essential simplifications.

Experiments and theory show that in the case of an explosion there arises on the boundary of the region of disturbed motion of the gas a spherical shock wave which propagates quickly into the gas at rest. The pressure p_1 in the gas at rest has been included in the system of determining parameters (8.7). The effect of the pressure p_1 on the disturbed motion of the gas may only manifest itself through the boundary conditions at the shock wave. By (6.33) and (6.34), these conditions, solved with respect to the characteristics of the motion of the gas behind the front of the wave, may be rewritten in the form

$$v_2 = \mathscr{D}\left(\frac{2}{\gamma+1} - \frac{2\gamma}{\gamma+1}\frac{p_1}{\rho_1\mathscr{D}^2}\right),$$

$$\rho_2 = \frac{\gamma+1}{\gamma-1}\frac{\rho_1}{1+\dfrac{2\gamma p_1}{(\gamma-1)\rho_1\mathscr{D}^2}}, \tag{8.10}$$

$$p_2 = \frac{2}{\gamma+1}\rho_1\mathscr{D}^2\left[1 - \frac{(\gamma-1)p_1}{2\rho_1\mathscr{D}^2}\right].$$

By the conditions of the problem, one has here

$$v_{\tau 1} = v_{\tau 2} = 0\,, \quad v_{n2} = -v_2\,, \quad v_{n1} = -\mathscr{D}$$

and

$$a_1 = \sqrt{\gamma p_1/\rho_1}\,.$$

In (8.10), at the initial stage of a strong shock wave, the non-dimensional quantity $\gamma p_1/\rho_1 \mathscr{D}^2$ is very small compared with $\gamma p_2/\rho_1 \mathscr{D}^2$, since the pressure p_2 behind the front of the wave is significantly larger than the initial pressure p_1.

The inequality $p_2 \gg p_1$ characterizes the sharp pressure jump at the front of the wave, i.e., a strong explosion wave. Neglecting small quantities of order $a_1^2/\mathscr{D}^2$, Conditions (8.10) may be simplified for a strong shock wave and replaced by the approximate conditions

$$\begin{aligned} v_2 &= \frac{2}{\gamma+1}\mathscr{D}\,, \\ \rho_2 &= \frac{\gamma+1}{\gamma-1}\rho_1\,, \\ p_2 &= \frac{2}{\gamma+1}\rho_1 \mathscr{D}^2\,. \end{aligned} \tag{8.11}$$

The conditions (8.11) represent the conditions on a strong shock wave. When the pressure in front of the jump is zero ($p_1 = 0$), these conditions coincide with the exact conditions; this corresponds to a neglect inside the region bounded by the explosion wave of the initial energy in comparison with the energy released instantaneously during the explosion at the centre of symmetry.

If one uses in an analysis of the solution and for an effective solution of the problem of a strong point explosion instead of (8.10) Conditions (8.11), then the pressure p_1 (and, consequently, also τ) is excluded from Table (8.7) of determining parameters. Hence it follows that the gas motion for a strong point explosion is self-similar.

The distribution of the characteristics of the disturbed motion of a gas behind the front of an explosion wave in the case of a strong explosion may be written in the form

$$
\begin{aligned}
v &= v_2 \mathscr{V}(\gamma, \lambda), \\
p &= p_2 \mathscr{P}(\gamma, \lambda), \\
\rho &= \rho_2 \mathscr{R}(\gamma, \lambda),
\end{aligned}
\tag{8.12}
$$

where v_2, p_2 and ρ_2 are defined by (8.11).

Obviously, the radius r_2 and the velocity of the front of the explosion wave $\mathscr{D} = dr_2/dt$ are determined by the quantities

$$\rho_1, \gamma, E, t$$

among which the constant γ is non-dimensional and the parameters ρ_1, E and t have independent dimensionalities; it is impossible to form from them non-dimensional combinations. On the basis of the Π-theorem, one has now for r_2 and $\mathscr{D}$ the formulae

$$
\begin{aligned}
r_2 &= k\left(\frac{E}{\rho_1}\right)^{\frac{1}{5}} t^{\frac{2}{5}}, \\
\mathscr{D} &= \tfrac{2}{5} k\left(\frac{E}{\rho_1}\right)^{\frac{1}{5}} t^{-\frac{3}{5}} = \tfrac{2}{5} k^{\frac{5}{2}} \sqrt{\frac{E}{\rho_1}} \frac{1}{\sqrt{r_2^3}},
\end{aligned}
\tag{8.13}
$$

where k is some abstract constant which depends only on γ. The exponents of the dimensional quantities are determined from the condition that r_2 must be a quantity with the dimension of a length.

Formula (8.13), which has been obtained with the aid of dimensional analysis from the formulation of the problem without effective solution of the entire problem, determines the law of motion of the shock wave. For $t=0$, one finds that $r_2=0$ and the velocity $\mathscr{D}$ of the shock wave is infinite. As time passes, the radius r_2 grows and the velocity $\mathscr{D}$ drops. Obviously, the assumption of a strong shock wave is only applicable in the time interval for which $\mathscr{D} \gg a_1$. It follows from the general theory that the velocity $\mathscr{D}$ of propagation of the front of a shock wave by the particles is always larger then the velocity of sound in front of the wave.

It follows from (8.8) for λ and from (8.13) that one may take for λ the quantity

$$\lambda = r/r_2, \tag{8.14}$$

changing the region of disturbed motion to the interval (0, 1).

The problem of the determination of the disturbed motion reduces to the determination of the functions $\mathscr{V}(\gamma, \lambda)$, $\mathscr{P}(\gamma, \lambda)$, $\mathscr{R}(\gamma, \lambda)$ and the constant $k(\gamma)$. The conditions at the shock front acquire the simple form

$$\mathscr{V}(\gamma, 1) = \mathscr{P}(\gamma, 1) = \mathscr{R}(\gamma, 1) = 1 \,. \tag{8.15}$$

At the centre of symmetry, the condition of the absence of a source of mass reduces to the condition that the velocity of the gas particles must vanish, i.e.,

$$\mathscr{V}(\gamma, 0) = 0 \,. \tag{8.16}$$

The condition that the total energy in the region of disturbed motion of a gas must be constant assumes the form

$$E = \int_0^{r_2} \left(\frac{v^2}{2} + \frac{1}{\gamma - 1} \frac{p}{\rho} \right) \rho 4\pi r^2 \, \mathrm{d}r \,.$$

On the basis of (8.12) and (8.14), this condition becomes

$$1 = \frac{32\,\pi k^5}{25(\gamma^2 - 1)} \int_0^1 (\mathscr{V}^2 \mathscr{R} + \mathscr{P}) \lambda^2 \, \mathrm{d}\lambda \,. \tag{8.17}$$

It may be used in the computation of k.

Obviously, the functions $\mathscr{V}$, $\mathscr{P}$ and $\mathscr{R}$ satisfy ordinary differential equations. These equations must be solved in the interval $0 \leqslant \lambda \leqslant 1$, where the conditions (8.15) and (8.16) are the boundary conditions at the two ends of the interval.

This problem will not be solved here.[1)] It will only be noted that one may, for an effective solution of the problem without operation on the actually written down ordinary differential equations and with the aid of the reasoning of dimensional analysis, write down two finite integrals of the system of ordinary differential equations, and thus obtain in simple finite form the exact solution of the problem of the strong point explosion. Figure 60 gives the graphs of these solutions.

Motion of a rigid body in an infinite mass of viscous incompressible fluid. Among the most fundamental central problems of continuum mechanics, one has the problems of the motion of bodies inside fluids and gases, of disturbed motions of fluids and gases due to bodies and of the

1) A complete study and solution is contained in L. I. Sedov, *Methods of similarity and dimensional analysis in mechanics,* Academic Press, 1959.

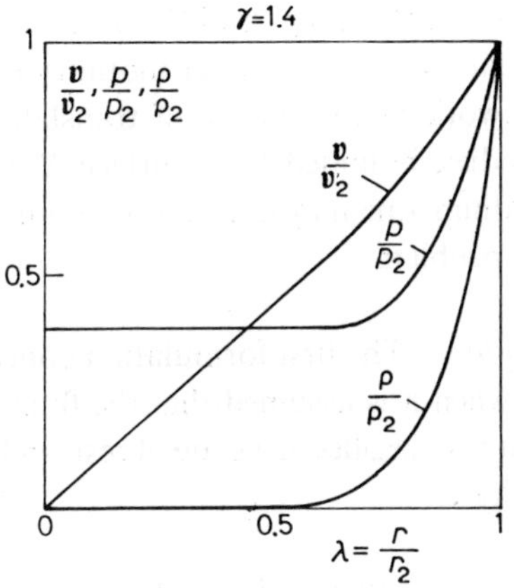

Fig. 60. Distributions of pressure, velocity and gas behind the front of the shock wave of a strong point explosion.

forces of their interaction with bodies.

At the boundaries of bodies, in contact with an external mobile continuous medium, there arises a system of forces of interaction. Great practical significance attaches to the properties of these forces of interaction, to their dependence on the laws of motion of the bodies, on the geometrical shapes and other peculiarities of the moving system of bodies. In technical problems, connected with the computation of the motions of all possible objects and devices in water and air, with the equilibrium of all possible technical structures, for example, of houses and towers, dams and pipes, etc., the forces of interaction of these objects and buildings with the surrounding media are most important. From a general formulation of these problems and dimensional analysis, one may derive certain general results which are important for the methods of analysis of different objects and for the execution of experiments.

Consider the basic schematic treatment of the problem of translation with constant velocity of an absolutely rigid body inside a weightless incompressible fluid, which is in contact with the body and fills all the space external with respect to the surface Σ of the body. Assume that the shape of Σ is arbitrary, but fixed, and that all geometrical dimensions of Σ are determined completely by specification of only one characteristic dimension d. In practice, this is an important typical case of the motion of a rigid body; at the same time, it is also necessary to consider the problems of motion of systems of deforming "rigid" and other bodies, of motions which are not translatory, of accelerated motions, etc.

Within the framework of the theory of translatory motion inside a medium of finite bodies, bounded by a surface Σ which represents the boundary of the medium, one may consider two fundamental equivalent formulations of the problem.

Absolute motion. The first formulation concerns the problem of "absolute" motion, when it is assumed that the fluid or gas at infinity in front of the body is at rest, i.e., lies in a state of rest, and the body translates with constant velocity $\mathbf{v}$.

Conversion of a motion. The problem of stream line flow. A second formulation corresponds to the problem of translatory flow of a fluid or gas around a fixed body, when the velocity $-\mathbf{v}$ at infinity in front of the body is constant.

By the principle of Galilei–Newton, addition of the translatory velocity $-\mathbf{v}$ to the entire system (external medium and body) is equivalent to the transition from one inertial system to another. Consequently, in these two formulations, all interaction forces are identical, the relative velocity field in the problem of the stream line flow is obtained by adding at all points to the vector of the absolute velocity the vector $-\mathbf{v}$.

Such an equivalence forms the basis for many experimental methods of investigation of this problem. In place of experiments with bodies in flight or swimming in water, bodies have been held fixed in wind tunnels, in flumes and other devices. Obviously, the requirement of complete equivalence of the problems of flight and stream line flow is linked in practice to a neglect of the influence of other bodies, in particular, of the walls of aerodynamic wind tunnels, tubes, etc. In cases in which this effect is not small, it must be taken into account in a special manner.

In the general formulation of a problem, the conditions at infinity are formulated for the state and velocity in front of the body. This circumstance is linked to the fact that in a study of steady motion, as the limit of an infinitely long continuing unsteady motion, the path pursued by a body in the limit is infinitely long. Therefore, at infinity, in the wake of a body, the motion of a liquid or gas is, in general, disturbed. One has to deal with such a position in the theory of wings with finite span, in the theory of ship motions and in many other cases.

System of the parameters which determine the velocity and stress fields in the case of the motion of a body in a viscous fluid. Next, consider the motion of a body in a viscous incompressible homogeneous fluid. It follows from the general equations of Chapter IV that the mechanical properties of a viscous incompressible fluid are completely determined by two constants: the density ρ and the viscosity μ. Two, chemically different, viscous incompressible fluids with identical $\rho=$const. and $\mu=$ const. are indistinguishable from the point of view of mechanics. The problem of steady flow of a viscous incompressible fluid around fixed bodies presents itself as a problem of the integration of the Navier–Stokes equations for conditions of adherence of the fluid to the surface of the body and for the condition that the velocity $-\boldsymbol{v}$ and the pressure p_∞ at infinity in the counterflow are specified. For the non-linear Navier–Stokes equations, this mathematical problem is very difficult; not even a particular exact solution of this problem for any body of the simplest shape is known.

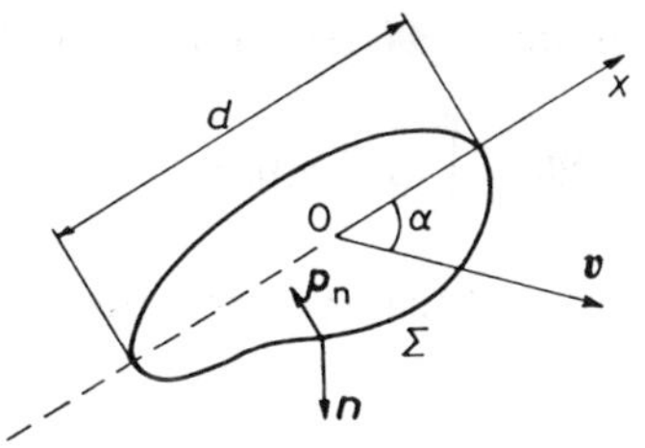

Fig. 61. The problem of the motion of a body in a fluid.

Nevertheless, there are a number of theoretical deductions and experimental data which make possible estimates of the character of the forces of interaction and of the properties of the fluid flow, and to obtain some ideas on the influence of the shape of a body on the magnitude of the forces of interaction between the fluid and the body.

In a system of coordinates, linked to a body, the steady velocity field of relative or absolute motion of a viscous incompressible fluid and likewise the distributions of pressure and internal viscous stresses are obtained as functions of the system of parameters

$$\rho, \mu, d, v, \alpha, \beta, p_\infty, x, y, z\,, \tag{8.18}$$

where α and β are the angles which determine the orientation of the vector of the constant speed of translation of the body with respect to the coordinate system fixed in the body.

Lift force and resistance. The fluid exerts on each element of the surface Σ of a body a surface force $\boldsymbol{p}_n d\sigma$ (by assumption, mass forces will not be taken into consideration). From the point of view of the resultant effect of the action of these elementary forces on a body, considered as rigid, great significance attaches only to the total force $\boldsymbol{P}$ (resultant vector) and the total moment $\mathfrak{M}$, determined by the integrals

$$\boldsymbol{P} = \int_\Sigma \boldsymbol{p}_n d\sigma\,, \qquad \mathfrak{M} = \int_\Sigma (\boldsymbol{r} \times \boldsymbol{p}_n) d\sigma\,. \tag{8.19}$$

The resultant force vector $\boldsymbol{P}$ may be presented in the form of the sum

$$\boldsymbol{P} = \boldsymbol{W} + \boldsymbol{A}\,.$$

The force $\boldsymbol{W}$, parallel to the velocity $\boldsymbol{v}$, if it is pointed in the opposite direction to $\boldsymbol{v}$, is referred to as resistance; the force $\boldsymbol{A}$, perpendicular to the velocity $\boldsymbol{v}$, is called lift. The forces $\boldsymbol{W}$ and $\boldsymbol{A}$ and the moment $\mathfrak{M}$ may be determined on the basis of theoretical calculations with the aid of direct or indirect computations of the integrals (8.19) or with the help of experiments with measurements of forces by weighing, for example, in wind tunnels or in special flumes or during experiments of a different kind.

In experiments and in theory, the forces $\boldsymbol{W}$ and $\boldsymbol{A}$ may be considered as quantities determined by the parameters.

$$\rho,\ \mu,\ d,\ v,\ \alpha,\ \beta,\ p_\infty\,. \tag{8.20}$$

A change of the pressure p_∞ does not affect the total force. It is readily seen that in the determination of the resultant forces the magnitude of the pressure p_∞ at infinity is not essential. In fact, the pressure enters into the Navier–Stokes equations for an incompressible fluid only by means of its derivatives with respect to the coordinates; therefore, on addition of a constant to the pressure, while the velocity field remains unchanged, one solution goes over into another solution. Hence it is seen that the pressure in a flow is an additive function of the pressure p_∞ at infinity. Note that this deduction is true only for incompressible fluids. On the other hand, it follows from the theorem of Gauss–Ostrogradskii that for each closed surface Σ bounding a volume V, one has

$$-\int_{\Sigma} p_{\infty}\boldsymbol{n}\,\mathrm{d}\sigma$$

$$= -p_{\infty}\int_{\Sigma}[\boldsymbol{i}\cos(\boldsymbol{n}, x)+\boldsymbol{j}\cos(\boldsymbol{n}, y)+\boldsymbol{k}\cos(\boldsymbol{n}, z)]\,\mathrm{d}\sigma$$

$$= -p_{\infty}\int_{V}\left(\frac{\partial \boldsymbol{i}}{\partial x}+\frac{\partial \boldsymbol{j}}{\partial y}+\frac{\partial \boldsymbol{k}}{\partial z}\right)\mathrm{d}\tau = 0$$

$$-\int_{\Sigma}\boldsymbol{r}\times p_{\infty}\boldsymbol{n}\,\mathrm{d}\sigma$$

$$= -p_{\infty}\int_{\Sigma}[\boldsymbol{r}\times\boldsymbol{i}\cos(\boldsymbol{n}, x)+\boldsymbol{r}\times\boldsymbol{j}\cos(\boldsymbol{n}, y)+\boldsymbol{r}\times\boldsymbol{k}\cos(\boldsymbol{n}, z)]\,\mathrm{d}\sigma$$

$$= -p_{\infty}\int_{V}\left[\frac{\partial(\boldsymbol{r}\times\boldsymbol{i})}{\partial x}+\frac{\partial(\boldsymbol{r}\times\boldsymbol{j})}{\partial y}+\frac{\partial(\boldsymbol{r}\times\boldsymbol{k})}{\partial z}\right]\mathrm{d}\tau = 0$$

Hence it follows that the assumption of incompressibility permits to exclude from Table (8.20) of determining parameters the quantity p_{∞}. From the remaining six parameters, one may form only the three non-dimensional parameters

$$\alpha,\ \beta,\ \mathsf{R} = \rho v d/\mu\,. \tag{8.21}$$

The angles α and β characterize the direction of the velocity $\boldsymbol{v}$ of the body with respect to Σ. If the body is a sphere, then the angles α and β are inessential; in other cases, the orientation of the velocity vector $\boldsymbol{v}$ relative to the body is essential and therefore the angles α and β are essential parameters. The non-dimensional number R is known as Reynolds number. Reynolds number plays a fundamental role in all phenomena which are connected with the viscosity of fluids and gases.

It is easily verified that the combinations $\rho d^2 v^2$ and $\mu d v = \rho d^2 v^2/\mathsf{R}$ have the dimension of force. On the basis of the Π-theorem, one may now write

$$W = c_W(\alpha, \beta, \mathsf{R})\rho d^2 v^2\,,$$

$$A = c_A(\alpha, \beta, \mathsf{R})\rho d^2 v^2\,,$$

$$\mathfrak{M} = c_{\mathfrak{M}}(\alpha, \beta, \mathsf{R})\rho d^3 v^2\,. \tag{8.22}$$

The coefficients c_W, c_A and $c_{\mathfrak{M}}$ depend only on α, β and the Reynolds number $\mathbf{R}$.[1] The determination of these coefficients for bodies of various shapes as functions of the arguments stated represents one of the principal problems of theoretical and experimental aerodynamics and hydrodynamics.

Today, there exist many data on these coefficients for a large number of different bodies encountered in practice.

The influence of the viscosity of the fluid on the motion and, in particular, on the resistance and lift forces is only manifested through Reynolds' number. It follows from Formula (8.21) for Reynolds' number that the influence of the viscosity, manifesting itself through the coefficient of viscosity μ, is closely linked to the density ρ, the velocity v and the linear scale d. For different abstract functions, the effect of an increase in the viscosity is equivalent to a decrease either of the linear dimensions or of

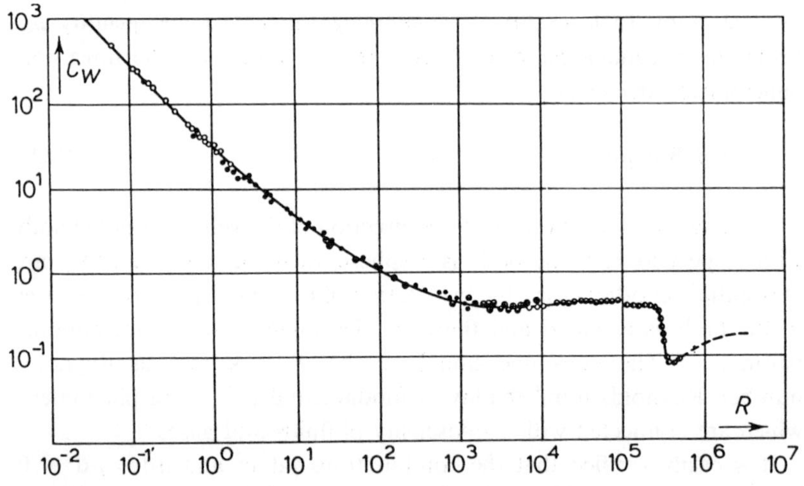

Fig. 62. Coefficient of resistance of a sphere $c_W = W(\frac{1}{4}\pi d^2 \frac{1}{2}\rho v^2)^{-1}$ as function of Reynolds' number $\mathbf{R} = \rho v d/\mu$, where d is the diameter of the sphere. ○ Schiller-Schmiedel, 1928 ● Lebster, 1924 ⊗ Allen, 1900 ◒ Göttingen, 1921 ◐ Göttingen, 1926 --- ram pressure tube 1922/29.

[1] Note that, for the sake of simplicity, only coefficients for the moduli of the force A and the moment $\mathfrak{M}$ have been quoted; in practice, one must consider analogous coefficients for the components of these vectors.

the velocity or of the density of a fluid. Obviously, an increase in the scale of a body or of the velocity of a motion for fixed viscosity is equivalent to a decrease of the viscosity for fixed dimensions and velocity.

Figure 62 shows experimental data on the influence of Reynolds' number on the coefficient of resistance c_W for a sphere for different ranges of values of Reynolds' number. Experimental data, obtained in different liquids and in air, fall well on to the same curve. For small values of Reynolds' number, the equation of this curve has the form $c_W = c/\mathbf{R}$; Fig. 63 shows

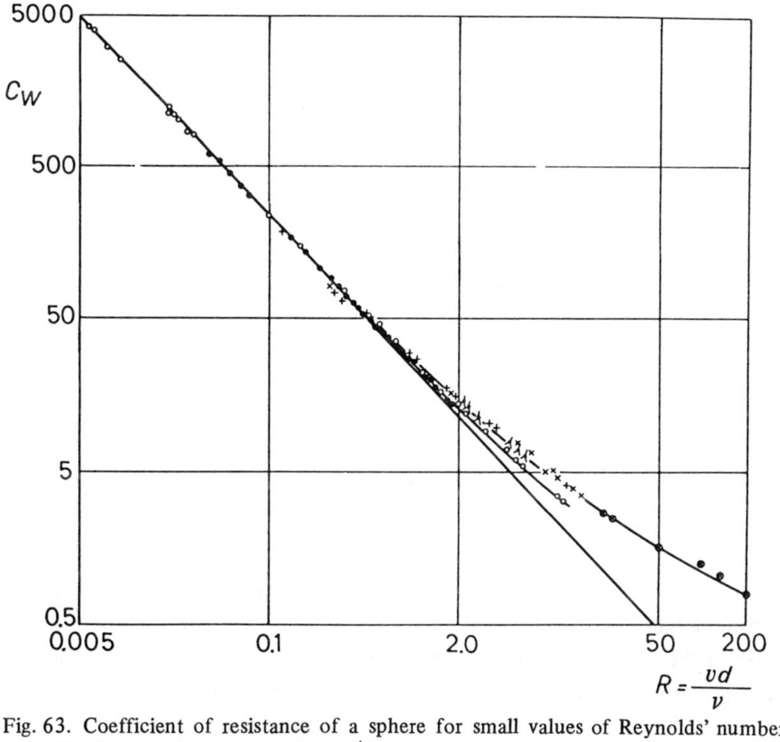

Fig. 63. Coefficient of resistance of a sphere for small values of Reynolds' number. Logarithmic scale along abscissa ($\nu = \mu/\rho$).

Arnold:
- ● small balls made of Rose alloy in molten butter,
- ○ small balls made of wax in alcohol,
- × air bubbles in water,
- \+ air bubbles in aniline,

Allen:
- λ small balls made of paraffin in aniline,
- ⊗ amber balls in water.

the corresponding straight line, since the abscissa has a logarithmic scale. Figure 64 presents examples of typical curves of the dependence of the coefficients c_W and c_A for a wing as functions of the angle of attack. i.e., of the angle of inclination of the velocity of motion of the wing to its profile.

With the aid of (8.22) and of the dimensionless coefficients c_A and c_W, determined by single experiments, for example, for the motion of a

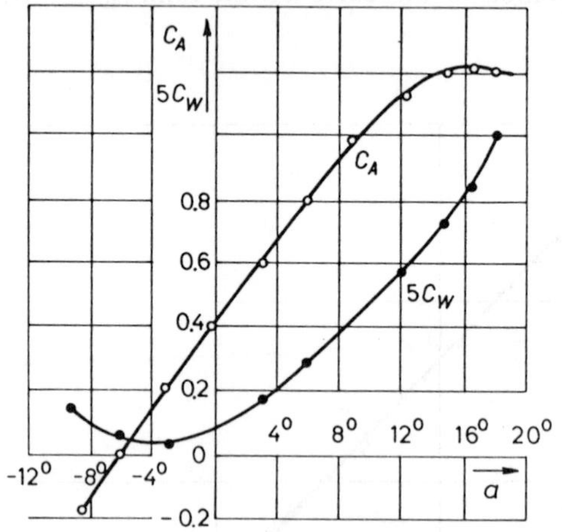

$$C_A = \frac{A}{\frac{\rho v^2}{2} S} \; ; \; C_W = \frac{W}{\frac{\rho v^2}{2} S}$$

Fig. 64. Typical curves showing the dependence of the lift coefficient c_A and the resistance coefficient c_W for a wing at an angle of attack α. The area of the planform of the wing is S.

given body in water, one may compute in other cases, in which no experiment has been performed, the resistance and lift forces for another body of the same geometrical shape, but with other dimensions, moving in other fluids or even in air, if the compressibility of the air may be neglected. Obviously, it is necessary for this type of computations to have available and utilize data for c_W and c_A for the same values of the angles α and β and for identical values of Reynolds' number $R = \rho v d/\mu$. In preliminary experiments, one must obtain data for c_W and c_A for the necessary ranges of the non-dimensional arguments α, β and R which are determined by the conditions of the applications to practical problems.

Motions in a viscous fluid for low Reynolds numbers. Consider the case of slow motion in a viscous fluid, corresponding to low Reynolds' numbers. A lowering of Reynolds' number corresponds to an increase in the coefficient of viscosity μ and of the internal stress forces, caused by the viscosity. If one neglects inertia forces in comparison with viscous forces, then this is equivalent to an assumption relating to the inessential nature of the density ρ as a determining parameter. Neglect of the density in the Navier–Stokes equations corresponds to an omission of the acceleration term. The corresponding mathematical problems and their solution constitute the approximate Stokes' theory.

In this case, the resistance W for translatory motion of a body of given shape with constant velocity v is determined by the parameters

$$\mu,\ d,\ v,\ \alpha,\ \beta\,. \tag{8.23}$$

Since it is impossible to form from μ, d and v non-dimensional combinations, it follows that

$$W = c(\alpha, \beta)\mu d v = \frac{c}{\mathsf{R}}\rho d^2 v^2\,, \qquad c_W = \frac{c(\alpha, \beta)}{\mathsf{R}}\,. \tag{8.24}$$

Consequently, for low Reynolds' numbers, the resistance force (and, analogously, the lift force) is proportional to the first power of the velocity of the motion of the body, to the linear dimension d and the coefficient of viscosity μ. The non-dimensional coefficient c depends only on the direction of the velocity of the body relative to its surface. For a sphere, the coefficient c is constant, as may be shown by one single experiment. Theoretical computations for the sphere yield $c = 3\pi$, if d is the diameter of the sphere. For bodies of arbitrary shape, it follows from the first formula (8.24) that $c_W = c(\alpha, \beta)/\mathsf{R}$, a formula which gives the dependence of c_W on Reynolds' number for small Reynolds' numbers. Experiments well confirm this formula and make it possible to state the largest values of Reynolds' number, depending on the shape of the body, for which (8.24) is practically completely applicable.

The resistance law (8.24) may, for example, be applied to a description of the process of settling of small particles in a fluid. However, for motions of submarines in water, of aircraft in air and even of motorcars, on account of their large dimensions and velocities, the corresponding Reynolds' numbers are larger and therefore (8.24) is unrealistic.

Motion in a viscous fluid for very large Reynolds numbers. In an ideal fluid, one has $\mu=0$; therefore an ideal fluid corresponds to an infinitely large value of Reynolds' number R. Large values of R may be considered for $\mu\neq 0$; however, then $vd\to\infty$, i.e., one is dealing with bodies of large dimensions and, generally speaking, at large velocities. In an ideal fluid, the resistance and lift forces are determined by the parameters

$$\rho,\ d,\ v,\ \alpha,\ \beta\,, \qquad \mathsf{R}=\infty\,. \tag{8.25}$$

Consequently, in an ideal fluid for stationary flow, one must satisfy the formulae

$$W=c_W(\alpha,\beta)\rho\, d^2 v^2\,, \qquad A=c_A(\alpha,\beta)\rho\, d^2 v^2\,, \tag{8.26}$$

which show that in an ideal fluid or, approximately, in a viscous fluid for $\mathsf{R}\to\infty$, the resistance and lift forces are proportional to the square of the velocity v^2, to the characteristic area d^2 and the density of the fluid. In many important cases, these relationships, derived only from the formulation of a problem with the aid of the Π-theorem, well correspond to experiment and always agree exactly with theoretical calculations within the framework of the hydrodynamics of an ideal incompressible fluid.

Ship resistance. In the case of motion of a ship on the surface of water, part of the boundary of the fluid is a free surface. The disturbed motion of the water in this case depends on the properties of its heaviness, due to which the surface of water covers itself with waves. The lift and resistance during the motion of a ship depends on the property of heaviness of the water. Therefore, in the case of translatory motion of a ship with constant horizontal velocity and fixed orientation with respect to the free surface of the fluid which fills all of the lower half-space, the system of determining parameters is given by the table

$$\rho,\ \mu,\ g,\ d,\ v\,. \tag{8.27}$$

Out of these parameters ($n=5$, $k=3$), one may construct the two independent non-dimensional combinations

$$\frac{\rho v d}{\mu}=\mathsf{R}\,, \qquad \frac{v}{\sqrt{gd}}=\mathsf{F}\,, \tag{8.28}$$

where F is referred to as Froude's number.

Froude's number is an essential non-dimensional parameter in all cases in which one has v, d and g in the table of determining parameters. The influence of the gravitational force is taken into consideration and manifests itself for non-dimensional quantities through Froude's number.

The formula for the resistance of the body of a ship may be written in the form

$$W = c_W(\mathsf{R}, \mathsf{F})\rho S v^2 , \tag{8.29}$$

where S is a characteristic area, which is, as a rule, taken equal to the wetted area of the external surface of the body, when the ship is in equilibrium on the surface of the water. The determination of the dependence of c_W on R, F and the geometrical shape of the body is a basic problem of naval hydrodynamics.

The non-dimensional parameters R and F are not only encountered as characteristic arguments in the problems discussed above; they are also employed and play important roles in many other problems in which one finds among the determining parameters the quantities ρ, μ, g, v and d, i.e., generally speaking, always when the property of viscosity and heaviness of a medium is essential. In particular, Reynold's number has fundamental significance in problems of the motion of viscous fluids through tubes.

Problem of the motion of a body in an ideal gas. The above formulation of the basic problems of the steady motion of a body in an infinite mass of fluid will now be modified in the following manner: Assume that the external medium is an ideal perfect gas. Take the property of compressibility into account and let the processes in each particle be adiabatic, reversible in the region of continuous motion and irreversible for a passage of the gas particles through a condensation front the presence of which in the flow is admitted.

In the problem of the flow around a fixed body, there are specified at infinity the pressure p_∞, the density ρ_∞ and the flow velocity v_∞. In this case, the distribution of the characteristics of state and motion of the gas are determined by the parameters

$$\gamma = \frac{c_p}{c_V}, \ p_\infty, \ \rho_\infty, v_\infty, \alpha, \beta, d, x, y, z \ . \tag{8.30}$$

Select for the space coordinates x, y, z a system which is fixed in the body. It is seen from (8.30) that $n = 10$ and $k = 3$. Obviously, all non-dimensional unknown quantities are determined by the six non-dimensional parameters

$$\gamma,\ v_\infty/a_\infty = \mathrm{M}_\infty,\ \alpha,\ \beta,\ x/d,\ y/d,\ z/d \tag{8.31}$$

with

$$a_\infty = \sqrt{\frac{\gamma p_\infty}{\rho_\infty}},$$

where, as before, α and β are the angles specifying the orientation of the counter flow with respect to the body, a_∞ is the velocity of sound at infinity in the counter flow and M_∞ the Mach number. In the problem under consideration, the Mach number M_∞ plays a role which is analogous to that of Reynolds' number and Froude's number in the problems considered above.

In particular, the linear dimension d may not be among the parameters (8.30), for example, in a study of the problem of the flow around an infinite wedge or cone, when the origin of the coordinate system is taken at the vortex of the cone or wedge. Then there arises again the property of self-similarity; instead of the three non-dimensional arguments x/d, y/d, z/d, one obtains the two non-dimensional arguments y/x and z/x.

In the problem under consideration, the formulae for the resistance and lift forces have the form

$$W = c_W(\alpha, \beta, \gamma, \mathrm{M}_\infty)\rho d^2 v_\infty^2, \qquad A = c_A(\alpha, \beta, \gamma, \mathrm{M}_\infty)\rho d^2 v_\infty^2. \tag{8.32}$$

Obviously, in an ideal compressible gas, the dependence of the forces W and A on the square of the velocity v is violated because of the influence of the Mach number. Formulae (8.32) are true for subsonic ($\mathrm{M}_\infty < 1$) as well as for supersonic ($\mathrm{M}_\infty > 1$) velocities of the counter flow. In the case of supersonic velocities, one may have condensation jumps in the flow. The functions $c_W(\alpha, \beta, \gamma, \mathrm{M}_\infty)$ and $c_A(\alpha, \beta, \gamma, \mathrm{M}_\infty)$ may be determined by computations on the basis of solutions of a hydrodynamic problem or with the aid of experiments in shock tubes, using special gas dynamic devices, or in free flight.

Problem of Boussinesq. Consider the classical problem of Boussinesq in the theory of elasticity. Let there be given a homogeneous elastic medium which obeys Hooke's law and fills a half-space. On the plane which forms the free surface of the elastic half-space, one has, by assumption, no external surface forces.

The initial, undeformed state of the half-space corresponds to a state in which any external loading and internal stresses are absent. A deformed state is brought about by application of a concentrated force $\mathscr{P}$ at some point of the free surface of the elastic half-space. For the sake of simplicity, assume that therefore $\mathscr{P}$ is perpendicular to the plane of the boundary in the undeformed state Fig. 65. It is required to determine the equilibrium state of stress and deformation in the elastic half-space under the assumption that the displacements tend to zero as one moves from the point O to infinity. Obviously, the solution of this problem possesses axial symmetry with the axis of symmetry passing through the force vector.

Consider this problem within the framework of the linear theory of elasticity with the linearized boundary condition at the free surface

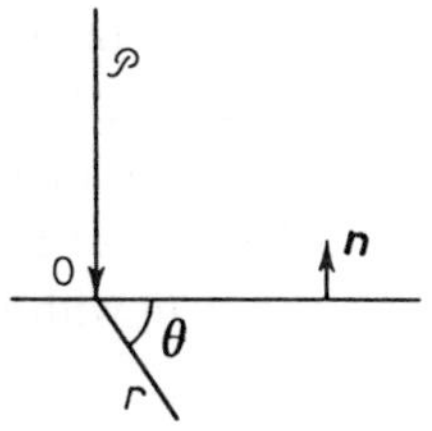

Fig. 65. On the problem of Boussinesq.

transferred to the original undisturbed boundary plane. This boundary condition has the form

$$p_n = 0$$

at all points of the free surface except at O where p_{nn} becomes infinite in such a manner that

$$p_{nn} = \mathscr{P}\delta(r)\,, \tag{8.33}$$

where $\delta(r)$ is Dirac's δ-function.

It is obvious from the equations of motion, the supplementary condition (8.33), and the condition at infinity that the displacement vector and the components of the internal stress tensor are determined by the parameters

$$\sigma,\, E,\, \mathscr{P},\, r,\, \theta\,, \tag{8.34}$$

where r and θ are polar coordinates with centre at O in planes passing through the force vector $\mathscr{P}$, σ is Poisson's coefficient and E Young's modulus. The equilibrium equations are partial differential equations in the two independent variables r and θ. One may form from the five parameters (8.34) only the three independent non-dimensional combinations

$$\sigma, \theta, \frac{\mathscr{P}}{Er^2}, \tag{8.35}$$

since Young's modulus has the dimension of pressure. Hence it follows that all unknown non-dimensional quantities depend on the three parameters (8.35).

Important significance attaches to the linearization of the problem. Since the magnitude of the force $\mathscr{P}$ enters into the boundary condition in a linear manner, it is obvious that the displacement and internal stress fields depend linearly on $\mathscr{P}$, i.e., all unknown components of the displacement vector and all components of the internal stress tensor are simply proportional to the parameter $\mathscr{P}/Er^2$. Therefore, knowing the dimension of the unknown quantities, one may determine completely their dependence on the radius vector r. For example, one may write down for the displacement vector $\boldsymbol{w}(\mathscr{P}, E, \sigma, r, \theta)$ with the dimension of length the expression

$$\boldsymbol{w} = \frac{\mathscr{P}}{Er^2} r\boldsymbol{f}(\sigma, \theta), \tag{8.36}$$

where $\boldsymbol{f}(\sigma, \theta)$ is some vector which depends only on one single variable coordinate.

Substitution of (8.36) into the equilibrium equations leads to ordinary differential equations for the unknown non-dimensional functions. These considerations greatly simplify the problem. The required solution is readily obtained by integration of the corresponding ordinary differential equations.

§9. Similarity and modelling of phenomena

Modelling and physical similarity. The theory of dimensionality and similarity is very important for the modelling of various phenomena.

Modelling is the replacement of the study of a phenomenon occurring in nature by a study of an analogous phenomenon on a model of smaller or larger scale, as a rule, under special laboratory conditions. The fundamental significance of modelling is that one may obtain from results of experiments with models required information on the character of effects and on different quantities connected with a phenomenon under natural conditions.

In a majority of cases, modelling is based on a study of physically similar phenomena. An investigation of a natural phenomenon of interest is replaced by a study of a physically similar phenomenon which is more convenient and more easily realized. Mechanical or, in general, physical similarity may be considered as a generalization of geometrical similarity. Two geometrical figures are similar, if the ratios of all corresponding lengths are identical. If the coefficient of similarity, i.e., the scale is known, then simple multiplication of the dimensions of one geometrical figure by the scale factor yields the dimensions of the other geometrical figure, similar to it.

Two phenomena are physically similar, if through given characteristics of the one one may obtain the characteristics of the other phenomenon by simple computations which are analogous to the transition from one system of dimensional units to another. In order to realize these computations, one must know the "transfer scale".

Numerical characteristics of two different, but similar, phenomena may be considered as numerical characteristics of one and the same phenomenon expressed in two different systems of dimensional units. For each set of similar phenomena, all corresponding non-dimensional characteristics (non-dimensional combinations of dimensional quantities) have identical numerical values. It is readily seen that the converse is likewise true, i.e., if all corresponding non-dimensional characteristics for two motions are identical, then the motions are similar. The set of mechanically similar motions determines itself the regime of the motions.

Similarity of two phenomena may at times be understood in a wider sense by assuming that the above definition relates only to a certain special system of characteristics completely determining a phenomenon and permitting the computation of any other characteristics which, however, it is impossible to obtain by simple multiplication by a scale factor corresponding to the transition from one to the other "similar" phenomenon. For example, in this sense, any two ellipses may be assumed to be similar, if one uses Cartesian coordinates directed along their principal axes. By the computations stated one can obtain the Cartesian coordinates of the points of any ellipse from those of the points of any other ellipse (affine similarity).

In order to conserve similarity during modelling, one must observe certain conditions. However, in practice, quite often these conditions, ensuring similarity of a phenomenon as a whole, are not fulfilled; then there arises the question concerning the magnitude of the accuracy (scale effect) which occurs during a transfer to nature of results obtained on models.

Similarity criteria. After establishment of a system of parameters which determine the subdivision of classes of phenomena, it is not difficult to arrive at similarity conditions for two phenomena.

In fact, let a phenomenon be determined by n parameters several of which may be non-dimensional. Further, assume that the dimensionalities of the determining variables and physical constants are expressed in terms of k dimensionalities of these parameters with independent dimensionalities ($k \leqslant n$). Obviously, one may then construct from the n quantities $n-k$ independent non-dimensional combinations. All non-dimensional characteristics of a phenomenon may be considered as functions of these $n-k$ independent non-dimensional combinations composed of determining parameters. Consequently, one may always indicate among all non-dimensional quantities composed of the characteristics of a phenomenon a certain base, i.e., a system of non-dimensional quantities which determine themselves all the remaining ones.

A definite corresponding formulation of the problem of classes of phenomena contains phenomena which are not, in general, similar between themselves. A separation from it of a subclass of similar phenomena may be realized with the aid of the following condition.

For similarity of two phenomena, it is necessary and sufficient that the numerical values of the non-dimensional combinations, forming the base, in these two phenomena be identical.

The conditions that the bases of the abstract parameters constituted from given quantities determining a phenomenon be constant are referred to as similarity criteria.

If the conditions of similarity are fulfilled, then it is necessary to know for an actual computation of all characteristics of a natural phenomenon from data on the dimensional characteristics on models the transition scale factors for all corresponding quantities. If a phenomenon is determined by n parameters k of which have independent dimensionalities, then for quantities with independent dimensionalities the transition scale factors may be arbitrary and must be specified with due consideration to the conditions of a problem, and for experiments also with consideration of the experimental conditions. The transition scale factors for all remaining dimensional quantities are readily obtained from formulae expressing the dimensions of each dimensional quantity in terms of the dimensions of the k quantities with independent dimensionalities for which the scale factors are fixed by the conditions of an experiment and the formulation of a problem.

Similarity for the flow of a viscous incompressible fluid around bodies. In the problem of flow of a viscous incompressible fluid around a body, all non-dimensional quantities are determined by three parameters: the angles α and β and Reynolds' number R. The conditions of physical similarity, i.e., the similarity criteria are thus given by

$$\alpha = \text{const.}\,, \quad \beta = \text{const.}\,, \quad \mathsf{R} = \rho v d/\mu = \text{const.}\,,$$

where it must be understood that the constant values of α, β and R express the equality of these quantities in different similar (corresponding) phenomena.

In order to model phenomena, results of experiments with a model may be carried over into nature only for identical α, β and R. The first two conditions are readily achieved in practice. It is more difficult to satisfy the third condition ($\mathsf{R}=$const.), especially in those cases when the body in the flow has large dimensions, as for example the wing of an aircraft. If a model is smaller than in nature, then conservation of Reynolds' number demands either an increase in the basic flow velocity, which, as a rule, is practically impossible, or, in essence, a change of the density and viscosity of the fluid. In practice, these circumstances lead to great difficulties in the study of aerodynamic resistance. The necessity for Reynolds' number to remain constant has brought about the construction of gigantic open aerodynamic tunnels in which air may be blown past full size aeroplanes, and likewise of tunnels of the closed type in which compressed, i.e., denser air is circulated at high velocities.

Special theoretical and experimental studies have shown that in a number of cases of aerodynamically well shaped bodies Reynolds' number noticeably influences only the non-dimensional coefficients of frontal resistance and sometimes very weakly the non-dimensional lift coefficient and certain other quantities which play a very important role in different practical questions. Consequently, the difference in the Reynolds numbers between model and large scale conditions is sometimes not essential.

Similarity in the flow around bodies taking into account compressibility of the gas. So far the similarity conditions have been stated for motions of bodies when the compressibility of air is not taken into account; the compressibility effect is inessential for velocities which are small in comparison with the velocity of sound.

In the aerodynamics of flight at large subsonic and supersonic velocities, the influence of compressibility manifests itself in the first place through the Mach number. In the above formulation of the problem of motion in an unlimited mass of gas, under adiabatic conditions, of an aerofoil or, generally speaking, of a body of any shape, the system of abstract determining parameters is given by (8.31). The parameters which determine the global characteristics of the gas flow or the characteristics of motion and state at characteristic points are

$$\alpha, \beta, \gamma, \mathsf{M}_\infty = v_\infty / a_\infty .$$

In modelling, one must ensure that these parameters have the same values in nature and in the model experiment. Obviously, the constant γ and Mach number M_∞ are the most essential similarity criteria. The values of γ are linked to the choice of the gas properties. For one and the same gas, the condition $\gamma=$const. is fulfilled automatically. The constancy of M_∞ may be ensured by the basic conditions for which an experiment is designed.

On the difficulties of modelling. By changing different parameters in modelling, one encounters in model experiments different effects which are inessential for a natural phenomenon of interest in the formulation of the problem introduced. For example, for steady air flows around wings at low velocities, the Mach number is small; hence the compressibility of air is unimportant. In this case, the basic similarity criterion is Reynolds' number. In order to model flows in aerodynamic wind tunnels using air, the desire to conserve the value of $\mathsf{R}=\rho v d/\mu$ when d is decreased leads, in general, to the need to increase on a small model the velocity v. An increase of the velocity of the basic flow leads to an increase in M_∞. On small models, Mach's number will be larger than in nature and, although in nature the influence of compressibility may be inessential, on a model it may be accentuated, and hence similarity violated. Such circumstances represent the basic difficulties in the design of aerodynamic experiments and highlight a number of requirements which must be taken into consideration in the construction of wind tunnels.

As other examples of essential effects, which may appear in model experiments and may be absent in a corresponding study of natural phenomena, may serve cavitation which arises during the motion of bodies in water, and condensation of gas in experimental situations. The occurrence of these effects is connected with the lowering of the dimensional values of pressure and temperature in certain regions of a moving medium. (Cavitation, i.e., evaporation of water in regions of low pressure, and condensation of air in aerodynamic tunnels may be the result of very sharp temperature drops with adiabatic expansion of the gas particles in certain areas of the gas flow.) In order to eliminate cavitation in water, one must increase the "non-essential" external pressure p_∞ at infinity. For an elimination of condensation of a gas, it is necessary to increase in the basic flow the temperature T_∞ which is inessential from the point of view of similarity criteria in the original formulation of a problem. In this connection, there occurs in aerodynamic tunnels with large supersonic velocities significant heating of the working gas.

Modelling of ship motions. In order to model ship motions for a study of resistance and to answer many other hydrodynamic questions (spray formation, rolling motion, flooding during motion through an aroused sea, etc.) one must undertake hydrodynamic experimentation with small ship models in special towing tanks.

It follows from the above formulation of the problem and from (8.28) that for similarity between nature and models one must ensure that Reynolds' number R and Froude's number F are constant. However, it is easily seen that for water with $g=\text{const.}$ it is impossible to satisfy simultaneously the conditions

$$\mathsf{R} = \rho v d/\mu = \text{const.}\,, \quad \mathsf{F} = v/\sqrt{gd} = \text{const.}$$

and to decrease the dimension d. The requirements to keep R and F constant imply that the velocity of the model must be larger and smaller, respectively, than that of the full scale ship.

Modelling keeping Froude's number constant. Thus, strictly speaking, complete mechanical similarity for the modelling of ship motion when g and $\nu=\mu/\rho$ are conserved is impossible. However, deep insight into the essence of hydrodynamic phenomena has shown that in many problems the influence of Reynolds' number may be taken into account with the aid of supplementary calculations or simple experiments. In the hydrodynamics of ordinary vessels, fundamental significance attaches to Froude's number F, and therefore modelling is performed in such a manner that F is kept constant.

Modelling of states of equilibrium of elastic structures. Next, consider the problem of modelling of states of equilibrium of elastic structures. Let there be given some structure made of homogeneous material, for example, a girder of a bridge. The elastic property of the isotropic material is determined by two parameters: Young's modulus $E[kg/m^2]$ and the non-dimensional Poisson coefficient σ. Consider geometrically similar structures and construct the table of determining parameters.

For a determination of all dimensions of the model, it is sufficient to specify some characteristic dimension d. If in the state of equilibrium under consideration the weight of the structure is essential, then the specific weight of the material $\gamma=\rho g[kg/m^3]$ must appear as a determining parameter. Beside the force of the weight of a part of a structure, it is acted upon, as a rule, still by external loading distributed in some definite manner over the structure. Let the magnitude of this loading be determined by the force $\mathscr{P}[kg]$. Thus, the system of determining parameters is

$$\sigma, E, d, \rho g, \mathscr{P} .$$

In this case, one has $n=5$, $k=2$; consequently, the base for mechanically similar states of elastic equilibrium consists of the three non-dimensional parameters

$$\sigma, E/\rho g d, \mathscr{P}/Ed^2 ,$$

which for similarity must be constant for the model and the large scale structure. In order to meet these conditions, all deformations must be similar. If the model is n times smaller than the full structure, then all displacements on the model will be n times less than in the full scale object.

If a model and the full scale structure are made of one and the same material, then the values of ρ, σ and E are the same for both; therefore one must satisfy, in order to achieve mechanical similarity, the condition

$$gd = \text{const}\,.$$

Under normal conditions, one has g=const.; consequently, one must have d=const., in order to achieve mechanical similarity, i.e., the sizes of the model and the full scale structure must be identical. In other words, modelling is impossible for constant g.

Modelling using a centrifugal table. A change in g may be achieved artificially by rotating a model at constant angular velocity on a so called centrifugal or rotating table. For sufficiently small dimensions of the model and large radius of rotation, the centrifugal inertia forces acting on the elements of the model may be assumed to be parallel. Using rotation about a vertical axis, one finds that in a state of relative equilibrium of the model (with respect to the rotating table) there act on it constant mass forces, analogous to the gravitational forces, but with a different acceleration. By the choice of the angular velocity of rotation, one may obtain any larger values of the acceleration.

In recent times, rotating tables have been constructed which are employed for studies on models of different processes[1)] which occur in soils.

Consider the stress τ $[kg/m^2]$ arising during deformation of an elastic structure under the action of its weight and a given load distribution. One may understand by τ the maximum value of any stress component or, in general, some stress component acting at a definite point of a structure.

The combination τ/E is non-dimensional; therefore one may write

$$\tau/E = f(\sigma,\ E/\rho g d,\ \mathscr{P}/Ed^2)\,.$$

If a model and its full scale structure are made of the same material, one

1) The condition $E/\rho g d$ = const. must be satisfied for modelling of processes in which side by side with other essential parameters one has the parameters ρ, g and E. Therefore, in all such cases, one may model with the aid of a rotating table.

has $E = \text{const.}$; hence for mechanically similar states, stresses at corresponding points will be identical.

If it is assumed that the state of stress is mechanically similar and that failure is determined by the values of the maximum stress, then it is obvious that in the model and the full scale object failure occurs at corresponding points.

If the external loading is high, and the own weight of the structure is small, so that it may be neglected, then the parameter $\gamma = \rho g$ (and, consequently, also the parameter $E/\rho g d$) is unimportant. In this case, the preceding ratio assumes the form

$$\frac{\tau}{E} = f\left(\sigma, \frac{\mathscr{P}}{Ed^2}\right),$$

and the similarity conditions consist only of the two equalities

$$\sigma = \text{const.}, \quad \frac{\mathscr{P}}{Ed^2} = \text{const.}$$

Hence it follows likewise that in order to model preserving material properties one must change the external loading proportionally to the square of the linear dimensions.

Small scale structures are stronger. Let l denote the change in length during deformation of some element of an elastic system. For structures of a definite large class, there holds true the relation

$$l/d = \varphi(\sigma, \rho g d/E, \mathscr{P}/Ed^2).$$

In a number of cases, it is directly clear from physical considerations that the quantity l/d is increased with an increase of the specific weights of structural elements, i.e., for increases of the parameter $\rho g d/E$.

Next, select two geometrically similar structures of different dimensions, made of the same material (E and σ identical). Assume that the external loading is changed proportionally to the square of the dimensions, i.e., that

$$\mathscr{P}/Ed^2 = \text{const.}$$

Obviously, in this case, the parameter $\rho g d/E$ increases with an increase in the dimensions of a structure; consequently, mechanical similarity is violated. In structures of smaller dimensions, relative deformations will be smaller; therefore such structures will be stronger. However, this deduction is only true when the specific weight $\gamma = \rho g$ of a material plays an essential role. If γ is unimportant and $\mathscr{P}/Ed^2 = \text{const.}$, then the relative deformations have the same values for bodies in different scales.

Consider still the case when γ is inessential and it is known that for a given structure the ratio l/d increases as the external load $\mathscr{P}$ is increased. If the external loading is proportional to the cube of the linear dimensions, then obviously for small scale structures the ratio l/d will be less than for large scale structures. Consequently, in this case, a decrease in dimension increases the strength.

In certain cases, modelling may be achieved by means of experiments with phenomena which are deliberately asimilar, when certain non-dimensional parameters $\Pi_1, \Pi_2, \ldots$ have different values for the model and the full scale object, but, at the same time, the form of the dependence of the unknown non-dimensional quantities on the determining non-dimensional parameters $\Pi_1, \Pi_2, \ldots$ is known beforehand from supplementary arguments. In these cases, one must keep for modelling purposes only those non-dimensional parameters constant the dependence of which is unknown.

At times, this method of modelling may be used when the form of the dependence of the unknown quantities on $\Pi_1, \Pi_2, \ldots$ brings out a working hypothesis which may be confirmed or disproved already after performance of a model study. As has been stated above, an example of such modelling is presented in a number of cases by modelling for different values of Reynolds' number when its effect on the unknown characteristics is inessential; however, this very method may also be applied in those cases when Reynolds' number is essential, but the dependence on it is known beforehand.

A study with the aid of models presents often the only possible means for an experimental investigation and solution of most phenomena which take place in the course of tens, hundreds, or even thousands of years; in model tests, a similar phenomenon may last only hours or days. Such a situation arises in the modelling of phenomena of infiltration of oil. One may also encounter the opposite case of a study of extraordinarily fast proceeding natural phenomena which one wants to replace by a similar model phenomenon taking much longer time.

Modelling is a crucial scientific problem which has, in general, principal and acknowledged significance, but it must be considered only as an initial basis for an original problem which consists of the effective determination of laws of nature, of the search for general properties and characteristics of different classes of phenomena, of the development of experimental and theoretical methods of investigation and solution of various problems and, finally, of the derivation of systematic materials, methods, rules and suggestions for the solution of concrete practical problems.

Nonlinear tensor functions of several tensor arguments [1)]

V. V. Lokhin and L. I. Sedov

Many fundamental geometrical and physical concepts are represented by scalar or tensor quantities. The mathematical formulation of a wide variety of laws of a geometrical or physical nature is accomplished with the aid of scalar or vector relations.

The tensorial expression of equations permits the formulation of laws which are independent of the choice of coordinate systems. Tensor characteristics and tensor equations have additional invariant properties and special peculiarities when the geometric or physical phenomena, objects, laws, and properties admit some symmetry.

Methods are developed below for automatically taking symmetry properties into account both in linear and nonlinear problems by suitable defining parameters which are associated with the basic assumptions in the formulation of the problem under study. Appropriate conclusions are arrived at concerning the effects of symmetry by the use of methods which are analogous to those developed in the closely related theory of similarity and dimensional analysis [1].[2)]

1) PMM, Vol. 27, No. 3, 1963, pp. 393–417.

2) Numbers in square brackets [ ] refer to the literature at the end of this appendix.

The present work is devoted to the solution of two basic problems.

(a) It is shown that the properties of textured media and crystals can be specified with the aid of tensors. Simple systems of tensors are actually determined as parametric geometrical quantities which define and specify the symmetry properties for all seven types of oriented media and all 32 classes of crystals.

(b) The general form is determined for the expression of tensors of arbitrary order when these tensors may be regarded as functions of a system of arguments consisting of a number of scalars and several independent tensors of various orders.

Both problems are intimately related to the consideration of the system of coordinate transformations which generate some symmetry group.

Symmetry problems play a fundamental role in physics. The specialization of the forms of functions and of tensors of various orders which are invariant under suitable symmetry groups is investigated in many works. The appropriate conclusions are applied and have contributed to the discovery of new effects in a number of different applications.

A summary of the basic data for different concrete examples is contained in a book by Nye [2]. Detailed references to the earlier literature may be found in the same book.

In algebra, a general theory is developed for obtaining and describing the properties of polynomial scalar invariants under finite transformation groups. These polynomials are formed from the components of tensors and vectors. It has been shown [3] that for every finite orthogonal group G there always exists an integral rational basis (integrity basis) of invariant polynomials. This integrity basis is a finite number of scalar invariant polynomials formed from the components of given tensors and vectors in such a way that any invariant polynomial formed from the same components can be expressed in terms of them.

An integrity basis forms a system of invariants with respect to the finite number of transformations of the group G. It is apparent, however, that its elements, polynomials in the components of given tensors, are not, in general, invariant under any arbitrary coordinate transformation, although such invariants are included in the basis. The number of elements of an integrity basis, which depends only on the group and on the choice of given tensors and vectors, is generally larger than the number of independent variable components of the given system of tensors and vectors. Therefore, the elements of an integrity basis are, in general, functionally dependent.

The actual construction of an integrity basis for the groups associated with oriented media and crystals has been carried out in works by Döring [4], Smith and Rivlin [5]. Pipkin and Rivlin [6], and Sirotin [7, 8]. It is shown below that, in order to construct tensor functions, it is necessary and sufficient to use a complete system of functionally independent simultaneous invariants [9, 10] formed from the components of the tensors which specify the symmetry groups and the other tensor arguments.

The construction of examples of scalars and tensors with specified symmetry is given in papers by Smith and Rivlin [5, 11, 12], in the book by Bhagavantam and Venkatarayudu [13], and in the works of Jahn [14], Shubnikov [15, 16, 17], and Sirotin [7, 8, 18, 19]. In a paper by Koptsik [20], various tensors of a physical nature are considered. He defines the symmetry of a crystal as "the intersection group of the symmetries of the existing properties of a crystal which are observed at a given instant" [20, p. 935].

Tensors which are functions of tensor arguments are considered in the case of second-order tensors. In this case, functional relations between tensors lead to functional relations between square matrices. In this area, the fundamental results reduce to the Cayley–Hamilton formula and to its generalizations to several matrix arguments [21–24, 25–28] (second-order tensors). However, basically, these generalizations consider only polynomial functions of matrices and components of tensors.

§ 1°. Fundamental concepts

As is well known, tensors may be regarded as invariant objects which are independent of the choice of coordinate system and which may be defined by the scalar components on a suitable basis. A tensor basis may be introduced in various ways; in particular, the polyadic product of the base vectors of a coordinate system in some manifold-space can always be taken as a basis.

For the sake of simplicity, consider only tensors in three-dimensional space. Let x^1, x^2, x^3 be coordinates of a point of the space and $\mathbf{э}_1, \mathbf{э}_2, \mathbf{э}_3$

be the vectors of a covariant basis.[1] Denote a tensor of order r by $\boldsymbol{H}$ and its components in the coordinate basis $\mathbf{з}_1, \mathbf{з}_2, \mathbf{з}_3$ by $H^{\alpha_1 \dots \alpha_r}$. In this paper, use will be made of the representation of the tensor $\boldsymbol{H}$ in the form of the sum

$$\boldsymbol{H} = H^{\alpha_1 \dots \alpha_r} \mathbf{з}_{\alpha_1 \dots \alpha_r}, \tag{A.1.1}$$

where summation occurs with respect to all indices $\alpha_1, \dots, \alpha_r$, which assume the values 1, 2, 3. In the general case, Formula (A.1.1) contains 3^r linearly independent terms, each of which may be considered as a special tensor.

Note that different continuous manifolds and the corresponding different base vectors can be introduced for a single coordinate system. For the same coordinates x^i and the same components $H^{\alpha 1 \dots \alpha_r}$, it is possible to consider different tensors corresponding to the various bases. In particular, such manifolds may be considered as different states of a given medium having a concomitant Lagrangian coordinate system which moves and deforms with time [26]. Cases are also possible where for a given Lagrangian coordinate system the various manifolds which correspond have different metrics. Thus, it is possible to consider simultaneously different tensors with given components which are the same, but on different bases and in different spaces, some of which may be Euclidean and others non-Euclidean (Kondo, Kröner, Bilby, and others).

The theory below will be developed for tensors in metric spaces. Denote the distance between two points with the coordinates x^i and $x^i + \mathrm{d}x^i$ by $\mathrm{d}s$. Let the quantity $\mathrm{d}s^2$ be defined by the formula $\mathrm{d}s^2 = g_{\alpha\beta}\,\mathrm{d}x^\alpha\,\mathrm{d}x^\beta$. The matrix $\|g_{ij}\|$ forms the covariant components of the fundamental metric tensor $\boldsymbol{g}$. The inverse matrix $\|g^{ij}\|$ gives the contravariant components. The contravariant base vectors $\mathbf{з}^i$ are determined from the formulae $\mathbf{з}^i = g^{i\alpha}\,\mathbf{з}_\alpha$.

The following formula are valid for the fundamental metric tensor $\boldsymbol{g}$:

$$\boldsymbol{g} = g_{\alpha\beta}\mathbf{з}^\alpha \mathbf{з}^\beta = g^{\alpha\beta}\mathbf{з}_\alpha \mathbf{з}_\beta = \delta^\alpha_\beta \mathbf{з}_\alpha \mathbf{з}^\beta, \tag{A.1.2}$$

where δ^i_j is the Kronecker delta.

Raising and lowering of the subscripts of the components of the various tensors is accomplished with the aid of the g_{ij} and g^{ij}. Formula (A.1.1) can be presented in the form:

[1] The coordinate system is arbitrary.

$$H = \sum_{s=1}^{p} k_s H_s \,, \tag{A.1.3}$$

where k_s are scalars and H_s are certain tensors of order r. Below it will always be assumed that the tensors H_s are linearly independent. It is obvious that $p \leqslant 3^r$.

Let the components of the tensor H be functions of the components of the m tensors

$$T_\varkappa = T_\varkappa^{\alpha_1 \ldots \alpha_{\rho_\varkappa}} \mathbf{э}_{\alpha_1} \ldots \mathbf{э}_{\alpha_{\rho_\varkappa}} \,, \qquad (\varkappa = 1, \ldots, m) \,, \tag{A.1.4}$$

the functions remaining the same, independently of the choice of the coordinate system. The integers $\rho_1, \ldots, \rho_m$ determine the orders of the tensors $T_\varkappa$. In the general case, $\rho_1, \ldots, \rho_m$ are different and are not equal to r. By definition, the tensor H will then be called a function of the tensors $T_1, \ldots, T_m$. The tensors $T_\varkappa$, among which there may be both variable and constant tensors, are the tensor arguments of the tensor function H.

If it is possible to form 3^r linearly independent tensors H_s of order r from the tensors $T_\varkappa$, then the tensor H will satisfy (A.1.3), in which the scalars k_s will depend only on the simultaneous invariants of the system of tensors $T_\varkappa$ and possibly on given additional scalar arguments.

Below, only those tensor functions will be considered for which the tensor g is included among the tensor arguments $T_\varkappa$.

The tensors H_s can be constructed from the tensors $T_\varkappa$ with the aid of two tensor operations: multiplication and contraction.[1)] The operation of contraction with respect to any two indices is always possible by virtue of the presence of the tensor g among the tensor arguments. Any multiplication of several tensors leads to a tensor the order of which is equal to the sum of the orders of the factors. Contraction with respect to l indices lowers the order of a tensor by $2l$. Multiplication and an obvious

[1)] It can be shown that such construction of the linearly independent tensors H_s of order r from the tensors $T_\varkappa$, when the covariant and contravariant components of the metric tensors g_{ij} and g^{ij} are among the tensor arguments, and as a consequence of it, there exist the components of the tensors $T_\varkappa$ with all possible structures of indices, is always possible. This statement is valid also in other, more general cases. Here such a proof is not necessary, because the practical method is given below for all possible forms of the orthogonal symmetry in the three-dimensional space with the aid of multiplication and contraction of the tensor arguments of p linearly independent tensor H_s. Together with it, from the general theory it follows that any $p + 1$ tensors of order r, admitting the given symmetry, are linearly dependent.

contraction of a given tensor $\boldsymbol{T}$ with the components $T^{ikjl\cdots}$ by the tensor $\boldsymbol{S}$ with components $\delta^i_n \delta^j_m$ results in the tensor $\boldsymbol{T}^*$ of the same order with the components

$$T^{*ikjl\cdots} = T^{jkil\cdots}.$$

The tensor $\boldsymbol{T}^*$ is called an isomer of the tensor $\boldsymbol{T}$. The operation of interchange of indices can be reduced to multiplication by the fundamental tensor and contraction. By definition, a tensor obtained as a result of permutation of several indices is also called an isomer of the tensor $\boldsymbol{T}$.

Methods are given below to construct general formulae of the type (A.1.3) for tensor functions. To this end, it is required to construct a linearly independent tensor basis $\boldsymbol{H}_s (s = 1, \ldots, p)$ in terms of the tensor arguments (A.1.4). The construction of the basis $\boldsymbol{H}_s$ from the defining tensors will be accomplished with the aid of the operations of multiplication and contraction.

§ 2°. Symmetry groups of tensors

The contravariant components $A^{\alpha_1 \ldots \alpha_r}$ of the tensor $\boldsymbol{A}$ admit the symmetry group G which is specified by the coordinate transformation matrices[1)]

$$\| a^{i}_{\cdot j} \| \left(a^{i}_{\cdot j} = \frac{\partial y^i}{\partial x^j} \,, \; y^i = y^i(x^j) \right) ,$$

if for each matrix of the group G

$$A^{i_1 \ldots i_r} = A^{\alpha_1 \ldots \alpha_r} a^{i_1}_{\cdot \alpha_1} \ldots a^{i_r}_{\cdot \alpha_r} . \tag{A.2.1}$$

If the fundamental tensor $\boldsymbol{g}$ admits a group, the group is called orthogonal. In other words, the transformation matrices of orthogonal groups satisfy the equivalent systems of equations

$$g^{ij} = g^{\alpha\beta} a^{i}_{\cdot \alpha} a^{j}_{\cdot \beta}, \quad g_{ij} = g_{\alpha\beta} a^{\alpha}_{\cdot i} a^{\beta}_{\cdot j} . \tag{A.2.2}$$

It is easy to verify that, if the group G is orthogonal, then the components

1) For the sake of simplicity, the enumeration of the elements of the matrices of the group G is omitted, so that $a^{i}_{\cdot j}$ is written instead of $a^{i}_{(\nu)\cdot j}$, where $\nu = 1, \ldots, h$ and h is the number of elements of the group G.

of the tensor A having any structure of the indices are invariant[1)] under the coordinate transformations generating the group G, provided that the condition (A.2.1) is met for the contravariant components of A. Therefore, for orthogonal transformations, it is possible to speak simply of symmetry of a tensor or of invariance of all its components relative to the group G.

The set of all orthogonal transformations under which a tensor $\boldsymbol{A}$ is invariant forms the symmetry group of the tensor $\boldsymbol{A}$. The symmetry group of a tensor may consist of only the identity transformation. For an arbitrary second-order tensor (non-symmetrical, $A^{ij} \neq A^{ji}$), the symmetry group consists of two elements: the identity transformation and the transformation of central inversion. For an arbitrary symmetric second order tensor, the symmetry group coincides with the group of self-transformations of a general ellipsoid. If the tensor ellipsoid is an ellipsoid of revolution, the symmetry group is infinite. A spherical (isotropic) tensor of second order has a symmetry group which coincides with the full orthogonal group of transformations, just like the fundamental tensor $\boldsymbol{g}$.

Consider several tensors $\boldsymbol{T}_1, \ldots, \boldsymbol{T}_m$ and denote their respective symmetry groups by $G_1, \ldots, G_m$. The group G which is formed by the intersection of the groups $G_1, \ldots, G_m$ is called the symmetry group of the set of tensors $\boldsymbol{T}_1, \ldots, \boldsymbol{T}_m$. It is not difficult to see that the tensor $\boldsymbol{H}(\boldsymbol{T}_1, \ldots, \boldsymbol{T}_m)$ will admit the symmetry group G. This follows from the fact that the components of the tensor $\boldsymbol{H}$ are functions of the components of the tensors $\boldsymbol{T}_i$ which are invariant with respect to the group G. Therefore the components of the tensor $\boldsymbol{H}$ will also be invariant with respect to the group G. In this connection, it is obvious that the symmetry group of a tensor, which is obtained as a result of the operations of multiplication and contraction of several tensors, will either coincide with the intersection of the symmetry groups of the component tensors or will possess greater symmetry and contain this intersection as a subgroup.

If the tensor $\boldsymbol{H}$ admits the symmetry group G, then the number of linearly independent terms p in (A.1.3) is, in general, less that 3^r. For a given group G and for a tensor of given order r, the number p can be computed using the theory of group characters [13, 14, 30]. Tables

2) If the group G is not orthogonal, it does not follow from (A.2.1) that the components of the tensor $\boldsymbol{A}$ with another structure of the indices are invariant.

suitable for the symmetry groups of oriented media and crystals are given in [13, 14, 18].

If a tensor $\boldsymbol{H}$ of odd order admits only the trivial group G consisting of the identity transformation, the number of terms is $p=3^r$; in this case, the tensor has the most general form. If the tensor $\boldsymbol{H}$ is of even order, its symmetry group consists of at least two elements: the identity transformation and central inversion. For symmetry groups consisting of only central inversion and the identity transformation, one has $\boldsymbol{H}=0$ for odd r and, therefore, $p=0$. For even r, $p=3r$ and, in this case, the tensor of even order has the most general form.

The scalar coefficients k_s in (A.1.3) are, in the general case, functions of the common invariants of the tensors $T_1, \ldots, T_m$ and of any number of given scalars (e.g., temperature, concentration, etc.). Some of the common invariants may be constant parameters, others may be variable.

The complete system of common invariants [9, 10] of the system of tensors $T_1, \ldots, T_m$ will be noted by $\Omega_1, \ldots, \Omega_N$. It follows from the completeness of the system of invariants that every invariant J, formed from the components of the system of tensors $T_1, \ldots, T_m$, satisfies the functional relation

$$J=f(\Omega_1, \ldots, \Omega_N)\,.$$

By definition, the invariants Ω_i retain their values in their same forms as functions of the components for any of the transformations of coordinates. These invariants can be obtained with the aid of the operations of multiplication and contraction. In this case, the invariants are homogeneous polynomials [9, 10] in the components of the tensors $T_1, \ldots, T_m$.

Assume that among the tensors $T_1, \ldots, T_m$ the tensors $T_\nu, \ldots, T_m$ $(1<\nu\leqslant m)$ are constant parametric tensors. Let the set of tensors $T_\nu, \ldots, T_m$ admit the finite symmetry group G^*.

Fix the values of the components of the tensors $T_\nu, \ldots, T_m$ given in the coordinate system x^i. After this is done, the invariants Ω_i reduce to ω_i, which are functions only of the components of the tensors $T_1, \ldots, T_{\nu-1}$. The equations $\omega_i = \Omega_i$ are true only in the coordinate system x^i.

In other coordinate systems, these equations are not, in general, satisfied. However, the equations will be satisfied for all coordinate transformations determined by the group G^*, since for these transformations all the tensors $T_\nu, \ldots, T_m$ are invariant. The quantities ω_i will not, in general, be invariant under any arbitrary coordinate transformation. It is clear that some ω_i, namely those which depend only on the components of the tensors $\boldsymbol{T}_\nu, \ldots, \boldsymbol{T}_m$ or only on the components of the tensors $\boldsymbol{T}_1, \ldots, \boldsymbol{T}_{\nu-1}$, will not depend on the coordinate transformation. It is obvious that all the quantities ω_i, as functions of the components of the tensors $\boldsymbol{T}_1, \ldots, \boldsymbol{T}_{\nu-1}$, may be regarded as invariant with respect to the group G^*. Thus, the invariant coefficients k_s in (A.1.3) will be functions of the Ω_i. The quantities k_s may be considered as functions of only the invariants ω_i under coordinate transformations in the group G^*.

The invariants ω_i are analogous to the invariants of an integrity basis. The quantities ω_i coincide with an integrity basis for proper choice of the complete system of invariants Ω_i. In the general case, variable, functionally independent invariants have special significance. Functionally independent invariants can be selected in various ways.

The actual construction of the tensors $\boldsymbol{H}_s$ in terms of specified defining tensors $\boldsymbol{T}_1, \ldots, \boldsymbol{T}_m$ is always possible and suitable general methods will be demonstrated by examples.

The linear independence of the tensors $\boldsymbol{H}_s$ can be established directly on the basis of geometric considerations or by verification with the aid of the appropriate determinants or by other general methods. In particular, the tensors $\boldsymbol{H}_{s_1}$ and $\boldsymbol{H}_{s_2}$ are linearly independent, if they are orthogonal or the symmetry groups of $\boldsymbol{H}_{s_1}$ and $\boldsymbol{H}_{s_2}$ do not coincide, since otherwise these two tensors would be proportional, which would contradict their conditions of symmetry. However, tensors which have the same symmetry group may be linearly independent.

Let the symmetry group G_s correspond to the tensor $\boldsymbol{H}_s$.

In a number of cases, it is convenient and advantageous [19] to choose the tensors $\boldsymbol{H}_s$ so that

$$G_1 \supseteq G_2 \supseteq G_3 \supseteq \ldots \supseteq G_p .$$

It is apparent that it is always possible to take as the first q $(q \leqslant p)$ linearly independent tensors the tensors $\boldsymbol{H}_1, \ldots, \boldsymbol{H}_q$, which depend either only on the fundamental tensor $\boldsymbol{g}$ or on $\boldsymbol{g}$ and the third-order tensor $\boldsymbol{E}$:

$$\boldsymbol{E} = |g^{ij}|^{\frac{1}{2}} (\mathfrak{z}_1\mathfrak{z}_2\mathfrak{z}_3 - \mathfrak{z}_1\mathfrak{z}_3\mathfrak{z}_2 + \mathfrak{z}_2\mathfrak{z}_3\mathfrak{z}_1 - \mathfrak{z}_2\mathfrak{z}_1\mathfrak{z}_3 + \mathfrak{z}_3\mathfrak{z}_1\mathfrak{z}_2 - \mathfrak{z}_3\mathfrak{z}_2\mathfrak{z}_1) . \tag{A.2.3}$$

These tensors correspond to isotropy with respect to the full or the proper orthogonal group. The isotropic tensors $\boldsymbol{H}_1, \ldots, \boldsymbol{H}_q$ of order r are well-known from the literature (3, 9, 10, 30]. In three-dimensional space, the maximum number q for isotropic tensors of order r equals [30]

$r=1$	2	3	4	5	6	7	8	9	10
$q=0$	1	1	3	6	15	36	91	232	603

All isotropic tensors of order r are isomers of the tensor H_1, where

$$H_1^{\alpha_1\ldots\alpha_r} = g^{\alpha_1\alpha_2}\ldots g^{\alpha_r\alpha_{r-1}} \qquad (r=2k),$$

$$H_1^{\alpha_1\ldots\alpha_r} = E^{\alpha_1\alpha_2\alpha_3} g^{\alpha_4\alpha_5}\ldots g^{\alpha_{r-1}\alpha_r} \qquad (r=2k+1).$$

The number q is equal to the number of different, linearly independent isomers of the tensor H_1, taking account of the symmetry of the components of the tensor g^{ij}. If the number r is odd, then $q=0$ for the full orthogonal group. All tensors of odd order which are invariant under the full orthogonal group reduce to zero. Tensors of odd order which are invariant with respect to the orthogonal group of proper rotations, with $\Delta=|a^{i}_{.j}|=1$, can be non-zero only for $r\geqslant 3$. For $r=3$, one has $H_1=E$ and, therefore, $q=1$.

The presence of symmetry of tensor functions with respect to some group of permutations of the indices will, generally speaking, decrease the numbers p and q. Formulae for tensor functions with certain symmetries with respect to some indices are always easily obtained from the general formulae by using the operations of symmetrization and alternation on the proper indices, retaining in the process only the linearly independent terms.

§3°. Tensors which specify the geometric symmetry of oriented media and crystals [29]

A medium is called isotropic, if all its properties at each point are invariant under the group of orthogonal transformations. One can distinguish between the following two types of isotropic media:

(1) media, isotropic with respect to the full orthogonal group of coordinate transformations with $\Delta=\pm 1$;

(2) media, isotropic with respect to the group of rotations with $\Delta=+1$ (gyrotropic media).

It is easily seen that in the first case the symmetry properties are completely characterized by the fundamental tensor $\boldsymbol{g}$. The condition of invariance of the components of the tensor $\boldsymbol{g}$ can be considered as the condition which defines the infinite class of all real matrices which are elements of the full orthogonal group.

The group of rotations with $\Delta = +1$, which defines gyrotropic media, is a subgroup of the full orthogonal group. This subgroup can be singled out by supplementing the condition (A.2.2) by the additional requirement of invariance of the components of the tensor $\boldsymbol{E}$ defined by (A.2.3). Therefore the infinite set of elements of the group of rotations is determined completely by the condition of invariance of the tensors $\boldsymbol{g}$ and $\boldsymbol{E}$. These two tensors may be considered as the tensors which determine the group of rotations with $\Delta = +1$.

Below, use will be made of the abbreviated symbols proposed by Shubnikov [15, 16] as notation for symmetry groups. According to these rules, the full orthogonal group is denoted by the symbol $\infty/\infty \cdot m$ (the generating elements of the group are: intersecting axes of infinite order and a reflecting plane of symmetry m). The group of relations corresponds to the symbol ∞/∞.

Results are given in §2° on the general form of tensor functions for tensors of any order when isotropy is present, i.e., when the arguments are only $\boldsymbol{g}$ or $\boldsymbol{g}$ and $\boldsymbol{E}$.

The simplest example of an anisotropic medium is the oriented medium. A medium will be called oriented if all its properties at each point are invariant under an infinite orthogonal group containing rotations by an arbitrary angle about some axis. Obviously, the symmetry groups of oriented media are subgroups of the full orthogonal group. A simple analysis shows that only seven different types of oriented media are possible, including the two types of isotropic media. The appropriate geometric illustrations for the different types of oriented media, the corresponding tensors and vectors which specify the symmetry groups of the oriented media are given in the table below. The correctness of these results is easily verified directly.

An anisotropic medium with a continuous or discrete structure is called a crystal, if it is possible to introduce a system of triply periodic Bravais lattices (with the same periods in the various lattices in a fixed coordinate system) having the same geometric properties as the medium under consideration. The set of Bravais lattices with given periods can admit finite point symmetry groups. The form of these groups depends on the structure of the set of lattices being examined and on the elementary parallelepiped of periods.

As is well-known [2, 16], there are only 32 different symmetry classes of crystals described by finite point groups. In the following table, the characteristic data are presented for all 32 crystal classes; the corresponding geometric figures illustrate each symmetry group. The unit vectors $\mathbf{e}_1$, $\mathbf{e}_2$, $\mathbf{e}_3$ form the orthogonal crystallographic basis. The orientation of this basis relative to the figure of symmetry of the crystal is indicated in the sketch. At the upper left of each box, the notation of the corresponding group, according to Shubnikov, is given. Moreover, each box contains symbols, used for a set of simple tensors which characterize and specify the given group. The formulae defining these tensors are also given.[1)]

Consider the tensors which determine the symmetries of the groups of the cubic system. It will be shown that the tensor $\boldsymbol{O}_h$ is invariant under a group of 48 transformations which give an isomorphic representation of the group $\bar{6}/4$, and that there are no other transformations under which the tensor $\boldsymbol{O}_h$ is invariant. For the proof, find all real transformations under which the tensor $\boldsymbol{O}_h$ is invariant.

The conditions of invariance of the contravariant components of the tensor $\boldsymbol{O}_h$ are equivalent to the following system of nonlinear algebraic equations for the nine elements of the transformation matrix $\| a^i_{\cdot j} \|$

$$a^\alpha{}_1 a^\beta{}_1 a^\gamma{}_1 a^\delta{}_1 + a^\alpha{}_2 a^\beta{}_2 a^\gamma{}_2 a^\delta{}_2 + a^\alpha{}_3 a^\beta{}_3 a^\gamma{}_3 a^\delta{}_3 = \begin{cases} 1 \\ 0 \end{cases} . \qquad \text{(A.3.1)}$$

The right-hand side is to be set equal to unity, if $\alpha = \beta = \gamma = \delta$, and equal to zero in the remaining cases. Now, setting $\alpha = \beta$ and $y = \delta$ for $\alpha \neq \gamma$, one obtains

$$(a^\alpha{}_1)^2 (a^\gamma{}_1)^2 + (a^\alpha{}_2)^2 (a^\gamma{}_2)^2 + (a^\alpha{}_3)^2 (a^\gamma{}_3)^2 = 0 \qquad (\alpha \neq \gamma) . \qquad \text{(A.3.2)}$$

It follows from (A.3.2) that

$$a^\alpha{}_i a^\gamma{}_i = 0 . \qquad \text{(A.3.3)}$$

Since the determinant $| a^\alpha{}_i | \neq 0$, it is seen from (A.3.3) that there is only one non-zero element in each column and in each row of the matrix $\| a^\alpha{}_i \|$.

1) In this table and in the sequel, powers of vectors are to be understood as dyadic or polyadic products.

Since

$$(a^{\alpha}{}_1)^4+(a^{\alpha}{}_2)^4+(a^{\alpha}{}_3)^4=1$$

for $\alpha=\beta=\gamma=\delta$, in accordance with (A.3.1), the following equality holds for each real non-zero element of the matrix $\|a^{\alpha}{}_i\|$:

$$a^p{}_q=\pm 1\,. \qquad \text{(A.3.4)}$$

Enumeration of all possible cases of (A.3.3) and (A.3.4) shows that the matrices consisting of the elements $(a^p{}_q)^2$, which are either equal to 1 or 0, can have the forms

$$\begin{Vmatrix}1&0&0\\0&1&0\\0&0&1\end{Vmatrix},\ \begin{Vmatrix}1&0&0\\0&0&1\\0&1&0\end{Vmatrix},\ \begin{Vmatrix}0&1&0\\1&0&0\\0&0&1\end{Vmatrix},\ \begin{Vmatrix}0&1&0\\0&0&1\\1&0&0\end{Vmatrix},\ \begin{Vmatrix}0&0&1\\0&1&0\\1&0&0\end{Vmatrix},\ \begin{Vmatrix}0&0&1\\1&0&0\\0&1&0\end{Vmatrix} \qquad \text{(A.3.5)}$$

A system consisting of only six matrices has been obtained. If, in accordance with (A.3.4), account is taken of the possibilities of different signs for the $a^p{}_q$, then each of the matrices (A.3.5) generates eight matrices for $\|a^p{}_q\|$. For example, the matrices corresponding to the first matrix of (A.3.5) are

$$\begin{Vmatrix}+1&0&0\\0&+1&0\\0&0&+1\end{Vmatrix},\ \begin{Vmatrix}+1&0&0\\0&+1&0\\0&0&-1\end{Vmatrix},\ \begin{Vmatrix}+1&0&0\\0&-1&0\\0&0&+1\end{Vmatrix},\ \begin{Vmatrix}+1&0&0\\0&-1&0\\0&0&-1\end{Vmatrix}$$

$$\text{(A.3.6)}$$

$$\begin{Vmatrix}-1&0&0\\0&+1&0\\0&0&+1\end{Vmatrix},\ \begin{Vmatrix}-1&0&0\\0&+1&0\\0&0&-1\end{Vmatrix},\ \begin{Vmatrix}-1&0&0\\0&-1&0\\0&0&+1\end{Vmatrix},\ \begin{Vmatrix}-1&0&0\\0&-1&0\\0&0&-1\end{Vmatrix}.$$

As is known, by the definition of the symmetry group of the cube $\overline{6}/4$, the system of matrices of the type (A.3.6) for each matrix of the system (A.3.5) forms the complete group of transformation matrices for symmetry of the cube of the group $\overline{6}/4$, and consists of $6\times 8=48$ orthogonal matrices. Thus, every matrix corresponding to a solution of the system of equations (A.3.1) must be one of the 48 matrices of the system (A.3.6). On the other hand, it is readily seen that the converse proposition is also

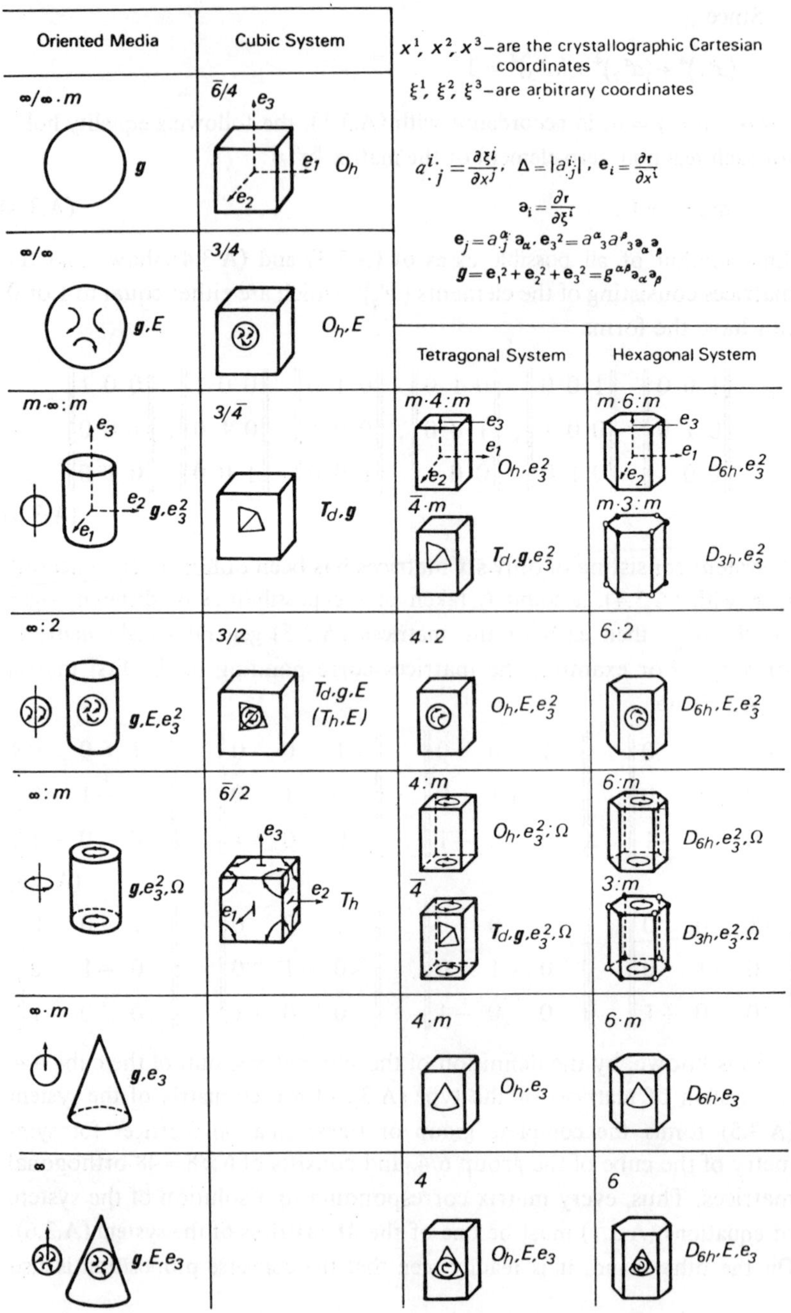
Oriented Media
Cubic System
x¹, x², x³ – are the crystallographic Cartesian coordinates
ξ¹, ξ², ξ³ – are arbitrary coordinates
a^i_j = ∂ξ^i/∂x^j, Δ = |a^i_j|, e_i = ∂r/∂x^i
э_i = ∂r/∂ξ^i
e_j = a^α_j э_α, e_3² = a^α_3 a^β_3 э_α э_β
g = e_1² + e_2² + e_3² = g^{αβ} э_α э_β
∞/∞·m
g
6̄/4
e3
e1
e2
Oh
∞/∞
g,E
3/4
Oh,E
Tetragonal System
Hexagonal System
m·∞:m
e3
e2
e1
g,e3²
3/4̄
Td,g
m·4:m
e3
e1
e2
Oh,e3²
4̄·m
Td,g,e3²
m·6:m
e3
e1
e2
D6h,e3²
m·3:m
D3h,e3²
∞:2
g,E,e3²
3/2
Td,g,E
(Th,E)
4:2
Oh,E,e3²
6:2
D6h,E,e3²
∞:m
g,e3²,Ω
6̄/2
e3
e1
e2
Th
4:m
Oh,e3²;Ω
4̄
Td,g,e3²,Ω
6:m
D6h,e3²,Ω
3:m
D3h,e3²,Ω
∞·m
g,e3
4·m
Oh,e3
6·m
D6h,e3
∞
g,E,e3
4
Oh,E,e3
6
D6h,E,e3

$$\boldsymbol{E} = \mathbf{e}_1\mathbf{e}_2\mathbf{e}_3 - \mathbf{e}_2\mathbf{e}_1\mathbf{e}_3 + \mathbf{e}_2\mathbf{e}_3\mathbf{e}_1 - \mathbf{e}_3\mathbf{e}_2\mathbf{e}_1 + \mathbf{e}_3\mathbf{e}_1\mathbf{e}_2 - \mathbf{e}_1\mathbf{e}_3\mathbf{e}_2$$
$$= \Delta(\mathbf{3}_1\mathbf{3}_2\mathbf{3}_3 - \mathbf{3}_2\mathbf{3}_1\mathbf{3}_3 + \mathbf{3}_2\mathbf{3}_3\mathbf{3}_1 - \mathbf{3}_3\mathbf{3}_2\mathbf{3}_1 + \mathbf{3}_3\mathbf{3}_1\mathbf{3}_2 - \mathbf{3}_1\mathbf{3}_3\mathbf{3}_2)$$

$$\boldsymbol{\Omega} = \mathbf{e}_1\mathbf{e}_2 - \mathbf{e}_2\mathbf{e}_1 = (a^{\alpha}{}_1 a^{\beta}{}_2 - a^{\alpha}{}_2 a^{\beta}{}_1)\mathbf{3}_{\alpha}\mathbf{3}_{\beta} = a^{\alpha}{}_1 a^{\beta}{}_2(\mathbf{3}_{\alpha}\mathbf{3}_{\beta} - \mathbf{3}_{\beta}\mathbf{3}_{\alpha})$$

$$\boldsymbol{O}_h = \mathbf{e}_1^4 + \mathbf{e}_2^4 + \mathbf{e}_3^4 = (a^{\alpha}{}_1 a^{\beta}{}_1 a^{\gamma}{}_1 a^{\delta}{}_1 + a^{\alpha}{}_2 a^{\beta}{}_2 a^{\gamma}{}_2 a^{\delta}{}_2 + a^{\alpha}{}_3 a^{\beta}{}_3 a^{\gamma}{}_3 a^{\delta}{}_3)\mathbf{3}_{\alpha}\mathbf{3}_{\beta}\mathbf{3}_{\gamma}\mathbf{3}_{\delta}$$

$$\boldsymbol{T}_h = \mathbf{e}_1^2\mathbf{e}_2^2 + \mathbf{e}_2^2\mathbf{e}_3^2 + \mathbf{e}_3^2\mathbf{e}_1^2\,,$$

$$\boldsymbol{T}_d = \mathbf{e}_1\mathbf{e}_2\mathbf{e}_3 + \mathbf{e}_2\mathbf{e}_1\mathbf{e}_3 + \mathbf{e}_2\mathbf{e}_3\mathbf{e}_1 + \mathbf{e}_3\mathbf{e}_2\mathbf{e}_1 + \mathbf{e}_3\mathbf{e}_1\mathbf{e}_2 + \mathbf{e}_1\mathbf{e}_3\mathbf{e}_2$$

$$\boldsymbol{D}_{3h} = \mathbf{e}_1^3 - \mathbf{e}_1\mathbf{e}_2^2 - \mathbf{e}_2\mathbf{e}_1\mathbf{e}_2 - \mathbf{e}_2^2\mathbf{e}_1\,, \quad \boldsymbol{D}_{3d} = \mathbf{e}_3(\mathbf{e}_1^3 - \mathbf{e}_1\mathbf{e}_2^2 - \mathbf{e}_2\mathbf{e}_1\mathbf{e}_2 - \mathbf{e}_2^2\mathbf{e}_1)$$

$$\boldsymbol{D}_{6h} = (\mathbf{e}_1^3 - \mathbf{e}_1\mathbf{e}_2^2 - \mathbf{e}_2\mathbf{e}_1\mathbf{e}_2 - \mathbf{e}_2^2\mathbf{e}_1)^2\,, \quad \boldsymbol{D}_{2h} = \lambda^{11}\mathbf{e}_1^2 + \lambda^{22}\mathbf{e}_2^2 + \lambda^{33}\mathbf{e}_3^2$$
$$= \lambda^{ij} a^{\alpha}_{.i} a^{\beta}_{.j}\mathbf{e}_{\alpha}\mathbf{e}_{\beta} = d^{\alpha\beta}\mathbf{e}_{\alpha}\mathbf{e}_{\beta} \quad (\lambda^{11} \neq \lambda^{22} \neq \lambda^{33} \neq 0, \quad d^{\alpha\beta} = d^{\beta\alpha})$$

$$\boldsymbol{C}_i = \boldsymbol{D}_{2h} + \omega^{ij}\mathbf{e}_i\mathbf{e}_j = \boldsymbol{C}^{\alpha\beta}\mathbf{3}_{\alpha}\mathbf{3}_{\beta}\,; \quad \omega^{ij} = -\omega^{ji} \neq 0$$

Trigonal System	Rhombic System	Monoclinic System	Triclinic System
$\bar{6}\cdot m$ 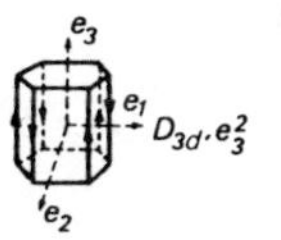	$m\cdot 2:m$ 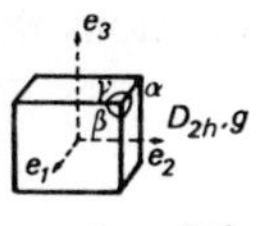		
$3:2$ 	$2:2$ 		
$\bar{6}$ 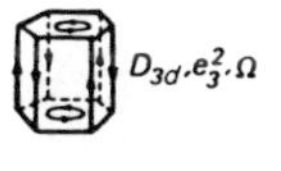		$2:m$ 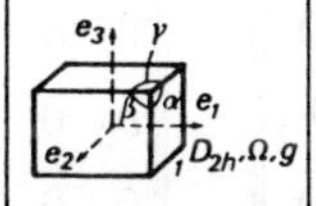	$\bar{2}$ 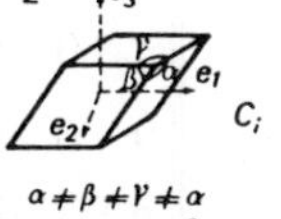
$3\cdot m$ 	$2\cdot m$ 		
3 		2 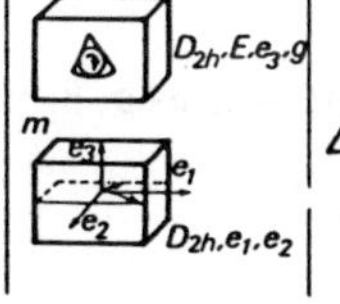	1 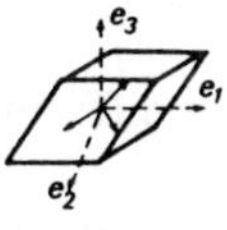

true: each matrix of the system of 48 matrices found provides a solution of the system of equations (A.3.1).

Next, find the matrices of the group of transformations under which the tensor $\boldsymbol{T}_d$ is invariant. The conditions of invariance of the contravariant components of the tensor $\boldsymbol{T}_d$ are equivalent to the following system of non-linear algebraic equations for the nine elements of the transformation matrix $\|a^i{}_j\|$:

$$a^\alpha{}_1 a^\beta{}_2 a^\gamma{}_3 + a^\alpha{}_2 a^\beta{}_1 a^\gamma{}_3 + a^\alpha{}_3 a^\beta{}_2 a^\gamma{}_1 + a^\alpha{}_1 a^\beta{}_3 a^\gamma{}_2 + a^\alpha{}_2 a^\beta{}_3 a^\gamma{}_1 + a^\alpha{}_3 a^\beta{}_1 a^\gamma{}_2 = \begin{cases} 1 \\ 0 \end{cases}. \tag{A.3.7}$$

The right-hand side of (A.3.7) should be set equal to unity, if α, β, γ are all different, and to zero, if at least one pair of α, β, γ are the same. Consider the equations of (A.3.7) for which $\gamma = \beta$. These equations have the form

$$a^\alpha{}_1 a^\beta{}_2 a^\beta{}_3 + a^\alpha{}_2 a^\beta{}_1 a^\beta{}_3 + a^\alpha{}_3 a^\beta{}_1 a^\beta{}_2 = 0 \qquad \begin{pmatrix} \alpha = 1, 2, 3 \\ \beta = 1, 2, 3 \end{pmatrix}. \tag{A.3.8}$$

Since $|a^i_{\cdot j}| \neq 0$, it follows from this system of equations that

$$a^\beta{}_i a^\beta{}_j = 0\,, \tag{A.3.9}$$

where β is any fixed index.

It may be concluded from (A.3.9) and the condition $|a^i_j| \neq 0$ that there is only one non-zero element in each row and each column of the matrix $\|a^i_{\cdot j}\|$. There are only six such matrices with different index structures on the non-zero elements:

$$\begin{Vmatrix} a_1 & 0 & 0 \\ 0 & b_1 & 0 \\ 0 & 0 & c_1 \end{Vmatrix}, \quad \begin{Vmatrix} a_2 & 0 & 0 \\ 0 & 0 & b_2 \\ 0 & c_2 & 0 \end{Vmatrix}, \quad \begin{Vmatrix} 0 & a_3 & 0 \\ b_3 & 0 & 0 \\ 0 & 0 & c_3 \end{Vmatrix}$$

$$\begin{Vmatrix} 0 & a_4 & 0 \\ 0 & 0 & b_4 \\ c_4 & 0 & 0 \end{Vmatrix}, \quad \begin{Vmatrix} 0 & 0 & a_5 \\ 0 & b_5 & 0 \\ c_5 & 0 & 0 \end{Vmatrix}, \quad \begin{Vmatrix} 0 & 0 & a_6 \\ b_6 & 0 & 0 \\ 0 & c_6 & 0 \end{Vmatrix} \tag{A.3.10}$$

Equations (A.3.7) with different indices α, β, γ yield

$$a_i b_i c_i = 1 \qquad (i = 1, \ldots, 6)\,. \tag{A.3.11}$$

It is easily seen that for orthogonal transformations, when

$$\sum_{\alpha=1}^{3} a^i_{\ \alpha} a^j_{\ \alpha} = \begin{cases} 1 \text{ for } i=j \\ 0 \text{ for } i \neq j, \end{cases} \tag{A.3.12}$$

one has

$$a_i = \pm 1, \quad b_i = \pm 1, \quad c_i = \pm 1\,. \tag{A.3.13}$$

In the general case, in order to obtain a representation of the symmetry group $3/\bar{4}$, the requirement of the invariance of the tensor $\boldsymbol{T}_d$ must be supplemented by the condition of invariance of the tensor $\boldsymbol{g}$, since only in this case will the conditions (A.3.12), which form part of the definition of crystal symmetry groups, be satisfied.[1)]

The system of matrices (A.3.10) together with the conditions (A.3.13) determines the 48 matrices of the symmetry group $\tilde{6}/4$. However, the additional equalities (A.3.11) select a subgroup of 24 matrices for which either $a_i = b_i = c_i = 1$ or two elements of the three numbers a_i, b_i, c_i are equal to -1. For instance, one obtains only four matrices from the first matrix of (A.3.10):

$$\begin{Vmatrix} +1 & 0 & 0 \\ 0 & +1 & 0 \\ 0 & & +1 \end{Vmatrix}, \quad \begin{Vmatrix} +1 & 0 & 0 \\ 0 & -1 & 0 \\ 0 & 0 & -1 \end{Vmatrix}, \quad \begin{Vmatrix} -1 & 0 & 0 \\ 0 & -1 & 0 \\ 0 & 0 & +1 \end{Vmatrix}, \quad \begin{Vmatrix} -1 & 0 & 0 \\ 0 & +1 & 0 \\ 0 & 0 & -1 \end{Vmatrix} \tag{A.3.14}$$

It is easily verified that the system of 24 matrices found, which represents the group $3/\bar{4}$, is the solution of (A.3.7), where this system of matrices for $|a^{i\cdot}_{\cdot j}| \neq 0$ forms the system of all real solutions of (A.3.7) under the condition that the matrices are orthogonal.

Now consider the conditions of invariance of the tensor $\boldsymbol{T}_h$. The following system of equations for $a^i_{\ j}$, the elements of the transformation matrix, is equivalent to the condition of invariance of the contravariant components of the tensor $\boldsymbol{T}_h$:

$$a^\alpha_{\ 2} a^\beta_{\ 2} a^\gamma_{\ 3} a^\delta_{\ 3} + a^\alpha_{\ 3} a^\beta_{\ 3} a^\gamma_{\ 1} a^\delta_{\ 1} + a^\alpha_{\ 1} a^\beta_{\ 1} a^\gamma_{\ 2} a^\delta_{\ 2} = \begin{cases} 1 \\ 0 \end{cases}, \tag{A.3.15}$$

1) It is easily verified that for $|\boldsymbol{e}_i| = 1$ equation $2\boldsymbol{g} = \boldsymbol{T}_d : \boldsymbol{T}_d$ holds, where the contraction is carried out with respect to two similarly located indices. However, it does not follow from this equation that $\boldsymbol{g}$ is invariant under the transformation (A.3.10) with (A.3.11).

where the right-hand side should be set equal to 1 for $\alpha=\beta=2, \gamma=\delta=3$; $\alpha=\beta=3$, $\gamma=\delta=1$; $\alpha=\beta=1$, $\gamma=\delta=2$, and to zero in all other cases. It follows from (A.3.15) that for $\alpha=\beta=1, \gamma=\delta=1,3$:

$$a^1{}_2 a^1{}_3=0,\ a^1{}_2 a^3{}_3=0,\ a^1{}_3 a^1{}_1=0,\ a^1{}_3 a^3{}_1=0,\ a^1{}_1 a^1{}_2=0,\ a^1{}_1 a^3{}_2=0; \tag{A.3.16}$$

for $\alpha=\beta=2$, $\gamma=\delta=1,2$:

$$a^2{}_2 a^1{}_3=0,\ a^2{}_2 a^2{}_3=0,\ a^2{}_3 a^1{}_1=0,\ a^2{}_3 a^2{}_1=0,\ a^2{}_1 a^1{}_2=0,\ a^2{}_1 a^2{}_2=0; \tag{A.3.17}$$

for $\alpha=\beta=3$, $\gamma=\delta=2,3$:

$$a^3{}_2 a^2{}_3=0,\ a^3{}_2 a^3{}_3=0,\ a^3{}_3 a^2{}_1=0,\ a^3{}_3 a^3{}_1=0,\ a^3{}_1 a^2{}_2=0,\ a^3{}_1 a^3{}_2=0. \tag{A.3.18}$$

It is seen from the 18 equations (A.3.16) to (A.3.18) and from the condition $|a^i{}_j|\neq 0$ that only one element in each row and each column of the matrix $\|a^i{}_j\|$ can be different from zero. If

$$a^1{}_1\neq 0, \text{ then } a^1{}_2=a^1{}_3=a^3{}_2=a^2{}_3=a^2{}_1=a^3{}_1=0,$$

and one obtains the matrices

$$\text{for } a^1{}_1\neq 0 \qquad\qquad \text{for } a^1{}_2\neq 0 \qquad\qquad \text{for } a^1{}_3\neq 0$$

$$\begin{Vmatrix} a^1{}_1 & 0 & 0 \\ 0 & a^2{}_2 & 0 \\ 0 & 0 & a^3{}_3 \end{Vmatrix}, \quad \begin{Vmatrix} 0 & a^1{}_2 & 0 \\ 0 & 0 & a^2{}_3 \\ a^3{}_1 & 0 & 0 \end{Vmatrix}, \quad \begin{Vmatrix} 0 & 0 & a^1{}_3 \\ a^2{}_1 & 0 & 0 \\ 0 & a^3{}_2 & 0 \end{Vmatrix}. \tag{A.3.19}$$

The three equations (A.3.15), when the right-hand side is equal to unity and $a^1{}_1\neq 0$, lead to

$$(a^2{}_2)^2(a^3{}_3)^2=1,\ (a^3{}_3)^2(a^1{}_1)^2=1,\ (a^1{}_1)^2(a^2{}_2)^2=1\,. \tag{A.3.20}$$

The real solutions of these equations and the equations which are obtained analogously for $a^1{}_2\neq 0$ and $a^1{}_3\neq 0$ are given by

$$\begin{aligned} &a^1{}_1=\pm 1, \quad a^2{}_2=\pm 1, \quad a^3{}_3=\pm 1, \\ &a^1{}_2=\pm 1, \quad a^2{}_3=\pm 1, \quad a^3{}_1=\pm 1, \\ &a^1{}_3=\pm 1, \quad a^2{}_1=\pm 1, \quad a^3{}_2=\pm 1. \end{aligned} \tag{A.3.21}$$

It follows from the values of a^i_j found that each of the matrices (A.3.19) splits up into eight matrices. One obtains a subgroup of the group of matrices $\bar{6}/4$, the subgroup consisting of altogether $3 \times 8 = 24$ orthogonal matrices. It is clear that the solution obtained satisfies the entire system of equations (A.3.15) and that every real solution is contained in the one just found.

The addition of the tensor $\boldsymbol{E}$ as a defining quantity results in the exclusion of matrices with $\Delta = -1$, since $\boldsymbol{E}$ is invariant only with respect to the group of proper rotations, for $\Delta = +1$. The set of two tensors $\boldsymbol{O}_h$ and $\boldsymbol{E}$ singles out a subgroup consisting of the 24 matrices with $\Delta = +1$ from the group of 48 matrices found for $\boldsymbol{O}_h$.

The set of tensors $\boldsymbol{g}$, $\boldsymbol{T}_d$, $\boldsymbol{E}$ also specifies a subgroup consisting of 12 matrices with $\Delta = +1$ from the 24 matrices which were found for the group of $\boldsymbol{g}$, $\boldsymbol{T}_d$. By actually singling out the proper matrices one can show that the transformation groups corresponding to the system of 12 matrices for the tensors $\boldsymbol{g}$, $\boldsymbol{T}_d$, $\boldsymbol{E}$ and that for the tensors $\boldsymbol{T}_h$, $\boldsymbol{E}$ coincide.

The equivalence of the tensors and the corresponding symmetry groups for the tetragonal system, indicated in the table, is a consequence of the following considerations. The symmetry group of the tetragonal system can be obtained as the intersection of the corresponding symmetry groups of crystals of the cubic system and symmetry groups of oriented media. Therefore the specification of the proper subgroups from the groups of the cubic system and of the oriented media may be accomplished by forming sets of tensors from the tensors which specify the corresponding cubic symmetry groups and those which specify groups for the oriented media. It is easy to see directly that the conditions of invariance indicated by the sets of tensors for each of the seven classes of the tetragonal system determines the group of transformation matrices of the corresponding symmetry group of these crystal classes.

In order to justify the choice of the tensors which specify the symmetries of the hexagonal and trigonal systems, one must consider the conditions of invariance of the components of the following pairs of tensors: $\boldsymbol{D}_{6h}$ and $\mathbf{e}_3^2$, $\boldsymbol{D}_{3h}$ and $\mathbf{e}_3^2$, $\boldsymbol{D}_{3d}$ and $\mathbf{e}_3^2$. The conditions of invariance of the dyad $\mathbf{e}_3^2$ selects only the matrices of the form

$$\left\| \begin{matrix} a^1{}_1 & a^1{}_2 & a^1{}_3 \\ a^2{}_1 & a^2{}_2 & a^2{}_3 \\ 0 & 0 & \pm 1 \end{matrix} \right\| \tag{A.3.22}$$

as admissible coordinate transformation matrices. It follows from the invariance of $\boldsymbol{D}_{6h}$ or $\boldsymbol{D}_{3h}$ or $\boldsymbol{D}_{3d}$ that $a^1{}_3 = a^2{}_3 = 0$.

If one requires invariance of the vector $\mathbf{e}_3$ instead of $\mathbf{e}_3^2$, one is led to transformation matrices of the form

$$\left\| \begin{matrix} a^1{}_1 & a^1{}_2 & 0 \\ a^2{}_1 & a^2{}_2 & 0 \\ 0 & 0 & +1 \end{matrix} \right\| \tag{A.3.23}$$

Since $\boldsymbol{D}_{6h}$, $\boldsymbol{D}_{3h}$ and $\boldsymbol{D}_{3d}$ are expressed in terms of the base vectors $\mathbf{e}_1$ and $\mathbf{e}_2$ only, the invariance of these tensors is related to the structure of the second-order matrices

$$D = \left\| \begin{matrix} a^1{}_1 & a^1{}_2 \\ a^2{}_1 & a^2{}_2 \end{matrix} \right\| . \tag{A.3.24}$$

In order to determine the structures of the matrices D, it is convenient to introduce a complex basis by means of the formulae

$$\mathbf{j}_1 = \mathbf{e}_1 + i\mathbf{e}_2, \quad \mathbf{j}_2 = \mathbf{e}_1 - i\mathbf{e}_2 .$$

In this basis, the tensors $\boldsymbol{D}_{3h}$, $\boldsymbol{D}_{6h}$ and $\boldsymbol{D}_{3d}$ take the form

$$2\boldsymbol{D}_{3h} = \mathbf{j}_1^3 + \mathbf{j}_2^3, \quad 4\boldsymbol{D}_{6h} = (\mathbf{j}_1^3 + \mathbf{j}_2^3)^2, \quad 2\boldsymbol{D}_{3d} = \mathbf{e}_3(\mathbf{j}_1^3 + \mathbf{j}_2^3) .$$

The conditions of invariance of these tensors in the real basis can be rewritten as conditions of invariance in the complex basis. If the transformation formulae of the complex basis have the form

$$\mathbf{j}_i = b^\alpha{}_i \mathbf{j}'_\alpha ,$$

then the relation between the matrices $\|a^i{}_j\|$ and $\|b^i{}_j\|$ is determined by the equations

$$\left\| \begin{matrix} a^1{}_1 & a^1{}_2 \\ a^2{}_1 & a^2{}_2 \end{matrix} \right\| = \left\| \begin{matrix} \frac{1}{2} & \frac{1}{2} \\ -\frac{i}{2} & \frac{i}{2} \end{matrix} \right\| \cdot \left\| \begin{matrix} b^1{}_1 & b^1{}_2 \\ b^2{}_1 & b^2{}_2 \end{matrix} \right\| \cdot \left\| \begin{matrix} 1 & i \\ 1 & -i \end{matrix} \right\| . \tag{A.3.25}$$

The condition of invariance of the tensor $\boldsymbol{D}_{3d}$ leads to the following system of equations for the $b^i{}_j$:

$$b^\alpha{}_1 b^\beta{}_1 b^\gamma{}_1 + b^\alpha{}_2 b^\beta{}_2 b^\gamma{}_2 = \begin{cases} 1 & \text{for } \alpha = \beta = \gamma , \\ 0 & \text{in the remaining cases.} \end{cases}$$

In expanded form, this system is equivalent to the equations

$$\begin{aligned}(b^1{}_1)^3+(b^1{}_2)^3&=1, \quad b^1{}_1(b^2{}_1)^2+b^1{}_2(b^2{}_2)^2=0\,,\\ (b^2{}_1)^3+(b^2{}_2)^3&=1, \quad b^2{}_1(b^1{}_1)^2+b^2{}_2(b^1{}_2)^2=0\,.\end{aligned} \tag{A.3.26}$$

All solutions of (A.3.26), satisfying the condition $|b^i_j| \neq 0$, are easily found from these equations. Since the a^i_j are real, it follows from (A.3.25) that $b^1{}_1=\bar{b}^2{}_2$ and $b^1{}_2=\bar{b}^2{}_1$. Hence one obtains six matrices for $\|b^i{}_j\|$:

$$\begin{Vmatrix}1 & 0\\ 0 & 1\end{Vmatrix}, \begin{Vmatrix}\varepsilon & 0\\ 0 & \varepsilon^2\end{Vmatrix}, \begin{Vmatrix}\varepsilon^2 & 0\\ 0 & \varepsilon\end{Vmatrix}, \begin{Vmatrix}0 & 1\\ 1 & 0\end{Vmatrix}, \begin{Vmatrix}0 & \varepsilon\\ \varepsilon^2 & 0\end{Vmatrix}, \begin{Vmatrix}0 & \varepsilon^2\\ \varepsilon & 0\end{Vmatrix}$$

$$\left(\varepsilon=\exp\frac{2\pi i}{3}\right). \tag{A.3.27}$$

The orthogonality of the corresponding matrices (A.3.22) is obtained automatically. On the basis of (A.3.27), (A.3.25) and (A.3.22), one can readily write down the 12 matrices corresponding to invariance of the tensors $\boldsymbol{D}_{3h}$, $\mathbf{e}_3^2$ which characterize the class $m\cdot 3:m$ of the hexagonal system. The invariance of $\boldsymbol{D}_{3h}$, $\mathbf{e}_3^2$ determines six matrices obtained from (A.3.23), (A.3.25) and (A.3.27) corresponding to the class $3\cdot m$ of the trigonal system.

The conditions of invariance of $\boldsymbol{D}_{3d}$ and $\mathbf{e}_3^2$ somewhat modify equations (A.3.26). The solution of the corresponding equations leads to a system of twelve matrices. The first six of these, which correspond to the invariance of $\mathbf{e}_3$, coincide with the matrices of the class $3\cdot m(\boldsymbol{D}_{3h}, \mathbf{e}_3)$, and the other six are obtained from the first by changing the signs of all the components of the matrices. The conditions of invariance of $\boldsymbol{D}_{6h}$ and $\mathbf{e}_3^2$ lead to matrices of the type (A.3.22). In the corresponding equations of type (A.3.26), ± 1 must be written instead of $+1$. As a consequence, the corresponding solution contains twelve matrices of the class $m\cdot 3:m$ and, in addition, the twelve matrices

$$\begin{Vmatrix}\tau^3 & 0 & 0\\ 0 & \tau^3 & 0\\ 0 & 0 & \pm 1\end{Vmatrix}, \begin{Vmatrix}\tau & 0 & 0\\ 0 & \tau^5 & 0\\ 0 & 0 & \pm 1\end{Vmatrix}, \begin{Vmatrix}\tau^5 & 0 & 0\\ 0 & \tau & 0\\ 0 & 0 & \pm 1\end{Vmatrix},$$

$$\begin{Vmatrix}0 & \tau^3 & 0\\ \tau^3 & 0 & 0\\ 0 & 0 & \pm 1\end{Vmatrix}, \begin{Vmatrix}0 & \tau & 0\\ \tau^5 & 0 & 0\\ 0 & 0 & \pm 1\end{Vmatrix}, \begin{Vmatrix}0 & \tau^5 & 0\\ \tau & 0 & 0\\ 0 & 0 & \pm 1\end{Vmatrix}. \qquad \left(\tau=\exp\frac{\pi i}{3}\right)$$

The corresponding real matrices are easily written out with the aid of (A.3.25).

The tensor parameters for all the remaining classes of the hexagonal and trigonal systems are easily obtained by considering the intersections of suitable groups the tensor characteristics of which have already been established. The reason for this fact is that the symmetry groups of these classes are subgroups of the symmetry groups investigated above.

As for the rhombic, monoclinic and triclinic systems, the tensorial characteristics indicated in the table are immediately apparent. It is clear that the corresponding sets of tensors, which specify the symmetry groups, are not uniquely determined.

In each case, another system of tensors, with a one-to-one relation to the system given in the table, may replace the latter. In particular, the number and powers of the tensors in a system need not remain the same. For example, instead of the tensors indicated in the table, the following correspondence of tensors and groups may be used:[1)]

$m\cdot 2:m$	$\mathbf{e}_1^2, \mathbf{e}_2^2, \mathbf{e}_3^2,$	2	$\mathbf{e}_1^2,$	$\mathbf{e}_2^2,$	$\mathbf{e}_3,$	$E,$
$2:2$	$\mathbf{e}_1^2, \mathbf{e}_2^2, \mathbf{e}_3^2, E$	m	$\mathbf{e}_1,$	$\mathbf{e}_2,$	$\mathbf{e}_3^2$	
$2\cdot m$	$\mathbf{e}_1^2, \mathbf{e}_2^2, \mathbf{e}_3,$	$\bar{2}$	$\mathbf{e}_1\mathbf{e}_2,$	$\mathbf{e}_1\mathbf{e}_3,$	$\mathbf{e}_2\mathbf{e}_3\,.$	
$2:m$	$\mathbf{e}_1^2, \mathbf{e}_2^2, \mathbf{e}_3^2, \Omega,$					

Each tensor of this system can easily be expressed in terms of the tensors given in the table. The inverse relations are immediately obvious.

The problem of determination of tensors which specify the symmetry groups of crystals and oriented media has been considered above. The inverse problem of determination of the orthogonal symmetry groups corresponding to a given tensor has been solved in important special cases.

§4°. Tensor functions of tensors characterizing the geometric properties of oriented media and crystals

General formulae of the type (A.1.3) are given below which are valid in arbitrary coordinates for the components of vectors A^i, second-order

1) Products and powers of vectors are to be understood as dyadic products.

tensors A^{ij}. third-order tensors A^{ijk}, and fourth-order tensors A^{ijkl} for oriented media[1)] and crystals. These tensors are functions of the tensor arguments in the table which determine the various symmetry groups.

Since the simultaneous invariants of the tensors which determine the symmetry groups are absolute constants, the invariant coefficients k_s ($s=1, \ldots, p$) are numerical constants or functions of certain scalars which may also be present in the list of defining quantities in addition to the specifying tensors.

Only p linearly independent terms are written out in the formulae. The choice of the terms may be changed; but in every other case a proper choice of terms can be represented as linear combinations of the terms written out in the formulae. The problem of selection of linearly independent tensors may prove to be important when using various supplementary hypotheses about the character of the functional relations (linear dependence of certain components, etc.). The known results [2] when the symmetry conditions

$$A^{ij}=A^{ji}, \quad A^{ijk}=A^{ikj}, \quad A^{ijkl}=A^{ijlk}, \quad A^{ijkl}=A^{jikl}, \quad A^{ijkl}=A^{klij}$$

are used, are easily obtained from the formulae given.

The above conditions are fulfilled when additional limitations are imposed on the invariant coefficients. Proper formulae are obtained from those presented by means of the operation of symmetrization.

Oriented media

Class $\infty/\infty \cdot m$ ($\boldsymbol{g}$)

$$A^{i}=0, \quad A^{ij}=kg^{ij}, \quad A^{ijk}=0, \quad A^{ijkl}=k_1 g^{ij} g^{kl}+k_2 g^{ik} g^{jl}+k_3 g^{il} g^{jk}.$$

Class ∞/∞ ($\boldsymbol{g}, \boldsymbol{E}$)

$$A^{i}=0, \quad A^{ij}=kg^{ij}, \quad A^{ijk}=kE^{ijk}, \quad A^{ijkl}=A^{ijkl}\ (\infty/\infty \cdot m).$$

Class $m \cdot \infty : m$ ($\boldsymbol{g}, \boldsymbol{B}=\mathbf{e}_{3}{}^{2}_{\cdot}$)

$$A^{i}=0, \ A^{ij}=k_1 g^{ij}+k_2 B^{ij}, \ A^{ijk}=0, \ A^{ijkl}=A^{ijkl}(\infty/\infty \cdot m)$$

1) Analogous formulae containing errors were published in [28]. The corrected formulae are given here.

$$+k_4 g^{ij} B^{kl} + k_5 g^{ik} B^{jl} + k_6 g^{il} B^{jk} + k_7 g^{kl} B^{ij} + k_8 g^{jl} B^{ik} + k_9 g^{jk} B^{il} + k_{10} B^{ij} B^{kl}.$$

Class $\infty:2$ $(\boldsymbol{g}, \boldsymbol{B}=\mathbf{e}_3^2, \boldsymbol{E})$

$$A^i=0, \quad A^{ij}=k_1 g^{ij}+k_2 B^{ij},$$

$$A^{ijk}=k_1 E^{ijk}+k_2 B^i{}_\alpha E^{\alpha jk}+k_3 E^{ij\alpha} B^k{}_\alpha, \quad A^{ijkl}=A^{ijkl}(m\cdot\infty:m).$$

Class $\infty:m$ $(\boldsymbol{g}, \boldsymbol{B}=\mathbf{e}_3^2, \boldsymbol{\Omega}=\mathbf{e}_1\mathbf{e}_2-\mathbf{e}_2\mathbf{e}_1)$

$$A^i=0, \quad A^{ij}=k_1 g^{ij}+k_2 B^{ij}+k_3 \Omega^{ij}, \quad A^{ijk}=0,$$

$$A^{ijkl}=A^{ijkl}(m\cdot\infty:m)+k_{11} g^{ij}\Omega^{kl}+k_{12} g^{ik}\Omega^{jl}+k_{13} g^{il}\Omega^{jk} +k_{14} g^{kl}\Omega^{ij}+k_{15} g^{jl}\Omega^{ik}+k_{16} g^{jk}\Omega^{il}+k_{17} B^{ij}\Omega^{kl}+k_{18} B^{ik}\Omega^{jl} +k_{19}\Omega^{ij}B^{kl}.$$

Class $\infty\cdot m$ $(\boldsymbol{g}, \boldsymbol{b}=\mathbf{e}_3)$

$$A^i=kb^i, \quad A^{ij}=k_1 g^{ij}+k_2 b^i b^j,$$

$$A^{ijk}=k_1 g^{ij} b^k+k_2 g^{ik} b^j+k_3 g^{jk} b^i+k_4 b^i b^j b^k,$$

$$A^{ijkl}=A^{ijkl}(\infty/\infty\cdot m)+k_4 g^{ij} b^k b^l+k_5 g^{ik} b^j b^l+k_6 g^{il} b^j b^k +k_7 g^{kl} b^i b^j+k_8 g^{jl} b^i b^k+k_9 g^{jk} b^i b^l+k_{10} b^i b^j b^k b^l.$$

Class ∞ $(\boldsymbol{g}, \boldsymbol{b}=\mathbf{e}_3, \boldsymbol{E})$

$$A^i=kb^i, \quad A^{ij}=k_1 g^{ij}+k_2 b^i b^j+k_3 E^{ij\alpha} b_\alpha,$$

$$A^{ijk}=k_1 g^{ij} b^k+k_2 g^{ik} b^j+k_3 g^{jk} b^i+k_4 b^i b^j b^k+k_5 \Omega^{ij} b^k +k_6 \Omega^{ik} b^j+k_7 \Omega^{jk} b^i,$$

$$A^{ijkl}=A^{ijkl}(\infty\cdot m)+k_{11} g^{ij}\Omega^{kl}+k_{12} g^{ik}\Omega^{jl}+k_{13} g^{il}\Omega^{jk} +k_{14} g^{kl}\Omega^{ji}+k_{15} g^{jl}\Omega^{ik}+k_{16} g^{jk}\Omega^{il}+k_{17} b^i b^j\Omega^{kl} +k_{18} b^i b^k\Omega^{jl}+k_{19}\Omega^{ij} b^k b^l \qquad (\Omega^{ij}=E^{ij\alpha} b_\alpha).$$

The cubic system

Class $\overline{6}/4$ $(\boldsymbol{O_h})$

$$A^i=0, \; A^{ij}=kg^{ij}, \; A^{ijk}=0, \; A^{ijkl}=A^{ijkl}(\infty/\infty\cdot m)+k_4 O_h{}^{ijkl}.$$

Class 3/4 $(\boldsymbol{O_h}, \boldsymbol{E})$

$$A^i=0,\ A^{ij}=kg^{ij},\ A^{ijk}=kE^{ijk},\ A^{ijkl}=A^{ijkl}(\bar{6}/4).$$

Class $3/\bar{4}$ $(\boldsymbol{g}, \boldsymbol{T_d})$

$$A^i=0,\ A^{ij}=kg^{ij},\ A^{ijk}=kT_d^{\ ijk},\ A^{ijkl}=A^{ijkl}(\bar{6}/4).$$

Class 3/2 $(\boldsymbol{g}, \boldsymbol{E}, \boldsymbol{T_d})$ or $(\boldsymbol{T_h}, \boldsymbol{E})$

$$A^i=0,\quad A^{ij}=kg^{ij},\quad A^{ijk}=k_1 E^{ijk}+k_2 T_d^{\ ijk},$$

$$A^{ijkl}=A^{ijkl}(\bar{6}/4)+k_5 T_h^{\ ijkl}+k_6 T_h^{\ iljk}+k_7 T_h^{\ ikjl}.$$

Class $\bar{6}/2$ $(\boldsymbol{T_h})$

$$A^i=0,\quad A^{ij}=kg^{ij},\quad A^{ijk}=0,\quad A^{ijkl}=A^{ijkl}(3/2).$$

The tetragonal system

Class $m\cdot 4:m$ $(\boldsymbol{O_h}, \boldsymbol{B}=\mathbf{e}_3{}^2)$

$$A^i=0,\quad A^{ij}=A^{ij}(m\cdot\infty:m)=k_1 g^{ij}+k_2 B^{ij},$$

$$A^{ijk}=0,\quad A^{ijkl}=A^{ijkl}(m\cdot\infty:m)+k_{11} O_h^{\ ijkl}.$$

Class $\bar{4}\cdot m$ $(\boldsymbol{g}, \boldsymbol{T_d}, \boldsymbol{B}=\mathbf{e}_3{}^2)$

$$A^i=0,\quad A^{ij}=k_1 g^{ij}+k_2 B^{ij},$$

$$A^{ijk}=k_1 T_d^{\ ijk}+k_2 T_d^{\ ij\alpha} B^{k.}_{.\alpha}+k_3 T_d^{\ ik\alpha} B^{j.}_{.\alpha},\ A^{ijkl}=A^{ijkl}(m\cdot 4:m).$$

Class $4:2$ $(\boldsymbol{O_h}, \boldsymbol{B}=\mathbf{e}_3^2, \boldsymbol{E})$

$$A^i=0,\ A^{ij}=k_1 g^{ij}+k_2 B^{ij},\ A^{ijk}=A^{ijk}(\infty:2),\ A^{ijkl}=A^{ijkl}(m\cdot 4:m).$$

Class $4:m$ $(\boldsymbol{O_h}, \boldsymbol{\Omega}=\mathbf{e}_1\mathbf{e}_2-\mathbf{e}_2\mathbf{e}_1, \boldsymbol{B}=\mathbf{e}_3{}^2)$

$$A^i=0,\quad A^{ij}=A^{ij}(\infty:m),\quad A^{ijk}=0,$$

$$A^{ijkl}=A^{ijkl}(\infty:m)+k_{20} O_h^{\ ijkl}+k_{21} O_h^{\ jkl\alpha}\Omega^{i.}_{.\alpha}.$$

Class $\bar{4}$ $(\boldsymbol{g}, \boldsymbol{T}_d, \boldsymbol{\Omega}=\mathbf{e}_1\mathbf{e}_2-\mathbf{e}_2\mathbf{e}_1, \boldsymbol{B}=\mathbf{e}_3{}^2)$

$$A^i=0, \quad A^{ij}=A^{ij}\,(\infty:m),$$
$$A^{ijk}=A^{ijk}(\bar{4}\cdot m)+k_4\,T_d^{ij\alpha}\,\Omega^{k.}_{.\alpha}$$
$$+k_5\Omega^{i.}_{.\alpha}\,T_d^{\alpha jk}+k_6\,\Omega^{i.}_{.\alpha}\,T_d^{\alpha j\beta}\,B^{k.}_{.\beta}, \quad A^{ijkl}=A^{ijkl}\,(4:m).$$

Class $4\cdot m$ $(\boldsymbol{O}_h, \ \boldsymbol{b}=\mathbf{e}_3)$

$$A^i=kb^i, \ A^{ij}=k_1g^{ij}+k_2b^ib^j, \ A^{ijk}=A^{ijk}(\infty\cdot m), \ A^{ijkl}=A^{ijkl}\,(m\cdot 4:m).$$

Class 4 $(\boldsymbol{O}_h, \boldsymbol{b}=\mathbf{e}_3, \boldsymbol{E})$

$$A^i=kb^i, \quad A^{ij}=k_1\,g^{ij}+k_2b^ib^j+k_3\Omega^{ij} \qquad (\Omega^{ij}=E^{ij\alpha}\,b_\alpha),$$
$$A^{ijk}=A^{ijk}(\infty), \ A^{ijkl}=A^{ijkl}(\infty)+k_{20}\,O_h{}^{ijkl}+k_{21}\,O_h{}^{jkl\alpha}\,\Omega^{i.}_{.\alpha}\,.$$

The hexagonal system

Class $m\cdot 6:m$ $(\boldsymbol{D}_{6h}, \boldsymbol{B}=\mathbf{e}_3{}^2)$

$$A^i=0, \ A^{ij}=k_1g^{ij}+k_2B^{ij}, \ A^{ijk}=0, \ A^{ijkl}=A^{ijkl}\,(m\cdot\infty:m).$$

Class $m\ 3:m$ $(\boldsymbol{D}_{6h}, \boldsymbol{B}=\mathbf{e}_3{}^2)$

$$A^i=0, \ A^{ij}=k_1g^{ij}+k_2B^{ij}, \ A^{ijk}=kD_{3h}{}^{ijk}, \ A^{ijkl}=A^{ijkl}\,(m\cdot\infty:m).$$

Class $6:2$ $(\boldsymbol{D}_{6h}, \boldsymbol{B}=\mathbf{e}_3{}^2, \boldsymbol{E})$

$$A^i=0, \ A^{ij}=k_1g^{ij}+k_2B^{ij}, \ A^{ijk}=A^{ijk}\,(\infty:2), \ A^{ijkl}=A^{ijkl}\,(m\cdot\infty:m).$$

Class $6:m$ $(\boldsymbol{D}_{6h}, \boldsymbol{B}=\mathbf{e}_3{}^2, \boldsymbol{\Omega}=\mathbf{e}_1\mathbf{e}_2-\mathbf{e}_2\mathbf{e}_1)$

$$A^i=0, \ A^{ij}=A^{ij}\,(\infty:m), \ A^{ijk}=0, \ A^{ijkl}=A^{ijkl}\,(\infty:m)\,.$$

Class $3:m$ $(\boldsymbol{D}_{3h}, \boldsymbol{B}=\mathbf{e}_3{}^2, \boldsymbol{\Omega}=\mathbf{e}_1\mathbf{e}_2-\mathbf{e}_2\mathbf{e}_1)$

$$A^i=0, \ A^{ij}=A^{ij}\,(\infty:m), \ A^{ijk}=k_1\,D_{3h}{}^{ijk}+k_2\,D_{3h}{}^{ij\alpha}\,\Omega^{k.}_{.\alpha},$$
$$A^{ijkl}=A^{ijkl}\,(\infty:m).$$

Class $6\cdot m$ $(\boldsymbol{D}_{6h}, \boldsymbol{b}-\mathbf{e}_3)$

$$A^i=kb^i, \ A^{ij}=k_1g^{ij}+k_2b^ib^j, \ A^{ijk}=A^{ijk}\,(\infty\cdot m), \ A^{ijkl}=A^{ijkl}\,(\infty\cdot m).$$

Class 6 ($\boldsymbol{D}_{6h}$, $\boldsymbol{b}=\mathbf{e}_3$, $\boldsymbol{E}$)

$$A^i=kb^i,\ A^{ij}=A^{ij}(\infty),\ A^{ijk}=A^{ijk}(\infty),\ A^{ijkl}=A^{ijkl}(\infty)\,.$$

The trigonal system

Class $\overline{6}\cdot m$ ($\boldsymbol{D}_{3d}$, $\boldsymbol{B}=\mathbf{e}_3{}^2$)

$$A^i=0,\ A^{ij}=k_1g^{ij}+k_2B^{ij},\ A^{ijk}=0,$$

$$A^{ijkl}=A^{ijkl}(m\cdot\infty:m)+k_{11}D_{3d}{}^{ijkl}+k_{12}D_{3d}{}^{jikl}+k_{13}D_{3d}{}^{kijl}+k_{14}D_{3d}{}^{lijk}.$$

Class $3:2$ ($\boldsymbol{D}_{3h}$, $\boldsymbol{B}=\mathbf{e}_3{}^2$, $\boldsymbol{E}$)

$$A^i=0,\ A^{ij}=k_1g^{ij}+k_2B^{ij},\ A^{ijk}=A^{ijk}(\infty:2)+k_4D_{3h}{}^{ijk},$$

$$A^{ijkl}=A^{ijkl}(m\cdot\infty:m)+k_{11}D_{3h}{}^{ij\alpha}E_{\alpha..}^{.kl}+k_{12}E^{\alpha ij}D_{3h..\alpha}^{kl.}$$

$$+k_{13}E^{\alpha ik}D_{3h..\alpha}^{jl.}+k_{14}E^{kj}D_{3h..\alpha}^{il.}\,.$$

Class $\overline{6}$ ($\boldsymbol{D}_{3d}$, $\boldsymbol{B}=\mathbf{e}_3{}^2$, $\boldsymbol{\Omega}=\mathbf{e}_1\mathbf{e}_2-\mathbf{e}_2\mathbf{e}_1$)

$$A^i=0,\ A^{ij}=A^{ij}(\infty:m),\ A^{ijk}=0,$$

$$A^{ijkl}=A^{ijkl}(\infty:m)+k_{20}D_{3d}{}^{ijkl}+k_{21}D_{3d}{}^{jikl}+k_{22}D_{3d}{}^{kijl}$$

$$+k_{23}D_{3d}{}^{lijk}+k_{24}D_{3d}{}^{ijk\alpha}\Omega_{.\alpha}^{l.}+k_{25}D_{3d}{}^{jik\alpha}\Omega_{.\alpha}^{l.}+k_{26}D_{3d}{}^{kij\alpha}\Omega_{.\alpha}^{l.}$$

$$+k_{27}D_{3d}{}^{lij\alpha}\Omega_{.\alpha}^{k.}\,.$$

Class $3\cdot m$ ($\boldsymbol{D}_{3h}$, $\boldsymbol{b}=\mathbf{e}_3$)

$$A^i=kb^i,\ A^{ij}=k_1g^{ij}+k_2b^ib^j,\ A^{ijk}=A^{ijk}(\infty\cdot m)+k_5D_{3h}{}^{ijk}\,.$$

$$A^{ijkl}=A^{ijkl}(\infty\cdot m)+k_{11}D_{3h}{}^{ijk}b^l+k_{12}D_{3h}{}^{ijl}b^k+k_{13}D_{3h}{}^{ikl}b^j$$

$$+k_{14}D_{3h}{}^{klj}b^i.$$

Class 3 ($\boldsymbol{D}_{3h}$, $\boldsymbol{b}=\mathbf{e}_3$, $\boldsymbol{E}$)

$$A^i=kb^i,\quad A^{ij}=A^{ij}(\infty),$$

$$A^{ijk}=A^{ijk}(\infty)+k_8D_{3h}{}^{ijk}+k_9D_{3h}{}^{ij\alpha}\Omega_{.\alpha}^{k.},\quad A^{ijkl}=A^{ijkl}(\overline{6}).$$

The rhombic system

Class $m\cdot 2:m$ $(\boldsymbol{D}_{2h}, \boldsymbol{g})$

$$A^i=0,\ A^{ij}=k_1 g^{ij}+k_2 D_{2h}{}^{ij}+k_3 D_{2h}{}^{i\alpha} D_{2h\alpha.}{}^{.j}\ \text{(Cayley–Hamilton)}$$

$$\begin{aligned}A^{ijk}=0,\ A^{ijkl}&=k_1 g^{ij}g^{kl}+k_2 g^{ik}g^{jl}+k_3 g^{il}g^{jk}+k_4 g^{ij}D_{2h}{}^{kl}\\&+k_5 g^{ik}D_{2h}{}^{jl}+k_6 g^{il}D_{2h}{}^{jk}+k_7 D_{2h}{}^{ij}g^{kl}+k_8 D_{2h}{}^{ik}g^{jl}+k_9 D_{2h}{}^{il}g^{jk}\\&+k_{10}g^{ij}M^{kl}+k_{11}g^{ik}M^{jl}+k_{12}g^{il}M^{jl}+k_{13}M^{ij}g^{kl}+k_{14}M^{ik}g^{jl}\\&+k_{15}M^{il}g^{jk}+k_{16}D_{2h}{}^{ij}D_{2h}{}^{kl}+k_{17}D_{2h}{}^{il}M^{jk}+k_{18}D_{2h}{}^{ij}M^{kl}\\&+k_{19}D_{2h}{}^{ik}M^{jl}+k_{20}M^{ij}D_{2h}{}^{kl}+k_{21}M^{ij}M^{kl}\qquad (M^{ij}=D_{2h}{}^{i\alpha}D_{2h\alpha.}{}^{.j})\,.\end{aligned}$$

Class $2:2$ $(\boldsymbol{D}_{2h}, \boldsymbol{E}, \boldsymbol{g})$

$$A^i=0,\ A^{ij}=A^{ij}\,(m\cdot 2:m),$$

$$\begin{aligned}A^{ijk}&=k_1 E^{ijk}+k_2 E^{ij\alpha}D_{2h\alpha.}{}^{.k}+k_3 E^{ik\alpha}D_{2h\alpha.}{}^{.j}+k_4 E^{ij\alpha}M^{k.}_{.\alpha}\\&+k_5 E^{ik\alpha}M^{j.}_{.\alpha}+k_6 D_{2h}{}^{i\alpha}E^{.j.}_{\alpha.\beta}M^{\beta k},\ A^{ijkl}=A^{ijkl}\,(m\cdot 2:m).\end{aligned}$$

Class $2\cdot m$ $(\boldsymbol{D}_{2h}, \boldsymbol{b}=\mathbf{e}_3, \boldsymbol{g})$

$$A^i=kb^i,\ A^{ij}=A^{ij}\,(m\cdot 2:m)=k_1 g^{ij}+k_2 b^i b^j+k_3 D_{2h}{}^{ij},$$

$$\begin{aligned}A^{ijk}&=k_1 g^{ij}b^k+k_2 g^{ik}b^j+k_3 g^{kj}b^i+k_4 b^i b^j b^k+k_5 D_{2h}{}^{ij}b^k\\&+k_6 D_{2h}{}^{ik}b^j+k_7 D_{2h}{}^{kj}b^i\ ,\quad A^{ijkl}=A^{ijkl}\,(m\cdot 2:m)\,.\end{aligned}$$

The monoclinic system

Class $2:m$ $(\boldsymbol{D}_{2h}, \boldsymbol{\Omega}=\mathbf{e}_1\mathbf{e}_2-\mathbf{e}_2\mathbf{e}_1, \boldsymbol{g})$

$$A^i=0,\ A^{ij}=A^{ij}(m\cdot 2:m)+k_4\Omega^{ij}+k_5\Omega^{i\alpha}D_{2h.\alpha}{}^{j.},\ A^{ijk}=0,$$

$$\begin{aligned}A^{ijkl}&=A^{ijkl}\,(m\cdot 2:m)+k_{22}g^{ij}\Omega^{kl}+k_{23}g^{ik}\Omega^{jl}+k_{24}g^{il}\Omega^{jk}\\&+k_{25}g^{kl}\Omega^{ij}+k_{26}g^{jl}\Omega^{ik}+k_{27}g^{jk}\Omega^{il}+k_{28}g^{ij}\Omega^{k\alpha}D_{2h\alpha.}{}^{.l}\\&+k_{29}g^{ik}\Omega^{j\alpha}D_{2h\alpha.}{}^{.l}+k_{30}g^{il}\Omega^{j\alpha}D_{2h\alpha.}{}^{.k}+k_{31}g^{kl}\Omega^{i\alpha}D_{2h\alpha.}{}^{.j}\\&+k_{32}g^{jl}\Omega^{i\alpha}D_{2h\alpha.}{}^{.k}+k_{33}g^{jk}\Omega^{i\alpha}D_{2h\alpha.}{}^{.l}+k_{34}D_{2h}{}^{ij}\Omega^{kl}\\&+k_{35}D_{2h}{}^{ik}\Omega^{jl}+k_{36}D_{2h}{}^{kl}\Omega^{ij}+k_{37}D_{2h}{}^{ij}\Omega^{k\alpha}D_{2h\alpha.}{}^{.l}\end{aligned}$$

$$+k_{38}D_{2h}{}^{il}\Omega^{j\alpha}D_{2h\alpha.}{}^{.k}+k_{39}D_{2h}{}^{kl}\Omega^{i\alpha}D_{2h\alpha.}{}^{.j}+k_{40}M^{kl}\Omega^{ij}$$
$$+k_{41}M^{ij}\Omega^{k\alpha}D_{2h\alpha.}{}^{.l}$$

Class 2 $(\boldsymbol{D}_{2h}, \boldsymbol{E}, \boldsymbol{b}=\mathbf{e}_3, \boldsymbol{g})$

$$A^i=kb^i,\ A^{ij}=A^{ij}(2:m),$$
$$A^{ijk}=k_1g^{ij}b^k+k_2g^{ik}b^j+k_3g^{jk}b^i+k_4b^ib^jb^k+k_5D_{2h}{}^{ij}b^k$$
$$+k_6D_{2h}{}^{ik}b^j+k_7D_{2h}{}^{kj}b^i+k_8\Omega^{ij}b^k+k_9\Omega^{ik}b^j+k_{10}\Omega^{kj}b^i$$
$$+k_{11}\Omega^{i\alpha}D_{2h\alpha.}{}^{.j}b^k+k_{12}\Omega^{i\alpha}D_{2h\alpha.}{}^{.k}b^j+k_{13}\Omega^{k\alpha}D_{2h\alpha.}{}^{.j}b^i,$$
$$A^{ijkl}=A^{ijkl}(2:m).$$

Class m $(\boldsymbol{D}_{2h}, \boldsymbol{b}=\mathbf{e}_1, \boldsymbol{c}=\mathbf{e}_2)$

$$A^i=k_1b^i+k_2c^i,\quad A^{ij}=k_1g^{ij}+k_2b^ib^j+k_3c^ic^j+k_4b^ic^j+k_5c^ib^j,$$
$$A^{ijk}=k_1g^{ij}b^k+k_2g^{ik}b^j+k_3g^{jk}b^i+k_4b^ib^jb^k+k_5g^{ij}c^k+k_6g^{ik}c^j$$
$$+k_7g^{jk}c^i+k_8c^ib^jb^k+k_9b^ic^jb^k+k_{10}b^ib^jc^k+k_{11}b^ic^jc^k$$
$$+k_{12}c^ib^jc^k+k_{13}c^ic^jb^k+k_{14}c^ic^jc^k,$$
$$A^{ijkl}=A^{ijkl}(2:m)=k_1g^{ij}g^{kl}+k_2g^{ik}g^{jl}+k_3g^{il}g^{jk}+k_4g^{ij}b^kb^l$$
$$+k_5g^{ik}b^jb^l+k_6g^{il}b^jb^k+k_7b^ib^jg^{kl}+k_8b^ib^kg^{jl}+k_9b^ib^lg^{jk}$$
$$+k_{10}g^{ij}b^kc^l+k_{11}g^{ik}b^jc^l+k_{12}g^{il}b^jc^k+k_{13}g^{kl}b^ic^j+k_{14}g^{jl}b^ic^k$$
$$+k_{15}g^{jk}b^ic^l+k_{16}g^{ij}c^kb^l+k_{17}g^{ik}c^jb^l+k_{18}g^{il}c^jb^k$$
$$+k_{19}g^{kl}c^ib^j+k_{20}g^{jl}c^ib^k+k_{21}g^{jk}c^ib^l+k_{22}c^ib^jb^kb^l$$
$$+k_{23}b^ic^jb^kb^l+k_{24}b^ib^jc^kb^l+k_{25}b^ib^jb^kc^l+k_{26}g^{ij}c^kc^l$$
$$+k_{27}g^{ik}c^jc^l+k_{28}g^{il}c^jc^k+k_{29}c^kc^lc^ic^j+k_{30}b^ib^jb^kb^l$$
$$+k_{31}g^{jk}c^ic^l+k_{32}b^ib^jc^kc^l+k_{33}b^ib^kc^jc^l+k_{34}b^ib^lc^jc^k$$
$$+k_{35}c^ic^jb^kb^l+k_{36}c^ic^kb^jb^l+k_{37}c^ic^lb^jb^k+k_{38}b^ic^jc^kc^l$$
$$+k_{39}c^ib^jc^kc^l+k_{40}c^jc^ib^kc^l+k_{41}c^ic^jc^kb^l\ .$$

If $\mathbf{e}_1$, $\mathbf{e}_2$ and $\mathbf{e}_3{}^2$ are taken as the defining tensors instead of $\boldsymbol{D}_{1h}$, $\mathbf{e}_1$ and $\mathbf{e}_2$, the last formula for fourth-order tensors may be replaced by

$$A_4=k^{ijkl}\mathbf{e}_i\mathbf{e}_j\mathbf{e}_k\mathbf{e}_l+k^{\alpha\beta 33}\mathbf{e}_\alpha\mathbf{e}_\beta\mathbf{e}_3\mathbf{e}_3+k^{\alpha 33\beta}\mathbf{e}_\alpha\mathbf{e}_3\mathbf{e}_3\mathbf{e}_\beta+k^{33\alpha\beta}\mathbf{e}_3\mathbf{e}_3\mathbf{e}_\alpha\mathbf{e}_\beta$$
$$+k^{3\alpha 3\beta}\mathbf{e}_3\mathbf{e}_\alpha\mathbf{e}_3\mathbf{e}_\beta+k^{3\alpha\beta 3}\mathbf{e}_3\mathbf{e}_\alpha\mathbf{e}_\beta\mathbf{e}_3+k^{\alpha 3\beta 3}\mathbf{e}_\alpha\mathbf{e}_3\mathbf{e}_\beta\mathbf{e}_3$$
$$+k^{3333}\mathbf{e}_3\mathbf{e}_3\mathbf{e}_3\mathbf{e}_3\ , \qquad (*)$$

where the summation is carried out with respect to the indices $i, j, k, l, \alpha, \beta$, which take on only the values 1 and 2. A simple calculation shows that there are 41 terms in this formula; their linear independence is immediately apparent.

It is not difficult to see that for tensors of even order, in particular, for fourth-order tensors referring to the classes $2:m$, 2 and m of the monoclinic system, the corresponding tensor parameters may be replaced by the same system of tensors $\mathbf{e}_1$, $\mathbf{e}_2$ and $\mathbf{e}_3{}^2$. The same formulae may therefore be used. Thus, for all classes of the monoclinic system, Formula (*) is applicable to fourth-order tensors.

It is also easily seen that the fourth-order tensors for the rhombic system with 21 linearly independent terms can be obtained from Formula (*) in which the term with $i=j, k=l$; $i=k, j=l$ and $i=l, j=k$ and $\alpha=\beta$ must be taken.

Thus, it is clear that in the construction of general formulae for tensor functions it is sometimes advantageous to change suitably the original basis of arguments for the particular cases at hand.

The triclinic system

Class $\bar{2}$ (C_i)

$A^i=0$, A^{ij} is the most general case with nine components,
$A^{ijk}=0$, A^{ijkl} is the most general case with 81 components.

Class 1 $(\mathbf{e}_1, \mathbf{e}_2, \mathbf{e}_3)$

All tensors have the most general form if symmetries are absent.

§5°. Tensor functions for oriented media and crystals with additional tensor arguments

Let it now be assumed that, besides the tensors which specify the geometric properties of oriented media or crystals, there are other tensors among the defining quantities or independent arguments. It is apparent that in this case the symmetry groups of the set of defining parametric tensors are suitable groups or subgroups of oriented media or crystals. Subgroups which specify the symmetry of a crystal are different from the symmetry group of the new sidering oriented media. If other tensors are adjoined to those which determine a crystal symmetry, either some crystal symmetry group will be obtained again or the symmetry group will reduce to the identity transformation.

All subgroups of a given crystal symmetry group are contained among the 32 crystal groups. Therefore, upon addition of other tensors to those which specify the symmetry of a crystal, the symmetry group of the new set of arguments will also belong to one of the 32 crystal groups.

A decrease in the number of linearly independent components of the tensors defined in the general case can occur only in the presence of some corresponding symmetry. It is apparent that simplifications of crystals will take place when the set of defining parameters admits a non-trivial symmetry group. After the determination of the type of crystal symmetry group which is appropriate for a set of tensor arguments, one of the formula of §4° can be used to determine the structure of the components of the tensor function which has been defined.

Thus, it is possible to use the formulae of §4° to determine the structure of tensor functions for crystals in the general case. In order to ascertain the nature of the appropriate formulae, it is first necessary to investigate the symmetry properties of the set of given arguments. For crystals, this is equivalent to representing the defining tensors in terms of tensors which characterize the crystal classes, as indicated in the table.

The argument above permits to analyze a large number of special cases easily, when the supplementary tensors are special or have a special form in the crystallographic axes.

When additional tensors are present, the scalars k_s are, in the general case, functions of the common invariants of the supplementary tensors and the tensors which specify the symmetry of the oriented media or crystals. Supplementary tensors can give rise to variable simultaneous invariants. In general, the number of functionally independent invariants is equal to the number of functionally independent components of the variable tensors. In certain special cases, the number of functionally independent components can be smaller.

It is possible to select, in the general case, the scalar invariants ω_i, in terms of which the k_s are defined, so that they retain their values for the different variable tensors which are equivalent from the point of view of symmetry of oriented media or crystals. These arguments, which are determined in a fixed coordinate system, may differ from the invariants Ω_i for arbitrary coordinate transformations, but coincide with them ($\omega_i = \Omega_i$) in the given fixed coordinate system.

§6°. On the Riemann curvature tensor and a generalization of Schur's theorem

The theory which has been developed above is directly related to all mathematical and physical laws which are formulated as vector or tensor equations and which, to some extent, are connected with geometric symmetry properties.

There are a great many important applications, for example, Hooke's law for oriented media and crystals, piezoelectric and optical effects, etc. As an example, consider the Christoffel-Riemann curvature tensor R_{ijkl}. As is known, this tensor is anti-symmetric with respect to interchange of the indices i and j for the indices k and l, and is symmetric with respect to interchange of the pairs of indices ij and kl. In the case of three-dimensional space, there are only six independent components of R_{ijkl} which may take on arbitrary independent values. These six components determine the six components of the symmetric second-order tensor K^{mn}, which may be introduced by

$$K^{mn} = E^{ijm} E^{kln} R_{ijkl} , \tag{A.6.1}$$

whence

$$R_{ijkl} = \tfrac{1}{4} E_{ijm} E_{kln} K^{mn} . \tag{A.6.2}$$

As is well-known [31], the components of the curvature tensor satisfy Bianchi's identity

$$\nabla_r R_{ijmn} + \nabla_m R_{ijnr} + \nabla_n R_{ijrm} = 0 ,$$

where the indices m, n and r are all different and $\nabla \varkappa$ is the notation for covariant differentiation with respect to the coordinate $x^\varkappa$. It may easily be seen that Bianchi's identity is equivalent to the following identity in the components of the tensor K^{mn}:

$$\nabla_\alpha K^{m\alpha} = 0 \,. \tag{A.6.3}$$

If the curvature tensor admits a symmetry of some type at points of Riemannian space, then, on the basis of the theory developed above, it is easy to write out the general formulae which determine the components of R_{ijkl} and K^{mn} in terms of the tensors which specify the corresponding symmetry group.

For instance, for symmetries of the type of oriented media, the following formulae are valid:

for the symmetry $\infty/\infty \cdot m$ and ∞/∞:

$$K^{mn} = kg^{mn} \;; \tag{A.6.4}$$

for the symmetry $\infty \cdot m$, $m \cdot \infty : m$, $\infty : 2$, $\infty : m$, ∞

$$K^{mn} = kg^{mn} + k_1 b^m b^n \;, \tag{A.6.5}$$

where b^m are the components of the unit vector directed along the axis of symmetry.

Analogous formulae can be written down in any case when the components of the tensor K^{mn} admit any finite symmetry group. For instance, for symmetry corresponding to any one of the five classes of the cubic system, one has:

$$K^{mn} = kg^{mn} \,. \tag{A.6.6}$$

Therefore, in this case, the tensor K^{mn} is spherical, just as in the case of complete isotropy. Corresponding formulae follow from (A.6.2) and (A.6.4)–(A.6.6) for the components of the tensor R_{ijkl}.

From (A.6.4) and Bianchi's identity (A.6.2), one has

$$g^{m\alpha} \nabla_\alpha k = 0. \tag{A.6.7}$$

Equation (A.6.7) expresses a well-known theorem of Schur, according to which isotropy of the curvature tensor at each point implies the constancy of the curvature in the whole space. Indeed, one has

$$k = \text{const.}$$

from (A.6.7)

A generalization of Schur's theorem is contained in the proof given above. This generalization consists of the fact that it is not necessary to require complete isotropy of the curvature at each point of the space for

Schur's theorem to hold. It is sufficient that at each point the symmetry conditions of the group 3/2 be satisfied, i.e., that the components of the tensors K^{mn} or R_{ijkl} be invariant under the 12 transformations of the symmetry group 3/2.

If the curvature is determined at each point by constant collinear vectors b^i, then Bianchi's identity gives

$$\nabla^\lambda k + b^\lambda b^\mu \nabla_\mu k_1 = 0. \qquad \text{(A.6.8)}$$

Equations (A.6.8) are a system of equations imposed on the curvature for the corresponding Riemannian spaces.

The above considered construction of the tensor function $\boldsymbol{H}$ in terms of the tensors $\boldsymbol{T}_1, \boldsymbol{T}_2, \ldots, \boldsymbol{T}_k$ was reduced to the introduction of the functional dependence between components $\boldsymbol{H}$ and $\boldsymbol{T}_s$, which retain their form in any coordinate system.

As an additional generalization of this conception, one can take into account that the conception of tensor is closely linked to the used co-ordinates systems, which are defined by the basis vectors. For each tensor one can introduce the canonical matrix and the corresponding canonical triad of the unit basis vectors, which in such a way can be considered likewise as, generally speaking, non-single value functions of a given tensor.

The authors express their gratitude to Iu. I. Sirotin with whom conversations helped to clarify matters in crystal physics, a branch of science which was new to them.

BIBLIOGRAPHY TO APPENDIX I

[1] Sedov, L. I., *Methods of Similarity and Dimensionality Mechanics.* Gostekhizdat 4th. ed. 1957.

[2] Nye, J., *Physical Properties of Crystals* 1960.

[3] Weyl, H., *The Classical Groups.* IL, 1947.

[4] Döring, W., Die Richtungsabhängigkeit der Kristallenergie. *Annalen der Physik,* 7. Folge, Bd. 1, Heft 1–3, S. 104–111, 1958.

[5] Smith, F. G. and Rivlin, R. S., The anisotropic tensors, *Quarterly of Applied Mathematics,* Vol. 15, No. 3, pp. 308–314, 1957.

[6] Pipkin, A. C. and Rivlin, R. S., The Formulation of Constitutive Equations in Continuum Physics, Part I. *Archive for Rational Mechanics and Analysis,* Vol. 4, No. 2, pp. 129–144, 1959.

[7] Sirotin, Iu. I., Anisotropic tensors. DAN *SSSR,* Vol. 133, No. 3, pp. 321–324, 1960. Vol. 133, No. 3, pp. 321–324, 1960.

[8] Sirotin, Iu. I., Integrity bases of tensor invariants of the crystallographic groups. *DAN SSSR*, Vol. 51, 1963.

[9] Gurevich, G. B. *Foundations of the Theory of Algebraic Invariants.* Gostekhizdat, 1948; Noordhoff 1964.

[10] Mal'tsev, A. I., *Fundamentals of Linear Algebra.* Gostekh-teoretizdat, Moscow, 1956.

[11] Smith, F. G. and Rivlin, R. S., The strain-energy function for anisotropic elastic materials. *Trans. Amer. Math. Soc.*, Vol. 88, No. 1, pp. 175–193, 1958.

[12] Smith, F. G., Further Results on the Strain-Energy Function for Anisotropic Elastic Materials, *Archive for Rational Mechanics and Analysis*, Vol. 10, No. 2, pp. 108–118, 1962.

[13] Bhagavantam, S. and Venkatarayudu, T., *The Theory of Groups and its Application to Physical Problems.* IL, 1959.

[14] Jahn, H. A., A Note on the Bhagavantam Suryanarayana Method of Enumerating the Physical Constants of Crystals. *Acta Crystallographica*, Vol. 2, Part 1, pp. 30–33, 1949.

[15] Shubnikov, A. V., Flint, E. E. and Bokii, G. G., *Fundamentals of Crystallography.* Izd. AN SSSR, 1940.

[16] Shubnikov, A. V., *Symmetry and Antisymmetry of Finite Configurations.* Izd. AN SSSR, 1951.

[17] Shubnikov, A. V., On the symmetry of vectors and tensors. *Izv. AN SSSR, ser. fiz.*, Vol. 13, No. 3, pp. 347–375, 1949.

[18] Sirotin, Iu. I., Group tensor spaces. *Kristallografia*, Vol. 5, No. 2, pp. 171–179, 1960.

[19] Sirotin, Iu. I., The construction of tensors with specified symmetry. *Kristallografiia*, Vol. 6, No. 3, pp. 331–340, 1961.

[20] Koptsik, V. A., Polymorphic phase transitions and crystal symmetry. *Kristallografiia*, Vol. 5, No. 6, pp. 932–943, 1960

[21] Spencer, A. J. M. and Rivlin, R. S., The theory of matrix polynomials and its application to the mechanics of isotropic continua. *Archive for Rational Mechanics and Analysis*, Vol. 2, No. 4, pp. 309–336, 1959.

[22] Spencer, A. J. M. and Rivlin, R. S., Finite integrity bases for five or fewer symmetric 3 × 3 matrices. *Archive for Rational Mechanics and Analysis*, Vol. 2, No. 5, pp. 435–446, 1959.

[23] Spencer, A. J. M. and Rivlin, R. S., Further results in the theory of matrix polynomials. *Archive for Rational Mechanics and Analysis*, Vol. 4, No. 3, pp. 214–230, 1960.

[24] Spencer, A. J. M. and Rivlin, R. S., Isotropic Integrity Bases for Vectors and Second-Order Tensors, Part I. *Archive for Rational Mechanics and Analysis*, Vol. 9, No. 1, pp. 45–63, 1962.

[25] Spencer, A. J. M., The Invariants of Six Symmetric 3 × 3 Matrices. *Archive for Rational Mechanics and Analysis*, Vol. 7, No. 1, pp. 64–77, 1961.

[26] Sedov, L. I., *Introduction to Continuum Mechanics.* Fizmatgiz, 1962.

[27] Lokhin, V. V., A system of defining parameters which characterize the geometric properties of an anisotropic medium. *DAN SSSR*, Vol. 159, No. 2, pp. 295–297, 1963.

[28] Lokhin, V. V., The general forms of relations between tensor fields in a continuous anisotropic medium the properties of which are described by vectors, second-order tensors and anti-symmetric third-order tensors. *DAN SSSR*, Vol. 149, No. 6, pp. 1282–1285, 1963.

[29] Sedov, L. I. and Lokhin, V. V., The specification of point symmetry groups by the use of tensors. *DAN SSSR*, Vol. 149, No. 4, pp. 796–7 7, 1963.

[30] Liubarskii, G. Ia., *The Theory of Groups and its Application in Physics*. Gostekhizdat, Moscow, 1957.

[31] Rashevskii, P. K., *Riemannian Geometry and Tensor Analysis*. Gos. Izd. Tekh. Teoret. Lit., Moscow, 1953.

Appendix II

Models of continuous media with internal degrees of freedom[1)]

L. I. Sedov

As is known, there is a need in modern physics and mechanics for the construction, analysis and utilization of new models of bodies with complicated properties. At present there is taking place actual development of macroscopic theories which require investigation not only of gas motions, but also of the motion of rigid deformable solids in close interaction with the physico-chemical processes occurring within a given particle and in its interactions with the neighbouring particles of a body and with external objects. The world literature of recent years contains numerous theoretical papers in which new types of generalized forces and equations of state are introduced. Most of these studies are based on formal mathematical constructions.

The construction of new theories is intimately connected with the introduction of new concepts as determining and unknown characteristics. It also involves quantities which are defined mathematically to describe the properties of space and time, the positions and states of substantial body particles and fields. These new concepts and mathematical entities make it possible to isolate the determining quantities from the general laws of motion and physico-chemical processes.

[1)] This paper has been read at the opening session of the Third All-Union Congress on Mechanics (January 25, 1968) and published in PMM, Vol. 32, 5, 1968.

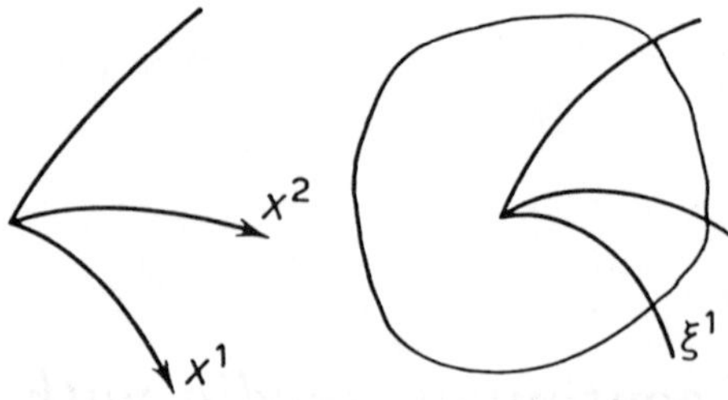

Fig. 66.

In order to consider these matters more specifically, examine the general formulation of the problems of the construction of models to describe broad classes of motions and processes in the mechanics of continuous media.

Some examples of basic characteristic quantities will be stated first. Physical investigation of the motion of material continua entails the use of the concepts of time and of a three- or four-dimensional metric space; it always requires two coordinate systems (Fig. 66),[1)] namely the observer's coordinate system x^1, x^2, x^3, x^4 and the corresponding Lagrangian system $\xi^1, \xi^2, \xi^3, \xi^4 = t$. In Newtonian physics, one can always assume that $x^4 = \xi^4 = t$ and consider absolute time as a scalar variable. The coordinates ξ^1, ξ^2, ξ^3 fix the positions of individual particles. In general, both coordinate systems are curvilinear.

In metric Riemannian space, the length element is given by

$$ds^2 = g_{ij} dx^i dx^j = \hat{g}_{ij} d\xi^i d\xi^j . \tag{A.1}$$

[1)] Some authors hold the view that the mechanics of moving continuous material media can be constructed by means of a single Cartesian coordinate system without significantly limiting generality. This point of view, which is reflected in certain texts and conveyed to students in all sincerity by their teachers, is incorrect and hinders a proper understanding of mechanics and its problems. Confusion is bred, on the one hand, by the fact that the mechanics of deformable bodies is usually concerned with linear problems in which one can assume the observer's system coincides with the concomitant system. On the other hand, it is encouraged by the fact that the metric of the concomitant Lagrangian coordinate system in the theory of liquids and gases manifests itself only through density. At the same time, it is often forgotten that even though all substantial characteristics such as velocity, acceleration, strain rate tensor, etc., are introduced with the aid of the observer's coordinate system, the notion of the concomitant coordinate system is still essentially involved.

The components of the tensor g_{ij} determine the metric and are the basic characteristics of space and time. In Newtonian mechanics and the special theory of relativity, the tensor g_{ij} is Euclidean and pseudo-Euclidean, respectively; its components are supplemented by the observer at his own discretion solely through his choice of the coordinate system x^1, x^2, x^3, x^4.

In the general theory of relativity, the tensor g_{ij} is determined from equations expressing physical principles. The invariant differential quantities which give the properties of the metric tensor g_{ij} of a four-dimensional Riemannian space can be taken as first and very important examples of nonclassical physical unknowns of a new type.

The basic unknown relationship in the observer's system which determines the motion of a medium is the law of motion represented by the four functions

$$x^i = x^i(\xi^1, \xi^2, \xi^3, \xi^4). \qquad (i = 1, 2, 3, 4) \tag{A.2}$$

In addition to the functions $x^i(\xi^k)$, it is convenient to introduce the following derivatives as determining arguments for various physical functions:

$$x^i_j = \frac{\partial x^i}{\partial \xi^j}, \nabla_{k_1} x^i_j, \ldots, \nabla_{k_1}\nabla_{k_2} \ldots \nabla_{k_p} x^i_j, \ldots \quad (p = 1, 2, 3, \ldots). \tag{A.3}$$

Here the symbol ∇_k denotes the covariant derivative with respect to ξ^k where the first derivatives x^i_j can be regarded for fixed values of the subscript j as vector components over the index i. These vectors determine the components of the velocity vector, the corresponding rotations, and the components of the strain tensor

$$\hat{\varepsilon}_{ij} = \tfrac{1}{2}(\hat{g}_{ij} - \mathring{g}_{ij}) = \tfrac{1}{2}(g_{pq} x^p_i x^q_j - \mathring{g}_{ij}).$$

in the comparison of a given position of a body with some imagined "initial position".

Here $\mathring{g}_{ij}(\xi^1, \xi^2, \xi^3, \xi^4)$ denote the components of the metric tensor corresponding to the "initial position" which is introduced by some convention based on physical considerations. In the simplest special cases, the initial position is introduced as an "unchanging rigid body" the three-dimensional spacial part of which coincides with the given deformable body at some "initial" instant [1].[1]

Together with the law of motion (A.2), one must introduce the variable parameters μ^A and their gradients (covariant derivatives) of various orders

$$\mu^A = \mu^A(\xi^1, \xi^2, \xi^3, \xi^4)\,, \; \nabla_{k_1}\nabla_{k_2}\dots\nabla_{k_q}\mu^A, \dots$$

$$(A = 1, 2, \dots, N;\; q = 1, 2, 3, \dots)\,. \qquad \text{(A.4)}$$

For those additional parameters μ^A, one can take:

the entropy; the concentrations of various components in a mixture;

the components of the tensors of residual strain and dislocation density[2] $\varepsilon_{ij}^{(p)}, S_{ij}$,

the components of the electromagnetic potential vector A_i for the electromagnetic field tensor

$$F_{ij} = \frac{\partial A_i}{\partial x^j} - \frac{\partial A_j}{\partial x^i}\,,$$

defined in the appropriate inertial coordinate system by the matrix (*cf.*, for example [3])

$$F_{ij} = \left\| \begin{matrix} 0 & B^3 & -B^2 & cE_1 \\ -B^3 & 0 & B^1 & cE_2 \\ B^2 & -B^1 & 0 & cE_3 \\ -cE_1 & -cE_2 & -cE_3 & 0 \end{matrix} \right\|,$$

where c is the velocity of light, E_1, E_2, E_3 are the components of the electric stress vector, and B^1, B^2, B^3 are the components of the magnetic induction vector;

the components of the magnetization and polarization tensor $\mathscr{P}_{ij} = \frac{1}{2}(F_{ij} - H_{ij})$, where the H^{ij} are given by the matrix

[1] Numbers in square brackets refer to references at the end of this Appendix.

[2] The theory of dislocations is presently being developed by refinement and generalization of the theory of plasticity through the addition of new parameters (*cf.*, for example, [2]).

$$H^{ij} = \begin{Vmatrix} 0 & H_3 & -H_2 & -D_1/c \\ -H_3 & 0 & H_1 & -D_2/c \\ H_2 & -H_1 & 0 & -D_3/c \\ D_1/c & D_2/c & D_3/c & 0 \end{Vmatrix}$$

with H_1, H_2, H_3 the components of the magnetic stress vector and D_1, D_2, D_3 the components of the electric induction vector;

the components of the internal mechanical moments of momentum m_{ik}, etc.

The variable parameters μ^A can be scalars, tensors, or spinors [4–6]. The presence of the variable parameters μ^A, which must be determined in accordance with (A.4) in the solution of problems means that the model of a continuous medium under consideration has internal degrees of freedom.

A characteristic and important feature of all microscopic models of deformable media and fields is the functional dependence of the unknown quantities for bodies of finite dimensions on the determining parameters. For example, for a deformable body of finite dimensions, the total internal energy U is always a functional of the functions $x^i(\xi^k)$ and $\mu^A(\xi^k)$.

In many practical cases, it is possible to assume the generalized property of additivity of the internal energy and to use for the total energy U a formula of the form

$$U = \int_m u(g_{ij}, x^i_j, \dots \nabla_{k_1}\nabla_{k_2} \dots \nabla_{k_r} x^i_j, \mu^A, \dots \nabla_{k_1} \dots \nabla_{k_s}\mu^A, S, K_B)\,\mathrm{d}m + U_0\,, \tag{A.5}$$

where m is the mass at rest, dm a mass element of the medium at rest, u the local internal energy per unit mass which is a physically determined function of the indicated arguments only, S is the entropy, and K_B ($B=1, 2, \dots$) are known functions of the coordinates ξ^i (a generalization of the specified physical constants). By a fundamental physical hypothesis, the local specific energy u at a given point does not depend on higher-order gradients[1)] which are not present among the arguments of u(r and s are

[1)] The possibility of having higher-order derivatives among the arguments of prescribed functions was already foreseen and predicted by Cauchy when he laid the ground work of the theory of elasticity. Limiting transitions from a discontinuum to a continuum in statistical theories indicate that the arguments of the specific internal energy u can generally include derivatives (A.3) of any order.

fixed numbers).

In the classical theory of elasticity, in three-dimensional Euclidean space, one has the simplest case when

$$u(\vartheta_{ij}, \varepsilon_{ij}, S, K_B)\,.$$

In the more complex new models[1] of continuous media, the arguments of the specific internal energy u also contain the additional physico-chemical characteristics μ^A and gradients of various orders of the quantities x^i_j and μ^A.

The presence of such gradients in the expression for the internal energy makes it necessary to reconsider the concepts concerning the equations of motion and processes, boundary and initial values, interaction mechanics, conditions at discontinuities, and many other matters.

The constant U_0, specially isolated and emphasized in (A.5), is entirely inessential in classical elasticity and is usually set equal to zero. In the more general case, the constant U_0 must be allowed for and cannot be regarded as an additive quantity for individual parts of a body when the body is, in fact, divided into separate parts. This is because any separation of a body into parts, any fragmentation, etc., involves losses of external energy. In a first approximation, the non-additivity of the total internal energy U can be allowed for by means of the constant U_0. Allowance for the variation of U_0 with changes in the body surface due to cracking, appearance and development of dislocations, and destruction is of paramount importance.

For elastic bodies with isolated singularities, it is possible to find changes in the constant U_0 for equilibrium processes from the total changes in elastic energy. The production or elimination of certain defects in the body through the action of internal processes or certain external influences requires energy for which the source may be the total internal energy of the body and known external energy inputs. In some cases, changes in U_0 are analogous to latent heat of fusion, or to phase-change energy, in general.

It must be noted[2] that further investigations of the strength of materials on a physical basis will be closely related to the analysis of changes in U_0.

1) In particular, recall the model of a fluid with bubbles [7].

2) *Cf.*, in particular, Vol. II.

The lack of finished theories and notable successes in the solution of problems on criteria of material strength can be attributed to the disregard of the quantity U_0. At the same time, advances in the theory of cracking of brittle bodies have been due largely to making allowance for changes in U_0.

In solutions of certain problems arrived at within the framework of the theory of elasticity, the theoretical stresses in certain small domains can increase without limit without noticeable or even local fracture. Because of this, fracture criteria based on the appearance of theoretical stresses in excess of limiting values in an elastic field are sometimes inadequate.

Fracture of various structural components and test specimen is generally a global phenomenon of the same character as instability of motion, impossibility of equilibrium, or impossibility of continuous motion.

Fracture criteria are generally not local in character. Nevertheless, global instability is often determined by entirely local conditions. One must bear in mind, however, that in many cases the corresponding local conditions may only be necessary, but not sufficient for loss of stability of equilibrium and for fracture of a given structure.

The problem of the construction of models of continuous media consists of the establishment of characteristic quantities and systems of functional or differential equations and various additional conditions which make it possible to formulate mathematical problems for the laws of motion $x^i(\xi^k)$ and the physico-chemical processes defined by the functions $\mu^A(\xi^k)$ for specific physical situations.

The problem of the construction of models of continuous media for known classes of real objects and real phenomena is one of the basic problems of physics. Solution of this problem must be founded on initial, universal and particular basic assumptions, on experimental data, and on the agreement of observations and experimental measurements with theoretical conclusions and computations within the limits of accuracy required in practice or implied by the meaning of a given problem.

The present paper contains a description, analysis, and elaboration of a general method which makes it possible to obtain complicated closed systems of equations and complicated supplementary boundary and other conditions for models of media with internal degrees of freedom from a minimum number of physical assumptions. The additional boundary and other conditions just mentioned are a means of specifying individual models and particular formulations of problems. The basic variational equation which is to be investigated and which constitutes the foundation of the present treatise is a simple and natural generalization of the variational principle of Lagrange. In many highly important cases, it coincides completely with the familiar applications and formulations of this principle [4, 6, 9, 1, 10].

As has been known for a long time, all of the basic equations of the theories of relativity, electrodynamics, analytical mechanics, thermodynamics of equilibrium processes, elasticity, hydrodynamics, and many other disciplines result from the application of the variational principle of Lagrange.

In many modern physical theories, this variational principle constitutes a working and, in essence, a unique initial investigative apparatus.

The following analysis will show that the variational equation of Lagrange for material continua and physical fields can be employed as a basis for all physical models not only of reversible phenomena, but also of irreversible phenomena.

The variational equation has made it possible to unify and synthesize on a common basis various phenomenological and statistical methods of the theory of irreversible processes in thermodynamics and mechanics. In particular, it has permitted the interpretation and evaluation of the associated law of residual plastic strains in mechanical plasticity within the framework of the existing thermodynamics of irreversible processes.

A new element of the following theory will be the use of the variational equation for:

1) description of irreversible phenomena actually occurring in continuous media;
2) the establishment of equations of state;
3) the establishment of kinetic equations;
4) the derivation of initial and boundary conditions;
5) the derivation of conditions at strong discontinuities ("jumps") inside a medium.

The important number of applications has been obtained during the foundations and introduction of models; in which instead of the three-dimensional models, the two- or one-dimensional models are constructed, when a motion or an equilibrium is obtained after the approximation of an unknown three-dimensional function by a corresponding known function in one- or two-dimensions maintaining additionally the unknown variable parameters.

In elaborating the modern theory of complicated macroscopic models of media and fields, it is important to bear in mind that even in Newtonian mechanics the description of phenomena with significant involvement of internal degrees of freedom on the basis of only the principal equation of Newtonian mechanics

$$ma = \mathbf{F} \tag{A.6}$$

is impossible.

Equation (A.6) is a sufficient basis for developing the analytical mechanics of a system of material points, the theory of absolutely solid bodies, the adiabatic theory of elasticity, the theory of motions of an ideal incompressible fluid, and certain other cases. It is already inadequate, however, for considering macroscopic thermal and electromagnetic effects.

Specifically, Equation (A.6) cannot serve as a basis for obtaining the macroscopic laws governing the growth of plastic strains, for the consideration of effects associated with the variation of continuously distributed dislocations, for taking account of various processes and effects associated with the macroscopic theories of electric polarization and magnetization of media and for many other purposes.

For example, the familiar equation for the moment of momentum of small particles or finite bodies does not follow from (A.6), but it is rather an independent fundamental equation derivable from the symmetry of natural laws with respect to the group of rotations. In fact, Equation (A.6) is a consequence of the symmetry of the laws of nature relative to the group of translations. For many classical models of continuous media, the differential equation of the moment of momentum reduces to the condition of the symmetry of the internal stress tensor or it is satisfied automatically, when the internal stress tensor is introduced as a characteristic to be determined from the general assumptions concerning the properties of a medium.

Note that the development of statistical theories for the derivation of macroscopic relations on the basis of (A.6) on a macroscopic level always involves some additional universal and particular assumptions which do not follow directly from (A.6).

Now consider the meaning of the basic variational equation which can be considered as the fundamental point of departure for macroscopic media with internal degrees of freedom.

For the sake of simplicity and greater generality, consider the following theory within the framework of the special theory of relativity, assuming that space-time is pseudo-Euclidean.

Experience and close examination indicate that the development of the theory using four-dimensional geometrically defined physical space-time and four-dimensional vectors and tensors is very convenient, natural, and quite necessary from a physical point of view in important cases.

In the observer's fixed coordinate system, complement the real motions and processes, described exactly or approximately by means of the piecewise continuous functions

$$x^i(\xi^k)\,,\quad \mu^A(\xi^k)\,,\quad S(\xi^k) \tag{A.7}$$

by some sufficiently broad class of piecewise continuous admissible functions which, by hypothesis, contains the system of functions (A.7):

$$\tilde{x}^i(\xi^k) = x^i(\xi^k) + \delta x^i\,,\quad \tilde{\mu}^A(\xi^k) = \mu^A(\xi^k) + \delta\mu^A\,,\quad \tilde{S}(\xi^k) = S(\xi^k) + \delta S \tag{A.8}$$

and, in view of the meaning of the quantities $K_B(\xi^k)$, assume that

$$\delta K_B(\xi^k) = 0\,.$$

The functions $\tilde{x}_i$, $\tilde{\mu}^A$ and $\tilde{S}$ are considered at points of some domain of a system of events of the four-dimensional volume V_0 in space-time bounded by the three-dimensional surface Σ_0. The construction to follow involves the assumption that in the class of permissible functions the variations δx^i, $\delta\mu^A$ and δS in the volume V_0 are continuous together with all their derivatives entering into the variational equations and that they are sufficiently arbitrary, while the variations δx^i_j, $\delta\nabla_k x^i_j$, ..., $\delta\nabla_k\mu^A$, ... etc., can be expressed in terms of the functions $x^i(\xi^k)$ and $\mu^A(\xi^k)$ for real phenomena, in terms of the variations δx^i and $\delta\mu^A$, and in terms of their derivatives with respect to the coordinates x^i.

The following are important new features of the theory to be developed:

1) the variations δx^i are defined as the components of a four-dimensional contravariant vector, and the variations $\delta\mu^A$ as the components of tensors of the same species as μ^A;

2) the variations δx^i and $\delta\mu^A$ and their derivatives can be different from zero and are to some extent arbitrary on the boundaries Σ_3 of the arbitrary volumes $V_4 \subset V_0$.

Write the fundamental basic equation in the form

$$\delta \int_{V_4} \Lambda \, d\tau + \delta W^* + \delta W = 0 , \tag{A.9}$$

where Λ is the density of the Lagrangian.

For a material medium, the function Λ may be given by[1)]

$$\Lambda = -\rho u(g_{ij}, x^i_j, \nabla_k x^i_j \dots \mu^A, \nabla_k \mu^A, \dots, S, K_B) , \tag{A.10}$$

where ρ is the scalar density (the ratio of the mass at rest to the three-dimensional volume in the concomitant coordinate system), and u is the internal energy per unit mass at rest in the concomitant coordinate system. In the special theory of relativity, the quantity u can be considered as a four-dimensional scalar. The first law of thermodynamics says that a function $u\rho \, d\tau$ can be introduced for any infinitely small physical particle.

Determination of the arguments and the form of the function u is the basic physical problem arising in the "concretization" of a model of a continuous medium. Stipulation of the internal energy as a function of its arguments always involves certain assumptions, some of which may appear to be very natural and self-evident.

In practice, the values of variable parameters can often be considered as characteristics of small perturbations. For this reason, the function u can be considered simply as a positive-definite form of the determining small variable parameters. In this case the problem of determining the function u reduces to the problem of determining the constant coefficients of the corresponding quadratic form. The determination of these coefficients is made easier by symmetry conditions [11, 12], and can be based on experimental data. In certain cases, the values of these coefficients can be related to molecular constants on the basis of a statistical theory (developed with assumptions which are universal and specific to a given model). Such coefficients are similar to Young's modulus and Poisson's coefficient which, in practice, can always be readily found by experiment; they can

1) To the variable integral over V_4 one can add terms allowing for the presence of the quantity U_0 in (A.5), which can generally vary due to the development of the boundary Σ_0 and of the discontinuity surfaces inside V_4. Such an additional term is not introduced in the basic variant of the theory presented below. In the arguments of the formula for Λ the entropy S has been isolated from the parameters μ^A. The arguments of the formula for Λ do not include gradients of the entropy S. The subsequent theory can be extended directly to the case where the entropy is not specially isolated, but is rather identified with one of the parameters μ^A entering into Λ together with its gradients of any order.

be computed by statistical means (on the basis of certain far-reaching assumptions). However, in the case of certain solids, the statistical values do not agree in general with those obtained experimentally. Agreement between theory and experiment is better for gases, but here too experimental verification of theoretical results is necessary. Nevertheless, statistical theories provide a means of gaining insight into certain relations between such coefficients which are not clear from phenomenological theories, i.e., into the relationship among the coefficients of heat conduction, viscosity, and diffusion.

In an inertial coordinate system, in Newtonian mechanics, Formula (A.10) can generally be replaced by

$$\Lambda = \rho(\tfrac{1}{2}v^2 - u),$$

where v is the velocity of the points of the continuous medium and u is a three-dimensional scalar, the internal energy.

In the theories already developed, the function Λ can be considered to be known both for the models of material media already defined and for the electromagnetic fields. In the general theory of relativity, the additive component of the quantity Λ associated with gravity is known and serves as a basis for determining the metric tensor g_{ij} representing the gravitational field. The various generalizations of the general theory of relativity, generally speaking, always involve a change in, or some other specification of, the density of the Lagrangian Λ.

It is important to note that from the physical point of view, one can say that a physical system has been specified or is known only if the internal energy or, respectively, the Lagrangian Λ, has been specified or determined [1, 10, 13–15].

Thus, the requirement of specifying the Lagrangian Λ as a function of the macroscopic variables in (A.9) is natural from a physical point of view. In satisfying this requirement, one can draw on the immense body of experience accumulated in the various branches of physics and in many experiments. The assumptions made in specifying the function Λ are always necessary and can be justified by various intuitive and other, generally simpler, assumptions.

In discussing the problem of specifying the function Λ, one can and must establish the most intimate possible contact between macroscopic theory, universal physical principles, experiment, and statistical theories.

Now examine the expression for the specified functional δW^* characterizing the external volume interactions in V_4 and surface interactions on Σ_3 of a given portion of the medium in V_4 with external fields and bodies and certain irreversible actions of neighbouring parts of a medium contiguous with the isolated volume V_4 along the surface Σ_3.

For adiabatic reversible processes, in the absence of external energy influxes inside V_4 and on the surface Σ_3, it is often possible to assume simply that

$$\delta W^* = 0 .$$

In the conservative systems of celestial mechanics, one can always assume that $\delta W^* = 0$.

In the general case of phenomenological theories where there are volume and surface energy influxes external to the medium under consideration and irreversible processes when the arguments of Λ include derivatives of various orders of $x^i(\xi^k)$ and $\mu^A(\xi^k)$ with respect to ξ^k or x^i, one can write down the following general expression for δW^{*}[1]:

$$\delta W^* = \int_{V_4} (\rho\theta\,\delta S - Q_i\,\delta x^i - M_A\,\delta\mu^A)\,d\tau$$
$$-\int_{\Sigma_3+S_\pm}\left(\sum_{j_1\dots j_p} Q_i^{kj_1\dots j_p}\nabla_{j_1}\dots\nabla_{j_p}\delta x^i + \sum_{j_1\dots j_q} M_A^{kj_1\dots j_q}\nabla_{j_1}\dots\nabla_{j_q}\delta\mu^A\right) n_k\,d\sigma . \tag{A.11}$$

In this expression, $S_\pm$ denotes the two sides of the three-dimensional surface S inside V_4 at which the characteristics of the motion can experience strong discontinuities, n_k are the components of the unit vector of the external normal to Σ_3, S_+ and S_-. The components $Q(Q_i, Q_i^{kj_1\dots j_p})$, $M(M_A, M_A^{kj_1\dots j_q})$ are some prescribed external generalized "forces". The quantity θ plays the role of the absolute temperature and can be regarded either as an unknown or as a prescribed quantity, depending on the circumstances. The variation in entropy δS in (A.11) has been introduced as a quantity independent of the variations δx^i and $\delta\mu^A$.

[1] In the four-dimensional approach, for determination of the sign of the scalar products, prescribing the influxes of energy, it is necessary to take into account the considerations carried out in the footnote on page 385.

Specification of the functional δW^* involves the problem of discriminating between internal and external interactions. For example, if the electromagnetic or gravitational field is considered as an external object, then the corresponding energy influxes for the electromagnetic ponderomotive and gravitational forces are present in the expression for δW^*; on the other hand, if these fields are given and the external influxes are included in the model of the medium, then, the corresponding total differentials are separable from δW^* and must be included in the expression for Λ. Upon transfer of the total differentials from δW^* into $\int \Lambda \, d\tau$, the meaning of Λ changes, and (A.10) can be replaced by another similar formula which contains the free energy or enthalpy instead of the internal energy, and which may contain other thermodynamic functions of state. For irreversible processes, the transfer of the complete term δW^* into Λ is impossible, since the variation δW^* is generally non-holonomic.

The definition of the components of the generalized mass and surface forces Q and M is a problem closely related to the theory of dissipative mechanisms. Solution of this problem necessarily entails various assumptions and contacts with the existing thermodynamics of irreversible phenomena. The determination of Q and M is analogous to the basic physical problem of Newtonian mechanics on the determination of laws for forces defined by Newton's equation, and in the present case by the variational equation (A.9). Consideration of the properties of the quantities appearing in the integrand of the expression for δW^* at the discontinuity surface $S_{\pm}$ can have special physical significance. The determination of δW^* involves a choice of the determining parameters x^i and μ^A, with determination of their variations and, in particular, with the property of continuity of the variations at discontinuities.

It is important to note that the determination or specification of the quantities Λ and δW^* serves to delineate common bases for the most varied models. This makes possible the use and synthesis of experience in various disciplines and the establishment of direct correlations between different theories. Moreover, additional means for the use of statistical considerations arise.

The variational[1)] equation (A.9) is closely-linked to the first and second law of thermodynamics. Let us consider this problem in detail, in the frame of the Newtonian mechanics. One can write the equation of energy conservation (the first law of thermodynamics) in the form (*cf.*, Equation (2.18), Chapter V):

$$\mathrm{d}\int_{V_3} \mathscr{E}\rho\,\mathrm{d}\tau = \int_{V_3} (\boldsymbol{F}_{\mathrm{mass}}\cdot \mathrm{d}\boldsymbol{r} + \mathrm{d}q^e + \mathrm{d}q^{**})\,\rho\,\mathrm{d}\tau + \\ + \int_{\Sigma_2} p_\alpha{}^{\beta} n^{\beta}\,\mathrm{d}x^\alpha\,\mathrm{d}\sigma \tag{A}$$

and the second law of thermodynamics (*cf.*, §5 Chapter V) in the form

$$\int_{V_3} \theta\,\mathrm{d}S\rho\,\mathrm{d}\tau = \int_{V_3} (\mathrm{d}q^e + \mathrm{d}q')\,\rho\,\mathrm{d}\tau\ . \tag{B}$$

Here $\mathscr{E}$ — the total energy calculated for the unit of mass, in general case θ plays the role of the absolute temperature, S — the specific entropy, $\mathrm{d}q'$ — the nonequilibrate heat, V_3 — the arbitrary inertial substantial volume of medium, and Σ_2 — the surface bounding V_3.

The equations (A) and (B) are written for the arbitrary three-dimensional volume V, and by definition one has

$$\mathrm{d}\int_{V_3} \mathscr{E}\rho\,\mathrm{d}\tau = \int_{V_3'} \mathscr{E}'\rho'\,\mathrm{d}\tau' - \int_{V_3} \mathscr{E}\rho\,\mathrm{d}\tau \tag{C}$$

where the quantities without stroke correspond to the instant t, and with stroke correspond to the instant $t + \mathrm{d}t$, and $\mathrm{d}t$ in the equation of energy conservation is the same for all particles. The relation (A), as the consequence of the equation of energy conservation, can be considered under the assumption, — that the quantity $\mathrm{d}t$, displayed by the increments of $\mathrm{d}x^i$, $\mathrm{d}q^e$ etc., is different for different particles and different instants of time t. However, in such a case the relation (A) is not the equation of energy, but represents the consequence of the equation of energy for the set of particles, occupying the volume V_3.

Consider an arbitrary four-dimensional volume V_4 in the space x^1, x^2,

[1)] Text in the square brackets was written especially for this book.

x^3, t. The section of this volume forms various three-dimensional volumes V_3 by the different surfaces $t = \text{const}$. Consider now the relations (A) for all kinds of V_3, corresponding to the different values of $t = \text{const}$.

Integrating the equality (A) with respect to the time, using the formula (B) and the expression (C), one can obtain the following relations representing themselves the consequences of the multitude of the relations of (A) type for the corresponding volumes V_3.

$$-\mathrm{d}\int_{V_4} \mathscr{E}\rho\, \mathrm{d}\tau\, \mathrm{d}t + \int_{V_4} (F_{\text{mass}} \cdot \mathrm{d}r + \theta\, \mathrm{d}S$$

$$-dq' + \mathrm{d}q^{**})\,\rho\, \mathrm{d}\tau\, \mathrm{d}t + \int_{\Sigma_3} p_\alpha^\beta n_\beta\, \mathrm{d}\sigma\, \mathrm{d}t = 0\,. \tag{D}$$

If one assumes, that during integration of the equality (A) all V_3 correspond to one and the same inertial substantial volume of a medium for the same $\mathrm{d}t$ for all particles, then the relation (D) is reduced to the usual equation of energy conservation for the finite interval of time.

As the fundamental natural assumption one takes that the definitions of Λ, δW^* and δW in the variational equations[1] are subjected to the

[1] It has been shown below, after giving Λ and δW, the quantity δW is determined; the establishment of the equations of state is reduced to it. The surface integral in ($\mathrm{A}.9_2$) (*cf.* page 600) is a part of δW; the contribution from δW into the sum $\mathrm{d}q^{(e)} + \mathrm{d}q^{**}$ can also exist.

When the specific form of functionals for Λ and δW^*, with aid of the physical theories and assumptions, is given then Euler's equations and the equations of state, in accordance with the canonical form (19) for δW, are fully determined.

From the other side, for given Euler's equations the functionals Λ, δW^* and δW are not determined unambiguously. As a matter of fact, if Λ is replaced by $\Lambda + \nabla_i \Omega^i$, where Ω^i – any four functions which depend on coordinates and unknown functions, then in the Cartesian system of coordinates one has

$$\nabla_i \Omega^j = \frac{\partial\Omega^1}{\partial x_1} + \frac{\partial\Omega^2}{\partial x_2} + \frac{\partial\Omega^3}{\partial x_3} + \frac{\partial\Omega^4}{\partial t}$$

and for the variation of the additional volume, integral of this term, using Green-Ostrogradski formula, one obtains

$$\delta\int_{V_4} \nabla_i \Omega^i\, \mathrm{d}\tau_4 = \delta\int_{\Sigma_3} \Omega^i N_i\, \mathrm{d}\tau_3\ .$$

Therefore, this variation can influence δW only, and correspondingly the equations of state and the conditions on jumps only. Consequently, the addition to the term $\nabla_i \Omega^i$ to Λ does not change Euler's equation.

condition that the equation (A.9) coincides exactly with the equation (D), when the variation of all quantities represents itself the real infinitesimally small increment of the determining parameters through the time dt, in other words, when the following equalities are valid

$$\delta x^i = \frac{dx^i}{dt}\, dt = v^i dt$$

$$\delta \mu^A = \frac{d\mu^A}{dt}\, dt = \mu^A\, dt \tag{E}$$

$$\delta S = \dot{S}\, dt\ .$$

Taking into account the circumstances one forms the basis for physical interpretation of various terms and functions appearing in the equation (A.9), when the variations are determined by the equalities (E). It follows from here, that in the general case for arbitrary, admissible by relations (for example by the condition $\delta(\rho\, d\tau) = 0$) of the variations δx^i, $\delta\mu^A$ and δS, the equations (A.9) can take the form:

$$-\delta \int_{V_4} \mathscr{E}\, \rho\, d\tau\, dt + \int_{V_4} (F_{\text{mass}} \cdot \delta r + \theta\, \delta S - \delta q'$$

$$+ \delta q^{**})\, \rho\, d\tau\, dt + \int_{\Sigma} p_{\alpha}^{\beta}\, n_{\beta}\, \delta x^{\alpha}\, d\sigma_2\, dt + \int_{V_4} \delta\Omega\, d\tau\, dt = 0\ ; \tag{A.9$_1$}$$

by $\delta\Omega$ one denotes the functional which turns identically into zero on the real motion-processes, $d\Omega \equiv 0$. In Newtonian mechanics, as well as in the relativity theory the quantity $\mathscr{E}$ can be considered as a four-dimensional scalar, and in addition, the virtual elementary work done by the external mass-forces of inertia must be consequently taken into account in the formula for δW^*, as one of the terms in the expression $(F_{\text{mass}} \cdot dr)$. In Newtonian mechanics, the virtual elementary work of the force of inertia on the virtual displacements δr of the infinitesimally small particle dm, moving with the acceleration a with respect to the inertial system of coordinates, can be written in two variants

$$
\begin{aligned}
-(\boldsymbol{a}\cdot \mathrm{d}\boldsymbol{r})\,\mathrm{d}m\,\mathrm{d}t &= -\delta\left(\frac{v^2\,\mathrm{d}m}{2}\right)\mathrm{d}t \\
&+\left[\delta v^2-\mathrm{d}\left(\boldsymbol{v}\cdot\frac{\partial \boldsymbol{r}}{\partial t}\right)\right]\mathrm{d}m\,\mathrm{d}t
\end{aligned}
\tag{F}
$$

or

$$
\begin{aligned}
(\boldsymbol{a}\cdot\delta\boldsymbol{r})\,\mathrm{d}m\,\mathrm{d}t &= \delta\left(\frac{v^2\,\mathrm{d}m}{2}\right)\mathrm{d}t-\frac{\partial \boldsymbol{v}\,\delta\boldsymbol{r}\,\mathrm{d}m}{\partial t}\,\mathrm{d}t \\
&-\frac{\partial\rho v^{\alpha}(v\cdot\delta r)}{\partial x^{\alpha}}\,\mathrm{d}\tau\,\mathrm{d}t\ .
\end{aligned}
\tag{G}
$$

The equality (G) was obtained using the equation of continuity of the form $\mathrm{d}m=\rho\,\mathrm{d}\tau_3=$ const., and consequently $\partial\rho/\partial t+\operatorname{div}\rho v=0$. The addition of the common multiplier $\mathrm{d}t$ does not change the essence of the matter, since $\delta\,\mathrm{d}t=0$.

On the basis of the equalities (F) and (G), the part of the expression for δW^* in (9) which is linked to the virtual work done by the forces to inertia, can be rewritten in the form given below. From (F), after integration, one obtains directly:

$$
\begin{aligned}
\delta W_a^* &= \int_{V_4}(-\boldsymbol{a}\cdot\mathrm{d}\boldsymbol{r})\,\mathrm{d}m\,\mathrm{d}t=-\delta\int_{V_4}\frac{v^2\,\mathrm{d}m}{2}\,\mathrm{d}t \\
&+\int_{V_4}\left[\delta v^2-\mathrm{d}\left(v\cdot\frac{\delta r}{\mathrm{d}t}\right)\right]\mathrm{d}m\,\mathrm{d}t\ .
\end{aligned}
\tag{F$'$}
$$

It must be emphasized, that for real processes the expression $\rho v^2 - \delta\,[v\cdot(\delta r/\mathrm{d}t)]$, and consequently the second integral, vanish identically and therefore the equality (F$'$) for real processes, after replacing δ by d, takes the form

$$
\mathrm{d}W_a^* = -\,\mathrm{d}T\,\mathrm{d}t\ ,
$$

where T — kinetic energy, i.e.,

$$T = \int_{V_3} \frac{\rho v^2}{2} \, dm \; .$$

If starting from the identity (G), after transformation of the corresponding ⟨⟨ volume ⟩⟩ integral into ⟨⟨ surface ⟩⟩ integral, one obtains

$$\delta W_a^* = \int_{V_4} (-a \cdot \delta r) \, dm \, dt = \int_{V_4} \frac{\rho v^2}{2} \, d\tau_4$$
$$- \int_{\Sigma_3} \rho v \cdot \delta r \, N_4 \, d\sigma_3 - \int_{\Sigma_3} v^\alpha (v \cdot \delta r) N_\alpha \, d\sigma_3 \; . \qquad (G')$$

Here N_i — components of the four-dimensional unit vector, normal[1)] to the boundary Σ_3 of the volume V.

The last two integrals can be included in the expression for δW.

In the version of theory, which uses the formula (F), the variation

[1)] Let the equation of the surface Σ_3

$$f(x^1, x^2, x^3, t) = 0 \; .$$

Then, obviously

$$N_\alpha = \lambda n_\alpha \qquad (\alpha = 1, 2, 3)$$

$$N_4 = - \lambda D \; ,$$

where n_α — components of the three-dimensional normal vector to the two-dimensional surface, bounding the three-dimensional volume of medium, D — velocity of motion of this surface,

$$\lambda = \sqrt{\frac{(\partial f/\partial x^1)^2 + (\partial f/\partial x^2)^2 + (\partial f/\partial x^3)^2}{(\partial f/\partial x^1)^2 + (\partial f/\partial x^2)^2 + (\partial f/\partial x^3)^2 + (\partial f/\partial t)^2}}$$

$$D = \frac{- \partial f/\partial t}{\sqrt{(\partial f/\partial x^1)^2 + (\partial f/\partial x^2)^2 + (\partial f/\partial x^3)^2}} \; .$$

The last integral in (G′) can be written in the form

$$\int_{\Sigma_3} v^\alpha (v \cdot \delta r) \, n_\alpha \, d\sigma \, dt \; .$$

$-\delta(u\rho\, d\tau)$ sums up with the variation $-\delta\, \rho(v^2/2)\, d\tau$ and as the quantity Λ, in the equation (A.9), appears the total energy with the minus sign

$$\Lambda \equiv -\rho u - \frac{\rho v^2}{2} = -\rho\left(\frac{v^2}{2} + u\right) = -\rho\, \mathscr{E}\ .$$

Additionally in δW_a^* remains the integral $\int_{V_4} [\, \delta v^2 - d(v \cdot \delta r/dt)\,] \times dm\, dt \neq 0$, which is exactly equal to zero for real processes, and the equation (A.9) has the form (A.9$_1$). In this case, the functional δW represents itself the total virtual influx of the external energy through the boundary of the volume V_4.

In the version of the theory, which uses the formula (G), δT sums up with $-\delta u$. Therefore, one can assume that

$$\Lambda = \rho\left(\frac{v^2}{2} - u\right) = L\ .$$

In this case, the Lagrangian L is not ⟨⟨ total energy ⟩⟩ with the minus sign. In the expression for δW in the version (G) there appear ⟨⟨ surface ⟩⟩ integrals from (G′), therefore, in this case δW does not coincide with the total primary influx of energy through the boundary V_4. The expression $\Lambda = \rho(v^2/2 - u) = L$ (i.e., variant (G)) are usually used in the analytical mechanics.

The advantage of the version (G) is that in the analytical mechanics it is not necessary to introduce additionally δW^*, when there are no internal nonpotential forces (work of the potential forces is included in u) beside the inertial forces, in other words, when the system is conservative. But in the general case $\delta W^* \neq 0$, and it must be also taken into account in the analytical mechanics, when the external forces are nonconservative or nonholonomic, and in the number of other examples. The basic variational equation (A.9) in the version (G) can be written in the form

$$\delta \int_{V_4} \Lambda\, d\tau_4 + \int_{V_4} [\, \theta \delta S + F_{\text{mass}} \cdot dr - \delta q' + \delta q^{**}\,]\, \rho\, d\tau_4$$

$$+ \int_{V_4} \delta\Omega_1\, d\tau_4 + \int_{\Sigma_3} [\, (p_\alpha^\beta - \rho v^\beta v_a) N_\beta - N_4\, \rho\, v_\alpha\,]\, \delta x^\alpha\, d\sigma_3\ , \qquad \text{(A.9}_2\text{)}$$

where

$$\delta\Omega_1 = \delta\Omega - \rho\,[\,\delta v^2 - \frac{\mathrm{d}}{\mathrm{d}t}\,(v_\alpha\,\delta x^\alpha)\,]\ .$$

If one takes into consideration that the physico-mechanical system itself represents the thermodynamical system with internal degrees of freedom and with irreversible processes, then the equation (A.9) and each of its term has the independent physical interpretation, and the equation in the infinitesimally small represents itself the equation in variation of balances of energy (more precisely, balances of power of transformations of energy). Therefore, the version (F), because of its clear physical interpretation (with preservation of four-dimensional scalar nature of each of its terms), is the most preferable.[1)]

In the general case, the Euler equations following from the variational equation (A.9) depends on the form of the functional $\delta\Omega$. The presence of the term $\delta\Omega$ is conditioned by the possibility of existence of ⟨⟨ generalized gyroscopic forces ⟩⟩ which appear in Euler's equaion but do not interfere in the equation of energy. These forces give the generalized work, different from zero on the mental increments of the defining parameters.

In general, the functional $\delta\Omega$ can be easily determined by comparing (A.9) and (D), when δW^* and the Lagrangian density Λ, and, consequently, the internal energy is given.

The problem of the general form of $\delta\Omega$ was considered in 1964. The Lorentz force, equal to $e/c(v \times \boldsymbol{H})$ perpendicular to the velocity of the moving charge for the real motions can be taken as the example of the gyroscopic force, when $\delta\Omega \neq 0$.

Obviously, the assumption of ⟨⟨ generalized gyroscopic forces ⟩⟩ represents itself the physical problem, which is solved below with the aid of the equation (A.9), after giving the function Λ and the functional δW^*. On the other hand, for the fixed physical relation (D), the functional $\delta\Omega$ and corresponding with it and with the equation of energy Λ and δW^* can differ because of the differences between ⟨⟨ generalized gyroscopic forces ⟩⟩.

Dissipative processes must and can be conveniently allowed for by means of an equation for the production of entropy in the laws of real motions and processes. The equation describing the variation of the entropy of particles is derived below from the Euler equations in the

[1)] The above considered versions of theory are investigated in detail in the work [21].

general variational problem (A.9). The positiveness of the growth of entropy due to irreversible internal processes must be ensured by the laws defining Λ and the generalized forces Q and M, for real phenomena.

In accordance with the basic meaning of (A.9), assume that the quantity δW is a surface integral over $\Sigma_3+S_\pm$. For the variations δx^i, $\delta\mu^A$ and their derivatives, not equal to zero on $\Sigma_3+S_\pm$, the variation δW can be determined from (A.9) in terms of $\delta\int\Lambda\, d\tau$ and δW^*.

If the quantity δW is defined on $\Sigma_0+S_\pm$ not only by (A.9) (for arbitrary δx^i, $\delta\mu^A$ and their respective derivatives), but also by external conditions, then, as will be shown below, this yields initial conditions, boundary conditions and conditions at the discontinuity.

If the variations δx^i and $\delta\mu^A$ and their derivatives of proper order are equal to zero on $\Sigma_3+S_\pm$, but arbitrary (linearly independent) inside V_4, then (A.9) yields the Euler equations[1)]

[1)] These equations have been obtained by equating the coefficients of ∂x^i, $\partial\mu^A$ and δS in the volume integral to zero, taking account of the equalities

$$\delta\Lambda = \partial\Lambda + \delta x^i\,\nabla_i\Lambda\,, \qquad \delta\mu^A = \partial\mu^A + \delta x^i\,\nabla_i\mu^A\,, \qquad \delta\, d\tau = \nabla_i\delta x^i d\tau\,.$$

In the equation (A.9) the variations δ of scalars, vectors and tensors are determined for the characteristic functions of the individualized particles, i.e., for constant ξ^1, ξ^2, ξ^3 and ξ^4.

The variations, denoted by ∂, are determined from the observer point of view at the system x^i, when the variation of the function is performed for the constant arguments x^i. The relation between the variations δ and ∂ has the form

$$\delta = \partial + \delta x^i\,\nabla_i\,.$$

In particular, by definition of $K_B(\xi^i)$ and $K_C(x^i)$, one has

$$\delta K_B(\xi^i) = 0 \quad \text{and therefore} \quad \partial K_B = -\,\partial x^i\,\nabla_i K_B$$

$$\partial K_C(x^i) = 0 \quad \text{and therefore} \quad \delta K_C = \delta x^i\,\nabla_i K_C\,.$$

For the partial forms of the variation $\delta_G T$ or $\partial_G T$ of any tensors T, components of which appear between the arguments of the scalar R, when the variation of the basis vectors $\mathfrak{z}_\alpha$ or $\mathfrak{z}_\alpha$ is performed only, one obtains that the following identities are valid

$$\delta_G R = 0 \quad \text{or} \quad \partial_G R = 0\,.$$

Such equalities can be useful for some intermediate transformations. For more details see [20].

$$\frac{\delta \Lambda}{\delta x_q^p} \nabla_i x_q^p + \nabla_s \left(\frac{\delta \Lambda}{\delta x_q^i} x_q^s \right) + \frac{\partial \Lambda}{\partial K_B} \nabla_i K_B + Q_i + M_A \nabla_i \mu^A = \rho \theta \nabla_i S \,, \tag{A.12}$$

$$\frac{\delta \Lambda}{\delta S} = \frac{\partial \Lambda}{\delta S} = -\rho\theta \,, \qquad \frac{\delta \Lambda}{\partial \mu^A} = M_A \,, \tag{A.13}$$

where $\delta\Lambda/\delta x_q^p$, $\delta\Lambda/\delta\mu^A$ and $\delta\Lambda/\delta S$ denote variational derivatives, e.g.,

$$\frac{\delta \Lambda}{\delta x_q^p} = \frac{\partial \Lambda}{\partial x_q^p} - \nabla_k \frac{\partial \Lambda}{\partial \nabla_k x_q^p} + \nabla_k \nabla_s \frac{\partial \Lambda}{\partial \nabla_s \nabla_k x_q^p} - \dots \,. \tag{A.14}$$

Multiplying (A.12) by x_4^i and summing over the index i, one obtains

$$\rho\theta \frac{dS}{d\xi^4} = Q_i \frac{\partial x^i}{\partial \xi^4} + M_A \frac{d\mu^A}{d\xi^4} + \frac{\partial \Lambda}{\partial K_B} \frac{dK_B}{d\xi^4} + \nabla_s F^s \,, \tag{A.15}$$

where

$$F^s = x_4^i x_p^s \frac{\delta \Lambda}{\delta x_p^i} \,, \qquad \frac{d}{d\xi^4} = \frac{\partial x^i}{\partial \xi^4} \nabla_i \,,$$

since

$$x_4^s \frac{\delta \Lambda}{\delta x_q^p} \nabla_s x_q^p + x_4^p \nabla_s \left(\frac{\delta \Lambda}{\delta x_q^p} x_q^s \right) = \nabla_s \left(x_4^i x_q^s \frac{\delta \Lambda}{\delta x_q^i} \right)$$

by virtue of

$$0 = \frac{\partial \Lambda}{\partial x_q^p} (x_4^s \nabla_s x_q^p - x_q^s \nabla_s x_4^p) = \frac{\partial \Lambda}{\partial x_q^p} \left(\frac{\partial^2 x^p}{\partial \xi^q \partial \xi^4} - \frac{\partial^2 x^p}{\partial \xi^4 \partial \xi^q} \right).$$

Equation (A.15) is the equation for the production of entropy in a particle, since, by hypothesis, the coordinate ξ^4 plays the role of time. In order to obtain the derivatives with respect to proper time $d\tau = (\hat{g}_{44})^{\frac{1}{2}} d\xi^4$, one must merely multiply both sides of (A.15) by $(\hat{g}_{44})^{-\frac{1}{2}}$.

The Euler equations contain the impulse and energy equations. Depending on the meaning of the parameters μ^A, the Euler equations also contain the Maxwell equations, chemical kinetic equations, and various other forms of equations for the required parameters μ^A, characterizing the internal degrees of freedom. It can be shown [2] that all existing macroscopic models of continuous media, including models of plastic media, can be obtained from the basis equation (A.9).

The Euler equations are, in general, partial differential equations the orders of which are related to the order of the derivatives entering into the arguments of the Lagrangian Λ. In the general case, this order can be fairly high.

If $\Lambda d\tau$ and δW^* are the four-dimensional scalars defined by (A.10) and (A.11), then after variation of the first integral and appropriate integration by parts, the basic equation (A.9) yields

$$\delta W = \int_{\Sigma_3 + S_\pm} \Big[\sum_{j_1 \ldots j_p} (P_i^{kj_1 \ldots j_p} + Q_i^{kj_1 \ldots j_p}) \nabla_{j_1} \ldots \nabla_{j_p} \delta x^i$$

$$+ \sum_{j_1 \ldots j_q} (N_A^{kj_1 \ldots j_q} + M_A^{kj_1 \ldots j_q}) \nabla_{j_1} \ldots \nabla_{j_q} \delta \mu^A \Big] n_k \, d\sigma + \int_{\Sigma_3 + S_\pm} \nabla_s \Omega^{sk} n_k \, d\sigma \,, \tag{A.16}$$

where $P_i^{kj_1 \ldots j_p}$ and $N_A^{kj_1 \ldots j_q}$ are certain quantities (tensor components) expressible in terms of Λ and derivatives of x^i and μ^A. These quantities result from transformations of the variation

$$\delta \int_{V_4} \Lambda \, d\tau$$

by integration by parts. These transformations do not yield unambiguous definitions of the components $P_i^{kj_1 \ldots j_p}$ and $N_A^{kj_1 \ldots j_q}$. This is because of the possibility of adding the first integral to the left-hand side of (A.16). This integral is identically equal to zero when Ω^{sk} is an arbitrary anti-symmetric tensor with continuous components having continuous first- and second-order derivatives at the points of the volume bounded by the surface $\Sigma_3 + S_\pm$.

This statement is obvious on the basis of the Gauss–Ostrogradskii theorem, since $\Omega^{sk} = -\Omega^{ks}$ implies that $\nabla_s \nabla_k \Omega^{sk} = 0$.

Any linear forms of the same character as those in the first terms of the integrand in (A.16) can be taken as the components of Ω^{sk}. It is clear that the formulae which yield expressions for the tensor components

$$P_i^{kj_1 \ldots j_p} + Q_i^{kj_1 \ldots j_p} \,, \qquad N_A^{kj_1 \ldots j_q} + M_A^{kj_1 \ldots j_q}$$

in terms of parameters characterizing the motion and state of the particles are not uniquely defined because of the arbitrary choice of Ω^{sk}.

This gives rise to the question of the ambiguity of the notion of the energy-momentum tensor, as well as to the question of the arbitrariness for specified Euler equations of the equations of state, in general, and of the fundamental notion of internal stresses, in particular.

The dependence of these tensor components in (A.16) on the determining parameters can be regarded and interpreted as equations of state of the physical medium. These equations constitute a generalization of Hooke's law.

Thus, an arbitrariness in the definition of the equations of state arises for a specified system of Euler equations. More detailed analysis shows that additional boundary and initial conditions at the strong discontinuities, which express physical interactions at the boundary of a body or at discontinuities inside a body, do not constitute a basis for an elimination of the above ambiguity of the equations of state.

For a specified system of Euler equations, it is possible to alter the density of the Lagrangian Λ by adding a divergent term. It is clear that this implies also a change in the equations of state. However, complete specification of the Lagrangian can be incorporated into the physical definition of a model of a continuous medium. Specification of the system of Euler equations does not provide the complete and necessary information on a concrete model of a medium.

The stresses are, of course, defined unambiguously once the equations of state have been established. But the whole significance of the ambiguity under discussion has to do with the fact that all the laws of motion and laws of processes of variations of the parameters μ^A in specific problems remain valid for certain other forms of the equations of state.

Note that the ambiguity under discussion is not related to the specific method used to establish the equations of state using the variational principle (A.9). The same situation arises in using the general heat influx equation of thermodynamics in differential form [14].

The significance of the ambiguity can be understood and explained on the basis of the following physical considerations.

It is a well-known fact that the problem of the internal stresses in an absolutely rigid body in motion has no definite solution. It is always possible to apply any system of internal forces, equivalent to zero, in such a body without being able to detect its presence or absence. The presence of such a system of forces can have no significance.

It is not difficult to convince that for any deformed body, analogously as for the absolutely rigid bodies, one can show many different systems of stresses, which do not influence the laws of motion, and because of that the presence or absence of such stresses is impossible to detect. For different systems of internal stresses the laws of motion and the supplementary conditions are the same, but the equations of state are different.

Obviously, the question of such a multi-valuedness does not arise when the equations of state are specified beforehand. However, in the problems of the construction of new models, when a system of equations of state is to be established, the problem of the possibility of a different choice of equations of state arises in the substance of the problem to be solved. This question may acquire special significance when the density of the Lagrangian depends on a sequence of gradients of the determining characteristics.

In order to illustrate the validity of this statement, consider the equations of the theory of elasticity for which the equations of state are given by

$$p^{ij} = \rho\left(\partial u/\partial \varepsilon_{ij}\right). \tag{A.17}$$

Instead of (A.17), choose other equations of state of the form

$$p^{*ij} = p^{ij} + \tilde{p}^{ij}, \qquad \tilde{p}^{ij} = \nabla_s \nabla_k N^{iksj} \qquad \left(N^{iksj} = -N^{ikjs}\right), \tag{A.18}$$

where the quantities N^{iksj} are, as indicated, antisymmetric in s and j, i.e., they constitute components of a tensor which in all problems depends in the same but arbitrarily specified manner on any of the parameters of state and any of their derivatives.[1)]

It is clear that all the laws of motion and deformation will be determined independently of $\tilde{p}^{ij}$, since the additional stresses $\tilde{p}^{ij}$ identically satisfy the equations of equilibrium

$$\nabla_j \tilde{p}^{ij} = 0$$

and since, moreover, for any volume V bounded by a closed surface Σ, one has

[1)] In the theory of elasticity, in problems of equilibrium, in the absence of external body forces, the solution for the stresses can also be represented in the form $p^{ij} = \tilde{p}^{ij}$. If Hooke's law or some other specific equation of state applies, however, the quantities N^{iksj} are functions of the coordinates and not universal functions (the same for all problems) of the strain characteristics. In particular, by (A.18), the components $\tilde{p}^{ij}$ are symmetric when $N^{iksj} = \omega^{ik}\,\omega^{sj}$, where ω^{pq} is any anti-symmetric tensor.

$$\int_V \nabla_j \tilde{p}^{ij}\,\delta x_i\,\mathrm{d}\tau = \int_\Sigma (\tilde{p}^{ij}\,\delta x_i + \nabla_k N^{iksj}\,\nabla_s\,\delta x_i)\,n_j\,\mathrm{d}\sigma =$$

$$\int_\Sigma \nabla_s(\nabla_k N^{iksj}\,\delta x_i)\,n_j\,\mathrm{d}\sigma = 0\,.$$

In this case, side by side with the stresses $\tilde{p}^{ij}$ there have been introduced the third order surface stresses[1] $\nabla_k\ N^{iksj}$. The normal components of the gradient ∇_s of the integrands in the surface integral, entering into the boundary condition, vanish identically at each point of Σ. The integral over any element $\Delta\sigma$ of Σ, by the Gauss–Ostrogradskii formula, leads to an integral over the contour Γ bounding this element. The interactions over $\Delta\sigma$ lead to interactions along Γ which may be considered to be internal. If $\delta x_i = 0$ on Γ, then the integral over $\Delta\sigma$ vanishes. Hence it follows that the additional stresses $\tilde{p}^{ij}$ and $\nabla_k N^{iksj}$ together do not contribute to the flux of the energy of the interactions (for arbitrary possible displacements δx_i) between adjoining particles, separated from each other by the elements of any surface Σ and, consequently, also with external bodies on the boundary Σ_0 of the body. Application of an analogous transformation to the components p^{ij}, determined by (A.17) in which u is the specific internal energy, leads to the energy equation in the presence of exchange of mechanical work between particles, separated by the same elements $\Delta\sigma$ of Σ for $\delta x_i \neq 0$. In this case an energy exchange between adjoining particles takes place only on account of the normal second order stresses p^{ij}. In this context, it must be emphasized that in more complicated models it is impossible to introduce internal surface interactions only into the ordinary second order stresses; therefore the reasoning above is of interest.

The variational condition (A.9) permits to explain more thoroughly the essence of the concept of the equation of state, of the boundary and initial conditions at strong discontinuities which do not follow from the differential equations without supplementary assumptions. It turns out that all of the conditions and equations listed are intimately interlinked and must be considered as a single complex.

The conclusions to follow are related to a transformation of (A.16) for δW such that the integrand contains only the variations δx^i and $\delta\mu^A$ and the covariant derivatives along the normal $\nabla_n^{(\alpha)}\delta x^i$ and $\nabla_n^{(\beta)}\delta\mu^A$, which are independent on $\Sigma + S_\pm$, with $\alpha, \beta = 1, 2, \ldots$. The fact is that the

[1] For more details see, Sedov L. I., ZAMP, 1969, Vol. 20, No. 5, pp. 653–658.

variations δx^i and $\nabla_j \delta x^i$ and also not all of the higher-order gradients $\nabla_{j_1}, \dots, \nabla_{j_p} \delta x^i$ on $\Sigma + S_\pm$ can be considered to be independent.

In the simplest particular cases, the appropriate transformations of (A.16) for obtaining the boundary conditions were carried out by Mindlin[1)] [17]. The corresponding particular transformations for obtaining the conditions at discontinuities were developed by M. V. Lur'e [18]. Assume that the surface $\Sigma_3 + S_\pm$ is smooth. A sufficient condition for this is that the surface S be smooth (since the volume V_4 and the chosen surface Σ_3 are arbitrary). The above transformations yield

$$\delta W = \int_{\Sigma + S_\pm} (\mathscr{P}_{i0}\,\delta x^i + \mathscr{P}_{i1}\,\nabla_n \delta x^i + \dots + \mathscr{P}_{i(r-1)}\,\nabla_n^{(r-1)} \delta x^i +$$

$$+ \mathscr{M}_{A0}\,\delta\mu^A + \mathscr{M}_{A1}\,\nabla_n \delta\mu^A + \dots + \mathscr{M}_{A(s-1)}\,\nabla_n^{(s-1)} \delta\mu^A)\,d\sigma\,, \tag{A.19}$$

where the components of the vectors $\mathscr{P}_{i0}, \mathscr{P}_{i1}, \dots, \mathscr{P}_{i(r-1)}$ and the components of the tensors $\mathscr{M}_{A0}, \dots, \mathscr{M}_{A(s-1)}$ are defined uniquely and are expressed in terms of $P_i^{kj_1\dots j_\nu} + Q_i^{kj_1\dots j_\nu}$ and $N_A^{kj_1\dots j_\nu} + M_A^{kj_1\dots j_\nu}$ which are not uniquely defined.

An important property of the vectors $\mathscr{P}_{i\alpha}$ and tensors $\mathscr{M}_{A\beta}$, defined at points of the elements $d\sigma$ on the boundary surface $\Sigma_3 + S_\pm$, is their dependence not only on the orientation of these elements, as in the case of ordinary stresses, but also on the curvature of these elements and other, more subtle differential geometric properties of the elements in question.[2)]

The true characteristics of a continuous medium are precisely the vectors $\mathscr{P}_\alpha$ and the tensors $\mathscr{M}_\beta$ which depend on the geometric singularities

[1)] V. Jelnorovich carried out the general transformation in four-dimensional space-time for any finite order of the variation of the gradients.

[2)] The transition from (A.16) to (A.19) is easily effected in the absence of edges or conic points on $\Sigma + S_\pm$; in the presence of such singularities, Formula (A.19) remains valid, but the value of the integral (A.19) must be considered as a limit along the smooth surface $\Sigma + S_\pm$ which tends to a surface with edges. Because the integrand in (A.19), which depends on the vector $\boldsymbol{n}$ and its tangential derivatives, has singularities and discontinuities, in the passage to the limit (to a three-dimensional surface $\Sigma + S_\pm$ with two-dimensional edges) there arise additional integrals taken over the two-dimensional surface with edges. These integrals can be written out, applying (A.16) which has no singularities on the edges directly to the surface with ribs. It is then necessary to convert to (A.19); in this transformation, the second integral of the divergent term, which vanishes for a smooth surface $\Sigma + S_\pm$, yields a readily computable non-zero integral over the edges in the case of a surface with edges.

of the areas on which interaction occurs, and on the determining parameters through the Lagrangian Λ and through $Q_i^{kj_1\dots j_\nu}$ and $M_A^{kj_1\dots j_\nu}$ which enter into the expression for δW^*. It is clear that the only combinations which matter in the formula for δW^* are those consisting of the $Q_i^{kj_1\dots j_\nu}$ and $M_A^{kj_1\dots j_\nu}$ which enter into the determination of $\mathscr{P}_{i\alpha}$ and $\mathscr{M}_{A\beta}$.

If the quantity δW is specified on a portion of the boundary Σ_0, then on the basis of (A.19), the arbitrariness of δx^i, $\delta\mu^A$ and their normal gradients on Σ_0, one obtains the following conditions at the points A on the portion of Σ_0 under consideration:

$$\mathscr{P}_{i\alpha} = f_{i\alpha}(A), \qquad \mathscr{M}_{A\beta} = g_{A\beta}(A) \tag{A.20}$$

$(i = 1, 2, 3, 4;\ A = 1, 2, \dots, N;\ \alpha = 0, 1, 2, \dots, r-1;\ \beta = 0, 1, 2, \dots, s-1)$,

where $f_{i\alpha}(A)$ and $g_{A\beta}(A)$ are, in general, given functions at the points A.

On the three-dimensional spatial portion of the boundary Σ_0, corresponding to t_0 = const., Equations (A.20) represent the initial conditions in the three-dimensional volume occupied by the body.

On the three-dimensional portion of Σ, formed by the two-dimensional boundary Σ_2 of the body and by the simultaneously varying time t, Conditions (A.20) can be considered as edge conditions at the boundaries of the variable three-dimensional volume occupied by a given body. Equations (A.20), on the instantaneous boundary t = const. $> t_0$, can generally be considered simply as relations defining the right-hand sides on the basis of the laws of motion, isolated by means of the initial and boundary conditions.

Now write down the conditions on the three-dimensional strong-discontinuity surface S situated inside the four-dimensional volume V_4 of a continuous medium. Assume that, on the basis of preliminary studies and appropriate hypotheses, all of the external interactions on the medium which are distributed over S are included in δW^* (for example, the variation of the "additive" constant u_0, and, specifically, heat release during chemical reactions at combustion or detonation fronts, or energy absorption at a different type of discontinuities along S can sometimes be considered as external influences; the same effects can be interpreted as internal processes, complicating and changing the density of the Lagrangian Λ, and, especially, isolating the variations of the corresponding additional surface integral over the surface of discontinuity S.

Assuming that the variations δx^i and $\delta\mu^A$ and all of their derivatives entering into δW are equal to zero on Σ_3, one obtains at the surface of discontinuity S

$$0 = \delta W_{S\pm} = \int_S [(\mathscr{P}_{i0}\,\delta x^i)_+ + (\mathscr{P}_{i0}\,\delta x^i)_- + \ldots + (\mathscr{P}_{i(r-1)}\nabla_n^{(r-1)}\delta x^i)_+ +$$

$$+ (\mathscr{P}_{i(r-1)}\nabla_n^{(r-1)}\delta x^i)_- + (\mathscr{M}_{A0}\,\delta\mu^A)_+ + (\mathscr{M}_{A0}\,\delta\mu^A)_- + \ldots$$

$$+ (\mathscr{M}_{A(s-1)}\nabla_n^{(s-1)}\delta\mu^A)_+ + (\mathscr{M}_{A(s-1)}\nabla_n^{(s-1)}\delta\mu^A)_-]\,d\sigma. \quad (A.21)$$

Applying in (A.21) to all quantities which depend on the direction of the normal to S, the same direction of the normal. From the definitions of $\mathscr{P}_{i\alpha}$ and $\mathscr{M}_{A\beta}$ and of the operator ∇_n^k one has

$$\mathscr{P}_{i\alpha}(n) = \mp\mathscr{P}_{i\alpha}(-n), \qquad \mathscr{M}_{\beta A}(n) = \mp\mathscr{M}_{A\beta}(-n), \qquad \nabla_n^{k-1} = \mp\nabla_{(-n)}^{k-1}, \quad (A.22)$$

where the minus sign corresponds to even, and the plus sign to odd values α, β and k.

As has already been noted, the basic condition of the class of admissible functions consists of the assumption that the required solutions and comparable functions in the volume V_4 are piecewise continuous, together with all their partial derivatives present in the basic variational equation (A.9). The basic significance of introducing the strong-discontinuity surface S inside the volume lies in the fact that the required solutions and the appropriately varied permissible functions experience discontinuities at the imagined intersection of the surface S.[1)] These discontinuities can be of various types: their character can depend, for example, both on the order and the form of the functions which experience discontinuities on S, or of their derivatives. For example, one can consider strong discontinuities of the crack type in which the required functions together with any of their partial derivatives are discontinuous, or discontinuities of the dislocation type in which small displacements normal to the surface S are continuous, while the displacements in the tangent plane to S are discontinuous on passing from one side S_+ to the other

1) In general, the magnitudes of the discontinuities of the unknown functions are also unknown. However, one may consider problems in which some of the discontinuities of the unknowns are specified in additional conditions.

side S_-, or discontinuities of the shock-wave type in classical gas dynamics, when all of the coordinates x^i (displacements) on S are continuous, while the derivatives $\partial x^i/d\xi^j$ can experience discontinuities.

When higher-order derivatives

$$\frac{\partial^k x^i}{\partial \xi^{j_1} \dots \partial \xi^{j_k}}$$

are present among the arguments of the function Λ, the number of possible types of strong discontinuities can become quite large.

Distinct cases are possible in the formulation and solution of specific problems in gas dynamics and in the simple theories of the mechanics of solids, when the type of surface discontinuity is specified, or the type of discontinuity is determined in the course of solution.

Correspondingly, in using variational equations, one must introduce or find classes of functions which must include the required solution.[1)] In particular, if one assumes that the class of admissible functions is defined by the following conditions at points of the surface S:

$$(\nabla_n^\alpha \delta x^i)_+ = (\nabla_n^\alpha \delta x^i)_-$$

$$(i = 1, 2, 3, 4;\ \alpha = 0, 1, \dots, r_1 - 1;\ r_1 \leqslant r), \quad \text{(A.23)}$$

where $(\nabla_n^\alpha \delta x^i)_+$, $(\nabla_n^\alpha \delta x^i)_-$ are arbitrary and independent for $\alpha = r_1$, $r_1 + 1, \dots r - 1$

$$(\nabla_n^\beta \delta \mu^A)_+ = (\nabla_n^\beta \delta \mu^A)_-$$

$$(A = 1, 2, \dots, N,\ \beta = 0, 1, \dots, s_1 - 1,\ s_1 \leqslant s),$$

while $(\nabla_n^\beta \delta \mu^A)_+$, $(\nabla_n^\beta \delta \mu^A)_-$ are arbitrary and independent for $\beta = s_1, \dots$ $s - 1$, then this defines on passage through the surface S the class of admissible functions $x^j(\xi^1, \xi^2, \xi^3, \xi^4)$, continuous together with their $r_1 - 1$ covariant partial derivatives, where the higher-order derivatives of these functions normal to S can have an arbitrary discontinuity. In addition to Conditions (A.23), assume here that all the quantities appearing in (A.9) are continuous on each side of the surface S for motion along the surface S. From the arbitrariness and independence of the quantites

[1)] Such assumptions are analogous to the very general assumptions about the continuity and differentiability of various functions in the mechanics of continuous media.

$\nabla_n^\alpha \delta x^i$ and $\nabla_n^\beta \delta\mu^A$, one finds, on the basis of (A.22) and (A.23), the following conditions at the discontinuity surface:

$$(\mathscr{P}_{i\alpha})_+ = (\mathscr{P}_{i\alpha})_- , \qquad (\mathscr{M}_{A\beta})_+ = (\mathscr{M}_{A\beta})_-$$

for $\alpha=0, 1, \dots r_1-1, \quad \beta=0, 1, \dots s_1-1$

$$(\mathscr{P}_{i\alpha})_+ = (\mathscr{P}_{i\alpha})_- = 0 , \qquad (\mathscr{M}_{A\beta})_+ = (\mathscr{M}_{A\beta})_- = 0 \tag{A.24}$$

for $\alpha=r_1, r_1+1, \dots r-1, \quad \beta=s_1, s_1+1, \dots s-1$.

Conditions (A.24) can be considered as conditions of continuity (the intersections of the world lines of the particles with the discontinuity surface S are conserved) of the quantities $\mathscr{P}_{i\alpha}$ and $\mathscr{M}_{A\beta}$ at the discontinuity surface S. This property of the quantities $\mathscr{P}_{i\alpha}$ and $\mathscr{M}_{A\beta}$ constitutes one of their important physical characteristics.

In more detailed solutions of the problems with discontinuous solutions and, especially, of problems which involve a varying boundary surface of discontinuity S (for example, with the propagation by particles of isolated dislocations within a medium, with the growth of cracks, and in other cases) it is possible to generalize the basic variational equation (A.9) and to introduce additional variations of the surface S or its edges with respect to the Lagrangian coordinates ξ^i.

Thus, in order to obtain additional relations corresponding to such complicated discontinuity phenomena in real bodies, it is generally necessary to complicate the variable functionals on the basis of the variational equation (A.9) by introducing additional terms into δW^* or $\delta \int \Lambda d\tau$, containing corresponding variations of the Lagrangian coordinates. This is due to the necessity of allowing for the special energy effects associated with the formation or possible propagation of various types of discontinuities by the particles of a medium.

The recent development of the general methods in the construction of specific models on the basis of (A.9) in the theory of continuous media and the media interacting with electromagnetic fields, are contained in the already published papers [19–22].

BIBLIOGRAPHY TO APPENDIX II

[1] Sedov, L. I., *Mathematical methods for the construction of new models of continuous media.* UMN, Vol. 20, No. 5, 1965.

[2] Berdichevskii, V. L. and Sedov, L. I., *Dynamic theory of continuously distributed dislocations. Its relation to the theory of plasticity.* PMM Vol. 31, No. 6, 1967.

[3] Sedov, L. I., *On the ponderomotive forces of interaction of an electromagnetic field and an accelerating moving material continuum taking into account finite deformations.* PMM Vol. 20, No. 1, 1965.

[4] Golubiatnikov, A. N., *A continuous medium with spinor and vector characteristics.* DAN SSSR, Vol. 169, No. 2, 1966.

[5] Jelnorovich, V. A., *Spinor as an invariant object.* PMM Vol. 30, No. 6, 1966.

[6] Jelnorovich, V. A., *Models of Media with Internal Mechanical and Electromagnetic Moments.* Collection of Papers to Celebrate the Anniversary of L. I. Sedov. Izd-vo "Nauka", Moscow, 1968.

[7] Kogarko, B. S., *On a method of a cavitating fluid.* DAN SSSR, Vol. 137, No. 6, 1961.

[8] Berdichevskii, V. L. *The construction of models of continuous media by means of a variational principle.* PMM Vol. 30, No. 3, 1966.

[9] Sedov, L. I., *On the energy-momentum tensor and macroscopic internal interactions in a gravitational field and in material media.* DAN SSSR, Vol. 164, No. 3, 1965.

[10] Sedov, L. I., *Variational methods of constructing models of continuous media, Symposia,* Vienna, June 22–28, 1966. Irreversible aspects of continuum mechanics. Springer-Verlag, 1968.

[11] Golubiatnikov, A. N., *Non-linear spinor functions.* DAN SSSR, Vol. 165, No. 2, 1965.

[12] Lokhin, V. V. and Sedov, L. I., *Non-linear tensor functions of several tensor arguments.* PMM Vol. 27, No. 3, 1963, (Appendix I of Volume 1).

[13] Sedov, L. I., *Introduction to the Mechanics of Continuous Media, Fizmatgiz,* Moscow, 1962.

[14] Sedov, L. I., *Some problems of designing new models of continuous media.* Proceedings of the Eleventh International Congress of Applied Mechanics. Munich, 1964, Springer-Verlag, 1966, pp. 9–19.

[15] Sedov, L. I., *A course in continuum mechanics.* Wolters-Noordhoff, 1971.

[16] Sedov, L. I. and Eglit, M. E., *Construction of non-holonomic models of continuous media for finite strain and for certain physico-chemical effects.* DAN SSSR, Vol. 142, No. 1, 1962.

[17] Mindlin, R. D., *Second gradient of strain and surface tension in linear elasticity.* Intern. J. Solids Structures, Vol. 1, No. 4, 1965.

[18] Lur'e, M. V., *Application of a variational principle to investigate discontinuities in a continuum,* PMM Vol. 30, No. 4, 1966.

[19] Sedov, L. I., *Thoughts on Science and Scientists.* Moscow, "Nauka", 1980.

[20] Sedov, L. I., *Selected Problems of Contemporary Mechanics,* Part I, (dedicated to the 50th Birthday of S. S. Grigorian), Part I, Moscow, Moscow Univ., 1981.

[21] Sedov, L. I., *Forms of energy and their transformations,* PMM Vol. 45, No. 6, 1981.

[22] Sedov, L. I., *On the description of dynamic properties of the gravitational field in vacuum,* PMM Vol. 44, No. 2, 1980.

Subject index